The Art of Paragliding

by Dennis Pagen

Learning Paragliding Skills for Beginner to Intermediate Pilots

white-tailed hawk

ILLUSTRATIONS BY DENNIS PAGEN

The Art of Paragliding

Learning Paragliding Skills for Beginner to Intermediate Pilots

Third Printing: March 2012
Printed in Canada
3 4 5 6 7 8 9 10 11 12 13 14

Published by Sport Aviation Publications
PO Box 43
Spring Mills, PA 16875 - USA
web site: www.lazerlink.com/~pagenbks
E-mail: pagenbks@lazerlink.com

Books by Dennis Pagen:

Understanding the Sky
Towing Aloft (by Dennis Pagen and Bill Bryden)
Hang Gliding Training Manual
Performance Flying
Secrets of Champions
Powered Ultralight Flying
Powered Ultralight Training Course

ISBN 13: 978-0-936310-14-1

ACKNOWLEDGEMENT

A work of this nature cannot be created by the author alone. There are countless individuals–instructors, students and pilots–who have helped develop paragliding and the awareness of proper instruction techniques. I am indebted to all. However, a number of individuals deserve special mention for direct input into ideas in this book. They are Larry Huffman, Jocky Sanderson, Lars Linde, Kari Castle, Steve Roti and Dixon White. Dixon (with additional input from instructors at his school Airplay) spent many hours reading, writing about and discussing fine points of flying and teaching. I cannot over-exaggerate my gratitude.

Next I must thank our hard-working editors, Dixon and Steve (again) as well as C.J. Sturtevant. They labored hard and long over a figureless manuscript to catch errors, elisions and gaffs. C.J., with her marvelous eye for grammar, spelling and correct structure keeps me from embarrassing myself in the face of astute readers.

The photographs in this book come from many sources and demonstrate the beauty of our fine sport. I appreciate the contributions of all the manufacturers who sent photos as indicated in the credits. I also wish to personally thank Jabe Blumenthal, David Levitt, Dave Masuda, Mike Eberle of North American Paragliding, Bruce Hamler of Golden Eagle Paragliding, Ornella Magnaguagno, Airwave, Greg Yarbenet, Brad Kushner and Paul Hamilton of Adventure Productions for the loan of their photograph collections. They have enhanced this book beyong telling.

Finally I wish to thank Claire Vassort who performed all the typesetting and composing. The attractive layout is her achievement and this book's success is as much due to her efforts as mine.

Dennis Pagen
Spring Mills, Pennsylvania
(January 2001)

PHOTO CREDITS

Front Cover – Flying in France – Courtesy of Jérôme Maupoint from the book Stolen Moments
Back Cover – Flying with friends – Courtesy of David Masuda
– Flying in France – Courtesy of Jérôme Maupoint

Courtesy of Airwave: Pages 10, 78, 90, 128, 205, 216, 217, 230, 235, 306, 324
Courtesy of Jabe Blumenthal: Pages 20, 22, 165, 24, 278, 291
Courtesy of Mike Eberle: 87, 106
Courtesy of Paul Hamilton: *x* (all photos), 350 (top and bottom)
Courtesy of Bruce Hamler: Pages 3, 8, 91, 138 (top), 156, 192, 204, 272
Courtesy of Brad Kushner: 82, 103, 112
Courtesy of Ornella Magnaguagno: 83, 135, 136, 189, 190, 215, 232, 253, 258, 260, 295, 297, 321, 323, 335, 354
Courtesy of Dave Masuda: Pages 4, 250, 281, 286
Courtesy of David Levitt: Pages 18, 45, 53, 55, 75, 84, 138 (bottom), 143, 243, 245, 248, 287, 288, 290, 299, 300, 311, 320, 325, 329, 331, 334
Courtesy of Greg Yarbenet: Page 7

ABOUT THE AUTHOR

Dennis Pagen has been flying paragliders since 1989, hang gliders since 1974 and ultralights since 1983. He wrote his first book Flying Conditions in 1975 and has written 13 other books since then, all related to flying for fun. He has had over 400 articles published in national and international magazines and probably will have more as long as inspiration occasionally drops by.

The author has been a school teacher and ski instructor in Leysin, Switzerland and began teaching hang gliding in 1975. He has been a member of the United Hang Gliding Association board of directors since that time and was instrumental in developing instructional and pilot rating programs. He put on the first Instructor Certification Program in 1976 and wrote the USHGA's instructor's manuals with a new edition in 1994. He is currently working on the USHGA's paragliding instructor's manual.

Flying has occupied his time perhaps more than it should over the years, but it has also been a dream come true. Flying was his quest ever since he watched hawks lazily circle up above the fields around his native Michigan home. He currently lives with his wife, Claire, in Central Pennsylvania. She, like Dennis, is a pilot and usually ends up showing him where the thermals are.

DISCLAIMER

WARNING: Paragliding, like other activities such as bicycling, skiing or driving a car involves motion in relation to the ground and other objects. This factor can result in injury or death to the participant if safe practices are not followed. This book describes all matters relating to paragliding with safety as the guiding principle. However, the unlimited variety of sites and conditions renders it impossible to describe all situations that a new pilot may encounter. Your instructor and this book can be your guide, but ultimately you, the pilot, are the determinate of your own safety. Do not attempt anything you think is unsafe. Your own safety is your responsibility, just as it is when you pilot an airplane or drive a car.

TABLE OF CONTENTS

PREFACE

For tens of thousands of years we humans have looked to the sky and envied the spontaneous flight of hawks and eagles soaring on outstretched wings. We longed for the freedom of the air. We dreamed of unlimited vistas and quiet hallways in the clouds. But such longing and dreams were frustrated until only the last century or so.

Human flight is now an everyday reality available to almost anyone of relatively sound mind, body and bank account. With the recent development of paragliding, flight is so convenient, so easy, so affordable that few with the desire to touch the sky are denied. If there ever was a form of flight to appeal to the masses it is this personal aviation realized by a seat and a wing which pops out of a backpack.

Paragliding is a young sport and as such is still undergoing development. It is this continued development that induced us to write this up-to-date and comprehensive manual. When we wrote the first paragliding book in English in 1990 we knew that the rapid refinement of the sport would require a rewrite. Some of the original information is outdated, and new designs and techniques are now in use. Indeed, so much has changed that an entirely new book is warranted. For that reason we have made the effort not only to acquire the most up-to-date information available, but also to attempt to perceive the future through our contacts with designers, instructors, manufacturers, competition pilots and organizations. This book is the fruit of that effort. Hopefully it will guide you on your lifelong pursuit of aerial adventure.

This book is not meant to be a substitute for expert instruction. It is designed for three things: First, it should be used as a classroom manual to guide you with groundschool material. Secondly, it will provide details to aid your practice of manipulating and maneuvering your wing. Finally, it should be used as a reference book to look up fine points of theory and technique as you progress beyond your classroom sessions. We hope we have succeeded in producing a manual that you will use over and over in your flying career. It is our wish to become your partner in the sky, providing you with words of encouragement, advice and pure joy as you progress on the wing. Some day you will fly into a certain situation and think "I read about this and the best thing to do is…" If this experience enhances your flying then our mutual efforts will be rewarded.

We emphasize effective learning, versatile skills and proper technique in this book, but most of all we emphasize safety. Safety means flying accident-free so the joys of flight are not diminished by abrupt interruptions. Safety comes from advancing your knowledge, honing your skills and perfecting your judgement. It is only by flying safely as we broaden our horizons that we develop most effectively over the long term. The goal is unbounded bird-like freedom.

In our modern, high-speed existence we sometimes forget how many opportunities we have to pursue our whims. With leisure time, money and the technology to conquer mountains, breathe under the sea, slide down slopes and vault through the sky, we are truly fortunate. Paragliding is the latest variation of wings for recreational flying. It is a fascinating sport. It is an engrossing sport. It is a liberating sport.

We welcome you to continue reading and take part in the quest to realize the oldest dream. You will change your outlook, your perspective and your sense of the possible. You will change your self.

Paragliding opens the door to many adventures at many sites around the world. Here pilots on modern canopies realize the flying dream.

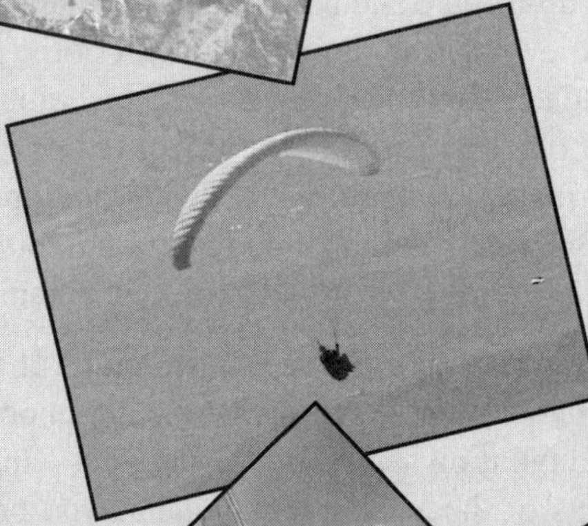

Chapter 1

Under a Rainbow
The Story of Paragliding

On a sunlit mountainside a paraglider pilot stands facing the warm breezes that waft up from below. The silence is broken only by the call of a hawk and the swish of wind in the pines. It is a glorious day.

The pilot lifts his arms then canters down the slope while his soft canopy arcs upwards and eases him into the sky. As the graceful craft banks to the left it slowly climbs in the updraft. The crescent wing changes perspective then the whole fantastic ensemble of glowing colors is lost in the brilliant sun. This impressionistic scene is the essence of pure flight.

Welcome to the world of free flight. Most readers are here to realize a life-long dream of floating like a bird over a pristine countryside. Perhaps you are simply thrilled by the exquisite image of multiple rainbow arcs flying in silent formation. Or you may be just curious to find out why the aerial sport of paragliding has stirred up such passion.

No matter what your motivation, we hope you will embark on the journey from inexperienced spectator to seasoned aeronaut. This journey is fun, fascinating and rewarding. The very thought of paragliding can charge your batteries. In fact, you're probably full-tilt to go, but first let's lay the groundwork so your skyward quest is made easier through understanding.

Paragliding and You

Paragliding looks easy, and in fact it is not physically taxing when compared to other outdoor sports such as rock climbing, kayaking or mountain biking. The amount of physical effort required to pursue paragliding is less and the effort-to-pleasure ratio is higher than for any other activity you can do in the upright position.

On the other hand, paragliding requires considerable study and practice in order to develop good judgement, understand the subtleties of the air and master the little control techniques that make for a great pilot. The learning process is fun, and that's what this book is all about. However, despite its great attraction, you may have some lingering doubts about the suitability of paragliding for you, or the suitability of you for paragliding. In order to address these doubts, we'll describe what it takes to fly a soft wing.

✓ *Whatever it Takes*

To begin, in the learning phase you must do a bit of hill climbing (unless you are learning by towing). This learning is so fun you won't mind the effort, but you have to be in good enough shape to slog up a hill a few times a day—not too high, not too fast, but you gotta get up there. The fact that some very senior citizens fly paragliders means that youthful athleticism isn't a requirement. Of course, if you are in great shape, lucky you—the climbs will be easier and you may be able to take more flights in a lesson.

> **Handicapped Training**
> Note: Handicapped individuals have learned to fly paragliders and discovered a new freedom. There are some schools which help handicapped people learn to fly with special arrangements to transport them up the training hill. If you have special needs, shop around for a school prepared to accommodate you.

Another useful attribute for a potential pilot is good spatial judgement. What this means is that you can judge distances, motions and orientations as you move around in three dimensions. Most of us can do this quite readily. If you can drive a car or ride a bike without having frequent near-misses, then you likely have reasonable spatial judgement. If you are a gymnast, diver, basketball player or involved in any other activities that require maneuvering in three dimensions, you'll probably have a leg up on this flying thing.

We mention spatial judgement because our experience has shown on rare occasions some students have a deficiency. They do not have the ability to judge how much they are turning and which way is up, for example. While such a deficiency can be a big drawback to learning to fly, it does not rule out flying altogether, since spatial judgement can be learned. In such a case it is necessary to find an instructor who is familiar with these matters and who will take extra time to help the student develop spatial skills (see Chapter 9).

✓ *Mental Aspects*

The final useful attributes are mostly mental. They are: good judgement, a concern for personal safety, the ability to make reasonable decisions, a positive attitude and the will or desire to fly. Good judgement can be taught and is part of any good school's program. One way to assess your judgement is to ask yourself "Am I accident-prone?" If the answer is yes, you may wish to take up bowling or some other activity where bad judgement or decisions do not have severe consequences. Many of us in sport aviation are adrenaline junkies in that we relish the rush and excitement of getting high and playing in three dimensions. However, to remain successful at our sport we must make careful judgements to maintain a margin of safety. A person who is accident-prone likely has either a low regard for personal safety or a carelessness that puts them at risk. In this case extra time must be spent on developing judgement skills.

A bit of self-confidence (but not too much) is important for learning since you will eventually be making decisions that greatly determine your safety. At the start, many of your decisions will be made for you by your instructor, but gradually you will begin to take charge of your own flight. If you already have good decision-making skills, fine. If not, they will be developed in the training program along with your judgement and flying skills.

Over-confidence is a deficit in aviation. Remember: There are old pilots, there are bold pilots but there are no old, bold pilots.

Finally, a positive attitude and the desire to learn help you progress because when you envision the rewards of becoming a pilot you have the motivation to learn effectively. Remember, we often subconsciously affect the outcome of our endeavors by the attitude we maintain. If you truly desire to become a versatile, experienced pilot you can. The personal rewards are beyond description.

✓ Fear and Danger

Paragliding, like skiing, rock climbing, scuba diving, kayaking and other adventure sports is not without some risk. In paragliding and other sports we are moving in relation to the ground and we can thus encounter that ground in a harsh manner. However, flying paragliders can be as safe as flying any aircraft, which is safer than driving a car in our opinion for various reasons. We fly more slowly than a car drives and we don't have to worry so much about the other drivers.

The key to safe flying, as in all of life, is risk management. In the course of your training you will learn how to make judgements or decisions based on many years of experience of pilots throughout the world. Your maturity, thought and care in duplicating these wise judgements will help assure your safety in flight.

The reward of beautiful vistas such as this Alaskan landscape is a motivation to overcome fear.

Fear is not an uncommon emotion in paragliding. Fear of heights or new experiences is natural and healthy. How we deal with fear is the important point. Fear that is manageable is an essential part of keeping us flying safely within our limits. When you learn to paraglide you will begin on the flat ground and gradually work your way higher. Your first solo flight at the training hill will be an exhilarating float across the ground. If you are learning to fly with a school that incorporates tandem training, your instructor will be in control. In either case your lessons will be tailored to minimize fear and maximize fun.

In later chapters we will discuss fear's role as we fly in more challenging situations. For now let us mention that quite a few pilots have an acute fear of heights, but when they are buckled in to their paraglider that fear goes away. Imagine that!

✓ Getting Started

Learning to fly is great fun, but it must be accomplished with a safe instructional program. The best way to proceed is to train with a respected and experienced instructor. Such individuals usually work for established schools which are often listed in the phone books of larger cities. Otherwise you can ask any pilot you encounter for the name and number of the nearest instructor or school, or you may be able to contact a national organization (see Appendix I) for information on schools in your area.

A good school prepares you for flight by familiarizing you with paragliding through videos and pictures or showing you actual flying at a site. You will probably sign waivers and provide some of your personal background so your instructor can best tailor the lessons to suit your needs. Then you will begin ground schools and actual lessons which start with tandem instruction or progress from flat-ground work to eventual flying and controlling the glider from various heights.

You will be provided with equipment (glider, harness and helmet) that is designed to suit your size and skill level. Most schools can accommodate large and small people of both sexes. If you are at the extremes in body size make sure the school can provide suitable equipment before going through the preliminaries.

Special Caution

You may on occasion see inexpensive used paragliders for sale in papers or from individuals. Be cautioned that old gliders are not necessarily airworthy because of line stretch or design faults. Often a great "deal" usually turns out to be useless or dangerous. Please resist the temptation to teach yourself or to learn from a friend. Many potential dangers have been detected and corrected through years of experience, and it is foolish to learn the hard way by repeating them. Any money spent on lessons will be saved on glider or body repair bills. If you do wish to purchase used equipment, have it checked out by an experienced professional.

The cost and the amount of time it takes you to become a proficient pilot varies greatly with location and your background. The better the weather in your location and the more time you have to take successive lessons, the faster you will progress. One thing that you can be sure of is that paragliding is about the least expensive way to fly and involves a minimum of official hassles.

Learning to paraglide is a great way to spend the afternoon.

✓ Organizations

At this time in its history, paragliding is well organized with local clubs, national organizations and an international body. These organizations are formed to share experiences, provide flying buddies, develop sites, preserve flying freedom and sanction competition. You can choose to get involved at any level you wish.

Birds of a feather flock together...

The strength of the sport in any country is the national organization. In most countries this is the body charged by the government with providing training and safety guidelines as well as pilot ratings. The English-speaking world's organizations are listed in Appendix I.

Most likely you will be required to join your national organization in order to have insurance coverage, even at the beginning stage. With this membership you will receive a magazine and begin progressing in the rating system. Your national organization can also put you in touch with your local club. Of course some countries and areas of large countries have more activity than others. If you can't locate a national organization, try the Internet—many pilots are connected and pilots know about others in different areas. It is very beneficial to have flying buddies so you can help each other develop. For that reason we urge you to seek out other pilots in your area.

Each country has its own rules regulating paragliding. In general these regulations are much more lenient than those for airplanes. For example, in the United States the Federal Aviation Administration (FAA) has allowed paragliding to be self-regulated as long as a national organization maintains certain safety-related programs. There are specific airspace rules which apply to all pilots and are discussed in Chapter 10.

One important part of the FAA requirements is glider safety standards. AFNOR in France and the DHV in Germany (which are combined under CEN standards) have developed programs for testing glider strength and stability. These recognized tests help assure that our aircraft are as safe as possible (see Chapter 11).

Two useful links:
- *ushga.org (U.S. organization)*
- *fai.org/hang_gliding (World organization)*

The international paragliding organization is the CIVL which stands for the French version of International Commission for Free Flight. This body meets annually and helps regulate competition, world records and some safety standards. As you see, paragliding is established around the world, and indeed, taken world-wide is one of the most popular ways to fly.

FROM PAST TO PRESENT

You are not alone in your dream to fly. Through ancient tales and myths we know that humans from the beginning of time have longed for the freedom of the skies. Cavemen and women probably stood on a mountain ledge and envied the eagle soaring over the wooded landscape on outstretched wings.

Certainly by the time writing was invented to help preserve thoughts through the ages, many were the tales of flight. The myths of winged gods such as

Mercury, Hermes and the Valkyries as well as fantastic birds like the Egyptian Phoenix, the Persian Hoama, the Indian Garuda and Sindbad's Roc all point to mankind's lust for the unique pleasures of flight. Tales of regular humans taking to the sky come down to us from long ago. The flying carpet fantasies of the Mid-east may have been inspired by pipe dreams, but reports of man-carrying kites from Tibet, China and Japan are likely based on fact.

✓ *Modern Times*

We all have heard the stories of Icarus and his father Daedalus as well as later accounts of the experiments of Leonardo Da Vinci and other would-be birdmen. But the real hero of human aviation is Otto Lilienthal who developed bat-like wings in Germany in the 1880's. He flew over 1000 flights from his man-made hill in a manner much like that of modern hang gliders. He directly influenced many other experimenters including the Wright brothers who developed practical aircraft a couple of decades later. Then war with its thrum and thrall of engines took over.

The first time a parachute-type device was used for other than a controlled descent was during World War I. The navy recruited a group of brave souls to be towed aloft on a parachute behind submarines to see what they could see. The sailors reportedly enjoyed the experience immensely except on the occasions when danger was spotted close by and the sub had to descend before they could reel the observer in. The dread factor must have been oppressive.

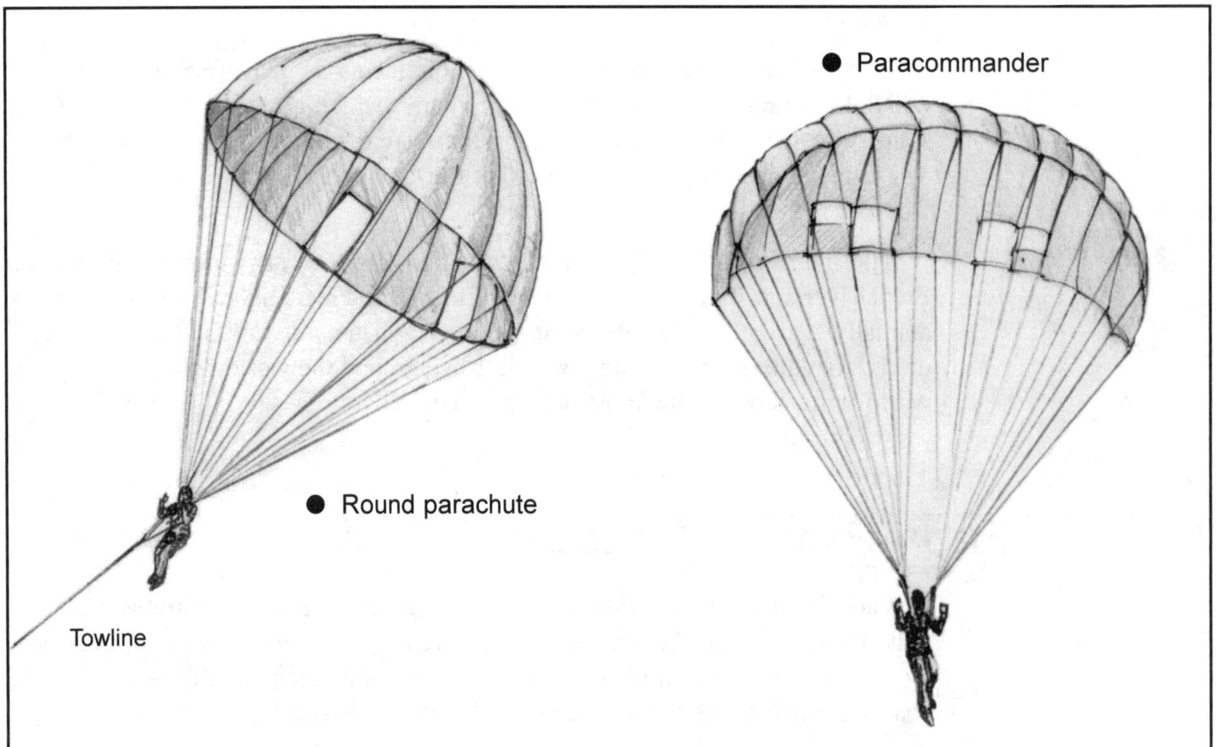

Figure 1-1: Early Parachute

This early experience was significant, for it demonstrated that a parachute could act like an aircraft, even though round canopies with little performance were used. After the first Great War, progress was made with the practice of land towing taking place in Germany and Holland. World War II put an end to such fun, but after this conflagration, there were over a thousand abandoned air bases throughout Europe and the U.S. These became ideal sites for parachute towing and the activity increased.

Things were far from ideal, however, for the round parachutes being used (see figure 1-1) required a lot of tow force to produce the necessary lift. Furthermore, they weren't very controllable. Landings after release from the towline were not highly predictable.

Figure 1-2: Square Parachute

Enter the Paracommander. This was a new style parachute with a more oblong shape (see figure 1-1) and cut-outs as well as longer panels and vents designed to propel the canopy forward in order to create more lift and allow excellent steering. This parachute was developed by Pioneer Parachute Corporation in the late 1950's and revolutionized sport parachute jumping as well as towing. The Paracommander would glide forward about 2.5 feet for every one foot of drop as opposed to falling almost vertically as did a conventional parachute.

The Paracommander allowed safer towing due to lesser speeds and tow forces required. Tourists soon found a new way to spend their Yankee dollars or pound sterling: taking rides high above the beaches of the pleasure capitals of the world under the gentle canopy of a Paracommander.

The next evolutionary step was provided by the National Aeronautics and Space Administration (NASA). Their experimentation in rocket recovery devices led them to design various controllable parachutes including Rogallo wings, which ultimately developed into hang gliders, and ram air parachutes which again revolutionized sport parachuting and ultimately developed into paragliding (see figure 1-2). Francis Rogallo is credited for his innovative development of both rudimentary hang gliders and paragliders.

Greg Yarbenet flying an early paraglider in 1974.

Eventually towing ram air parachutes became the passion of a group who formed the British Association of Parascending (BAPC) in the early 1970's. They called their sport parascending, naturally, for the main event was towing aloft and staying aloft. However, the competitive spirit soon made an appearance and the passengers turned into pilots by releasing from the towline and floating back to earth to test their skills at spot landing.

The problem with these ram air parachute or "squares" or "airfoils" (aerofoil in Britain) as they were called was that they were relatively fast and unforgiving of poor landing techniques as a host of ankle injuries proved. The quest was on for better performance and more gentle characteristics.

✓ Paragliding Realized

Greg Yarbenet, an experienced parachutist from Erie, Pennsylvania (USA) launched a modified parachute both from a slope and a tow vehicle in 1968. The canopy he used was very similar to the paragliders of mid-eighties vintage. His photos of the experience with the date printed in the margin leave no doubt to his prior claim as a progenitor of our sport. Possibly others made similar attempts in this era as well, but the sudden development of hang gliding in the early 70's diverted much of the attention from parachutes. However, the quest for performance in hang gliders rendered them too heavy for carrying up mountains on the pilot's back by the late 70's. Parachutes started looking better.

Instructor Bruce Hamler's first flight in 1988 near Anchorage, Alaska.

In 1978, French parachutists Jean-Claude Bétemps, André Bohn and Gérard Bosson refined the technique of running and launching from a slope at Mieussy, France. The practice soon attracted attention and Mieussy became the first Mecca of paragliding. It wasn't long before skiers and mountain climbers got into the act and the skies throughout the Alps were festooned with bright rainbow wings.

When word reached England of this new twist on parachuting, two manufacturers of square parachutes, John Harbot and Andrew Crowley were intrigued. They soon realized that less porous materials and larger canopies would help performance. This early experimentation led John Harbot to become one of the earliest test pilots, manufacturers, and instructors in paragliding. Another early manufacturer was Frenchman Laurent Kalbermatten who formed the company Ailes de K.

From this brief outline you can see that the development of paragliding was a slow process. As in most forms of aviation, it took a number of geniuses and dreamers who combined efforts to put together various ideas and experiences to bring the sport to its present state. The improvement of design progressed rapidly when paragliding took off, but eventually slowed as the limits of materials and the laws of nature were approached. Likewise the techniques of piloting and training greatly improved until very good standard methods were developed. We intend to pass these methods along to you with the note that even though advancement of paragliding has slowed its furious pace, there are still some design and technique innovations that will be coming along. But with a sound grounding of aerial fundamentals you will be prepared for flight into the next century.

How Do They Do That?

By now you know a little about the background of paragliding, but you may still wonder about the practicalities of actual flying. How do they take off? How do they stay up? Does the pilot have to work hard? The following description will answer most of your questions.

After checking the launch site and landing site for appropriate wind conditions and safe open areas, a paraglider pilot unpacks his or her canopy on launch. The canopy is spread, the lines checked to eliminate tangles and the harness is attached to the canopy lines. When everything is deemed ready the pilot begins running down the slope. The canopy arcs forward and inflates. With a few more steps the pilot is airborne! Life is a breeze.

Once in the air the pilot relaxes, sits back in the seat of the harness and steers the glider by pulling down on control lines—one on each side of the harness. The glider maintains its shape or inflation due to the air pressure forces. It is powered by gravity, so it always must be descending with respect to the air. However, if the air is rising fast enough the glider can rise too. By riding updrafts a pilot can stay aloft for many hours and choose to fly long distances. Flights of over 200 miles (320 km) have been achieved. Updrafts from winds deflected by a mountain or due to warm thermal currents are the most common forms of lift used by pilots flying silent wings.

When a glider nears the ground, it's time to land. The pilot uses the controls to set up a landing pattern that controls his position and height in relation to the chosen landing field. Just before touching the ground the pilot pulls both control lines to stop the forward speed, slow the descent and step to the ground gently. The canopy is then gathered and carried off the field while the pilot beams a happy smile.

Imagine yourself as the pilot of the glider in the opening paragraph of this chapter.

> *You are no longer thinking of the mundane things of this earth, but are soaring far and wide on personal wings. You feel the rush of air and the surge of lift. As you climb higher the horizon recedes and you swear you can see forever. You spot your brother, the hawk, and ease over to greet him. He watches warily but readily joins you in your circles of delight. Your spirit knows it has found its place and you are truly alive.*

This is not an imaginary tale after all, for most paraglider pilots have had similar experiences. You will too as you progress on your way to master mankind's simplest flying.

A pilot finds freedom in the French Alps.

Chapter 2

Beginning Wings and Gear

"You are as good as your equipment."
— Anon

Paragliding is flying reduced to the essentials. There is little between you and the sky and that's what makes the sport so enjoyable, so aesthetic. In its purest form, paragliding becomes a wedding between pilot and craft resulting in harmonious flight. Making your wing an extension of your body and mind is the goal.

An important part of this union is the glider and the other equipment that make this marvelous flying all possible. You must understand how to handle and care for your equipment properly so it doesn't let you down. In this chapter you'll learn the basic arrangement and care of equipment necessary for your early flying. Later, in Chapter 11 we will offer a lot more information about flying gear.

We begin with the paraglider itself, then progress to the harness and additional equipment such as helmets and instruments.

The Right Wing

Gliders used for training and beginning flight are different from those used later when more skill has developed. You wouldn't jump into a race car when you have first learned to drive, would you? Well maybe so, but with paragliding you must go through the fledgling stage and let your feathers develop. That means you should use an easy-to-fly glider initially.

What are the qualities of easy training gliders? The main thing is they are designed for easy inflation (rising above you) during launch, slow flying speeds, predictability, easy control, stability and resistance to turbulence in flight as well as ease of landing. We'll learn more about some design factors later, but here

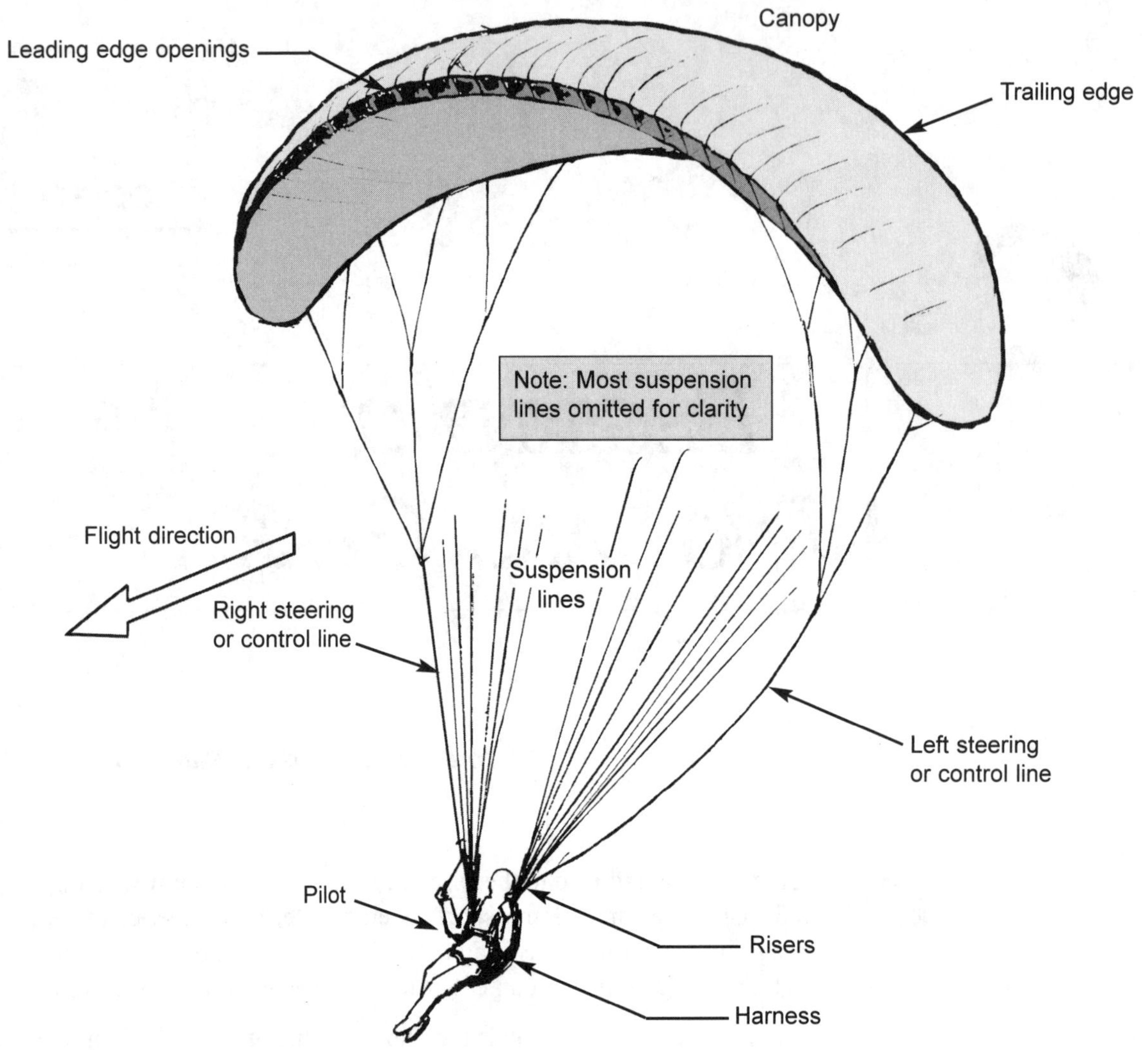

Figure 2-1: Overall View of Paraglider

we'll mention that light canopies, large cells, high inflation pressures, and other factors achieve the above desired characteristics.

Because easy training gliders are slower flying, they should not be flown in as much wind as higher performance models and they don't glide as far. However, they will still take you as high as you wish to fly. In fact, we recommend you learn all your skills well on a low-performance glider and don't rush the transition to a higher performance model. Usually, the higher the performance, the more a glider requires active flying in the air to maintain stability. It takes time to develop control skills, so you should move up in glider performance carefully.

Your instructor will supply the right size glider for you. If a glider is too small, takeoff and landing will require faster running. If it is too large, it will not maintain its stability in flight. This latter point is very critical and is an important safety factor. We recommend you use the school's equipment until you are ready to buy your own, then enlist the aid of an experienced dealer (your instructor in most cases) to guide your purchase. Be cautious of buying used equipment until you are experienced, for you may not know how to assess it for condition and size suitability (see Chapter 11 for a guide on used equipment). If a wing is not a training glider you won't be able to use it safely initially anyway.

✓ The Name Game

In the course of your learning, your instructor and the other pilots will refer to different parts of a paraglider. You should learn these parts so you can communicate. Start with the main terms like *canopy, leading edge, trailing edge, cells, openings, suspension lines, control lines, toggles, risers, carabiners, harness and straps.* Can you figure out some of these terms? As you progress in your flying, continue to learn more of the details of your flying ensemble.

Look at figure 2-1. Here we present the wing and pilot in flight to show the overall look of a paraglider and some of the main items.

Knowing the paragliding lingo lets you understand and communicate technique.

- The *canopy* is the arc of cloth wing above you. The canopy is usually made of nylon. The *leading edge* is the front part of the canopy which meets the air first.
- The *trailing edge* is the rear edge of the canopy.
- The *leading edge cell openings* are holes at the front of the canopy to allow the air to flow inside. The trailing edge is closed, so once the air fills the canopy it no longer flows but maintains a pressure that keeps the canopy inflated.
- The *suspension lines* are the main support cords which connect the canopy and harness. They are fastened to the lower surface with little sewn loops and may be removed for replacement if damaged.
- The *control or steering lines* run from the harness to the trailing edge of the canopy. There is a left one and right one; they are used to alter the shape of the rear of the canopy and thus control the glider. Often control lines are called brakes, but we take the lead of some instructors in discouraging the use of this term since it may lead to misconceptions.
- The *harness* is what the pilot sits in. It has straps to hold the pilot securely as well as containers for a parachute and other items.
- The *pilot* with the silly grin is... you!

✓ The Canopy

Now let's look closer at the canopy to see some special parts. Figure 2-2 shows the canopy arc with a couple of measurements. The width from tip to tip is known as the *span*. The distance from leading edge to trailing edge is the *chord* or chord length. We again have labeled the leading edge as the front margin of the wing and the trailing edge as the rear. The furthest out parts of a wing is known as the *wing tip*. Most paragliders have an area at the tip which is pulled downward; this area is called the *stabilizer*.

You can see that the wing is divided into sections called *cells*. Each cell is separated by an internal *rib*. The rib connects the top and bottom surface of the wing. Beginner paragliders usually have larger and fewer cells and cell openings—often around twenty. Higher performance models will have up to seventy small cells. Larger cells and openings help maintain inflation better and lead to easier launches.

The second part of the figure shows a cross-section of the wing as if we sliced it with a big cleaver. Here you can see the curved *upper surface* and the less curved *lower surface*. The suspension lines are attached to the lower surface and the rib or cell wall between the upper and lower surface maintains the wing's shape. Now

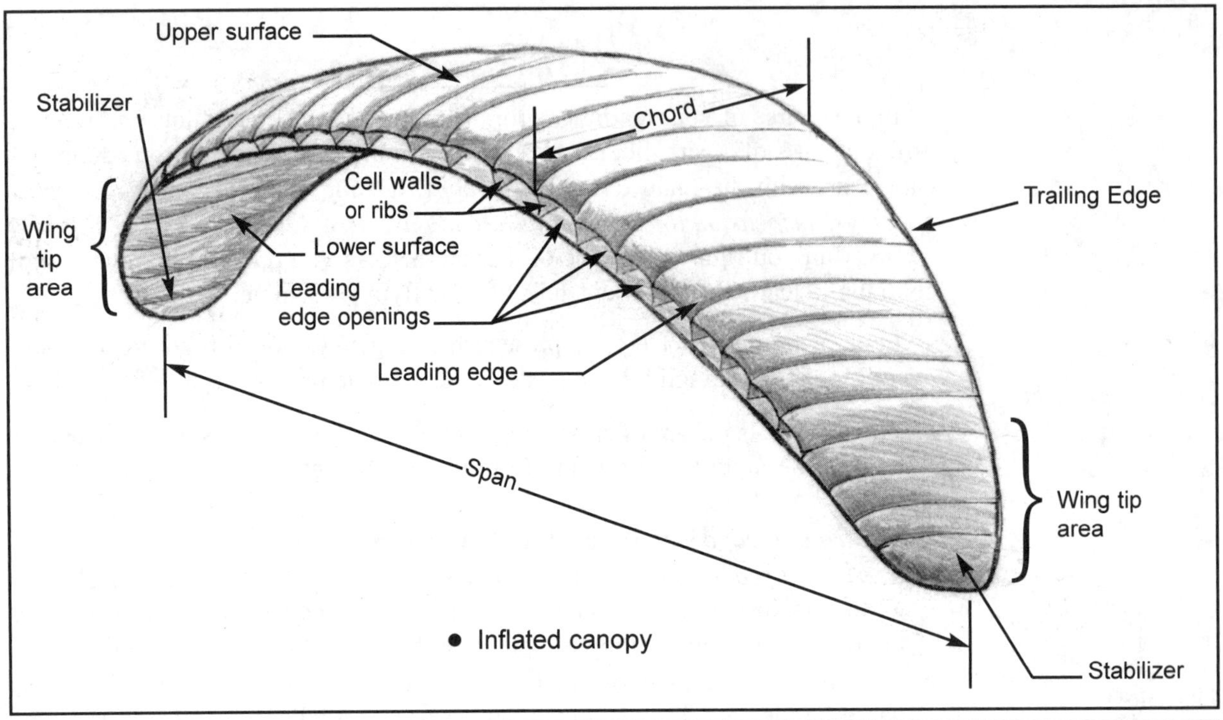

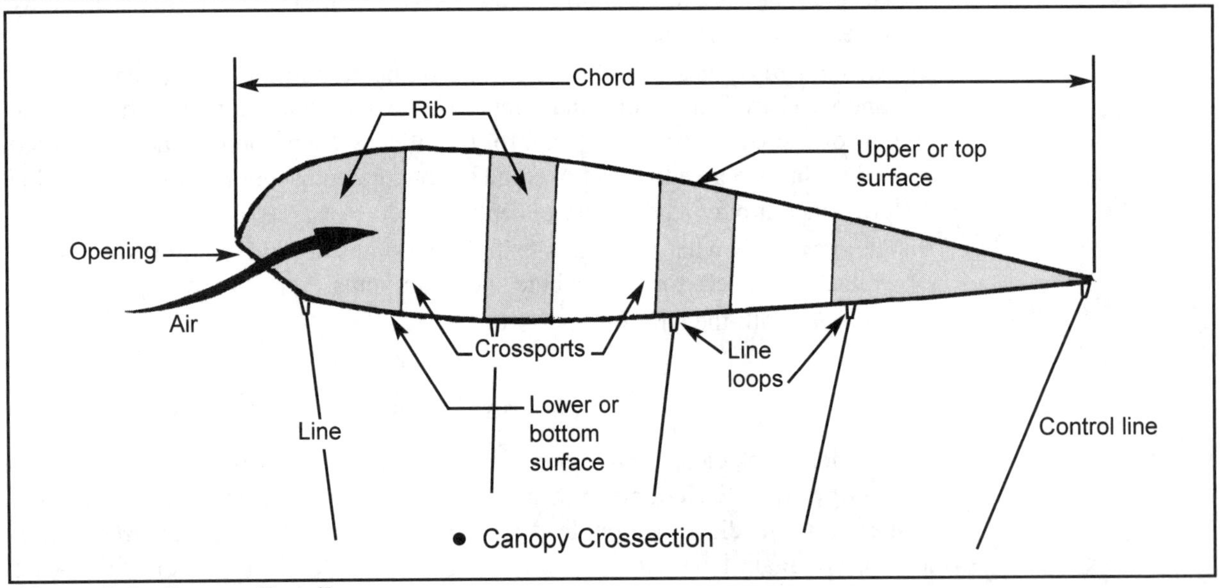

Figure 2-2: Canopy Details

look at the rib and notice the holes in it. These holes are known as *cross-ports*. They are one of the keys to paraglider design because they let the air that enters at the leading edge move outward to pressurize the entire wing and hold it inflated.

A cross-section shape with a curved upper surface as shown in the figure is called an *airfoil*. By definition, the chord of this airfoil is the distance from the furthest forward point to the rearward point as we indicate in the figure.

✓ *The Harness*

Finally, we present figure 2-3 which shows a close-up of the harness system. Note: this drawing depicts a beginning or training harness. More harness features are described in Chapter 11.

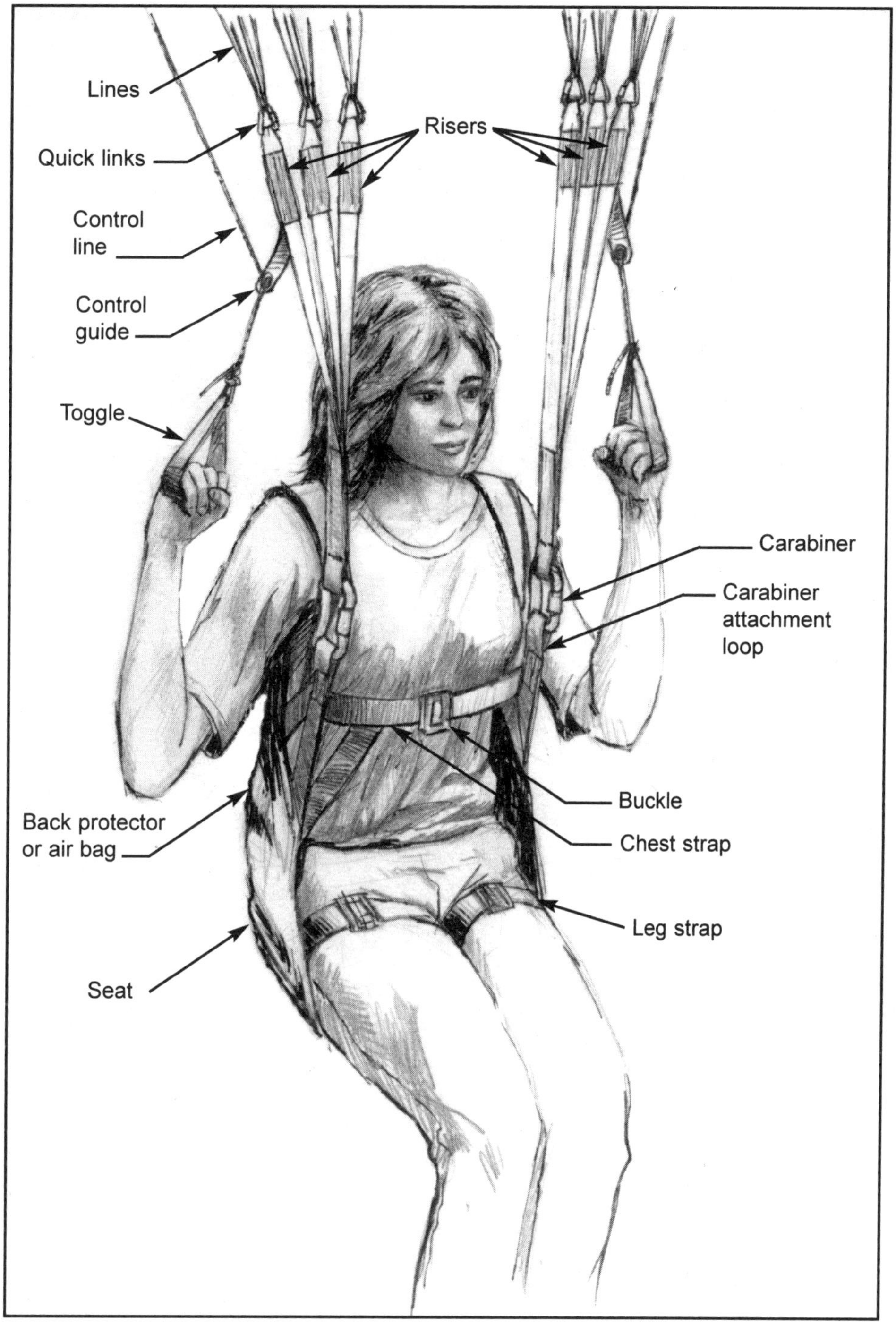

**Figure 2-3:
Harness Detail**

The main parts of the harness system are:

- The *seat*. Usually harnesses have a solid composite board padded for comfort.
- The *leg straps* are adjustable loops to keep you from sliding out of the seat.
- The *chest or waist straps*. This strap helps hold you in the harness and helps position the carabiners in relation to your body.
- The *carabiners* are metal loops which connect the risers to the harness. They may be opened to remove the risers from the harness and thus free it from the canopy. They lock closed for flying.

- The *risers* are short lengths of strap or webbing which connect the suspension lines to the carabiners. Their function is to organize the suspension lines and allow the pilot to selectively influence certain lines.
- The *quicklinks* are the small rings that attach the risers to the suspension lines.
- The *toggles* are the loops you hold on to that allow you to pull the control lines. The toggles normally have a snap or a magnet catch to attach them to their riser so they don't get tangled when not in use.
- The *control line guide* is a pulley or ring which keeps the control line in position for easy access.
- The *back protector* is a protective foam pad that curves under your rear and up your back. Some training harnesses have less elaborate pads for ease of learning. Usually a parachute compartment is incorporated in a harness, but a parachute is useless close to the ground, so you won't have one for your early flights.

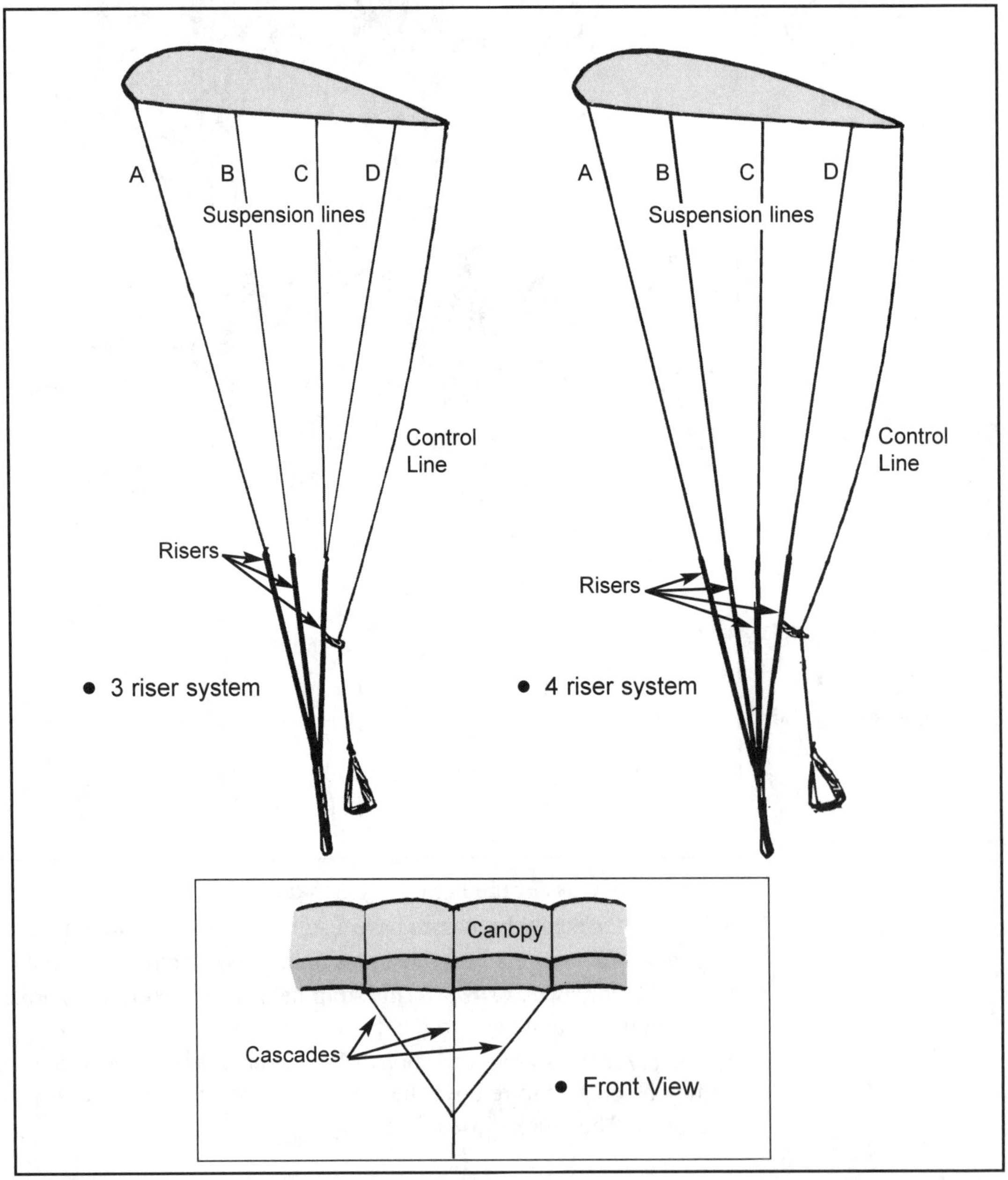

Figure 2-4: Three and Four Riser Systems

✓ Suspension Line Layout

The last thing we'll sort out on this subject is the position and identity of the suspension system. Various harnesses and canopies have slightly different arrangements in risers and lines. The most common setups use three or four risers per side as shown in figure 2-4. Most training gliders also use three or four risers so your transition to a more advanced canopy doesn't include figuring out a new riser system.

To insure continued line strength: never step on, jerk or kink a line.

Whatever number of risers a glider uses, they nearly all have four sets of lines running up to the canopy. These lines can be grouped according to where they meet the canopy. The ones in front are called the A lines and come off *the front risers* (one on each side of the harness). This understanding is important because we'll be using the term *front riser* when we learn launch procedures and other techniques as well. The next lines back are called *B lines*, then comes *C lines* and finally *D lines* at the rear of the canopy. The risers at the back are called *rear risers*. If a system uses four risers, they can be identified as A, B, C and D. But with three risers more than one set of lines must come off one riser, as shown in the figure, so we generally don't refer to B or C risers to avoid confusion.

Training gliders typically have 20 lines—that's 10 per side. For example, if three risers are used, the A risers may have 3 lines, center riser 4 and the rear riser 3, per side. More advanced gliders may have 30 to 40 suspension lines.

Here are two more things to notice: The D lines do not go to the very rear of the canopy. The last line back there is the control line. From figure 2-4 you can imagine how pulling on the control line will pull the rear of the canopy down like a flap or an aileron on an airplane. Also notice that the suspension lines split at the top to spread out along the canopy in order to distribute their load evenly. Typically a suspension line splits into three parts. These splits are called *cascades.*

Paraglider Construction

In order to understand the strength of our glider and how to care for it, we should first get to know a bit about its construction. We won't get too technical here, just provide a familiarization. In Chapter 11 we cover equipment in more depth because when you buy your personal gear you need to know more detail.

✓ The Canopy

The wing arcing through the sky is not just any old rag. It is made of rip-stop nylon with a special weave and resin coating that render it very strong with low porosity. Designers choose the cloth by weight which generally relates to the thickness of the cloth. Heavier material can be less porous, but may render the wing more difficult to inflate in light winds merely due to its additional weight.

Other materials used in a modern canopy include webbing and Mylar. Webbing loops form tabs to attach the suspension lines. Mylar is stiff sheeting of polyester which is sometimes placed near the leading edge to help keep the openings open. Mylar triangles are typically sewn in the front of the ribs for this purpose.

A glider ready to fly showing the arrangement of canopy, lines and harness

✓ The Lines

The lines themselves are typically made of Kevlar (trade name for an aramid fiber). Kevlar is low stretch and extremely strong. Some lines are made of Spectra (called Dyneema in Europe) which is another strong manufactured fiber. Spectra has some tendency to change length so it is often only used for short sections of line. It has better flex properties than Kevlar.

All lines are covered by a woven protective coating called a *sheath*. This coating is usually made of polyester and is not considered to be part of the strength of the line. The coating is colored and sometimes different colors are used on the different sets of lines to help identify them. Most suspension lines cascade or separate into smaller diameter lines near the canopy. Typically they are 2.1 mm diameter in the lower part and 0.9 to 1.2 mm in the upper cascaded section (line diameters are usually given in millimeters, even in English usage. Two millimeters equals .078 inches). Each line can hold at least 400 lbs (180 kg) when new. If you multiply this by 20 to 35 lines on a typical paraglider you can see that these wings are strong indeed.

✓ The Harness

The harness is usually made out of nylon pack cloth which is durable and resistant to moisture, dirt, abuse and insults. The webbing and risers are normally formed from Dacron webbing material. The quick links are steel and the carabiners are specially forged aluminum or steel with redundant strength.

All harnesses incorporate clips or clasps for easy entry and exit. They also have adjustable straps for big or little thighs, big or little shoulders. The adjustable attachments are usually obtained from the parachuting or climbing industry and are very strong and reliable. You will note that the harness webbing is sewn together in places to form loops or connections. The sewing uses Dacron thread, and on a properly constructed harness is incredibly strong.

When your body is suspended from the seemingly diaphanous wing, the spider web lines and the cloth swing seat, you can rest assured that it has been tested and will easily handle ten times your weight!

Paraglider Care

When you start out in a school your instructor generally takes care of packing, storing and the initial handling of the glider and associated equipment. However,

• Excessive heat

it is important for you to understand some basics of glider care so you do not cause unnecessary wear and tear on the training glider. In Chapter 11 we again go into more care detail so you are in the know and up to speed when you get your own wings. Figure 2-5 summarizes potential glider damage.

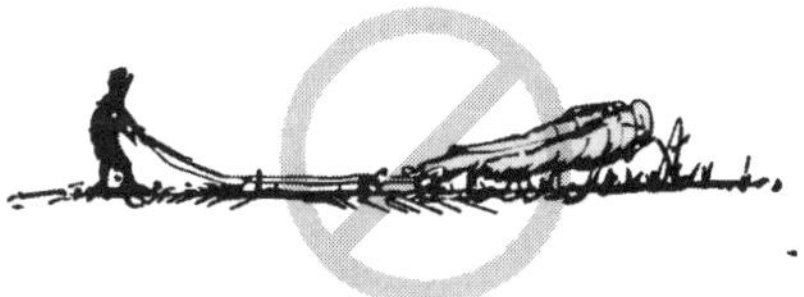

• Laying it on sharp objects

• Sitting on packed glider

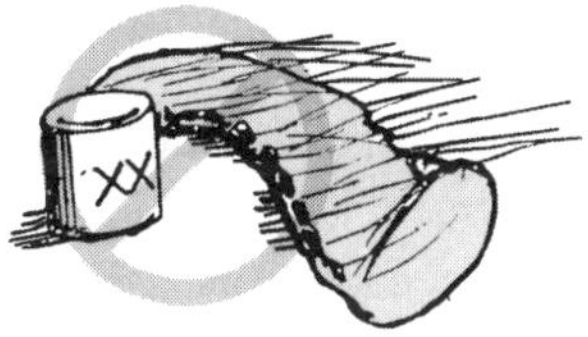

• Harsh chemicals and dirt

• Excessive exposure to sunlight

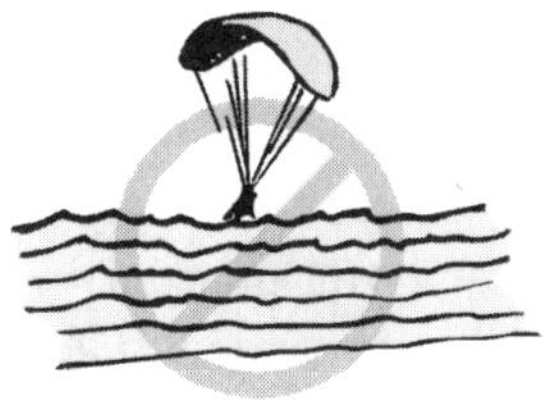

• Exposure to water and damp storage

• Pulling on snagged lines

Figure 2-5: Causes of Glider Damage

✓ *Environmental Damage*

The single item (besides hard use) that shortens a paraglider's life is ultraviolet (UV) radiation. That comes from the sun. nylon, Dacron, Spectra and Kevlar all deteriorate slowly in UV beams. For that reason we suggest you keep a canopy in its carrying pack or tote sack until you're ready to use it. When it is lying around, bunching it up is better than leaving it spread out, and putting it in the shade somewhere is better yet. We recommend having your canopy checked annually for porosity and strength by your dealer.

Heat is the second environmental threat to a glider's life. Paragliders should never be stored in enclosed cars or trunks that are left to bake in the tropical desert sun. They should also not be transported near an engine or exhaust that can heat them beyond ambient temperature. Treat a glider like you would a puppy and it will be your long-term friend.

Finally, moisture can deteriorate a glider, mainly the canopy. Always store it in a cool, dry place. Rain and a dunking in the pond will normally not hurt your wing—nylon is fairly impervious to moisture. However, if the canopy is stored wet, mildew may form that can do a nice job of rotting nylon. Dry it before storing. Also, a canopy filled with water and pulled out of a lake or carried can stretch due to the water weight. We provide more information in Water Landings in Chapter 10. Salt water causes additional problems and requires special attention which we cover in Chapter 11, page 282.

✓ *Handling Abuse*

When you first start out, you don't know what the limits of use or abuse are. We impress you with how strong a glider is, but it can also be vulnerable to localized damage if you concentrate force or wear at a point. Here's what to avoid:

First, *never* step on a part of your glider—not the canopy, the lines or harness. A rock or hard surface below the stepped on part can easily damage it. Lines are especially vulnerable to such damage. If you must get close to the canopy and step through lines, open up a space with your hand and step carefully between them.

In a similar vein, you should never sit on your packed glider. The buckles or other hard points can damage the soft stuff. Many pilots lean against their packed glider while waiting for the wind to make up its mind, but a full-body sit is ill-advised.

Be gentle with your equipment—your life depends on it!

Pulling hard or jerking on an individual line can damage it. If the line is caught on something—a stick or part of your gear—exercise patience and free it gently at the point of entanglement, even if you have to get out of the harness to do so. Often small sticks can snag in the lines. The only solution is to untangle and toss the sticks. If you merely try to shake them loose, they may drop and get tangled again.

Thorns and sharp sticks are particularly harmful to a canopy. Carefully inspect where you are going to lay out the canopy and avoid (or remove) sharp objects. The hard pull you give during launch can easily rip a canopy if it is resting on a sharp point.

Finally, be aware that beating and flailing the canopy on the ground—usually due to botched takeoff technique—takes its toll. Such treatment can break down the resin in the sail and eventually increase its porosity as well as cause more apparent physical damage. However, you are just learning to fly and will not be totally able to avoid contacting the ground with the canopy, sometimes hard. The price of your lessons includes this expected wear and tear on the equipment as well as expert training.

Take as good care of the training gliders you use as you possibly can, for it is wise to develop proper habits for the near future when you will have your own equipment.

Other Equipment

One of the beauties of paragliding is that it doesn't require much gear other than the glider and harness. In fact, for training that's it, other than a helmet, proper shoes and gloves. Here we'll include a few other items that we cover in detail later.

✓ Helmets

A helmet is a necessity for all your paragliding practice, *including* ground handling. When you are learning to handle your canopy your focus is often on the wing, not on where you're stepping. A trip and a tumble can send you on your head in some instances. Besides, it is good to get used to the helmet from the beginning so you are comfortable with the slight change in sound and field of view by the time you are ready for first flights.

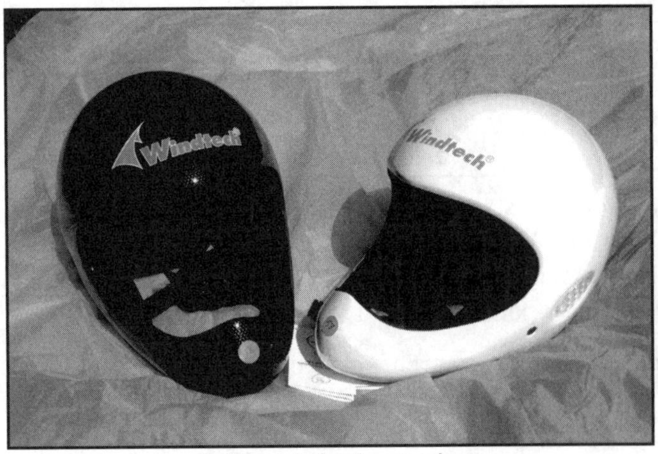

Full-face helmet examples

Your instructor will supply a helmet for your training. Later you will have to buy your own. A good hard-shell helmet with an inner crushable foam lining is the recommended type. The hard shell deflects blows and the inner foam provides shock absorption. At least

1 inch (2.5 cm) of foam should be present.

Some pilots prefer an open-face helmet, but a full-face helmet may offer more protection. The front mouthpiece of a full-face helmet may reduce your ability to feel the wind on your face. Snaps and other protrusions on a helmet tend to catch lines (especially when you are looping risers over your head for a reverse inflation—see Chapter 4), it is best to remove protrusions or purchase a helmet without them.

Hockey, cycling and climbing helmets rarely provide enough protection for flying. Motorcycle helmets tend to be too heavy. The helmets made specifically for paragliding are ideal. You can find them available through your school or in magazines devoted to the sport.

✓ Shoes and Gloves

Most paraglider pilots use boots that are designed specifically for the sport. Their main feature is stiff ankle support, like a light ski boot. Obviously you can't run in these boots as fast as you can with your Nikes, but you can run plenty fast enough for takeoff.

Most instructors don't expect you to purchase boots just to take lessons, but if you have a form of hiking boots that support the ankles, you may consider using them. However, you should not use boots with open lacing hooks, since these hooks readily catch the paraglider lines. Running shoes are also generally OK for the initial lessons, especially in a carefully controlled training situation. In any case, ask your instructor what he or she recommends. You don't want to wear heavy expedition boots but neither do you want to use sandals.

Gloves are advised even in warm weather to prevent abrasion to your hands from the risers during launch and the control toggles during flight. Of course, you can fly if you forget them, but a pair of light gloves when it's warm and heavy gloves when it's not is highly recommended.

✓ Instruments

In your beginning stages you will not use instruments. Your instruments are your eyes, ears and sense of touch. But as soon as you begin flying high, even in the training stages, you may have instruments strapped on by your instructor. Here's a brief description of what you may encounter.

Radios – We use radios very frequently in paragliding. Instructors guide students and pilots team fly with other pilots. In some countries the radio frequencies are HAM or business channels (e.g. U.S. and U.K.). In others it's aircraft or specially assigned channels (e.g. U.K. and Australia). Your instructor will have radios to use when necessary. Before purchasing your own radio you should find out the frequency range appropriate for your area of flying. If you get the wrong radio, it will end up as a very expensive paper weight or a door stop.

Airspeed Indicators – An airspeed indicator does what the name says: it tells you how fast you are flying through the air. The same instrument can also serve as a wind speed indicator on the ground. You hold it into the wind and it tells you how

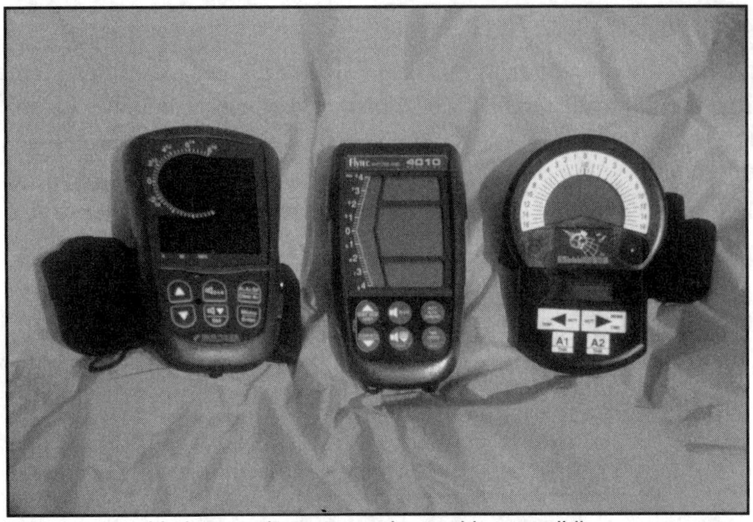

Various varios commonly used in paragliding

fast the wind is blowing. Use a wind meter like this as often as you can to help you learn to judge the wind speed by its feel on your body and its effect on the surroundings (trees, bushes, grass, etc.). Your instructor may hook an airspeed indicator on your harness to help you perceive your airspeed when you begin flying high.

Altimeter – An altimeter is an instrument that measures barometric pressure (the air's pressure is lower the higher you go) and tells you how high you are above a set point. An altimeter is often the first instrument a new pilot acquires.

Variometer – A variometer also measures the barometric pressure but it measures the rate of change of this pressure and thus it can tell how fast you are going up or down. Pilots use this instrument to tell how fast they are climbing in lift or how fast they are sinking. You do not need such an instrument in your training program, but as soon as you begin to learn soaring, a variometer is a valuable aid.

Note that most variometers today include an altimeter and often an airspeed indicator. There are many models to choose from with many varied features. Used instruments are usually available to help reduce costs. Enlist the advice of a trusted experienced pilot before investing in such items.

SUMMARY

This chapter is devoted to bringing you a little closer to your first experience of floating through the sky. Familiarization with your equipment helps you better understand the communications of your instructor. When he yells "pull the rear risers," you can comply immediately once you know where they are.

The main points to remember are the names of the canopy, leading edge, control lines, risers and harness. These are the items that will be repeatedly referred to at the hill. Also, don't forget the guidelines on care.

TEST YOURSELF *(Answers in Appendix II)*

1. What is the name of the straps that run from the carabiners to the suspension lines?
2. What do we call the front edge of a canopy?
3. What are the rearmost suspension lines called?
4. Name three sources of potential damage to a paraglider.

Chapter 3

Why It Flies

"You will be necessarily upborne by the air if you can renew impulse on it faster than the air can recede from the pressure."
— Samuel Johnson, 1759

Most of us have seen airplanes and birds powering their way through the sky. It is not hard to imagine that the propeller or flapping pulls them through the air so that the wings can develop lift and hold them aloft. But what about gliding craft with no engine such as sailplanes, hang gliders and paragliders? Why don't they fall? More mysteriously, how do they manage to climb?

We'll answer these important questions in this chapter. Here you will discover the secrets of how wings work in general and how a paraglider works specifically. The understanding you develop here is important throughout all of your flying experience for the more you understand what makes a paraglider fly, the more you will be able to avoid danger and excel as a pilot. This chapter is an introduction for beginner students. Later in Chapter 12, we will provide more details to go along with your expanded knowledge and skill.

A Paraglider Invades the Sky

A paraglider comes in a big wad that emerges from a bag, then lies on the ground like a limp jellyfish. There's not a bone in its body. But suddenly, magically it leaps into the air and behold, a semi-solid wing arcs above us to obey our bidding. Perhaps one of the greatest mysteries of all to people first seeing paragliding is how the wings inflate and stay inflated. Let's clear up that mystery.

✓ *Relative Speed and Ground Speed*

When a paraglider launch begins, the openings in the leading edge are pulled forward to catch the wind and fill the canopy up just like a plastic bag held up with the opening facing a breeze. Even if there is no actual wind, the paraglider fills with air because the pilot is moving it forward. The air flowing by the paraglider is called the *relative wind.* On takeoff it can be created by a wind moving across the ground, by moving the paraglider itself through the air, or a combination of both.

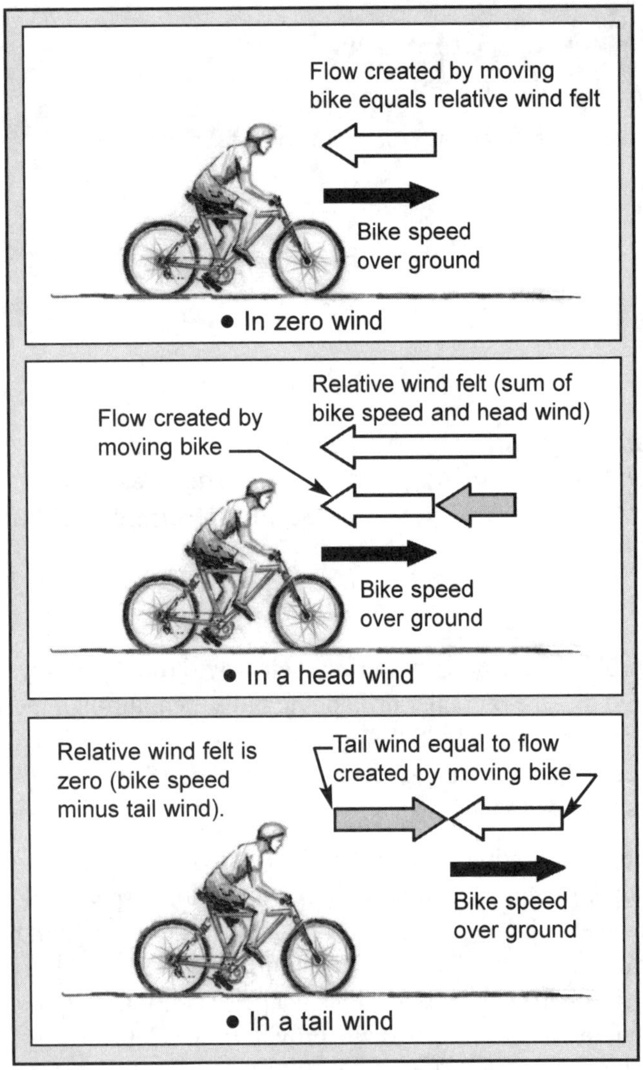

Figure 3-1: Relative Wind

To understand this concept, imagine riding your bicycle. If you are standing still in a wind you will feel a certain wind force. If there is no wind but you ride along, you again feel a wind. This wind is the relative wind you have created by your relative movement through the still air. Now if you pedal along into a wind blowing toward you, you will feel even more relative wind. On the other hand, if you travel with the wind behind you, you may adjust your speed so you feel no wind at all. The relative wind you feel is zero even though there is an actual wind blowing. Figure 3-1 helps you get the picture.

We should point out that a glider in wind is completely different from a bicycle. Whereas the bicycle is in contact with the ground and propels itself forward even in a head wind, the glider has no contact with the ground and is only propelled through the air an amount equal to its normal airspeed (we'll see below what this normal airspeed is). Therefore the glider's speed over the ground varies with the wind direction and strength in relation to the glider's flying airspeed.

This idea of relative wind is important for a couple of reasons. First all wings require a certain minimum relative wind or airflow in order to produce the necessary lift. Secondly, since our wings need a relative flow to inflate and lift us away from the hill, *we must always take off facing into the wind.* If for example, we try to take off with a wind behind us, we will encounter what we experienced on our bicycle with the wind behind us: no relative wind and therefore no inflation.

> **First Rule of Flight**
> Always takeoff and land aiming directly into the wind. Slight violations of this rule can be allowed when you gain a bit of experience. In general, pilots always launch within 25° of the wind direction.

We added the landing part in the rule not because it is necessary to fly into the wind to maintain inflation, but because flying into the wind slows our speed down in relation to the ground. In Chapter 5 and Chapter 8 we will get a better understanding of landing matters. For now we must develop the idea that the wind the glider's wing "feels" is the relative wind. Once the glider is up in the air it produces the necessary relative wind by gliding forward. Even with a tail wind our wings will glide and produce their own relative wind. We can summarize our discussion:

Speed in the Air

- The *relative wind* is the flow a wing "feels". Is is always *equal and opposite* to its flying *airspeed* and path.
- A glider's *groundspeed* is a combination of its *airspeed* and the *windspeed.*

Figure 3-2 should make these ideas clear. You will note that the groundspeed varies with the wind while the airspeed and relative wind always stay the same,

Figure 3-2: Airspeed, Windspeed and Groundspeed

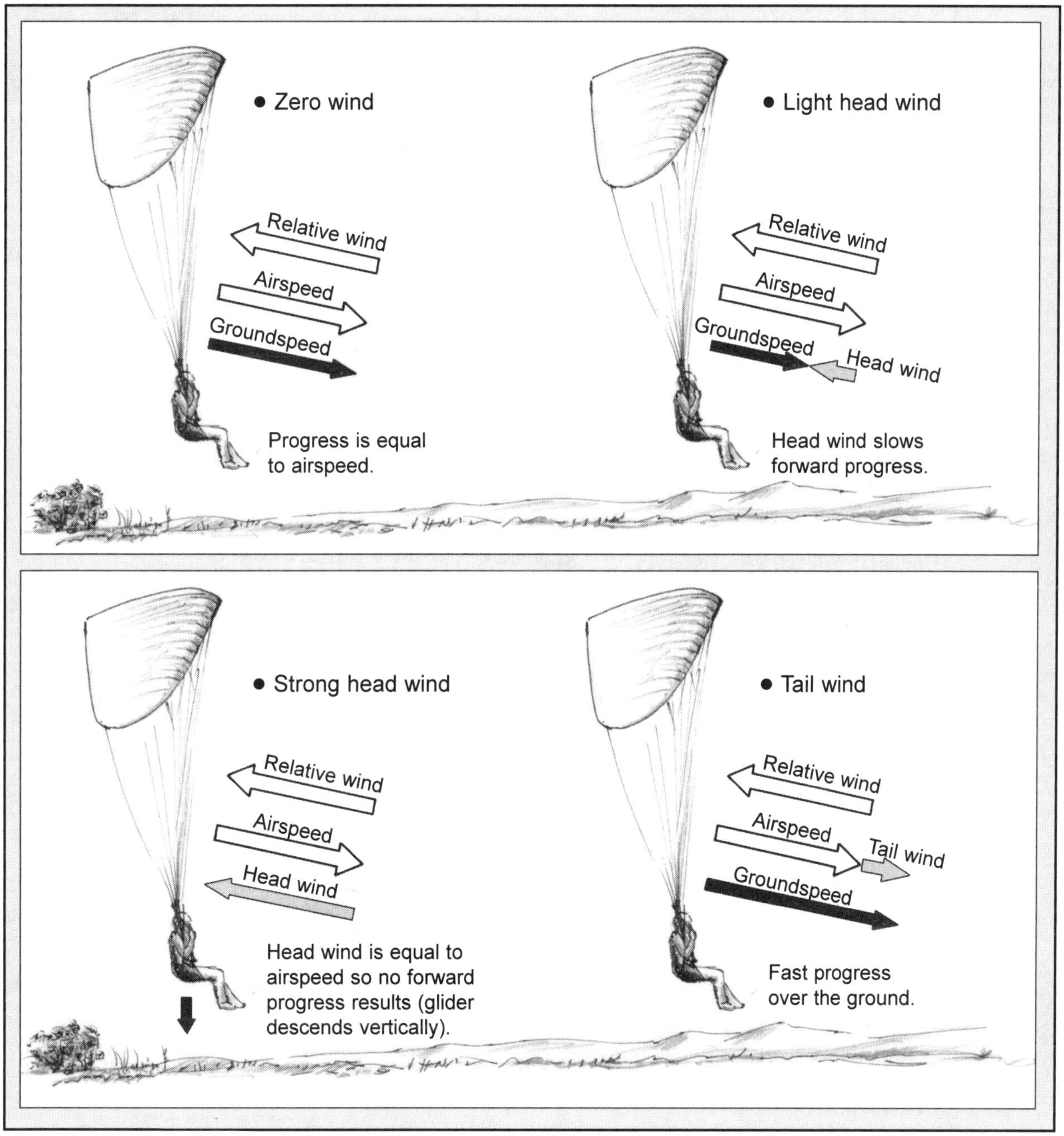

which is different from the bicycle situation. (We can vary the relative wind on our wings by making speed controls, but we aren't concerned with controls here, just the concept of relative wind.)

We have spent some time on the meaning of relative wind and groundspeed, for of all the principles that govern how wings fly, these are perhaps the most important practical matters as we shall see throughout our training. Concepts such as stalls, ground tracking, control, stability, takeoffs and landings all depend on relative wind or groundspeed or both. For example, now we can immediately see why we must takeoff and land facing into the wind: this is the flying direction which minimizes our ground speed.

✓ *Inflation*

Now let's continue the inflation process we began in the previous section. Follow figure 3-3 as we offer a brief description. As we see in (a), the wind entering the front of the canopy begins inflation and a little lift is produced by the air flowing over it (we describe lift and drag in the next section). If there is no wind, the same thing happens (at a slightly slower rate) since pulling the canopy forward creates a relative wind. Also, note that just the leading edge of the canopy is being pulled initially since it is laid out so the leading edge is furthest from the pilot.

Next we see in (b) that the process has continued. The glider continues to inflate and lift due mainly to the flow at the leading edge area. Now a force opposing the forward pull on the lines has set up. This force can be felt by the pilot as

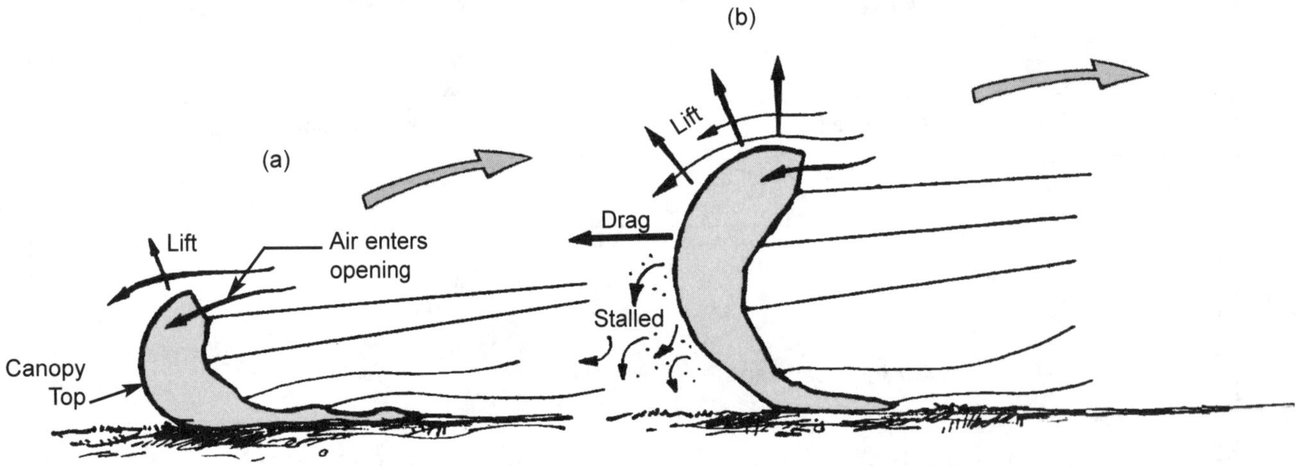

Figure 3-3: Glider Inflation

a resistance to his or her forward run. It is due mainly to the drag of the canopy. Note we have shown a stalled area. This is where the air is not flowing smoothly.

In (c) the entire canopy is off the ground. More lift is being created so it accelerates rapidly upward. Also the force opposing the pilot's forward run is at a maximum.

Finally, in (d) the canopy has nearly reached the apex of its arc above the pilot. Typically it reaches this point in 2 seconds. The drag force has greatly diminished, lift force has increased, and smooth airflow has been established over most of the wing. The pilot is ready to commit aviation.

✓ Forming a Wing

Let's take a closer look at what keeps a wing inflated and in good shape. First turn to figure 3-4. Here we see the airflow in flight above and below a cross-section of the wing similar to that shown in figure 2-2. We also see the flow entering the front openings. The air here doesn't flow continuously, but builds up pressure to a certain level then stops flowing. It initially flows into the canopy to inflate it then becomes stagnant. However, the pressure is maintained by the flow trying to get inside the canopy much like the airflow when you fill your car tire. Once the pressure in the tire equals that in the air hose the flow stops into the tire.

This inflation sequence shown on both pages indicates why there is a hard pull on the canopy when it is low to the ground. This drag force goes away once the canopy is aloft and the forces become mostly lift.

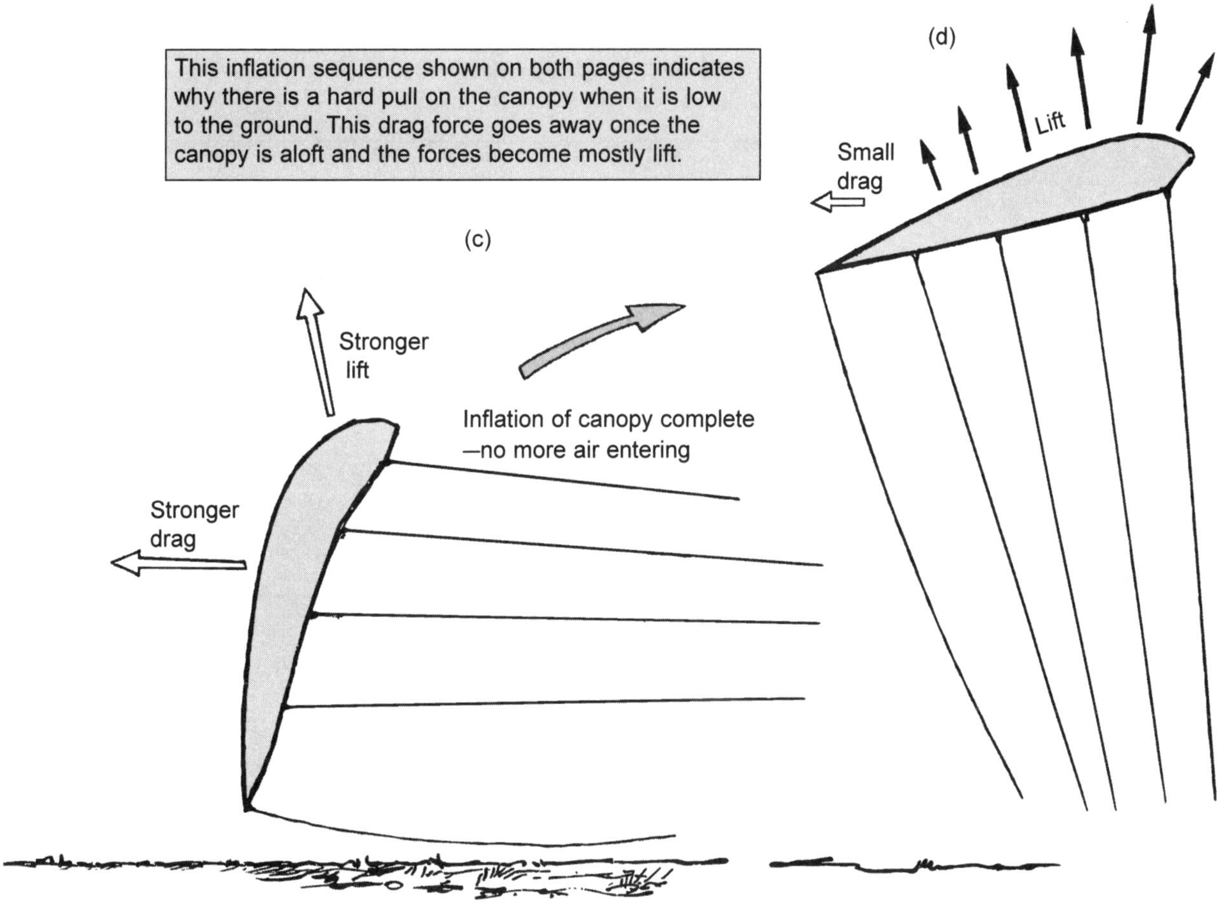

Figure 3-3 (continued): Glider Inflation

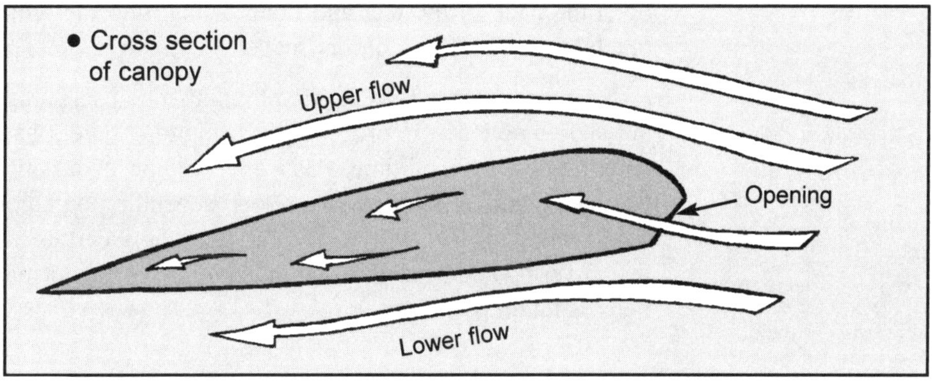

Figure 3-4: Canopy Pressurization

How *much* pressure builds up in the wing depends on how fast you are flying and how the cell openings are oriented to the flow. How *fast* the canopy pressure builds up depends on the size of the canopy, and the size of the openings. As figure 3-5 shows, the air entering the front of the wing moves to the back and then to the side to inflate the tips, which usually do not have cell openings. The air flows sideways through the crossports. Once inflation is complete the internal flow stops and the entire wing is at the same internal pressure.

The pressure inside the wing is known as *dynamic pressure* because it is caused by the air's molecules moving against the wing (or vice versa) due to the airflow. Another form of air pressure is called *static pressure*. This pressure is simply due to the weight of the atmosphere. The average static pressure of the air at sea level is 14.7 pounds per square inch (1.0 kg/cm^2). It is mainly the effects of dynamic pressure that keep our wings flying.

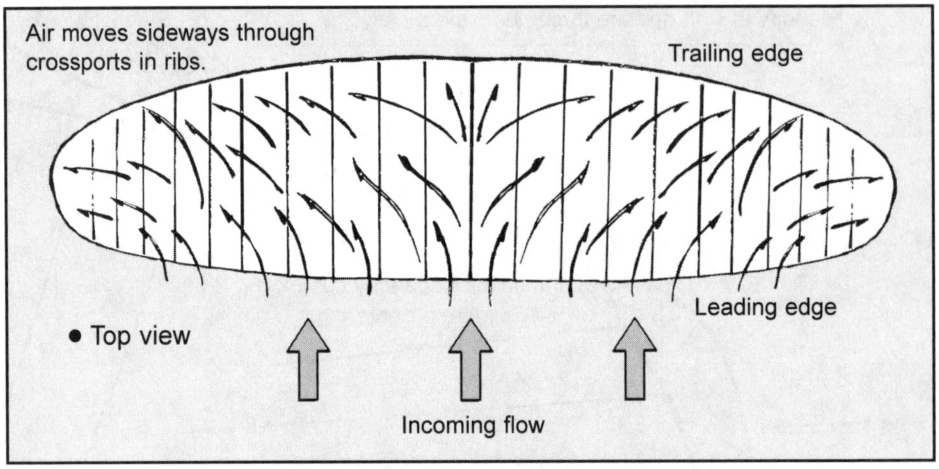

Figure 3-5: Initial Inflation Flow

✓ Maintaining Form

We are often asked why the rear of a paraglider is closed and not opened like a windsock. The easy answer is that if air was allowed to flow freely through the canopy interior, no pressurization would occur and the canopy would be much less inclined to maintain its shape, especially at the tips. You can see this effect on a windsock when it flaps in higher winds. Better yet, take a plastic bag and hold it open into the wind. It will fill with air and pull out to hold its shape. Now cut open the back of the bag and hold it open into the wind. You will see that it is much less inclined to hold its shape.

So internal pressure is the first thing that helps a paraglider wing hold its form. We should note that in order for the wing to deflate it has to expel air from inside which takes a bit of time. Also, the volume of air inside a paraglider weighs between 6 and 10 lbs (2.7 to 4.5 kg), depending on canopy size. This air weight combined with the canopy weight gives it a certain amount of inertia during maneuvering and in turbulence. In some ways the air mass inside the canopy can help maintain its shape.

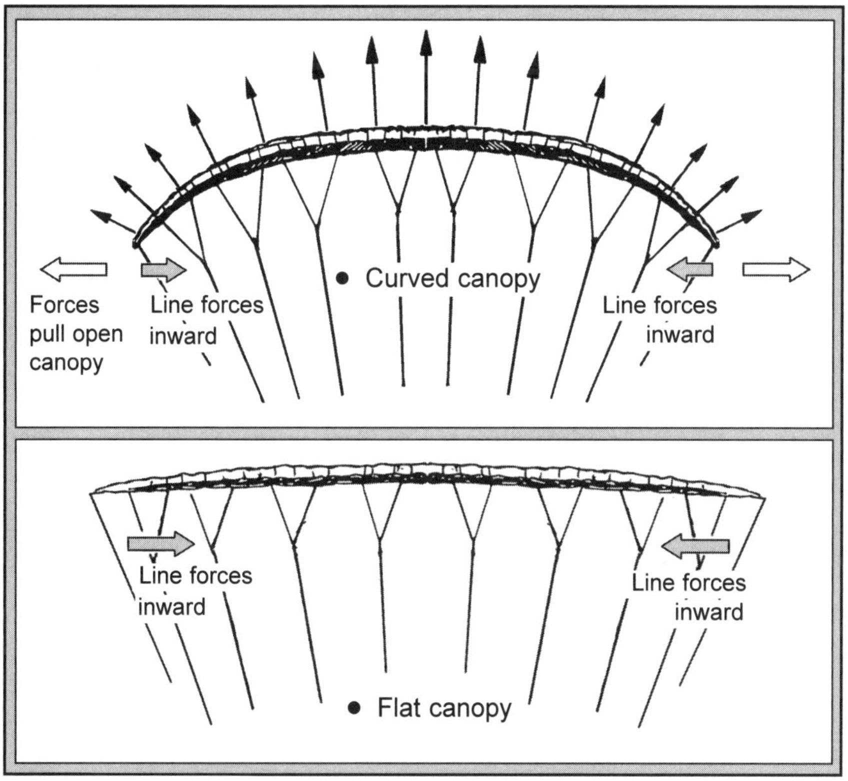

Figure 3-6: Outward Lift Helps Keep Canopy Open

Finally, the canopy is held open by its arced shape. Look at figure 3-6 which shows a canopy head on. We have drawn arrows to represent the lift forces on the wing. As you progress further towards the tips you see the lift forces angle outward more and more. This outward angling produces outward forces that serve to pull the tips outward. If a canopy was made to lie flat as shown, the slight inward force of the suspension lines or variations in the air would pull the tips in like an accordion.

Now we must turn our attention to the details of how our wings fly.

CREATING LIFT

Watch a big bird's slow wing beats and it's easy to imagine that it is rowing against the air. But suddenly it stops flapping, and still it glides along, barely losing altitude. It is apparent that something is holding the bird up. That something is a force we call lift, which is developed by the wings.

To understand how a wing produces lift, simply realize that air is a fluid just like water. If you run your hand through water, you feel a force as the water is deflected. In fact, the force you feel on your hand is exactly opposite to the force the water undergoes. You can feel the same effect in air, but you must move faster since air is so much less dense than water. Try holding your hand out a car window while driving at a moderate speed to feel lift force from the air.

✓ *The Airfoil*

A flat surface can be used to form a wing and deflect the air, as shown in figure 3-7. However, this flat surface is not very efficient because as it creates lift it also creates a lot of drag which retards the wing (see below).

Airfoil - A curved surface designed to deflect the air smoothly to produce as much lift with as little drag as possible. The cross-sectional shape of a wing is an airfoil.

Through much experimentation over the years, aircraft designers learned to model their wings after the birds and create an *airfoil*.

When the surfaces of the wing are curved, the air is gently coaxed over the airfoil without abrupt changes

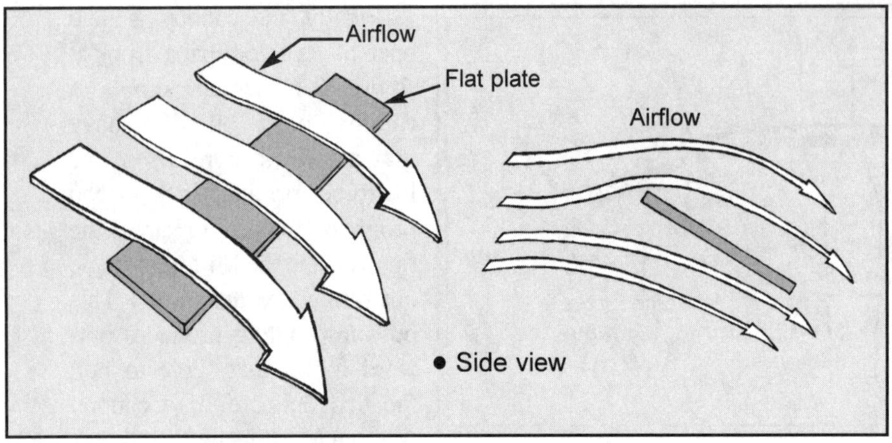

Figure 3-7: Airflow Around a Flat Plate

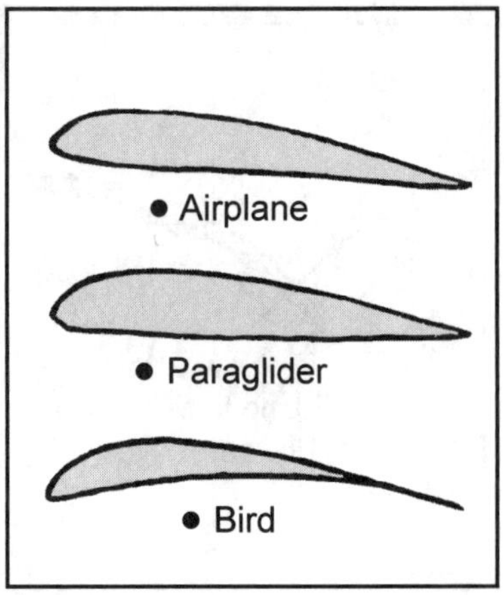

Figure 3-8: Different Airfoils

and thus creates less drag. The cross-sections of various airfoils are shown in figure 3-8. Note how the paraglider airfoil is a shape much like that of an airplane.

✓ Lift and Drag

Now we can see that the air gets deflected from both the upper and lower surfaces of the airfoil. The upper surface pulls the air down as it passes while the lower surface pushes it down. In fact, the airfoil plows some air ahead of it like the bow wake of a boat. This effect causes a slowing of the air below the wing and a speeding up of the air above the wing as the air moves out of the way of the passing wing. The result is more deflection forces (about 2/3 of the total) occurring on top of the wing than the bottom.

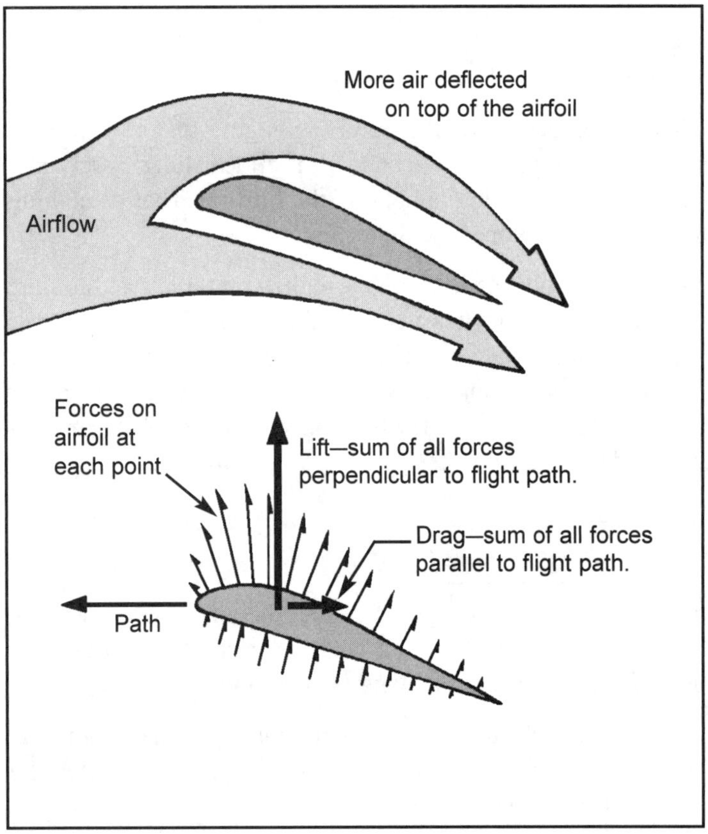

Figure 3-9: Forces in an Airfoil

If we realize that whatever force we impart to the air is equally imparted to the airfoil in the opposite direction (Sir Isaac Newton's third law), we can depict the force along our airfoil. This is shown in figure 3-9 with the top drawing showing the deflected airflow and the bottom drawing showing the forces the airfoil feels.

We now can define *lift* as the upward forces and *drag* as the rearward forces on our airfoil or wing. Think about it for a minute and you will see that we want to be creating as much lift as possible to suspend our airborne craft while producing as little drag as possible so our forward progress isn't impeded.

The Effect of Forces

We have been speaking as if our wings move along horizontally. This would be the case if we had a propeller to pull us along. However, if we are without an engine we must get our power from gravity. This isn't hard to understand if we imagine a toboggan being pulled down a slope by gravity. As figure 3-10 shows, gravity pulling downward is translated to forward motion by the action of the toboggan surface on the snow or the airfoil on the air.

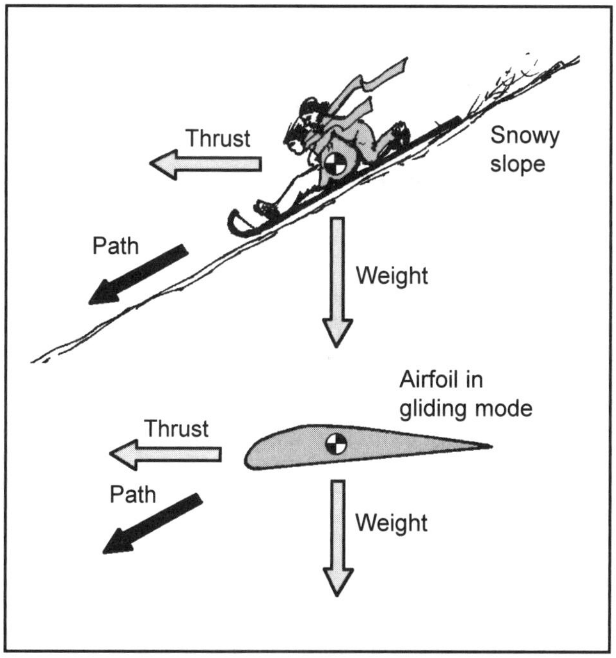

Figure 3-10: The Pull of Gravity

✓ *Angle of Attack*

Now we come to an important feature on our wings: unlike the toboggan that must run along the snow's surface, our wings can be angled in the air to alter our flight path and speed. The way we describe this effect is by the term *angle of attack.*

Angle of Attack – The angle a wing makes with the onrushing air. Since the overall flow the airfoil "sees" is exactly opposite its flight path, we can define angle of attack as the angle between the wing itself and the flight path.

Figure 3-11 illustrates an airfoil's angle of attack and shows various angles of attack. In a paraglider we control angle of attack (and thus speed) with the control lines. If we release both lines the canopy accelerates and the trailing edge is level. Our angle of attack is low as shown. However, as we start applying control force, we pull down the trailing edge. This change increases the angle of attack because the canopy retards a bit in relation to our body. In addition, the angle of the wing is measured with the straight line drawn from the trailing edge to the leading edge (defined in Chapter 2 as the chord). You can see from the figure that the more control line we apply the greater the angle of attack. Also, the more the airflow will be deflected as it passes over the wing.

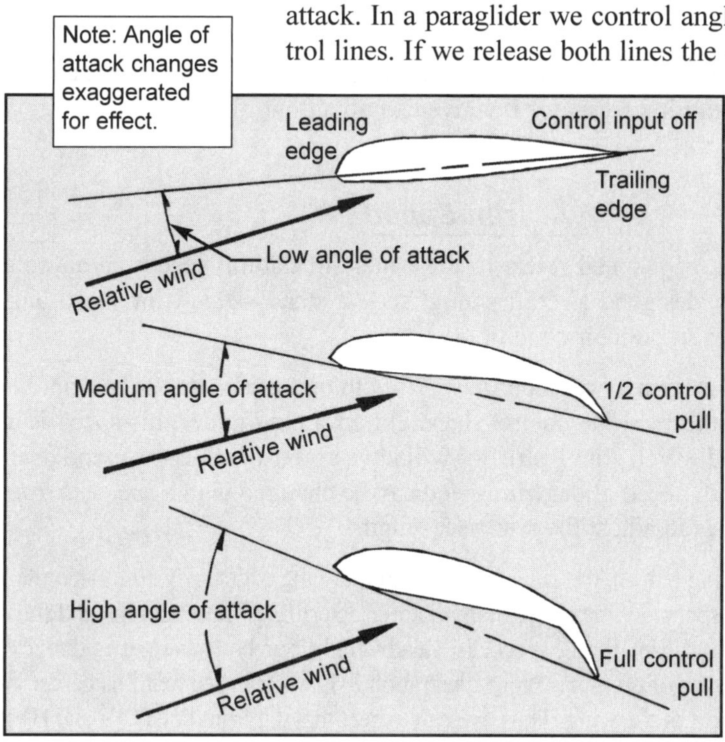

Figure 3-11: Different Angles of Attack

With a high angle of attack we fly slowly while a lower angle of attack results in faster flying speeds. Think again of the toboggan sliding faster down a steeper hill.

THE BALANCING ACT

In flight the forces on our wings are normally balanced. If this were not so, we would not be able to fly comfortably because we would be constantly making corrections to stay on top of things. Figure 3-12 shows the balance. The downward pointing arrow represents the weight of our body and glider—the total weight of the flying system at the *center of mass*. The vertical arrow is the total of all the lift and drag forces on the wing. We call this the *resultant* since it is the result of the net aerodynamic (caused by the airflow) forces. The location of the resultant is called the *center of pressure*, which makes sense when we recall that lift and drag are created by deflection pressures.

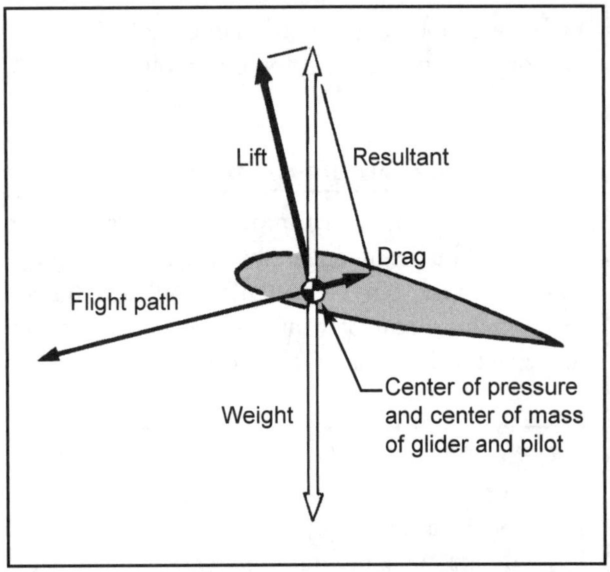

Figure 3-12: Forces Applied to Paraglider

We have also illustrated the lift and drag forces with arrows. Note that lift is not pointing directly up since we are not flying horizontally. Actually, since both lift and drag are part of the overall aerodynamic forces on the wing, we need to define them better. By convention:

> **Lift** is the sum of all aerodynamic forces *perpendicular* to our flight path.
> **Drag** is the sum of all the aerodynamic forces *parallel* and opposite to our flight path.

SPEED CONTROL

A paraglider is a small aircraft with great control possibilities. We can speed up, slow down and turn. First let us see how we can affect our glider's speed.

✓ Trim Speed

In most aircraft, trim speed refers to the hands-off natural speed the aircraft seeks. All properly designed aircraft should seek a steady-state trim speed and stay at that speed in smooth air conditions.

Pilots often refer to the normal flying position with the controls at eye level as "trim". Such a usage will confuse airplane and hang glider pilots. We advise you to avoid "hands off" flying in a paraglider.

With a paraglider we may consider *trim speed* to occur with the weight on the hands held comfortably on the control lines. On a training glider this speed is a normal flying speed which your instructor will have you experience over and over again. On a more advanced glider trim speed can be changed with a *speed stirrup* or *trimmers* (buckles to adjust the rear riser length).

When we are flying at trim, the angle of attack of the wing is determined by the relative length of the suspension lines. The manufacturer determines these lengths and thus the angle of attack and trim flying speed. In steady flight our body weight is directly below the wing's center of pressure. From illustration 3-12 we can see that this center is located in the front part of the wing. Thus, we can expect that the front lines (A's and B's) carry more weight than the rear ones. In fact, these lines are often of larger diameter.

✓ Changing Speed

In the course of your training and certainly later in your more advanced flying, you will be frequently changing your flying speed. By now you have probably surmised that you do this by applying a pull on your control lines. Here we will not go into too much detail about speed control, for later we discuss it with a view of how to achieve the best performance. Suffice it to say that you must learn a few important control positions initially, then later you'll develop a more thorough understanding of how to fine-tune your airspeed.

We have already discussed trim speed. That's as fast as you can go on a training/beginner glider. We'll call this position "off controls." The next speed you should consider is *maneuvering speed*. We give it this name because it's a good speed for flying close to the terrain and turning. This speed is set with the controls about at your eyes. We'll refer to the position as "quarter controls." Figure 3-13 illustrates this position and others.

Figure 3-13: Control Positions

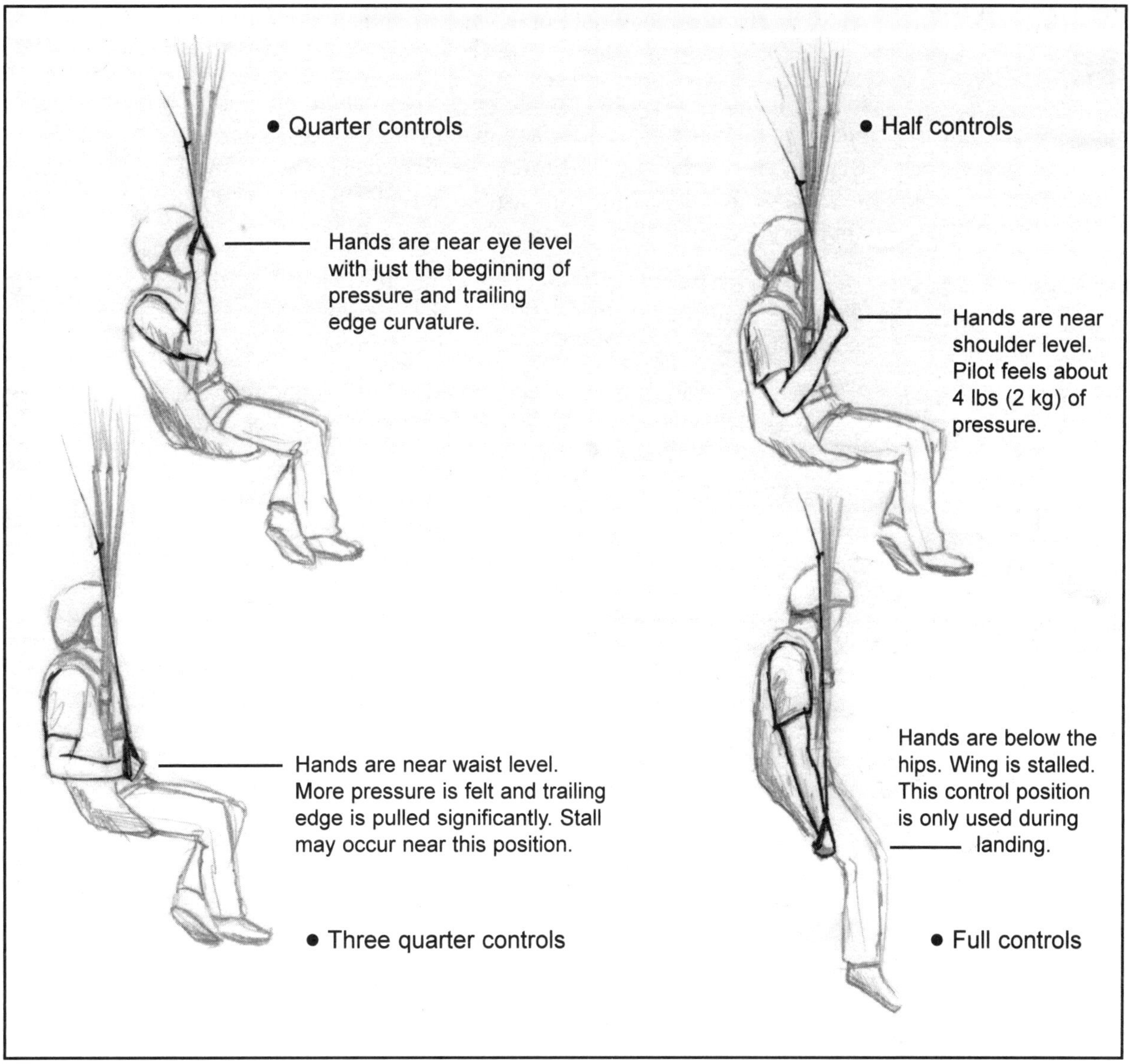

The "half-controls" position will be discussed in Chapter 9 when we speak of minimum sink rate flying. The "three quarter controls" position should not be used by novice pilots as it is dangerously close to a stall (see our next section). We illustrate the "full controls" position here because this position nearly stops the glider and is used for landings. Flying with half or more control pulled is known as being "deep in the controls" in paragliding lingo.

✓ Stalls

Stalls are major no-nos in most aircraft unless lots of altitude is available. A stall is even more serious in a paraglider because it usually results in a loss of the wing's shape and position above you. Only experts attempt stalls in very carefully controlled situations (over water usually).

What is a stall? Simply put, a *stall* occur when our angle of attack gets so high that the airflow cannot make the great change of direction required to follow the contour of the upper surface. As a result, the smooth flow separates from the wing, swirls or eddies result, drag increases greatly and lift dies away. Figure 3-14 illustrates some of these effects.

The fact that airspeed is related directly to angle of attack means that a stall occurs at a given airspeed for any given wing with the same loading (as long as we aren't accelerating). Therefore you will often hear pilots speaking of a glider's *stall speed.* For a paraglider this speed is typically from 13 to 15 mph (21 to 24 km/h) and possibly slightly less for a training glider.

A beginner paragliding pilot should not lower the controls below his or her shoulders (and should fly mostly with the controls at eye level) to avoid stalls, except during landings. We learn more about landings in Chapter 5.

When your paraglider stalls you typically lose control over your glider and lose altitude as well. Different types of stalls and stall recovery are covered in later chapters. For now we must be aware of stalls and avoid them during our training (and throughout our flying career). Fortunately, this is an easy thing to do for the most part.

⚠ **Caution:** *To avoid stalls, remember the first commandment of flight: Maintain thy airspeed.*

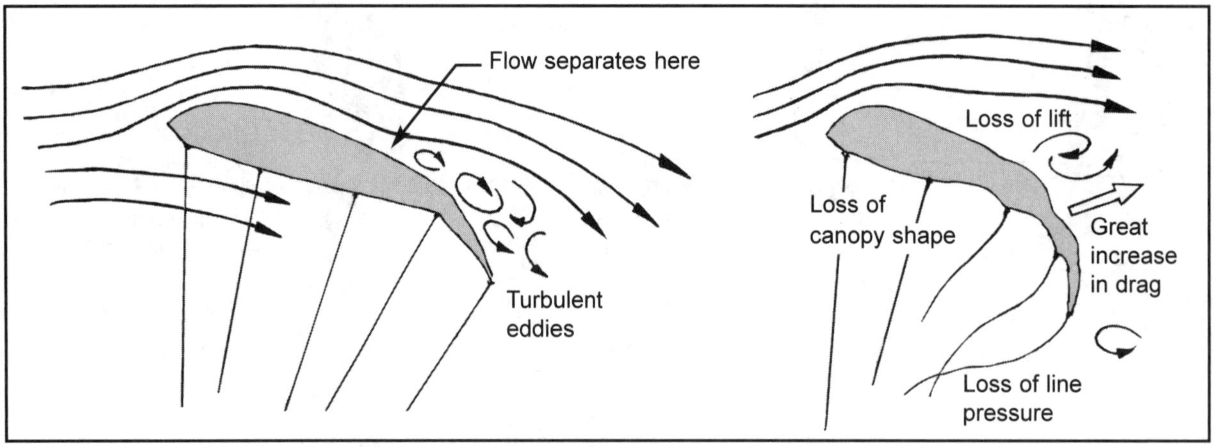

Figure 3-14: Stall of a Paraglider

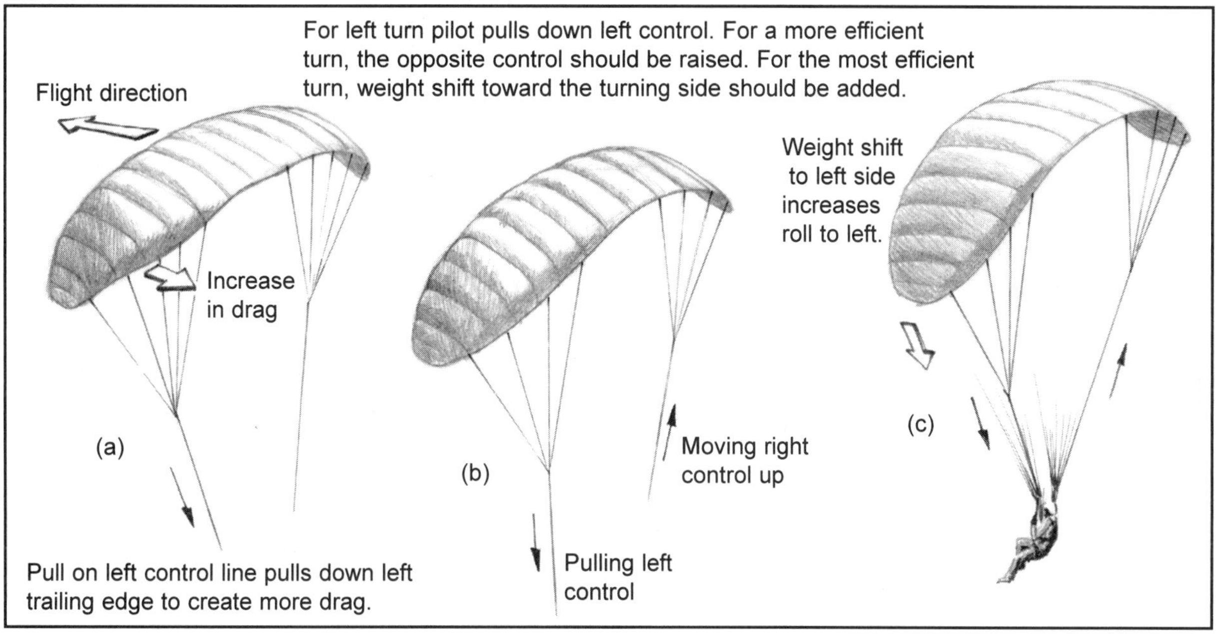

Figure 3-15: Turn Control Variations

✓ *A Good Turn*

Turns are fun and functional. Naturally, we want to be able to go where our heart desires and turns do the trick. Turns have a wide variety of techniques and practices which we'll learn in later chapters. Here we'll simply point out how a turn is produced on the glider so you understand the principle when you start doing them.

Figure 3-16: The Meaning of Bank Angle

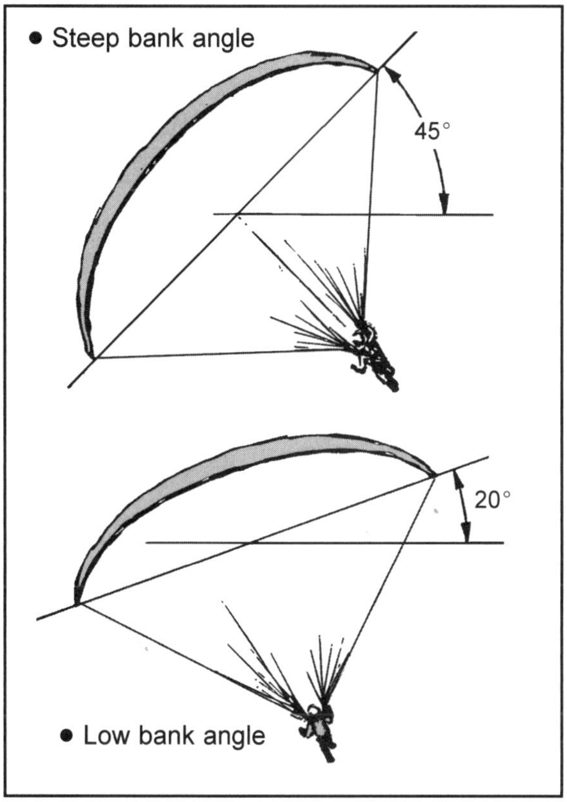

The simplest way to perform a turn is to maintain good airspeed (controls at eye level) and pull down on one toggle. Figure 3-15 (a) shows the effect on the glider. The wing with the pulled down trailing edge experiences more drag (a higher angle of attack) so it slows and you turn toward that side.

A more efficient turn or coordinated turn is accomplished if you pull down on one control toggle while raising the other as in (b). This procedure has the effect of allowing the outside wing to speed up for a smoother turn.

The final technique we'll mention is adding a weight-shift. This method involves twisting your hips so your weight sits on the buttock which is on the side you want to turn towards (c). The effect is to pull down the risers on that side while allowing the risers on the opposite side to rise a bit. The wing then more readily enters a *bank angle* and *rolls*. You will learn that you can turn a glider by weight shift alone, but weight shift combined with the dual control input method is the most efficient turn.

A bank and a roll (mentioned above) are essentially the same thing. When one wing tip goes up and the other down, the wing is banked and has rolled. Figure 3-16 shows what we mean by bank angle or roll with a rear view of a glider. Note that we measure the bank angle with respect to the horizon. Later, when we learn to turn we'll talk about "low banked turns" or "20° banked turns." You will then understand the lingo.

SUMMARY

We have begun our study of how paraglider wings work and behave in the air. We presented many ideas which may take some thought, but the reward is an easier grasp of the practical side of flying. Soon, when you learn to perform controls you will already understand how they work. Such a good start may accelerate your learning and at the very least let you know that your glider flies on the basis of the timeless laws of physics and is a dependable aircraft indeed.

There are many more things to learn about the glider's relationship with the air, but this knowledge will come later as you gain experience and we add insights a bit at a time. Learning such as this with its potential to produce real pleasure is not a chore. As you study, keep thinking how you will relate what you are learning to your aerial experience. Next we try our wings.

TEST YOURSELF (*Answers in Appendix II*)

1. The name of the airflow a glider "feels" in flight is ________________.
2. A very important concept in paragliding is the relationship between airspeed, windspeed and groundspeed. In a head wind our groundspeed will be (*more, less*) than our airspeed.
3. In a tail wind our relative wind will be (*higher, lower, the same*) as with no wind.
4. In a tail wind the wind will (*add to, subtract from*) our groundspeed.
5. Our canopies remain inflated through (pick all that applies):
    - a. Dynamic pressure
    - b. Static pressure
    - c. Head wind effects
    - d. The effect of the curved shape
    - e. Drag forces
    - f. Magical incantations
6. We control our airspeed by ________________________.
7. A stall occurs from too much ______________ and too little ______________.
8. Stalls are dangerous because of ________________________ and ________________________ .

Chapter 4

FLIGHT FAMILIARIZATION

" The centipede was happy quite, until a toad in fun,
Said, 'Pray, which leg after which?'
That worked her mind to such a pitch,
She lay distracted in the ditch, considering how to run."
— Anonymous

We have taken the time to prepare ourselves with a little theory and now are ready for the real thing: handling the glider. This is our first hands-on experience where we will be making the canopy fly. The learning is easy and fun, but we have to progress in little steps or the canopy is likely to end up at our feet in a tangle of synthetic line and cloth looking more like laundry than a wing.

But don't worry, in a few tries you'll be inflating the canopy with reliability and floating it as if it were a balloon. That's the beauty of paragliding—it's so easy to learn. But this ease of learning is a bit deceiving. The truth is you can learn to fly in a day, and no doubt you will be flying on your first outing. However, there are many safety-related matters to attend to that require time, thought and practice in order to master. So, as we begin our skill development, you'll notice two things: we will always emphasize the *gradual approach* and we strive to insure *safety through good judgement*. In fact, these two items will be a recurrent theme throughout all the practical skill development in this book. It is our goal to make them your guiding principle throughout your entire flying experience.

UNPACKING, LAYING OUT AND PACKING

The first part of practical flying is taking a glider out of its carrying bag and opening it out properly. In training, as in later flight, you normally lay the canopy out in position to launch, so here we'll learn the proper laying out practice. In addition, we'll learn how to check out the entire glider in a thorough manner which is called a preflight.

Your Goal: *To learn to properly open and stow a glider as well as lay it out, detangle lines and preflight it.*

✓ Spreading Your Wings

The first thing we do at a training site or flying hill is check the wind. Once we're satisfied it's fine (not too strong, not too gusty, not too cross) we locate a clear place to lay out our canopy with the correct orientation. Upright sticks and weeds or oversized rocks are problems because they don't allow the canopy to lie flat and may snag lines or fabric when you attempt to pull the wing up.

When you begin, you may not be able to judge how much space you need to spread your wing, so normally your instructor will help you. Keep in mind, however, a square of about 10 or 11 good paces on a side is the minimum. With practice you'll be able to easily judge the size of the necessary clear area.

Now open up the pack and pull out the canopy. The harness may be separate or it may be folded up in the wing. If it's separate, set it aside. Place the canopy at the top of its layout area and unfold it to its full chord length as shown in figure 4-1. The leading edge should be placed at the furthest downwind point in the clear area. You can identify the leading edge because it has the cell openings and

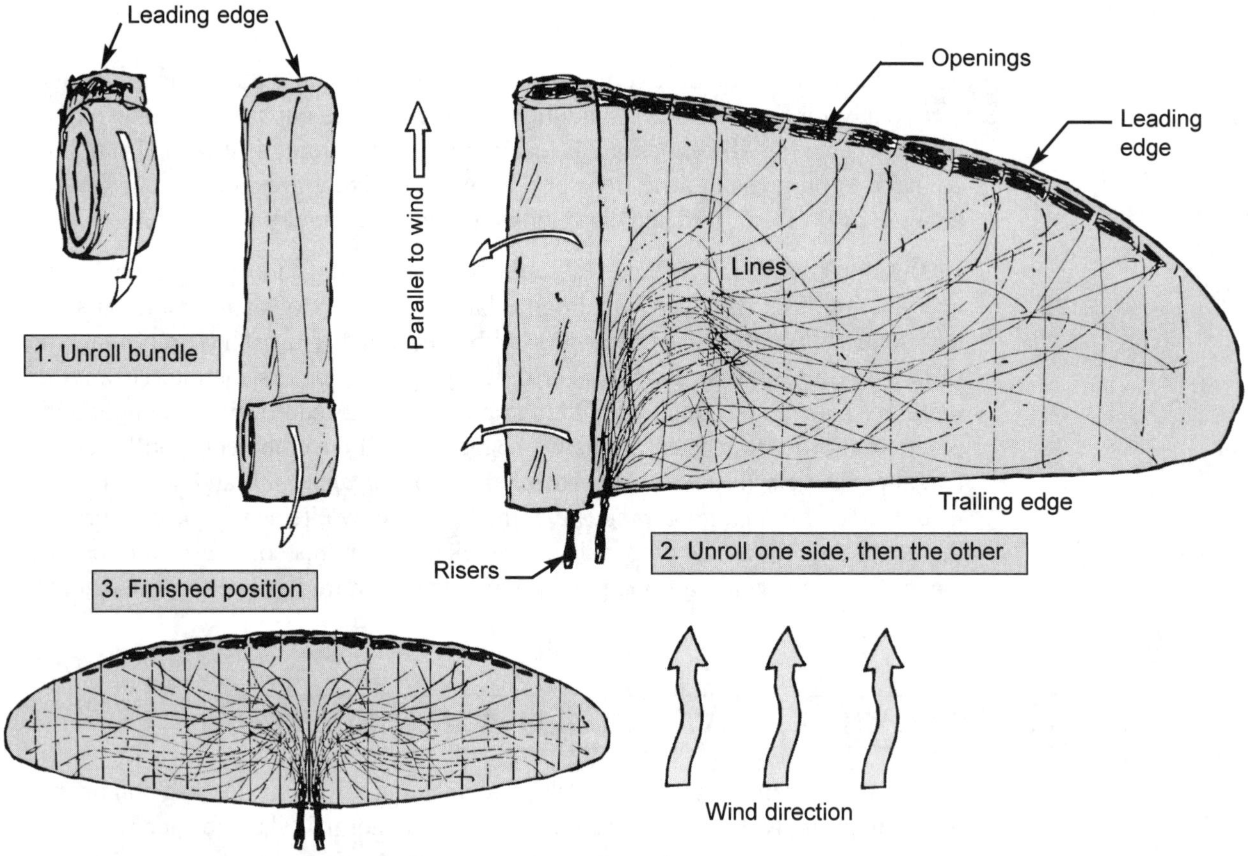

Figure 4-1: Laying out Canopy

it is normally the first part to open. Unfold the long bundle as shown and make sure the cigar-shaped package is oriented perfectly parallel to the wind. Next roll out one wing and then the other. From the figure you can see that the canopy is lying on its back (the upper surface) with the trailing edge upwind. This is the proper launch position.

Once the wings are spread-eagled on the ground, double-check to make sure the leading edge is furthest downwind (the direction away from your intended heading) and the trailing edge is upwind or facing the launch direction. You'll probably only make the mistake of opening up backwards once—it's embarrassing! This orientation will allow you to pull the canopy up and move into the wind.

✓ Clearing the Lines

Now look at the lines lying on the canopy. If the harness is still attached, gently lift it off and detangle any lines hooked on the harness protrusions. Walk the harness away from the canopy in the upwind direction until the lines are almost fully extended. If the harness is separate from the rest of the glider, locate the risers, lift them gently and extend them upwind nearly full length of the lines one at a time.

One of the best ways to insure clean inflations is to have clear lines.

Before we can proceed we must make sure all the lines are in proper order and not entangled. Several things can occur during the packing and unpacking. The lines may entangle amongst themselves (remember those fishing reel snarls when you were a kid) or they may catch up on part of the harness. If the harness is removed the risers may twist or loop through some of the lines.

The way to correct the above problems is to perform a line check. The simplest way to do this check is to stand facing the canopy with the risers of one side in your hand (with the harness attached, let it dangle). Move toward the canopy if necessary so you have enough slack and lift the A riser as shown in figure 4-2. How do you know what the A riser is? It's the one attached to the lines going to the canopy's leading edge. Since the leading edge is furthest away from you, the A lines will be the tightest. Also note that often the A riser will be identified by a color patch sewn on it near the quick link.

Although some gliders have color-coded lines, it is important for you to learn to sort out the risers by their relationships to one another whether they are slack, under tension or overhead. Finding the correct riser without having to look at it is an important skill, much like reaching for the stick shift of your car without looking. Begin learning this skill at once by thinking correctly about riser relationships: the A's in front, the rears in back and the rest in between.

First lift the A riser and check to make sure there are no twists or entanglement in the A lines. Occasionally some of the other lines may be looped over the As and be held in a sort of slip knot. Shake the A lines to try to remove any entanglement. If the knot doesn't shake out, lay the risers and lines on the ground and walk in to undo the knot by hand. Be careful—don't step on the lines!

Once the A lines are worry-free, lift all the risers and pull the control lines out from the bottom and to the side as shown in figure 4-2. Shake all the lines if neces-

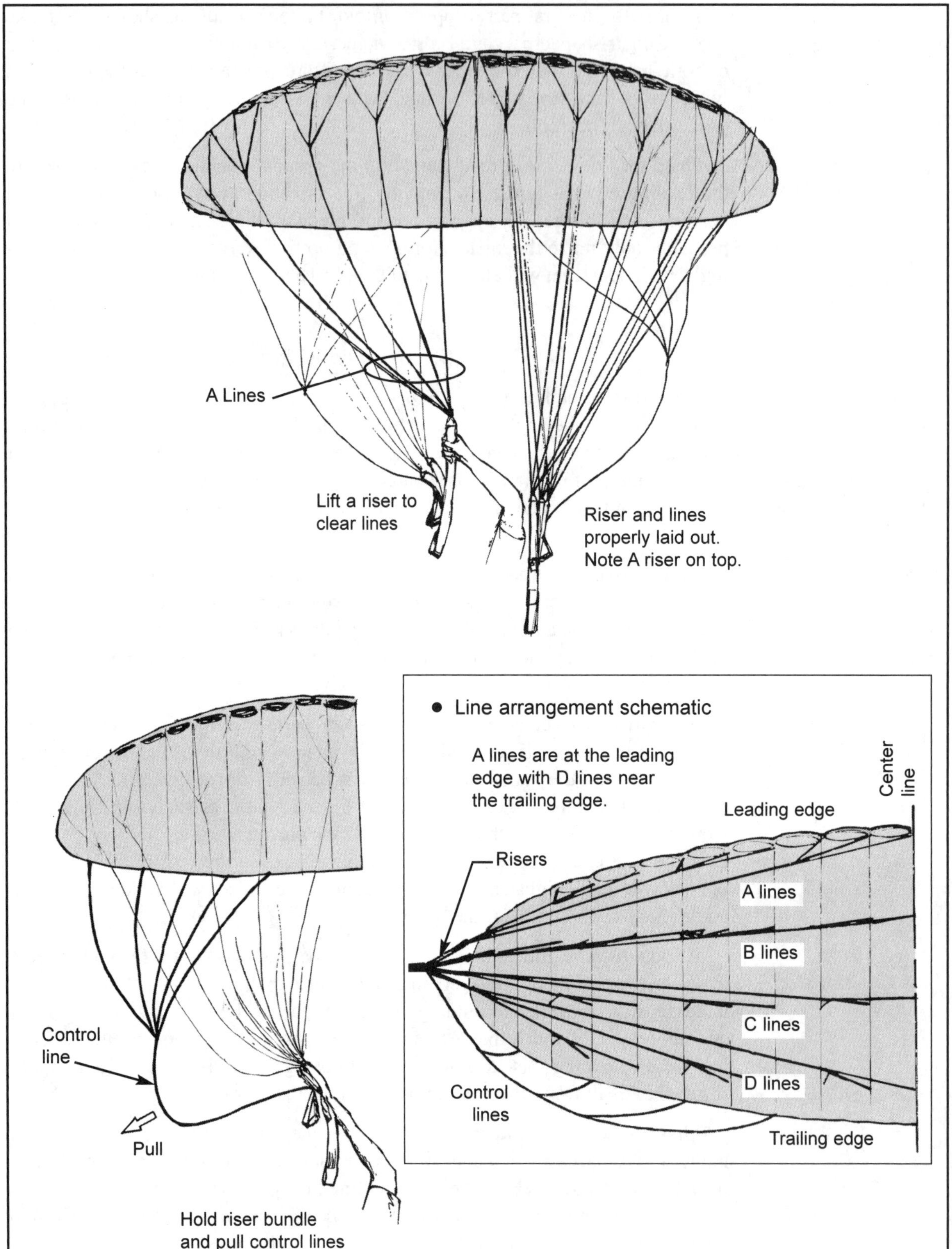

Figure 4-2: Line Clearing

sary to free the control lines. Again, if any unshakable snags exist, set things down and go fix them by hand. After you have cleared one side, lay the risers on the ground in front of the canopy with the A risers up and the A lines extended nearly full length as shown in the figure. Now repeat the process with the other side.

Line Clearing

- First clear the A lines by lifting and shaking the A riser.
- Second clear the control lines by lifting all the risers and pulling the line to the side.

In the process of clearing the A lines, you should look at the other lines and check to see that they are generally clear. If major knotting occurred in the A's, chances are there are knots in the others as well. You can check them by holding each riser up with the A riser, one after another. You can even put a checked riser over your head to hold it while you go on the check the next one.

The most serious problem occurs when the risers are packed free from the harness and they loop through some lines. The result is some lines or risers are twisted and some may not be. The whole problem can be confusing until you learn the simple technique:

Straightening Twisted Lines

Start with the A lines and separate them untwisted back to the riser. If the riser itself is twisted, the end of the risers has passed between some of the other risers as shown in figure 4-3. Hold the A riser up and let the rest dangle. It's usually clear where you have to pass the riser end to straighten the system. If it still looks confusing, pull up the next riser with the A riser and make sure it is straight. Hold these two and things should be apparent.

Sometimes the riser end can pass through only some of the lines from a riser. If this is the case, all the lines from that riser cannot be straightened. Again start with the A lines and guide them back to the riser. If they are twisted in different directions, separate them and it should be clear how the riser must be passed to straighten the lot. If the As are all twisted the same, go on to the next riser and make your assessment. Repeat the process until you find the source of the problem.

Hint Do not make a correction unless you are reasonably sure of the cure—don't make matters worse. With an assistant straightening the D lines (rearmost) while you lift the As, the problem and solution is often more clear.

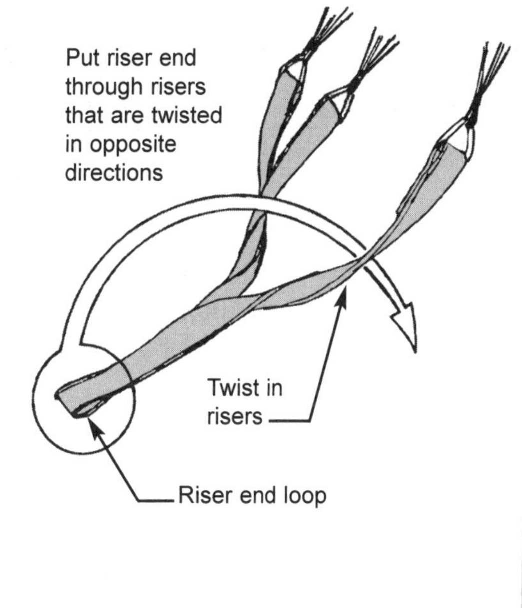

Figure 4-3: Untwisting Lines and Risers

Preventing twisted lines by stowing them carefully is the best policy (see our later discussion).

✓ Horseshoeing the Canopy

This is a little trick which can help ease your inflations, especially in light wind. The idea is to ensure that the center of the canopy inflates first so that the tips don't fold inward. The method involves laying the canopy out in a horseshoe shape as shown in figure 4-4.

If you think for a moment, you'll see that the horseshoe shape mimics the arced shape of the canopy in flight. If you were to lay the canopy straight on the ground, the suspension lines at the tips would be tighter than those in the center. Thus, the tips would pull up first.

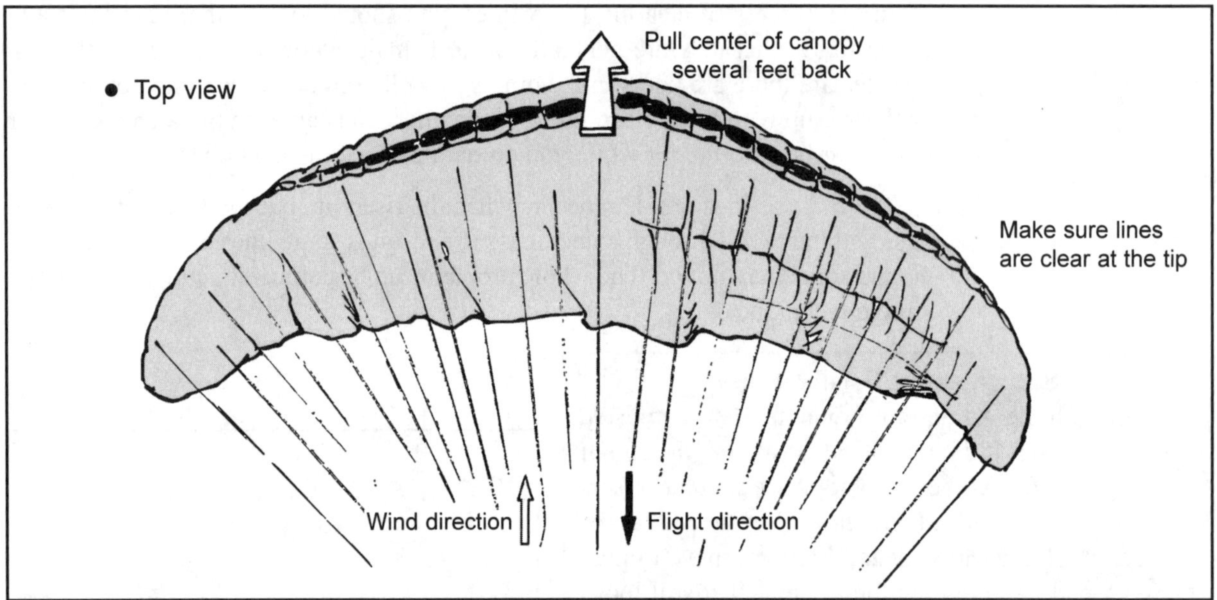

Figure 4-4: Horseshoeing the Canopy

To create the horseshoe shape, lay the canopy out normally then move the risers or harness closer to the canopy so you don't drag them. Now move to the leading edge of the canopy and pull it downwind (away from the harness) 2 to 3 feet (.6 to 1 m) at the center. Move out either side of the center and pull the canopy back a little to make a nice rounded horseshoe shape. The tips should stay where they were originally. While horseshoeing the canopy, make sure you don't cause a lineover (see Preflight below). Also, try to leave the center section of the leading edge standing up a bit (see Building a Wall under the heading Reverse Launches later in this chapter).

✓ Packing the Glider

Naturally we pack or stow the glider at the end of our flying. However, we will discuss it here because you may be practicing with a glider in ground school first, and at this point the arrangement of the canopy is fresh in your mind.

When you are ready to put your glider away, lay it out on its back (lower surface up) with the long axis of the canopy parallel to any significant wind as shown in figure 4-5. Next you must place all the lines on your canopy as shown. The easiest way to do this is to grab the riser bundle and whip the lines up on the canopy with a gentle motion. Often parts of the lines remain beyond the canopy edge, so

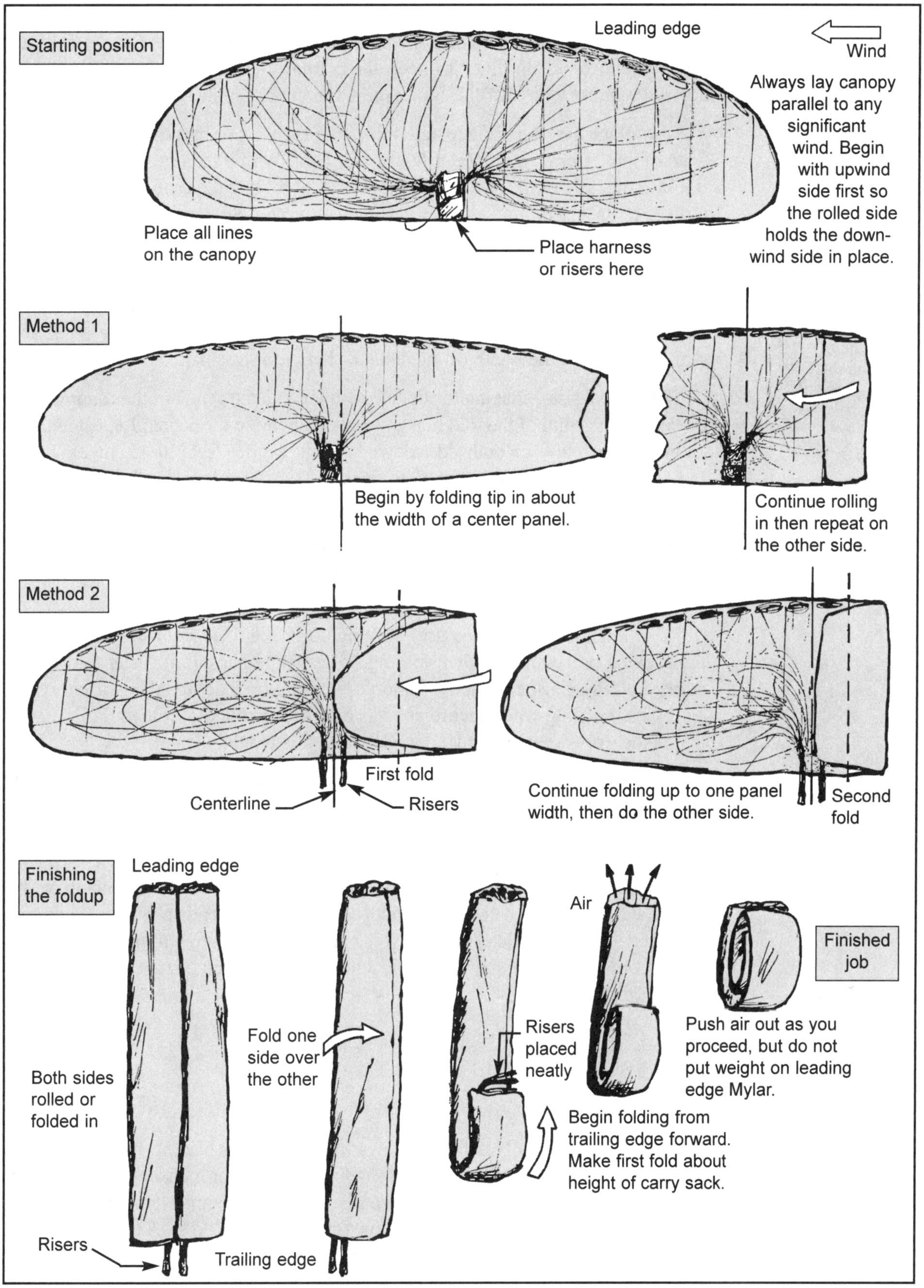

Figure 4-5: Folding the Canopy

simply move these stragglers onto the canopy. Try not to create too tight a wad of lines, for they will all be folded or rolled in the next steps. Note: On older gliders pilots would braid the lines to keep them from tangling. This procedure is ill-advised with the new stiff lines for excess flexing can reduce their life.

Now place the risers (if separated from the harness) or harness near the trailing edge of the canopy as shown in figure 4-5. Hint: With the harness attached, place it just to one side of the centerline since we'll be folding the canopy along the centerline. If the risers are separate, let them protrude beyond the trailing edge as shown in method 2. Keep them separate and straight to prevent line twists.

If the ground is rough or abrasive, open your glider pack and put the canopy on it for the final rollup.

Now pack up the canopy. There are two recommended methods: rolling it or folding it. *Rolling* consists of rolling the wing in from the tip in a roll width about equal to a cell width. Start by folding the stabilizer (tip) in an amount so the second fold is about the width of the distance between the line tabs.

Folding consists of taking the tip and walking it to the center of the canopy. Then go back to the point of the fold line and walk that part to the center. Repeat until the entire wing is folded on both sides down to about 1 to 1½ feet (30-45 cm) as shown. In windy conditions you can tuck the tip of the second side under the first side roll so it doesn't blow around.

A third way to fold a glider is the *accordian method.* This practice requires two people who are positioned at the center of the glider and pull each side in panel by panel and fold it like an accordian.

Which method should you use? Rolling keeps the lines from sliding around in the packing process better than folding. It is also perhaps a bit easier to do when alone. Folding is usually faster. Canopies with mylar leading edge stiffeners will suffer less bending when accordianed as opposed to rolled or folded. Follow your instructor's or owner's manual's guidance in this matter.

The next step is to fold one side over the other as shown in figure 4-5. Finally fold the long bundle from the trailing edge to the leading edge in lengths that are equal to the desired finish length (this is largely determined by the pack or bag length and is typically 2 to 2½ feet (60 to 75 cm). Always fold from the trailing edge to the leading edge so the air can escape the canopy interior. As you make these last folds, smooth out the bundle and press it down to evacuate trapped air or you will end up with too big a bundle (but don't crush the leading edge Mylar or sit on the canopy). If the harness is separated from the risers we suggest putting it in the carrying pack first so it somewhat protects the canopy.

REVERSE SETUP

Reverse inflating involves facing the canopy with your back to the wind as you inflate it. The benefit is that you can watch the canopy throughout the entire inflation process and can float it above you before you attempt a launch. This view allows you to check for a good inflation—no line snags, no sticks and no folds—before committing to the air. Even though we teach (preach) looking up to check the canopy during a forward launch it is difficult to assess and correct problems while you are running. Thus, with a forward inflation it is almost always necessary to abort the run when a malfunction is detected, especially on a steep slope.

The drawback of a reverse inflation is that you must turn around at some point while maintaining control. However, despite this drawback, we recommend that you use reverse inflations almost exclusively.

Your Goal: *To learn how to prepare your glider and your positioning for inflating and later launching from the reverse position.*

Special Note to Students and Instructors

In the process of learning to fly paragliders we adhere to the principle that learning correct procedures from the very beginning creates better and safer pilots in the long run. For that reason we teach launches in the reverse position from day one. With modern canopies, reverse inflations should be used for most launches, including calm wind launches. About the only time forward launches are needed are when running in deep snow or uneven footing, operating on skis, you are at high altitude, the launch is very shallow or an injury such as a weak knee would be compromised during the turning motion required with a reverse inflation.

No doubt reverse hookup of the risers to the harness and the actual turning and running process required with a reverse inflation are a bit more difficult to learn in comparison to a forward position. On the other hand, the inflation of the canopy and kiting control is usually easier since the student can see the canopy. It can be argued that the cross controls (right hand affects the canopy's left side and vice versa) when in the reverse position can be confusing, but the truth is, a person new to the paragliding experience has developed no great bias to the process and learns the proper controls readily.

The big advantage of this modern approach is enhanced safety. Pilots can handle gusts and stronger winds easier with reverse launches.

Pilot on the right performs a reverse inflation. The pilot on the left has inflated and turned to face forward.

✓ Putting on the Harness

Climbing into the harness seems like a no-brainer, but there is at least one important point to make. *Always* put the leg straps on as soon as you put your arms through the shoulder straps. The reason for this warning is that occasionally leg straps are forgotten which is very dangerous (see Emergency Procedures in Chapter 10). Start this good habit now and keep it throughout your flying career.

Figure 4-6 shows the general steps to putting on the harness. They are: shoulder straps, leg straps, then chest strap. The buckles used to attach the straps may be seat belt type push-in buckles, but on training gliders may be nesting rectangles. These items may be confusing if you haven't used them before. The figure shows how to insert them. Start with one side first, then pull the entire smaller rectangle through the larger one and square it up. Removing these buckles is the reverse procedure: pull some slack in the strap and feed one side of the buckle out first.

Make a tightness adjustment on the strap if necessary. Leg straps should be barely snug or a bit looser. You should feel comfortable for running. The chest strap should be adjusted so that the strap is tight when the upright straps that hold the carabiner loops are straight (not pulled in). Chest strap adjustment affects control and stability, and we cover it in Chapter 9.

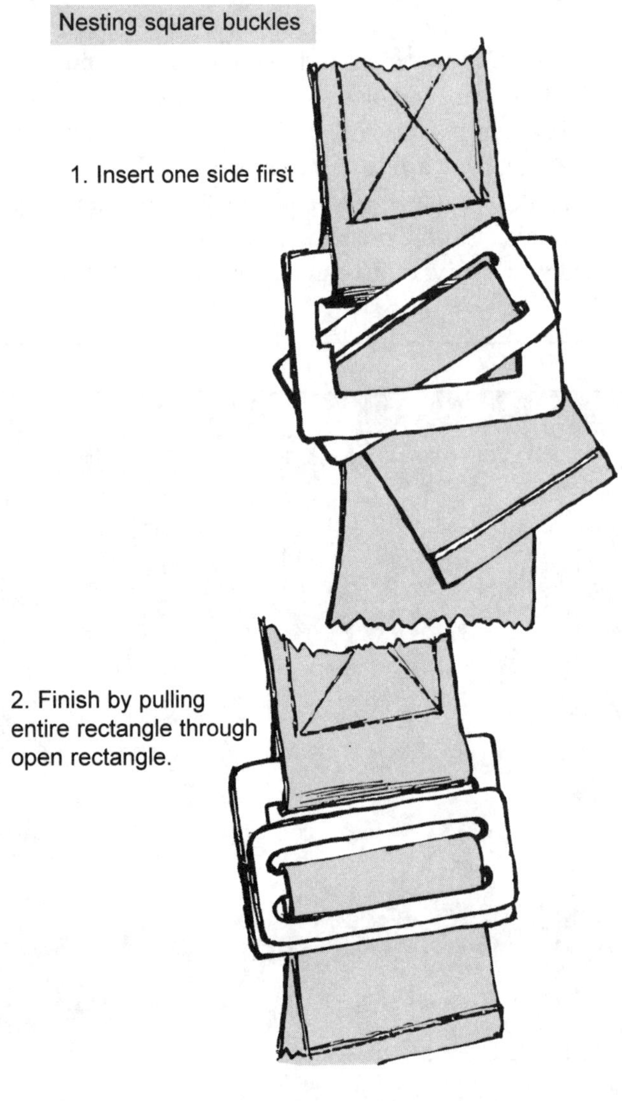

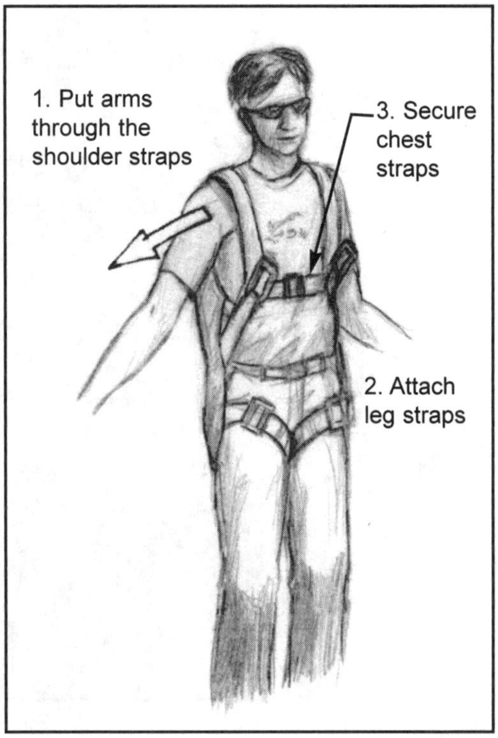

Figure 4-6: Putting on the Harness

Figure 4-7:
Reverse Inflation Turn Direction

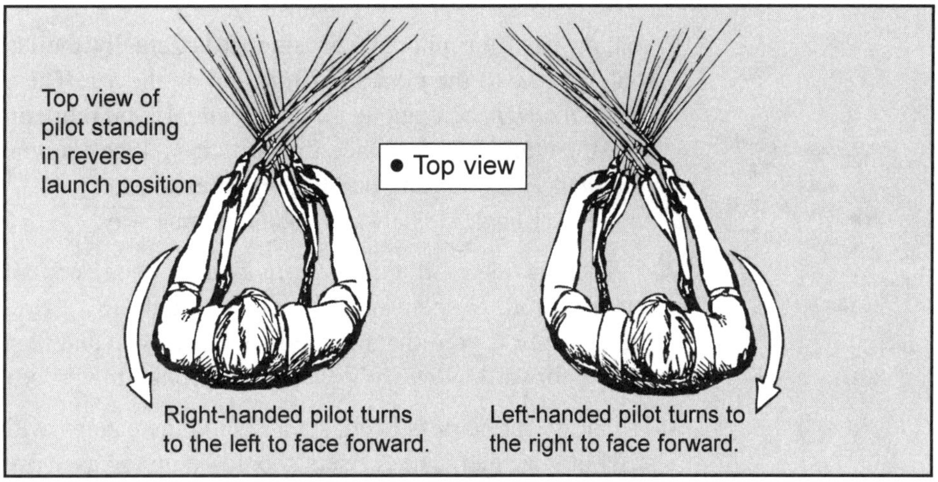

*Safety Tip:
Always put on and fasten your helmet before hooking into risers.*

✓ Attaching the Risers—Reverse Position

Once you have the harness on and the lines arranged, it's time to hook up the risers to the harness. Before you can begin, it is necessary to make a life decision: which way will you be turning to face forward and launch once the canopy is inflated above you? The turn direction you use determines how the risers are routed to the harness. We called it a life decision only half in jest because whichever way you learn is the way you should use throughout your flying career. The turnaround must be an automatic process that you do the same way every time.

We'll make the decision for you. If you are right-handed, always turn toward the left (counterclockwise when viewed from above) when you start from the reverse position and are rotating to face away from the canopy to begin your run. Conversely, left-handed pilots should turn right out of the reverse position (clockwise when viewed from above) as figure 4-7 shows.

**Figure 4-8:
Right-Handed Riser Arrangement**

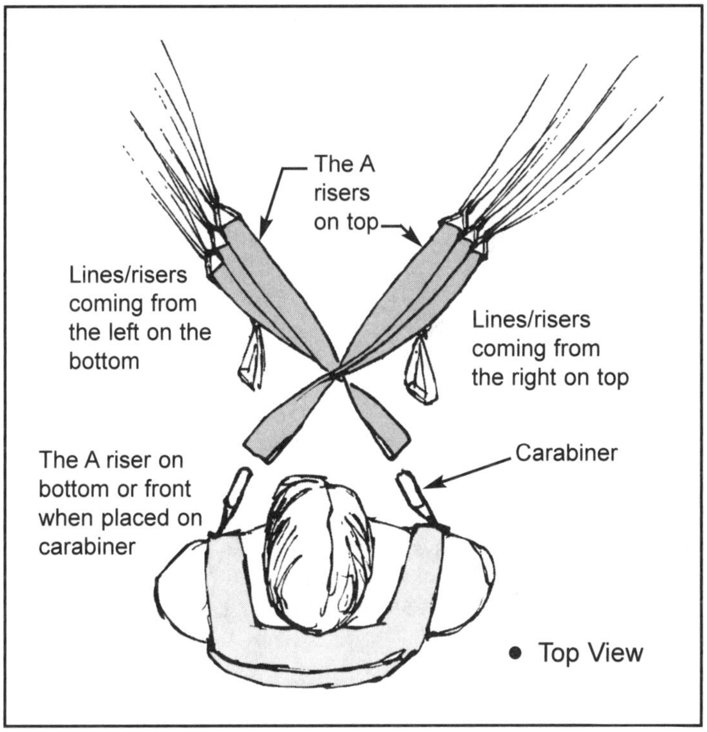

The reason for this right or left-handed bias is because later, if you learn to be a tandem pilot, you must turn to face the canopy and must rotate so a side-mounted parachute doesn't brush your tandem passenger. Since a 'chute should be mounted on the side of the dominant hand, a right hander will have the 'chute on the right and his or her left side will be next to the passenger as the turn is completed to the left. Simple logic, isn't it?

Now here are the steps for attaching the risers properly so everything is magically untangled when the canopy inflates above you and you turn forward to launch. We deal first with right-handers, then left-handers for simplicity.

Right-handed reverse attachment – The general arrangement is shown in figure 4-8. The illustration shows that the riser array coming from the pilot's right side (it is actually the canopy's left riser system) routes across to the pilot's *left side* and on the *top*. The other riser array routes to the pilot's *right side* and is on the *bottom*. If you think of the canopy lifting above you, then turning left to face forward, this all makes sense. Again you can see the importance of choosing one turning direction for your life-long flying experience so the riser hookup is always done the same way.

Right-handers should turn around left. Left-handers should turn around right when beginning their run from the reverse position.

There is one more thing you must do to the risers before hooking them to the harness. Both of them must be given a half (180°) twist to the left. Again, this makes sense if you imagine the riser above you and then turning your body 180° to face forward. Now we'll walk you through the process step by step.

1. Once your harness is on, stand facing the canopy with the risers laid out neatly at your feet. The A risers should be on top as shown in figure 4-9.

2. Now pick up the riser array on your *left* and give it a 180° twist to the left (counter clockwise). Attach it to the harness by looping it through the carabiner on your *right* side (see Carabiner Hookup below). If you have any doubt as to how the risers should orient, remember the poetic phrase:

   The A's are away, the rears are near.

Figure 4-9: Right-Handed Riser Hookup

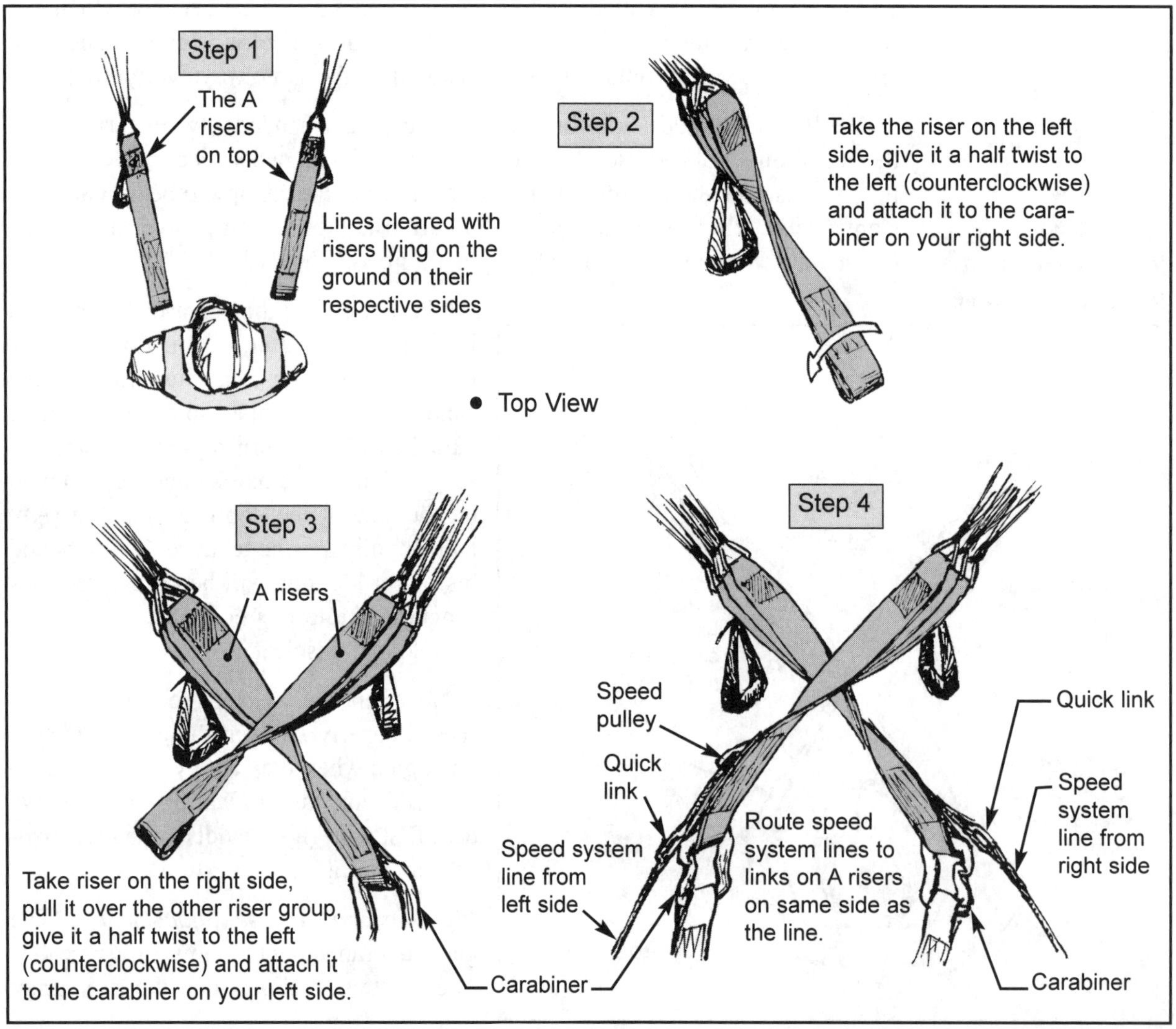

3. Now, lift the risers on your right and give them a 180° twist to the left (counter-clockwise) and route them *over* the previous riser array. Attach them to your harness at the left carabiner
4. Finally, in the future if you have a speed system to hook up, route the speed rope to the A riser that is on its same side (left speed rope on the left side riser) in the most direct manner (no twists or wraps around any other riser or line) as shown. Hook the rope to the link provided on the A riser. Note: With a front mounted parachute the speed rope *must* route *behind* the parachute bridle.

Left-handed reverse attachment – This process is the mirror image of the right-handed one. Figure 4-10 shows the overall look and the steps. In this case, the riser array on your *left* side (coming from the canopy's right half) is routed *over* the one coming from your *right* side. The top array goes to your right carabiner and the bottom array goes to your left. Again there is a half (180°) twist put in the riser array. This time it is to the right (clockwise).

Here's a stepwise account:

1. Start with the risers placed at your feet and the A risers on top. Pick up the riser array on your right and give it a 180° twist to the right as shown. Now attach

Figure 4-10: Left-Handed Riser Hookup

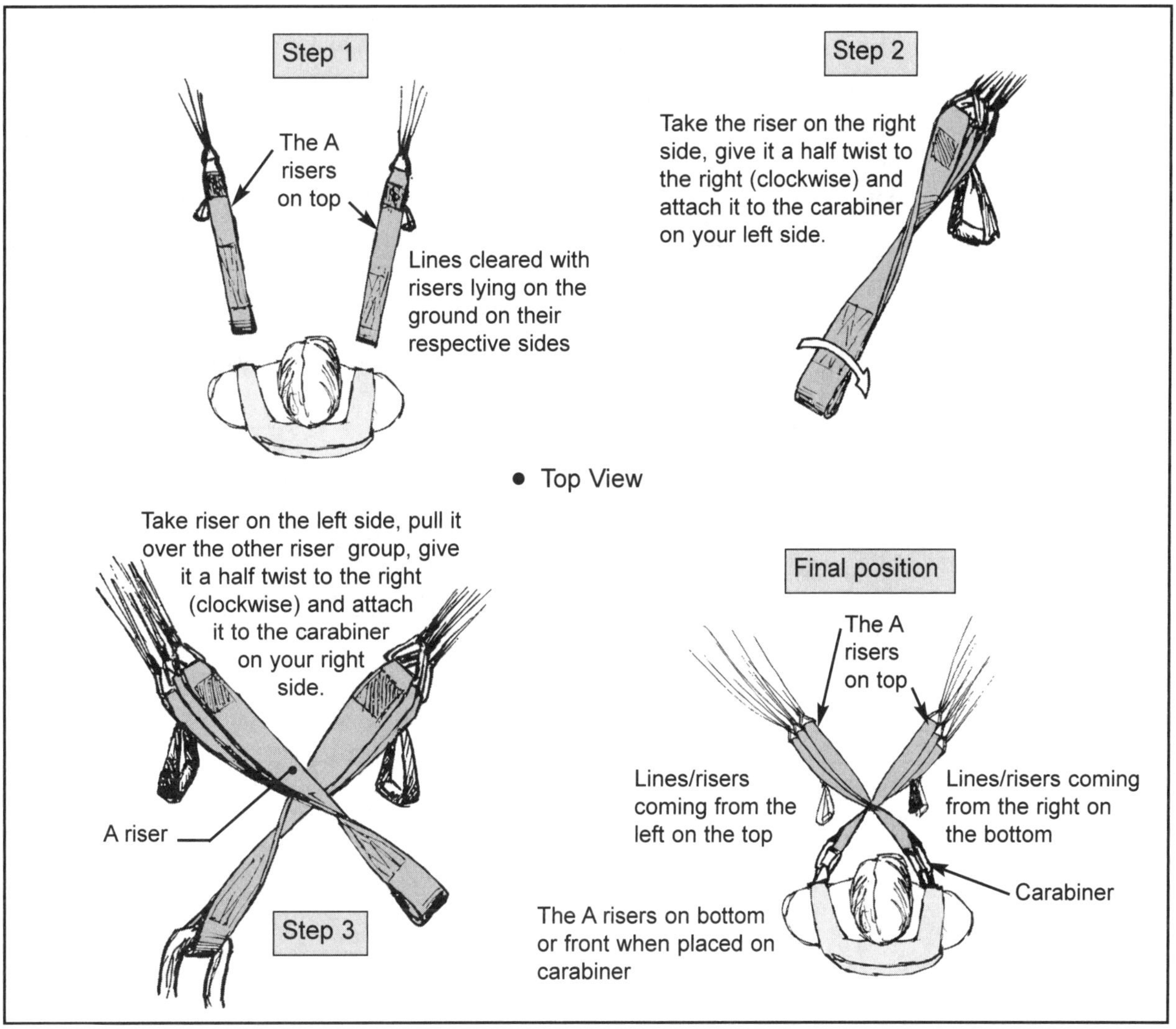

it to the carabiner on your left side (see Carabiner Hookup below).

2. Next, pick up the other array (at your left foot), route it over the previous array and give it a 180° twist to the right. Now attach it to the right carabiner. Again remember: *the As are away and the rears are near.*

3. If you have a speed system hookup, route the speed rope to the A riser that is on its same side (left speed rope on the left side riser) in the most direct manner (no twists or wraps around any other riser or line—see parachute bridle note on page 49.

When you are attaching your risers, it is important to take a step or two forward to provide slack in the A lines so you don't distort the canopy from its nice layout. Practice this hookup a few times and the process will become automatic—as long as you don't vary it.

✓ Carabiner Hookup

To help avoid confusing the way to hook up the riser loops to the carabiner, we provide figure 4-11. In general, carabiners are oriented at an angle when viewed from above as shown (not fore-and-aft or side-to-side). The riser loop should be put on the carabiner so the A riser is forward and to the outside (away) as shown. The rear riser is to the rearward and inside (near).

Once you have hooked the risers on the carabiners, check to make sure the carabiner gates are locked and the risers have one half twist in them from carabiner to canopy. To do this riser check, reach forward to straighten the lines and note the twist in the riser group. Check the left and right groups separately.

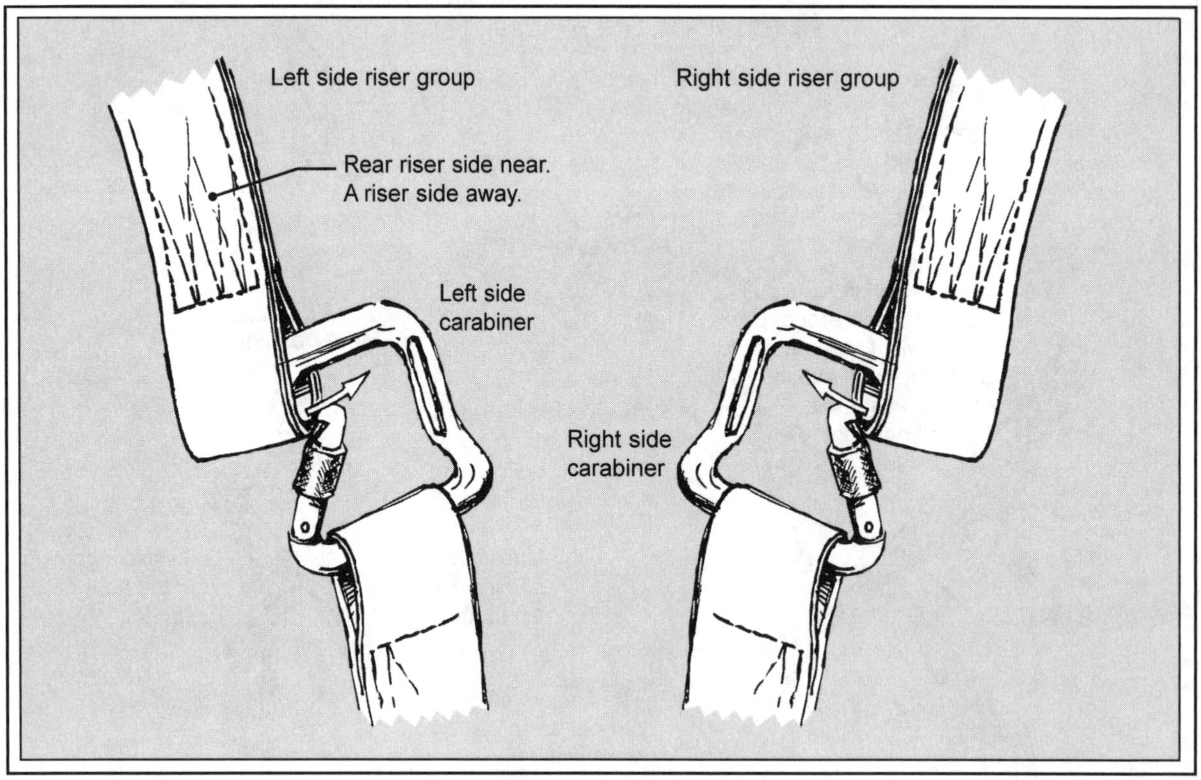

Figure 4-11: Attaching Risers

There are several types of carabiners used on paragliding harnesses. One type has a little lever that must be pressed to open the gate as shown in figure 4-12. Another type shown in the figure has a rotating spring lock. Rotate the cylinder and pull it down to open as we illustrate. Both of these types lock themselves when the gate is released.

There is some disagreement as to how the carabiners should be oriented. For towing they should have the gates facing rearward. However, normally the gates face forward although some pilots turn them around to avoid catching lines or other bits of equipment. Don't change the orientation on your training harness and do not remove the carabiner from the harness in normal use, only detach the risers if necessary.

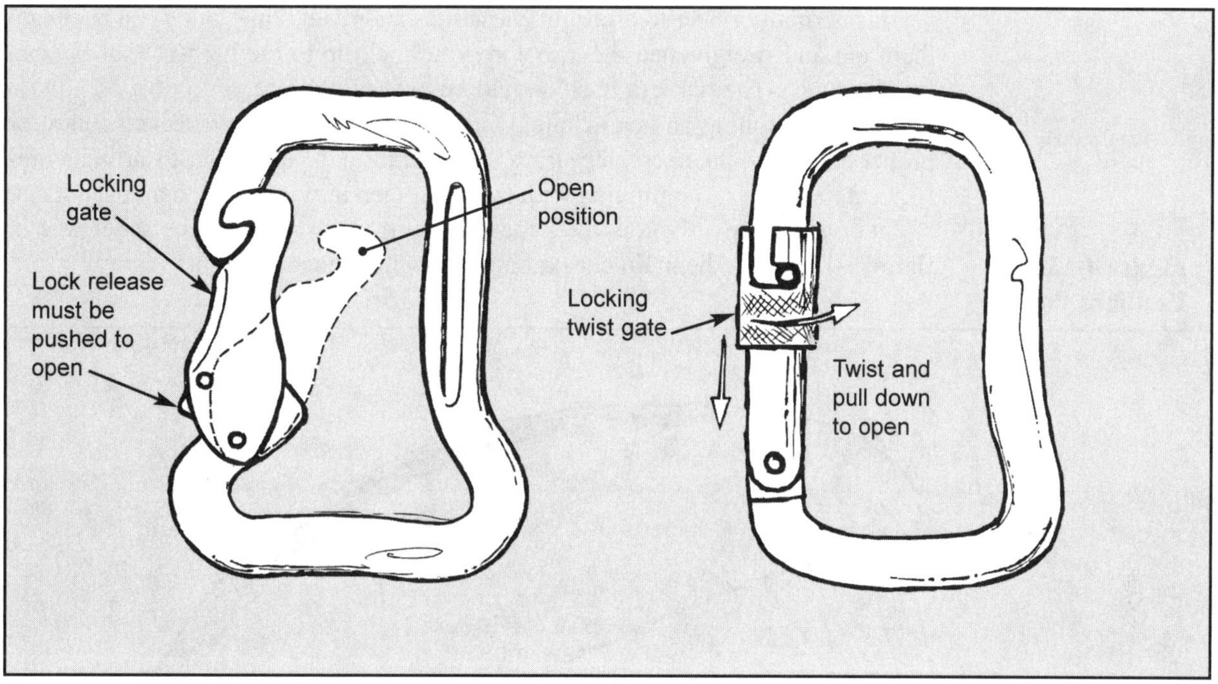

Figure 4-12: Carabiner Types

✓ *Preflighting*

A *preflight check* is an essential part of all safe flying. From this point on you should incorporate it into your layout and launch routine. A preflight check consists of checking all vital parts of the glider for integrity. The purpose of a preflight check is to inspect for wear and damage to your glider as well as to make sure you have completed all the necessary steps in preparing it for flight.

Safety Tip: *Finding and correcting an equipment problem on the ground is infinitely better than doing so in the air.*

We'll first describe a preflight check and then include a checklist which you can copy to use as a guide until you can perform the check flawlessly from memory.

Preflight Requirements

Thoroughness — You must look at *every* part in detail.
Organized system — Start at one point and go around the glider in a systematic manner. If you jump from point to point chances are you will miss something.
Uninterrupted process — Do not let a distraction stop your pre-flight. If you are interrupted you should start over.
Correct deficiencies — If something is wrong correct the problem. If it cannot be corrected at the site, do not fly—a few minutes of airtime are not worth compromising safety.

It is normally easier to preflight check the canopy and lines once you have laid them out and straightened them, before you hook up to the harness (so you can walk around). The harness itself should be checked before you put it on so you can see the webbing and parachute. Once it is on it should be checked again for proper hookup to the risers, leg straps, chest straps, etc. We suggest starting a preflight at the harness, continuing with the lines, then moving to the canopy. Set and maintain a routine so you don't forget any item of inspection. The schematic in figure 4-13 shows the main checkpoints. Now let's do a preflight.

Figure 4-13: Preflight Points

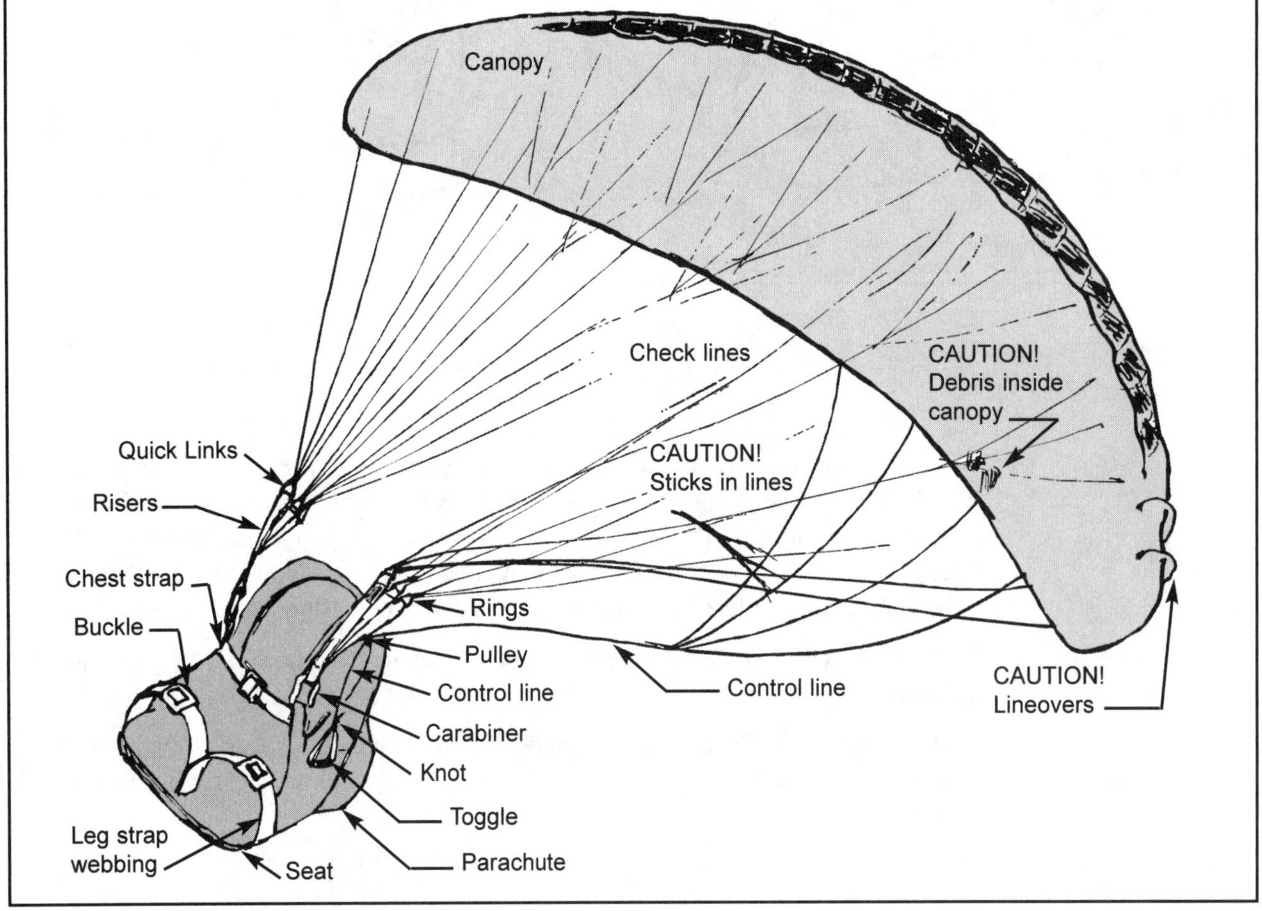

Instructor Dixon White shows a student proper reverse inflation positioning.

1) *Harness* – Start with the webbing straps—especially the leg straps. Are they in good shape and adjusted properly? Make sure there are no cuts or abrasions. At the same time check all buckles and connections. Check the general condition of the seat and cloth area that holds everything together.

Next check the carabiners. Are they locked? Check to make sure the risers are properly attached at the carabiners. Now check the risers to make sure they are not twisted and they have suffered no cuts or abrasions. Finally, check all the quick links at the top of the risers to make sure that they are closed (they should be protected with a rubber or plastic cover to keep the lines and risers oriented properly. At the same time, check the suspension line loops at the quick links to make sure they aren't worn.

2) *Reserve Parachute* – We categorize the parachute separately to emphasize its importance (note, training harnesses usually don't carry a parachute). Check to make sure the bundle looks normal. *Double-check* to make sure the closure pins are in place and the handle is secured to its Velcro in the proper position. Check the routing of the bridle.

3) *The Lines* – A preflight is not normally the place to check every single line in detail (see Line Care and Repair in Chapter 11). However, we should run our eye up all the lines and check for kinks, knots and twists. This check can actually be part of your line clearing, but be aware that you may be preoccupied with straightening the lines and not be inspecting them carefully. We suggest inspecting each riser's lines in turn *after* you have straightened them. Be especially on the lookout for frays, for it means that the outer coating is damaged and the inner structural part may be as well. Kinks are also serious matters, for it can mean that the inner line is damaged. You should replace any line that has obvious damage before you fly.

The control lines demand special attention. That's because if one of these lines develops a problem in flight, control is compromised. Check to make sure they are clear, of course, but also inspect them for damage. The obvious place for wear is where they pass through the pulley or ring at the rear riser. Check this point and the pulley for proper operation. Finally, check the toggle and especially its knot. It should have several knots to make sure it does not come untied. The stiff cord used does not hold knots well, so this check is especially important.

Finally, make sure that no sticks or weeds are entangling the lines. Pilots fondly call such sticks "death sticks," which is an over-dramatization, but you get the point.

4) *The Canopy* – The cloth on your canopy is very strong, but it can be damaged if it is dropped hard on a sharp object, or stepped on, or snagged as you take off. Perhaps the best time to inspect it is when you are opening it up or folding it

away. However, again you may be distracted in these procedures, so a complete walk around to look at the canopy is advised. The only way to check the top surface (which is against the ground) is to lift the canopy up in selective places and look under it. Look especially for small tears, pulled threads, separated sewing and signs of abrasion damage.

An ounce of prevention is better than a pound.

Also, check inside the canopy to make sure there isn't heavy debris (like small stones) inside the canopy which can stall the wing. These items typically migrate to the trailing edge and you'll have to lift the trailing edge up and shake them to the front to get them out.

Finally, check to make sure that there isn't a line over one the wing tips (known as a *lineover*). This problem is shown in figure 4-13. Normally you should catch this when you do the line checks, but a double-check is warranted for you would have a fouled canopy if you tried to launch with such an arrangement.

Now your glider is ready to fly. If any potential damage occurs between preflight and launch—such as stepping on the lines or snagging the canopy—pause to reinspect the areas of concern.

Don't forget your helmet. Check it for cracks and a secure chin strap which includes the buckle. Also make sure it's adjusted properly—loose helmets are unsafe.

Preflight Chart

☐ Harness straps	☐ Parachute bridle
☐ Leg straps	☐ Riser quick links
☐ Buckles	☐ All lines
☐ Harness pack cloth	☐ Control lines
☐ Seat	☐ Control line pulleys
☐ Both carabiners	☐ Control line toggles and knot
☐ All risers	☐ Canopy material for damage
☐ Back support	☐ Canopy debris
☐ Parachute handle	☐ Canopy layout (lineover)
☐ Parachute position	☐ Helmet

Copy this checklist and use it until you have memorized all the steps to a thorough preflight.

Reverse Inflations

Now that you have accomplished the preliminaries you are ready to get the glider in the air. Your instructor will choose the site and conditions for your first practice. Generally, you will start on flat ground with a light, gentle wind (ideally 5 mph or 8 km/h). The flat keeps you from getting airborne and the wind helps raise the canopy. Of course, it is sometimes hard to find the perfect training situation, so a slight slope in calm winds can substitute for the ideal conditions. If winds are more than light, your instructor may assist you by tugging your harness.

✓ Reverse Inflation Types

There are at least four types of reverse launch techniques that have been developed over the years. The differences have to do with which risers and controls you hold and whether or not your arms are crossed. We teach two methods in this book. Here we show the straight arm method. It is the easiest and safest method. Later, in Chapter 7, we will teach you a modified method for use in higher winds.

A pilot practices bringing the canopy up in the reverse position. Note the high force on the pilot with the canopy low to the ground.

The third method is with the arms crossed. This method is an unnecessary complication and is rarely used today. The fourth method is a method which was used earlier in the development of paragliding. It involves holding opposite controls and risers while reversed (right hand holds left A and left control) which requires letting go of the controls as you turn around forward. Releasing the controls at a critical time during launch can be dangerous if gusts occur. We do not recommend this method.

✓ Wind Sense

During the previous discussion on canopy layout we mentioned the importance of having the canopy placed perpendicular to the wind flow. We repeat that advice here. Later, when you are more skilled you may tolerate very slight deviations from this practice, but for now you'll greatly accelerate your success if the wind is straight into your canopy.

We also previously noted that the wind, even when it is light, can vary its direction slightly. Since it is difficult without assistance to move the canopy to angle it into the wind, it is important to exercise patience and wait for the proper wind. Of course, during early instruction your instructor and assistants may quickly angle the canopy for you, but soon you will be by yourself on launch.

When the wind does come straight in it is important to inflate immediately so you don't waste the good cycle. Prepare yourself for inflation *before* the wind straightens out so you can move as soon as it is reliably straight. Learn to watch the wind streamers while you are getting ready to get a feeling for the duration of the straight-in cycles. Develop a wind sense in this manner for it is valuable judgement material throughout all your flying.

Pro Tip: *To readily detect the wind direction right at the point of launch, tie a short (4" or 10 cm) piece of yarn to one of your toggles. Hold it away from your body and you have a sensitive wind indicator.*

✓ Initial Turning Around

When you first hook into your canopy you will be facing the glider and ready to inflate it. However, at the training hill you will occasionally be facing forward when the canopy drops and you will need to turn yourself around both to carry the canopy and prepare for the next inflation.

Naturally, you cannot simply do a turn, for one of the riser groups will be wrapped around your body. So here's what to do. Hold the risers and take a step or two back to allow some slack in the lines (don't step on them). Now, lift one of the riser groups (and lines) over your head with both hands as shown in figure 4-14. If you turn around towards the *right*, lift the *left* riser over your head. If you turn around towards the *left*, lift the *right* risers over your head.

Once you do this a couple of times it becomes easy. Be aware that the lines can catch on your harness or helmet as you turn around, so use both hands spread apart and duck a little or swing the lines out. Once you have turned you should check the positions of the risers and you'll see that they look exactly like the setup in the earlier figures 4-8 or 4-10.

Which way should you turn to face the rear? The answer is easy: the opposite way you will turn when coming out of the reverse position. So a right-hander should turn right to face the canopy and left once the canopy is inflated in order to face forward. A left-hander should turn left to face the canopy and right to face forward for launch.

Again we point out the importance of selecting a turn direction that remains the same throughout your flying career. The reason for this is that in the confusion of launch you don't want to make a mistake and turn the wrong way to face forward. If you do, your risers will be wrapped a full turn around each other and you may get picked off the ground in this mode. Yes, it happens occasionally (we'll deal with this emergency in Chapter 7 in Turn Around Emergency Procedure) so the best defense is to start with the best turn direction for you and *never* use the other.

Safety Tip: *Select a comfortable reverse launch turn direction and always use that direction so it becomes automatic.*

Figure 4-14: Turning Around from Forward to Reverse

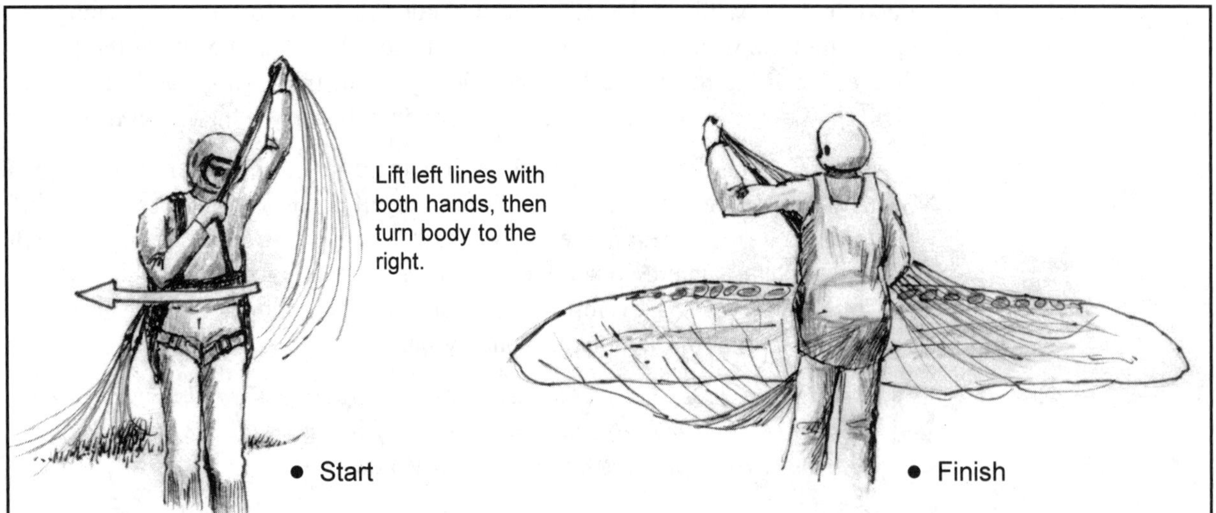

✓ Straight Arm Arrangement

You are nearly ready to practice inflating the canopy and lofting it over your head. The first thing you must do is grasp the controls and the A risers properly. Starting from the reverse position you learned before, let each riser dangle with the rear riser upward as shown in figure 4-15. The weight of the risers will hold them apart and the lines will be crossed further forward. In this bird-eye view we show both the right-handed and left-handed version.

Now grab the control toggles on each riser group with the hand on the same side as the riser group. You may be able to pop the toggles off their snaps by pulling the toggles away from your body, or you may have to hold the riser with the opposite hand to accomplish this feat.

Next grasp the risers. You will be holding and pulling only the A risers. The reason for this is that these risers are attached to the A lines going to the front of the canopy. By pulling only on the front of the canopy you will pull the openings into the air stream which inflates the canopy and causes it to generate lift like a wing.

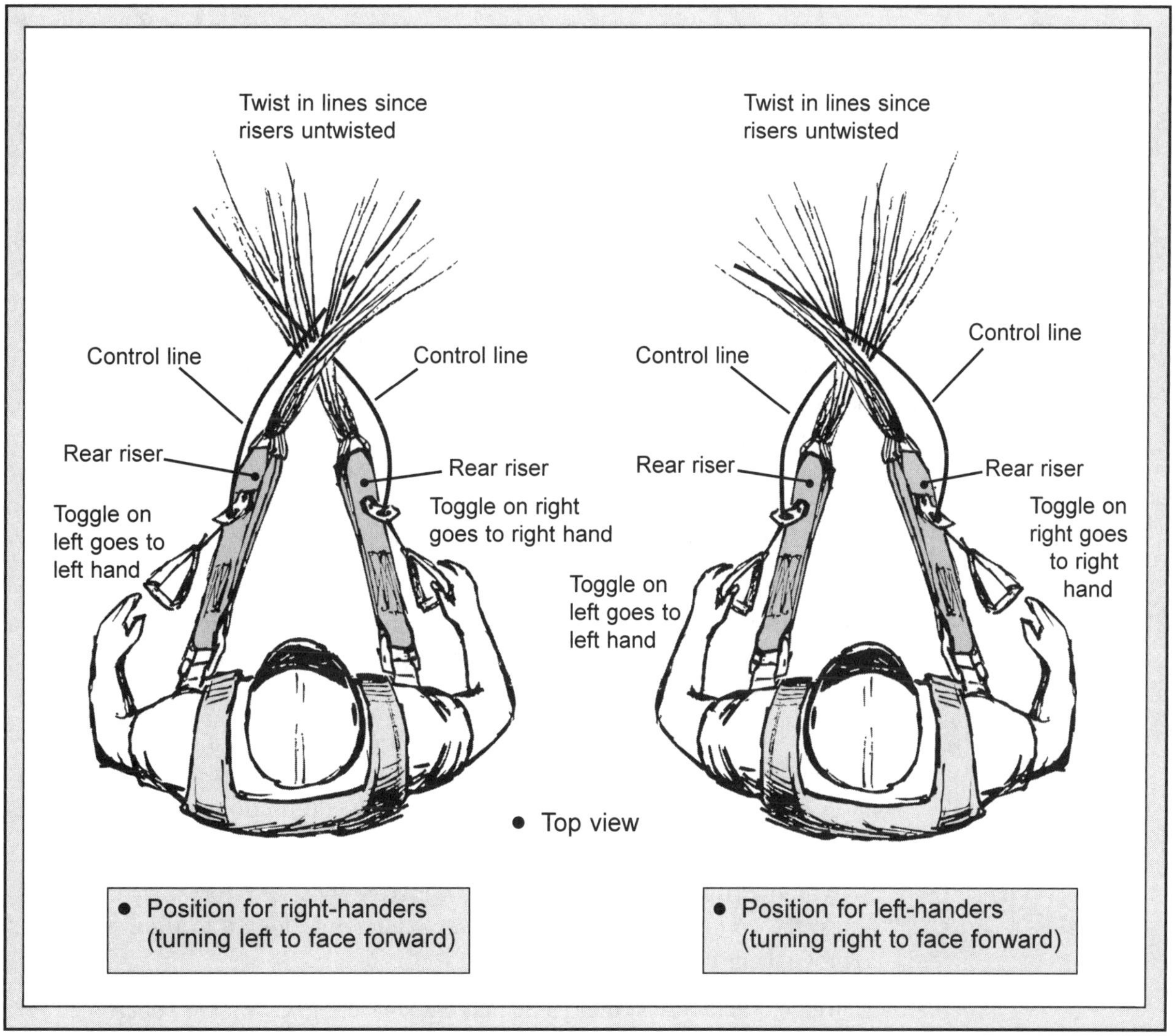

Figure 4-15: Grasping the Controls

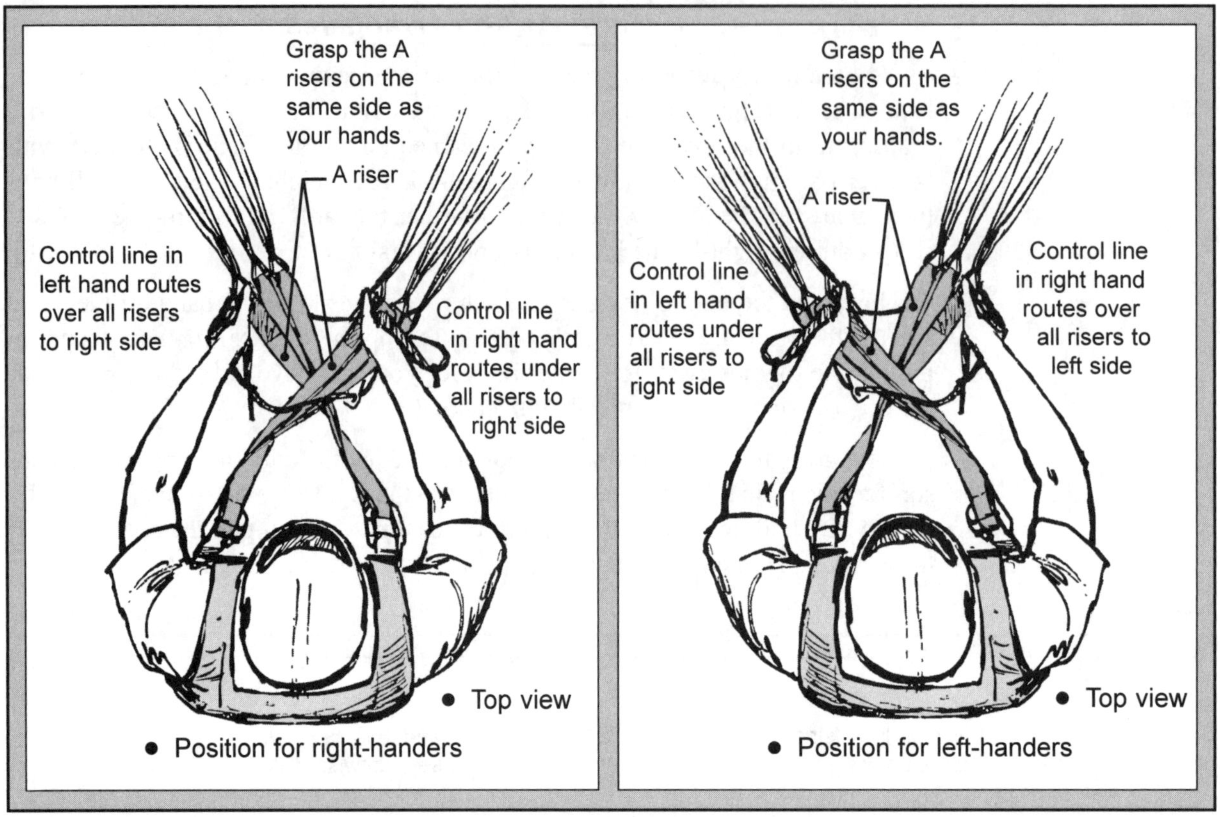

Figure 4-16: Grasping the Risers

Matters complicated to describe become easy once you practice them a few times.

Begin by uncrossing the lines, which crosses the risers in the proper sense as shown in figure 4-16. Next take the twist out of the lines which puts the twist in the risers. This process puts the A risers on top as shown. Now grasp the A riser on the left with your left hand and the A riser on the right with your right hand. The control toggles should remain in your hands and the control lines will be routed properly. Hold the A risers with your thumb on top at the quicklinks as shown. Your right hand should hold the control toggle routing to the risers crossing to the left while holding the A riser routing to the right. Your left hand holds the control toggle routing to the right and the A riser routing left as shown in the figure. Be sure that neither control line is lying over a forearm.

You can check for correctness of your position by making sure that the remaining risers drop away without their lines hanging over the A lines. This process is easier to do than describe. Once you do it correctly, follow the same procedure every time and you will have no problems.

When you let go of the rest of the risers in a group to hold only the As, they will droop and look like a confused mess. However, they sort themselves out as soon as you begin the inflation. To avoid being a confused mess yourself, focus on the few simple steps:

1. With the risers uncrossed and untwisted (the twists are forward in the lines) grasp the control toggles on the correct side (right toggle to right hand) and remove them from the risers.

2. Recross and retwist the risers, making sure the lines are then uncrossed and untwisted.

3. Grasp the A risers on each side without crossing your hands. For right-handers, both A risers will come out from their riser group to the left: for left-handers the A risers come out of their riser group to the right.

4. Make sure the A lines are not crossed over by other lines as you hold the A risers.

5. Double check each control line routing by making sure it passes either over or under an entire group of risers.

Arrangement Memory Aid

Right-Handers

- Turn forward from the reverse position towards the...... LEFT
- The riser group on top routes to your...................... LEFT
- Put a half twist in the risers when attaching to the ....LEFT
- Pull the A risers out to the side of the group to the ...... LEFT

Left Handers

- Turn forward from the reverse position towards the...... RIGHT
- The riser group on top routes to your .... RIGHT
- Put a half twist in the risers when attaching to the ....RIGHT
- Pull the A risers out to the side of the group to the .... RIGHT

✓ Building a Wall

One of the best techniques for effective inflations in wind is to start with a "wall." Building a wall consists of teasing the A risers with a few light jerks to pull up the leading edge a bit as shown in figure 4-17. In light winds only a low wall can be built and the leading edge will stay up mainly on the stiffness of the cell ribs. If the wall tumbles forward in light winds, it's exasperating, but you can often recover yourself by pulling the entire canopy forward from the trailing edge with a tug on both rear risers. Do this only if the canopy is on grass, or it will suffer abrasions.

Figure 4-17: Building a Wall

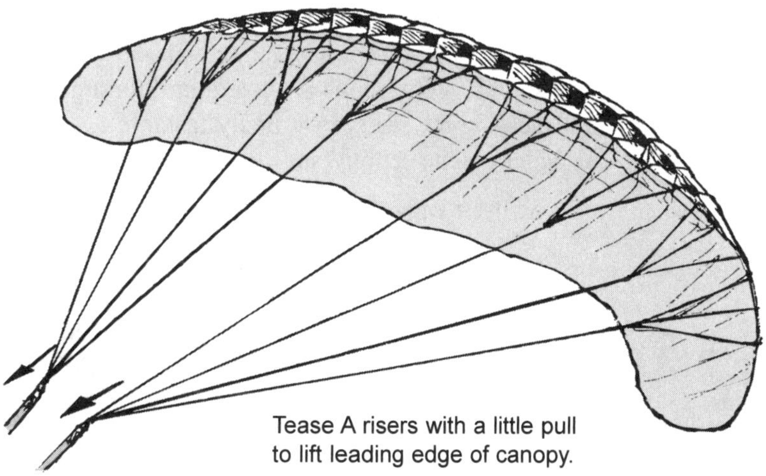

In higher winds the wall will hold its position and you should back up to provide just the required pressure to hold the wall.

Caution: *In stronger winds, be cautious of building a high wall or the canopy may inflate itself before you are ready.*

You can control the canopy on the ground in the reverse position in high winds or gusts with the control lines. In a sudden gust, brace yourself with your legs, lean forward, release the risers and pull both control toggles behind you fully as shown in figure 4-18. The glider will remain grounded. This procedure should be well-practiced in medium winds before you have to use it in stronger stuff.

If the glider starts to rise more than 3 feet (1 m) above the ground it's best to use a pull on the rear risers to bring it back down. The reason is, pulling on the controls on a fast rising canopy can result in the canopy lifting you off the ground.

A wall can be built even when the canopy is in a horseshoe shape on the ground. The benefit of building a wall is that the canopy has already begun to inflate so the launch process progresses more smoothly.

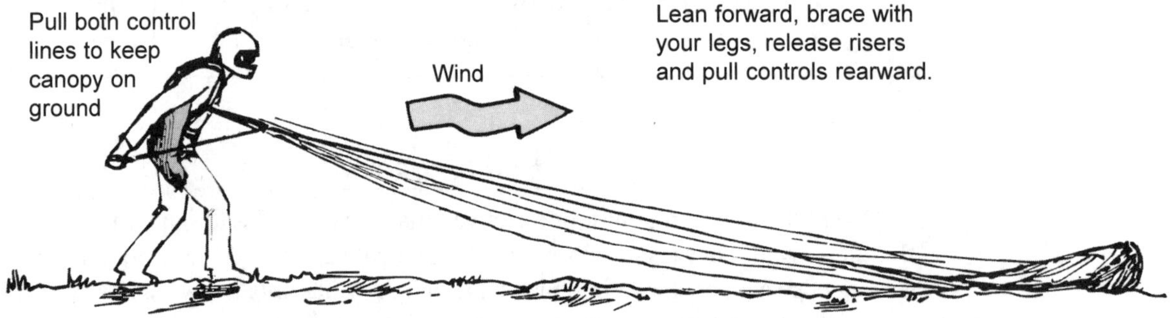

Figure 4-18: Gust and Wind Control Position

✓ *Reverse Inflation*

Now you're ready to learn how to raise the canopy over your head. The act is simple if we break it down into its elements. These are:

1. Body centered and square to the canopy
2. Holding the risers properly
3. Arms in position for an even, equal pull
4. A smooth, continuous backward pull through the harness
5. Releasing the risers in a timely fashion
6. Watching the canopy and controlling it in the right sense.

Now let's walk through the steps in more detail.

1. *Center and square* – You must begin by being in the center of the canopy. Some canopies have an identifying center mark (such as a logo or colored cell rib) or you can simply note where the middle of the left and right lines is on the canopy. Square means you stand on and follow a path which is 90° to the axis of the canopy as shown in the top view of figure 4-19. Start with one foot slightly in front of the other, shoulder width apart with your knees bent (add more knee flex the stronger the wind). Hips and shoulders should stay square to the canopy with most of the weight on your rear foot.

2. *Holding the risers* – We discussed the proper arrangement of the risers for inflation earlier. Your arms should be nearly extended with your hands on the A riser quick links as shown previously in figure 4-16. Hold the risers in a comfortable grip with your thumb on top as shown. Your hips and legs will be doing most of the work through the harness, so you don't have to latch on to the risers.

3. *Arm position* – Both arms should lift evenly upwards on the A risers as you move backwards.

4. *Smooth pull* – The force you exert on the canopy should be continuous and steady. In a wind, flex your knees more and lower your body to load the harness. It should feel like you are inflating the glider with your seat. In light winds, keep moving backwards until you turn around.

For a good reverse inflation arm motion, think of throwing a basketball over your head to someone behind you.

5. *Releasing the risers* – As the canopy rises the required pull on the A risers diminishes. If you pull too long, or pull downward once the canopy is aloft, you will move the canopy too far past you or pull the front down. Either way a front fold is likely.

6. *Watching and controlling the canopy* – As soon as the canopy begins inflating you should watch it to make sure it comes up evenly. As it comes up, check it for tangles in the lines, a folded tip or crookedness (one wing back). We'll see how to make corrections for a less than perfect inflation next. In stronger winds, control how fast the canopy rises by releasing the risers sooner.

Now let's practice. After a check of the wind over your shoulder, look at the canopy center and begin to back up vigorously as you lift upwards on the A risers with your arms outstretched. The canopy should rise above you in a graceful, smooth manner. If it does not, drop it back down by releasing the risers and pulling on both control lines. Reorganize the canopy and start again.

Lift both risers evenly to prevent one side from rising faster. If one side does lift faster, pull more on the opposite riser. Think of pulling the retarded side forward with the riser. With wind the canopy tends to inflate easier (and faster) than in calm air, so you can release the A risers in a gradual manner by the time the canopy is 60° up from the ground. Figure 4-20 shows the entire process.

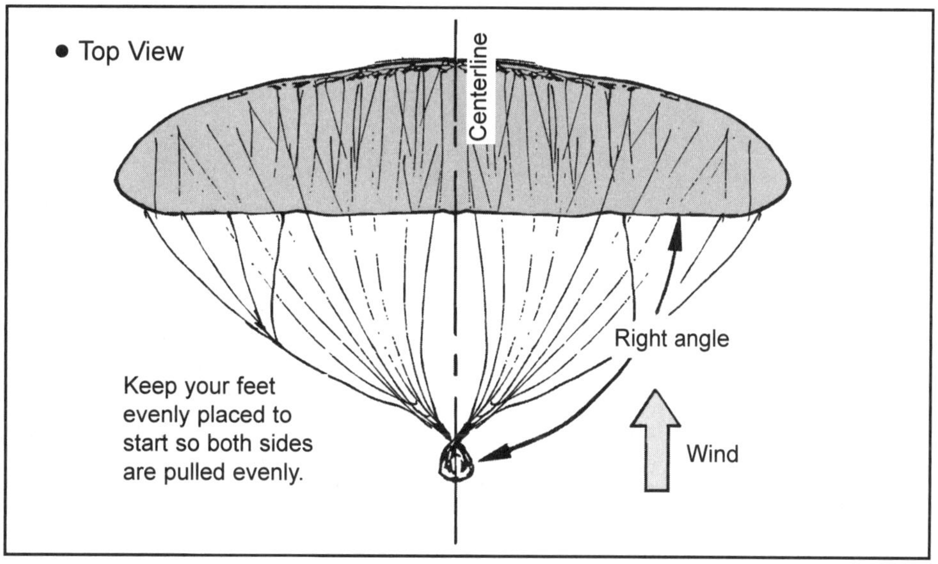

Figure 4-19: Proper Positioning for Inflation

Once you have released the A risers, use the control lines to keep the glider floating squarely above you. This is challenging because the control lines are crossed so that your left hand controls the right side of the canopy and vice-versa. Your first practice should be to overcome confusion by using the controls and learning automatic response. If you pull the controls too much the canopy will drop back and may lose inflation. You can correct for this effect by moving your hands up to stop the control pull, grasping the A risers and moving backwards.

In very light winds you'll have to keep moving backwards to maintain canopy inflation. In stronger winds you can stand still while maintaining back pressure. Once the canopy is stabilized, check it for integrity. Did you get a tip fold on the way up? If so, continue controlling the glider and shaking out such a fold as the canopy floats over your head. A few quick jerks of the control line on the folded side should remove the fold. You might have to hold a little control force on the opposite side to prevent it from moving forward.

If everything looks fine, keep your progress into the wind and in one quick, smooth motion turn around to run upwind. Did you turn the correct way? If not, practice makes perfect. It is important to develop these reverse launch skills until every move is automatic and correct.

If you are still reversed and you want to drop the canopy to try again, pull both controls down as far as you can to stall the canopy. An alternative to this action is to pull both rear risers evenly downward. Be cautious of pulling too hard on the rear risers for the canopy can slam into the ground. "Tenderizing" the canopy in this way shortens its life.

The four big mistakes to avoid when reverse inflating and launching are:

1. Confusing controls
2. Rotating the wrong way to face forward
3. Stopping your upwind progress as you turn back around.
4. Focusing on the canopy and not watching where you are going as you turn around.

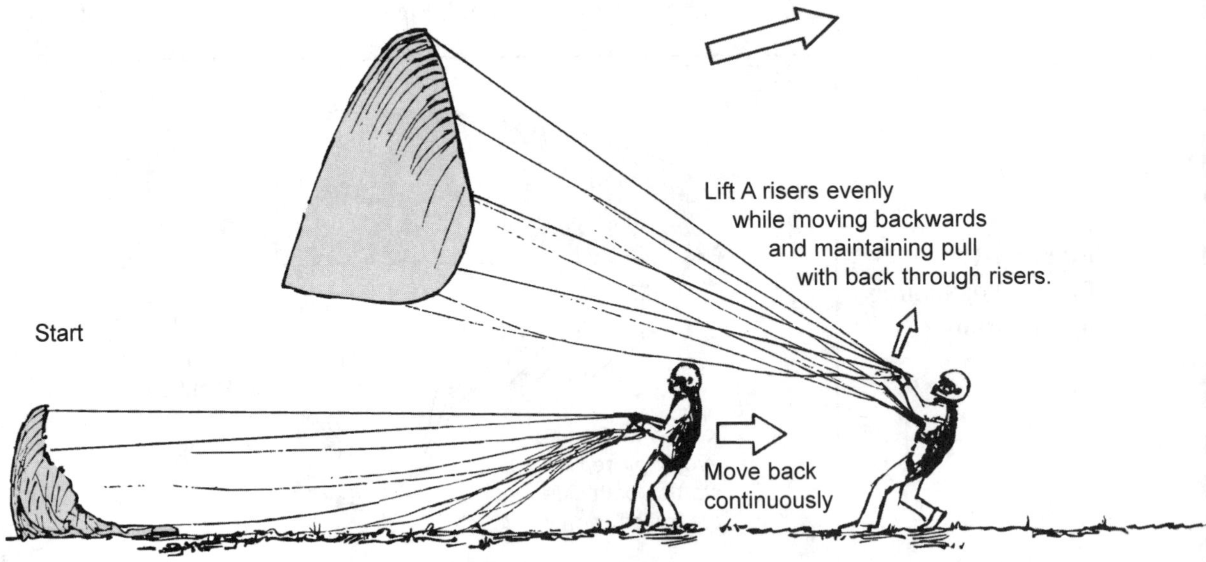

Figure 4-20: Reverse Inflation

A bit of practice in gentle conditions will prevent or cure these mistakes.

We spend a lot of time on reverse inflation technique because the better you are at producing consistent inflations, the more trouble-free your takeoffs will be. You simply must practice inflations until you have mastered the riser release timing, the canopy control and the process of rotating to run forward without pausing.

Skill Check: *Your reverse inflation can be considered accomplished when you can raise the canopy in a controlled fashion to hold a kiting position even in slightly varying conditions. Overshooting is not allowed!*

✓ Stopping and Dropping the Canopy

When you come to the end of your run (or patience) and are ready to drop the canopy, we always recommend you quickly turn around to reverse position just before dropping the canopy. This action is a safety factor in case a wind is present (remember all the force on the canopy when it was near the ground during inflation), and it sets you up in the correct position for another reverse inflation (even if you have to carry the glider).

Naturally you must turn in the correct direction. This direction is right for right-handers and left for left-handers. Note that this direction is opposite to the

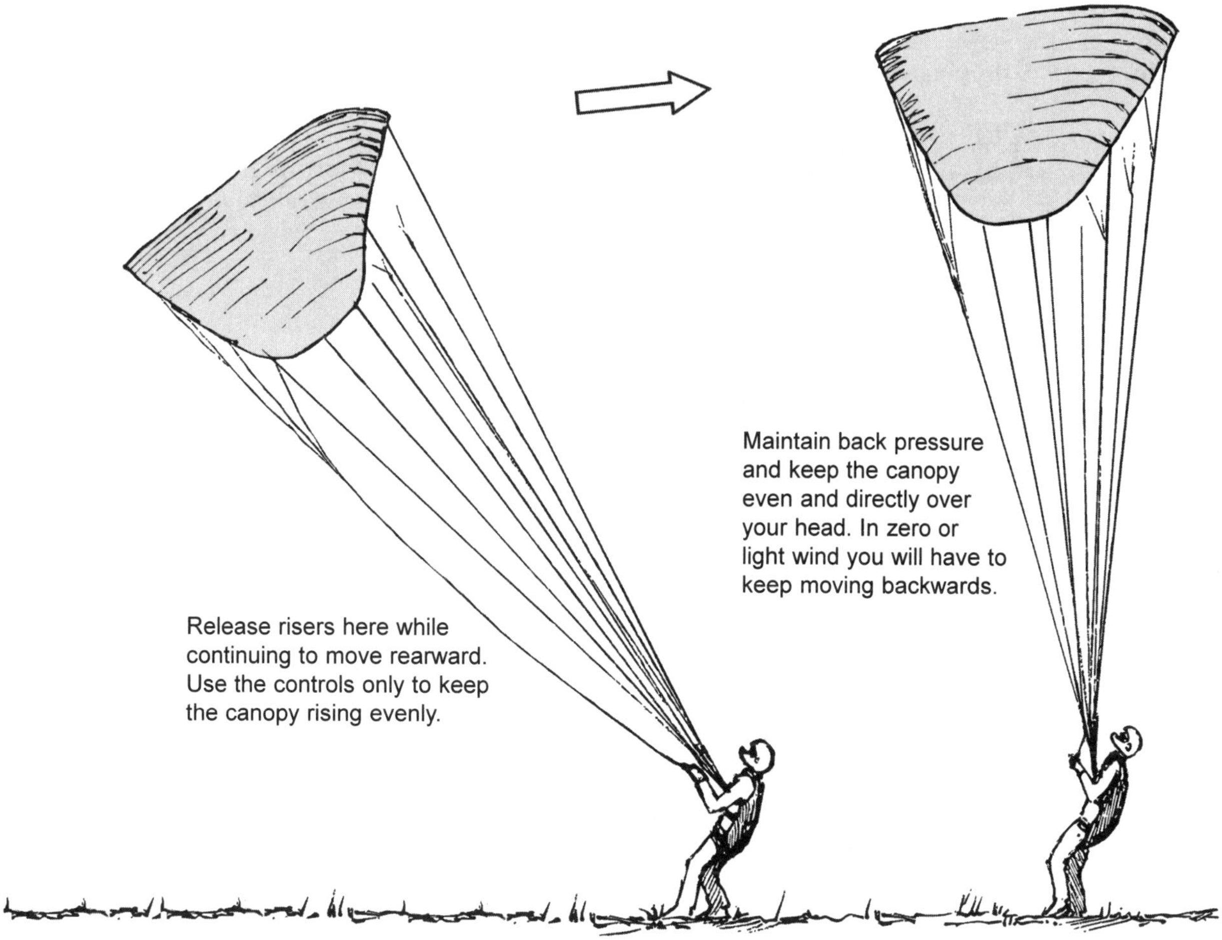

Figure 4-20: Reverse Inflation (continued)

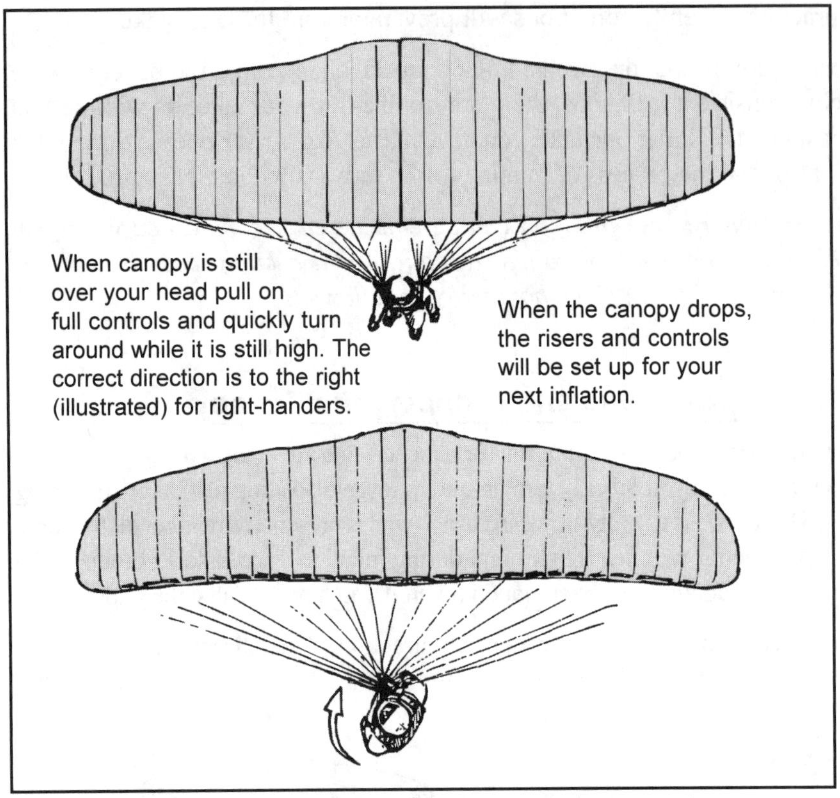

Figure 4-21: Dropping the Canopy While Turning

direction you turn to face forward after you have performed a reverse inflation. You knew that didn't you?

The turn around should be in one quick fluid motion. Try to keep moving in the original run direction if there is no wind so the canopy drops towards its rear or it may fall on or past you. Figure 4-21 illustrates this process. At the same time you turn around, pull on the controls to retard the canopy and drop it faster, but remember the warning we provided in the previous section. We discuss stopping the canopy in the forward position at the end of the Forward Inflation heading.

Once the canopy is stabilized on the ground there's one more important step:

Pro Tip: *As soon as you stop your glider and intend to carry it, take the control toggles off your hands and attach them to their snaps on the rear risers. This is an important procedure for preventing unnecessary tangling.*

✓ Sorting out a Canopy

Often in your early flying experience an errant wind or an error in control results in your canopy turning and falling with a twist or dropping in front of you into a seemingly tangled mass. Naturally it will have to be straightened out before you can proceed. Here are a few tricks to help you sort it out.

Learn to drop your canopy neatly and your next inflation will come sooner.

If the glider has flipped one wing over the other to lie on its under surface, it should appear as in figure 4-22. The upper surface of the canopy is facing upward which isn't normal when it's on the ground. If you are facing forward there will be a twist in the lines. To correct this situation, note which way the lines are twisted, then walk to the canopy to pull the correct tip up and over to the other side in order to straighten the lines. You can do this without getting out of the harness, but be careful and avoid stepping on lines. Gather all the lines in one hand and shake them out of your way as you proceed.

If you reversed your position before the canopy dropped you would normally have put a twist in the lines. But with the canopy itself flipping over, you either have no twist or a full twist in the lines as shown. In case of a full twist, proceed as before and flip the canopy over to result in only half a twist. You will then be set up properly for a reverse launch (assuming you turned to face rearward in the correct direction).

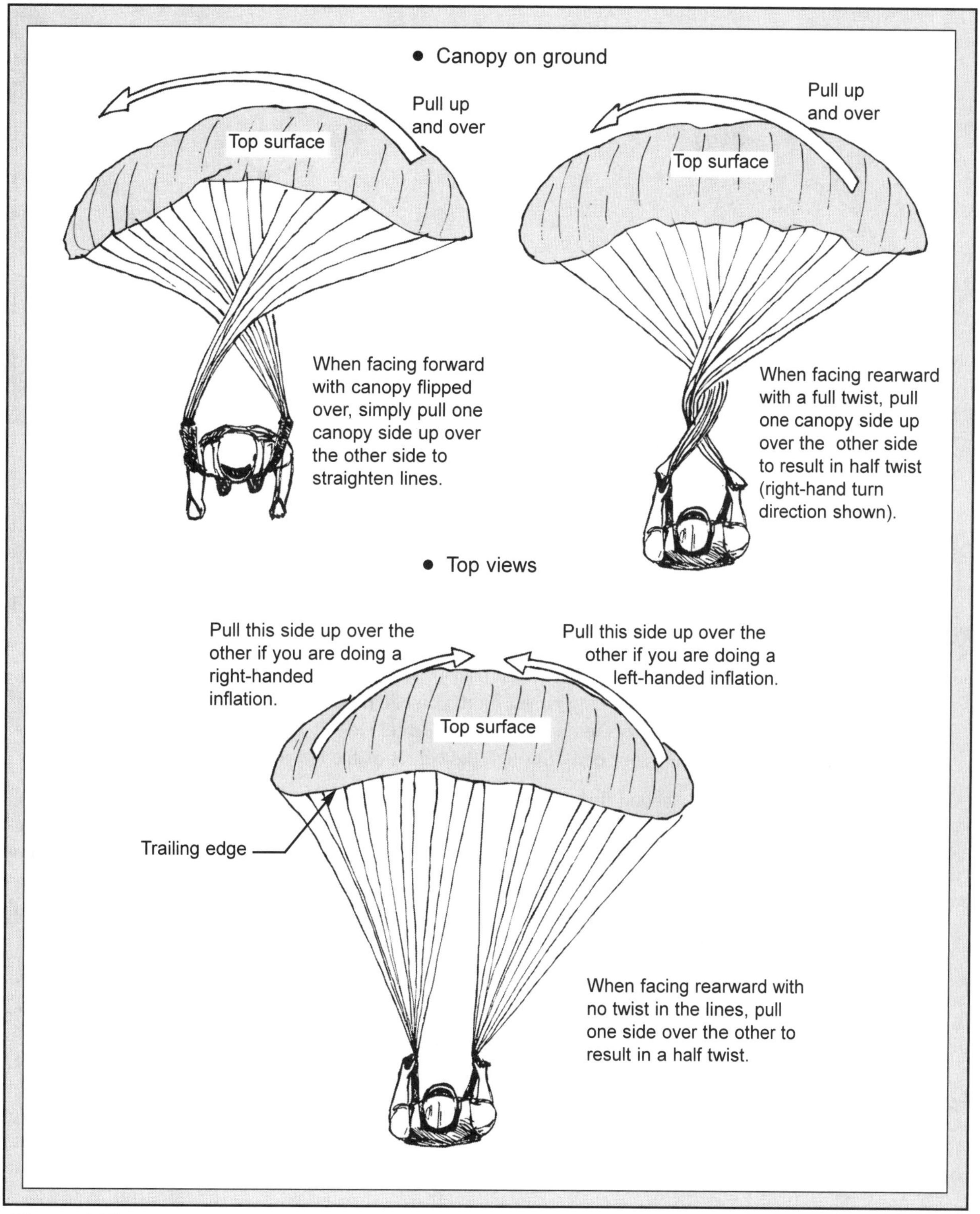

Figure 4-22: Righting the Canopy

If no twist is in the lines, you must be careful to flip the canopy over in the correct direction. This direction is to the right (clockwise) for a right-hander and to the left (counterclockwise) for a left-hander. Remember, a right-hander wants to end up with the riser group coming from his left on top and vice versa. The figure illustrates the correct directions.

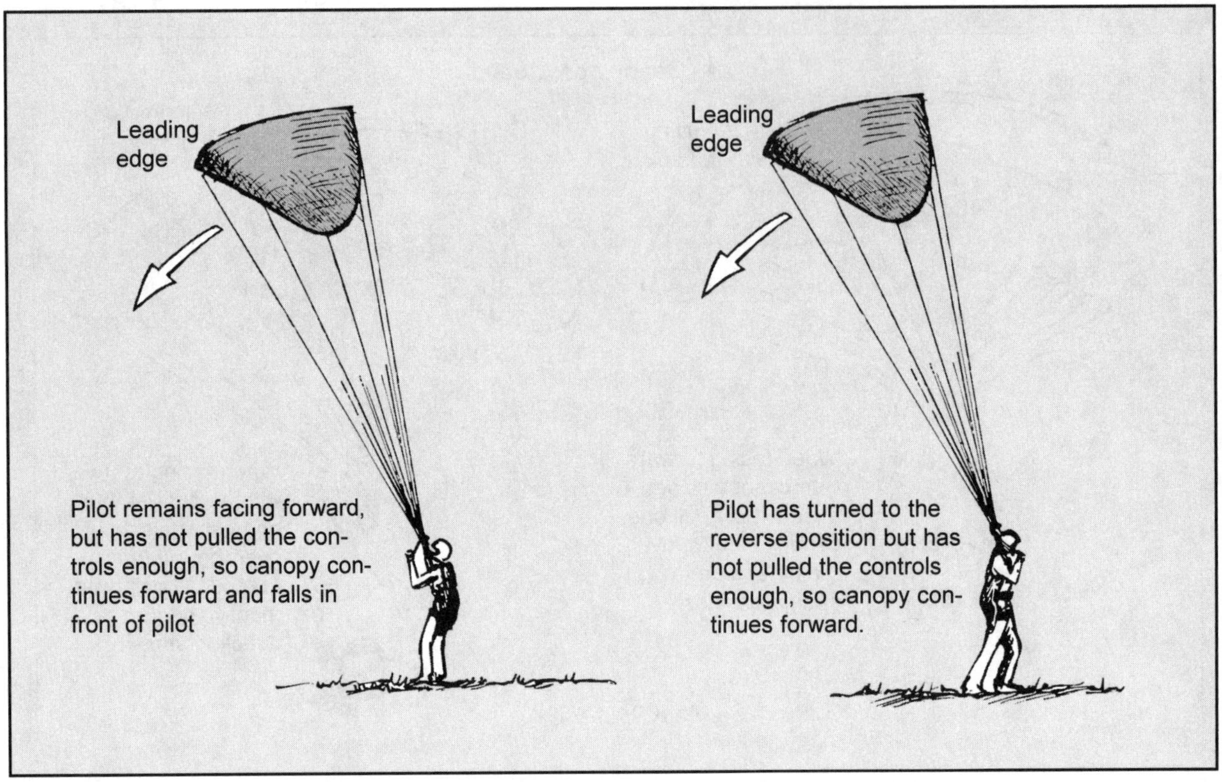

Figure 4-23: Canopy Falling Forward

If the canopy has traveled past you to deflate forward (as shown in figure 4-23), you have several recourses. First, you can have a friend pull the entire canopy over your head to be on its back on the correct side. In this case you must be very careful to pass your body through the center of the left and right lines.

The second thing you can do if you are facing the canopy (which means you did not turn around as the canopy dropped) is to flip one side over the other as in the lower drawing of figure 4-22. When you do this you are in a reverse setup since you started in the same position (facing the canopy with the top up and the lines untwisted). However, you are downwind of the canopy, so gather it up and take it to your launch point.

Good ground handling sets your glider up for easy carrying, relaunching or packing.

A third situation arises if you had turned to face backwards when the canopy was above you and it continued to go forward to fall behind you as shown. In this case there is a half twist in the lines behind you. You should gather the risers in your hands (left in left hand, right in right) and turn around while flipping one group of lines over your head, just as you do when you turn around for a reverse launch. Which way should you turn? Turn in the direction that untwists the lines (the lines that lie on top are the ones that go over your head). Now follow the procedure in the previous paragraph and everything will be correctly arranged.

Some of this discussion may appear confusing but after a few trials you'll readily use these techniques, or better yet, you'll avoid having to use them by improving your canopy control. Of course, if you get confused, you can always detach the risers from the harness and lay out the canopy then clear the lines.

✓ Carrying the Glider

During training you will often have to carry the glider back to your starting point. Also when you are flying higher, it is desirable to carry the canopy off the landing field immediately after landing. How can you do this effectively? The best way is to turn and face the canopy. If you are going to reverse inflate next, you can loop one side riser group over your head, but this procedure is somewhat difficult when the canopy is not neatly laid out. You can simply turn to face the canopy and let one riser group wrap around you.

Now grab *all* the risers and lines together and start gathering lines by sliding your forward hand up the lines. Use the other hand to hold the section of lines that you gathered in a loose loop as shown as figure 4-24. Walk towards the canopy as you proceed so you don't pull it over the ground. As you reach the canopy, lift it several times to gather it as you pull in more line until you have a big ball.

Now lift the whole mass with your forward hand and haul it over your shoulder as we see in the figure. Avoid dragging the canopy.

Pro Tip: *In windy conditions, keep the canopy over your downwind shoulder or the wind will blow it over your head and blind you. In hot conditions, do not place the canopy between your body and the sun, for it doesn't shade you, it acts like a greenhouse!*

When you walk back up a hill you'll find the leg straps often restrict your steps. That hill climbing business is hard enough without a crotch constrictor. *You must resist the urge to unbuckle the leg straps.* This act can and has led to taking off with the straps unfastened. This event is a true emergency with its own recovery procedures (see Chapter 10) which cannot be followed at a training hill. To ease your climbing either remove the harness or simply loosen the leg straps without removing them.

Caution: Never *unfasten your leg straps once you are in your harness until you are ready to take off the harness.*

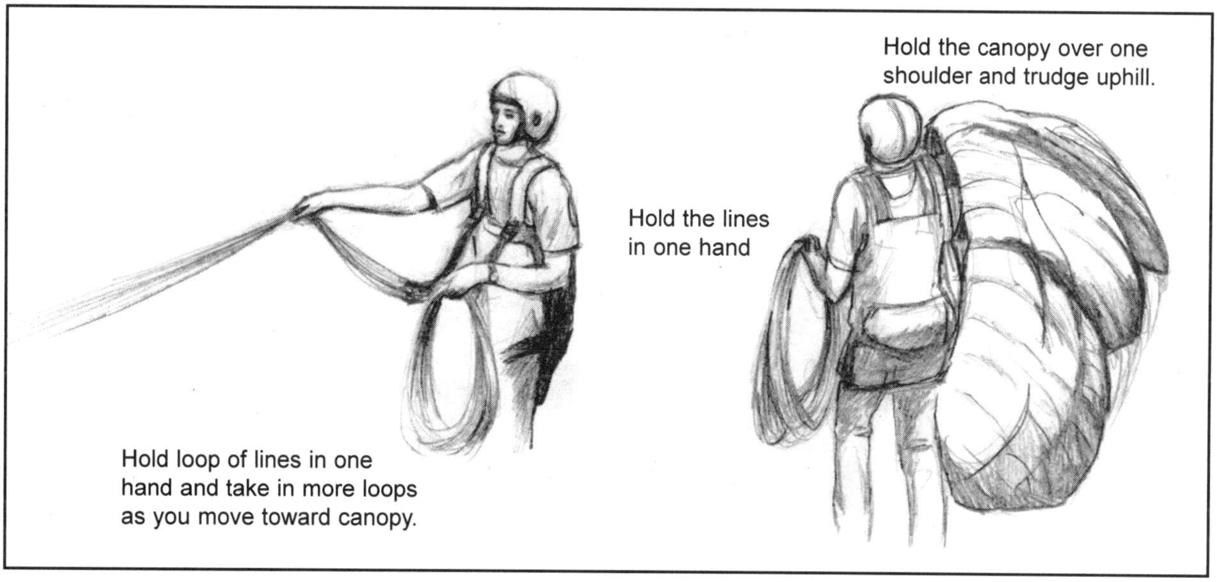

Figure 4-24: Gathering and Carrying the Canopy

✓ Self-Spreading

Once you have returned to your start point the canopy must be spread. As a beginner you normally step out of the harness and do this procedure bit by bit. Start by placing the harness in the proper position in front of the canopy with slack in the lines. Here are a few points of care: You are no longer unrolling the canopy but pulling it along the ground. To avoid getting it dirty or abrading it, lift it as much as possible and pull it carefully. Stop if it gets snagged and find the source of the problem.

Quite often, the lines (which are now hooked to the harness) catch on debris, rocks or stubble as you try to spread the canopy. This problem may warrant a couple trips to the harness to lift the risers, shake them and toss the lines out to the side. Proceed from one side to the next until the entire canopy is laid out. Horseshoeing the canopy (see Laying Out the Canopy) makes the process easier.

Later, when you are doing reverse inflation practice, you can even lay out the canopy by yourself without removing the harness. In fact, this technique is normal procedure when launching at crowded sites as we'll see later. Here's how to do it: Lay the bundle down at your desired departure point. Now release one or two loops of line from the hand holding the loops. With the other hand pull first one wing tip out as far as the lines allow, then the other as shown in figure 4-25. Continue in this manner as far as you can go until the lines are all stretched out. Be sure to get the center section well-positioned and the tips clear of any lineovers.

A premium spread leads to reliable inflations.

You will not be able to get the canopy entirely spread because you can't walk all the way to the tips, but a good horseshoe can be made. Now return to launch position and clear the lines in either the forward or reverse position. Shake out any apparent rat's nests in the lines (learn at the training hill what is a reasonable line layout). You're ready to go!

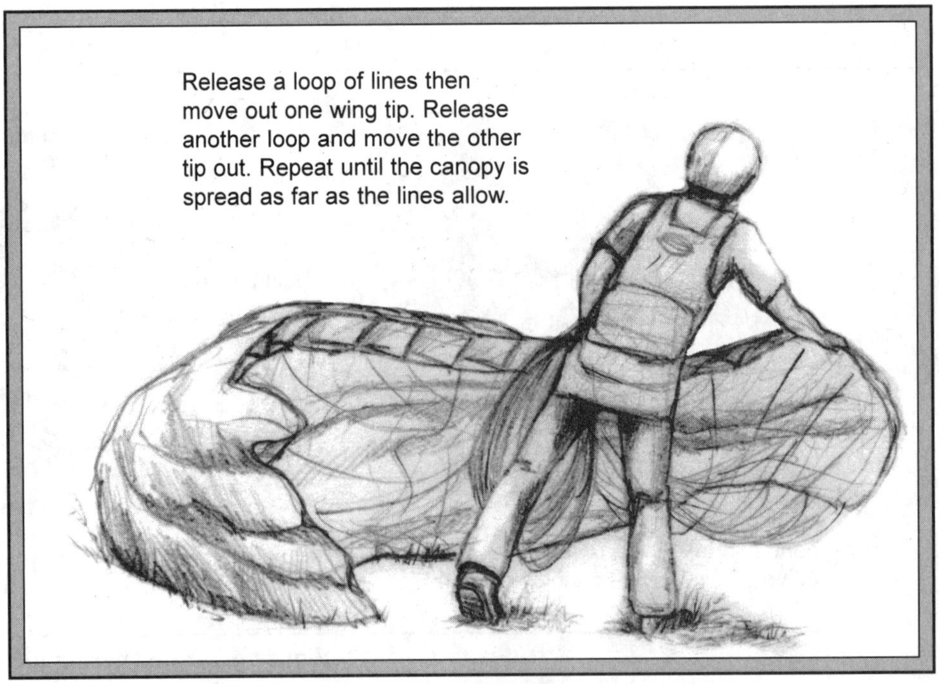

Figure 4-25: Self-Spreading the Canopy

Forward Inflation

Once you have mastered reverse inflations, forward inflating is relatively easy. The main difference is you can't see the canopy as it rises and you must look up to check it before actually launching. Your initial holding and positioning is also different.

The best conditions for learning forward inflations, just as with reverse inflations, is in a light (5 mph or 8km/h) wind. However, we can also learn forward inflations in zero wind because it's easier to keep a fluid motion when you are facing forward. A slight slope is permissible, even during your initial practice.

Forward inflations give you a better view of your initial run, but a poorer view of your canopy.

Your Goal: *To learn how to inflate consistently in various conditions while facing forward.*

✓ Attaching the Risers

Even when inflating or launching in the forward position, we recommend getting into your harness first, attaching the risers in the reverse position, then ducking under the appropriate set of lines to launch in the forward position. The reason for this technique is to avoid getting dragged backwards in winds later in your flying experience. Remember, good habits should be learned from the beginning.

If you are going to attach the harness facing forward (with it on or off), here are the procedures. Start with the two riser groups laid out on the ground 2 to 3 feet (.6 to 1 m) apart in front of the canopy. Make sure the lines are untwisted and the A riser is on top. Now stand between the risers facing away from the canopy and lean to one side to pick up a riser group. Keep its orientation and affix it to your carabiner with the A riser forward as shown in figure 4-26. The easiest way to do this initially is to pull the riser over your shoulder as shown. This method avoids twisting the riser or causing confusion. After you are familiar with the system, you can bring the riser under your arm while maintaining the same A riser-up orientation. Bend the loop of the riser down as shown to line up and hook to the carabiner. Once it's on, double-check to make sure the lines aren't twisted and the A riser is facing forward. Repeat this procedure with the other side. You should take a step back with the risers as you attach them so you don't pull on the canopy inadvertently.

Figure 4-26: Forward Facing Riser Hookup

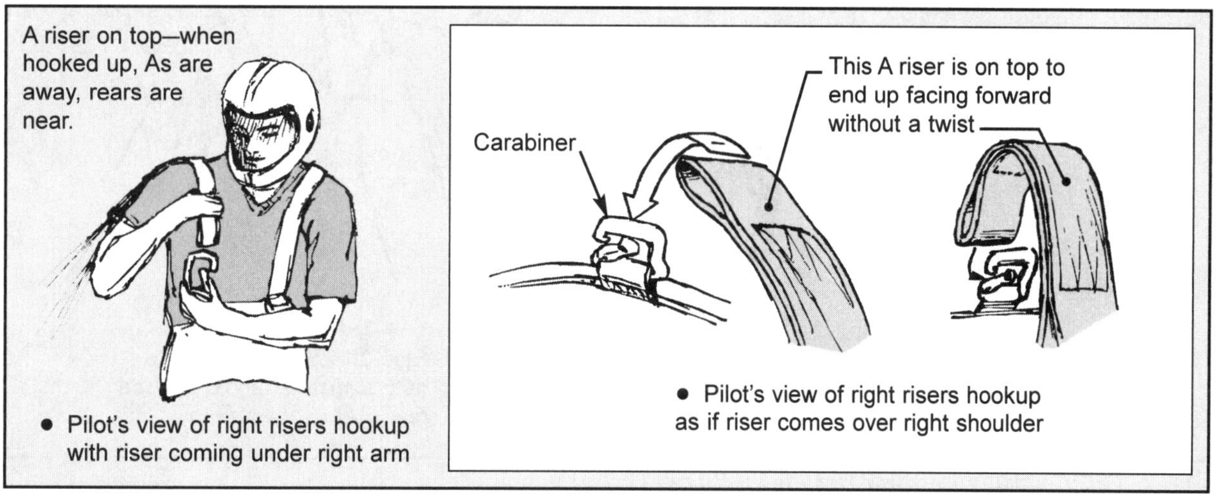

✓ Lofting the Canopy

Just as with reverse inflations we break forward inflation into its elements for clarity. These are:

1. Centered and square to the canopy
2. Holding the risers properly
3. Good arm position for an even, equal pull
4. A smooth, continuous run
5. Releasing the risers in a timely fashion
6. Looking up to check the canopy

Here are the steps in detail:

1. *Centered and square* – This positioning is the same as for the reverse inflation except you are facing away from the canopy. You must look over both shoulders to check that you are centered on the canopy.

2. *Holding the risers* – Now we position the risers properly for inflation. It is best to deal with the risers one at a time, at least when beginning. With all the risers hanging from the carabiner, locate a rear one (hint: it is the one with the control toggle attached to it). If you are working on the right side, lift it to full length with your left hand, unsnap the control toggle and put this in your right hand in a natural control position. Now drape this rear riser over your right arm to lie in back of you as shown in figure 4-27.

Repeat this draping process with the next-to-rear riser until you come to the A riser. With the A riser, hold it at its full length and then grip it with your thumb and fingers in the manner shown in figure 4-28. You should be holding the riser (without a twist) at or just below its attachment to the quick link. The finished position is shown in the

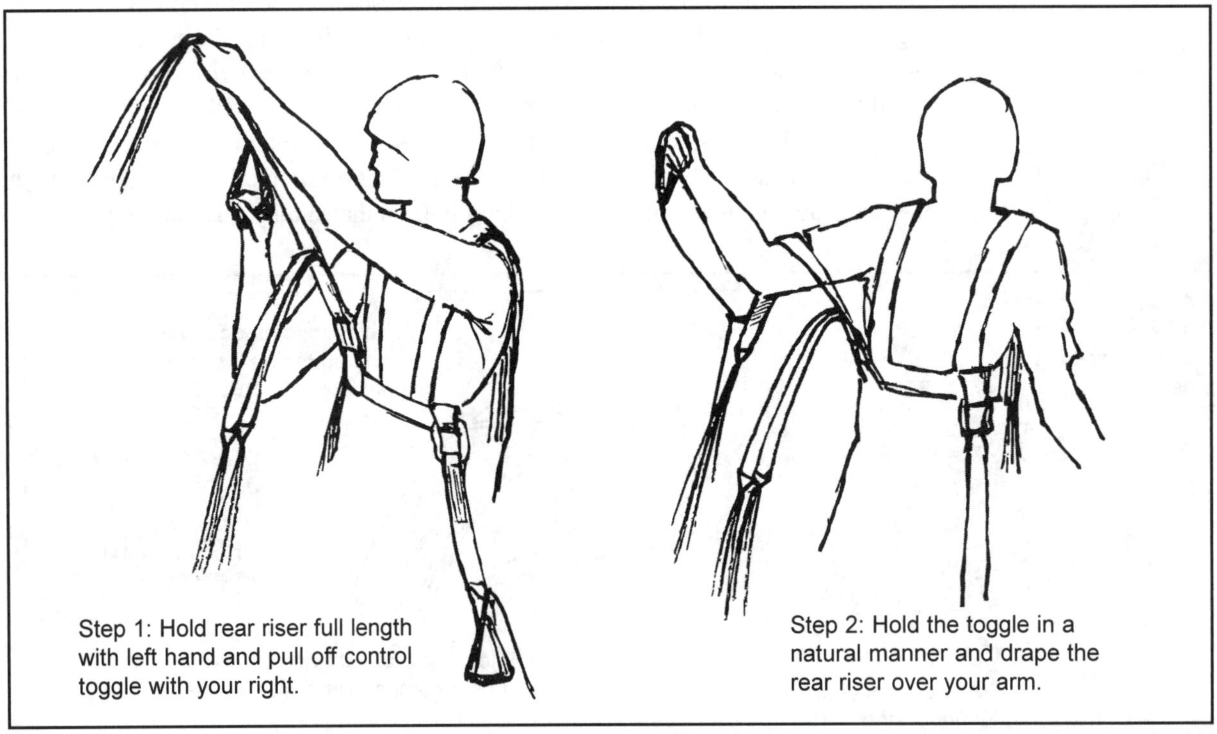

Figure 4-27: Arranging the Risers for a Forward Launch

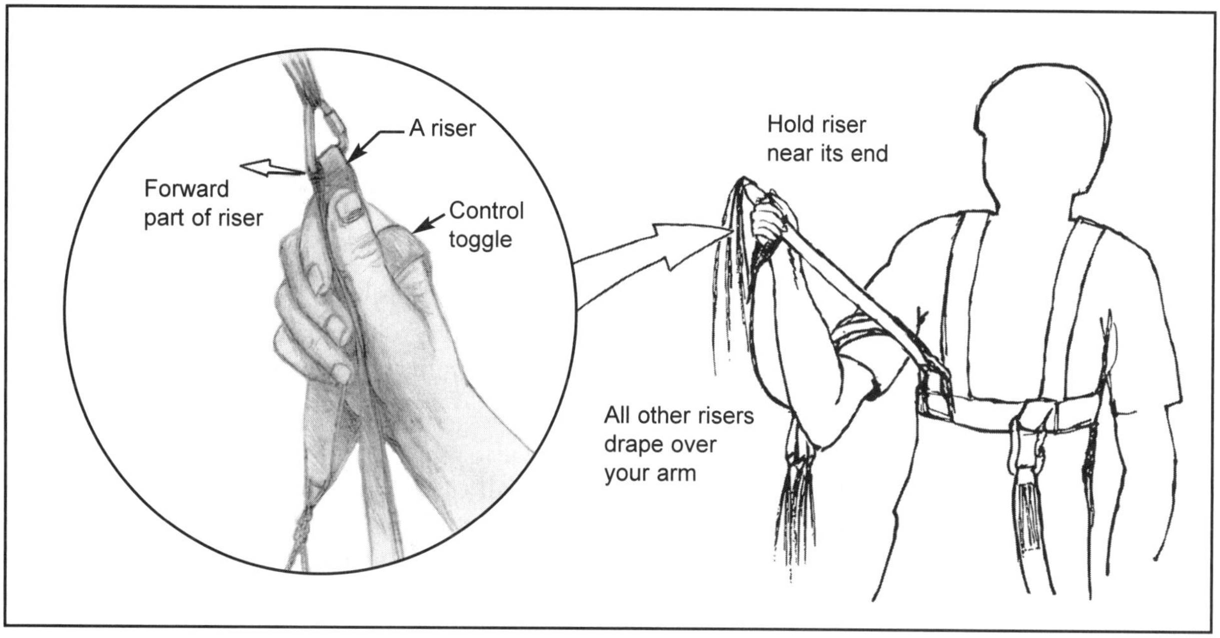

Figure 4-28: Riser Arrangement

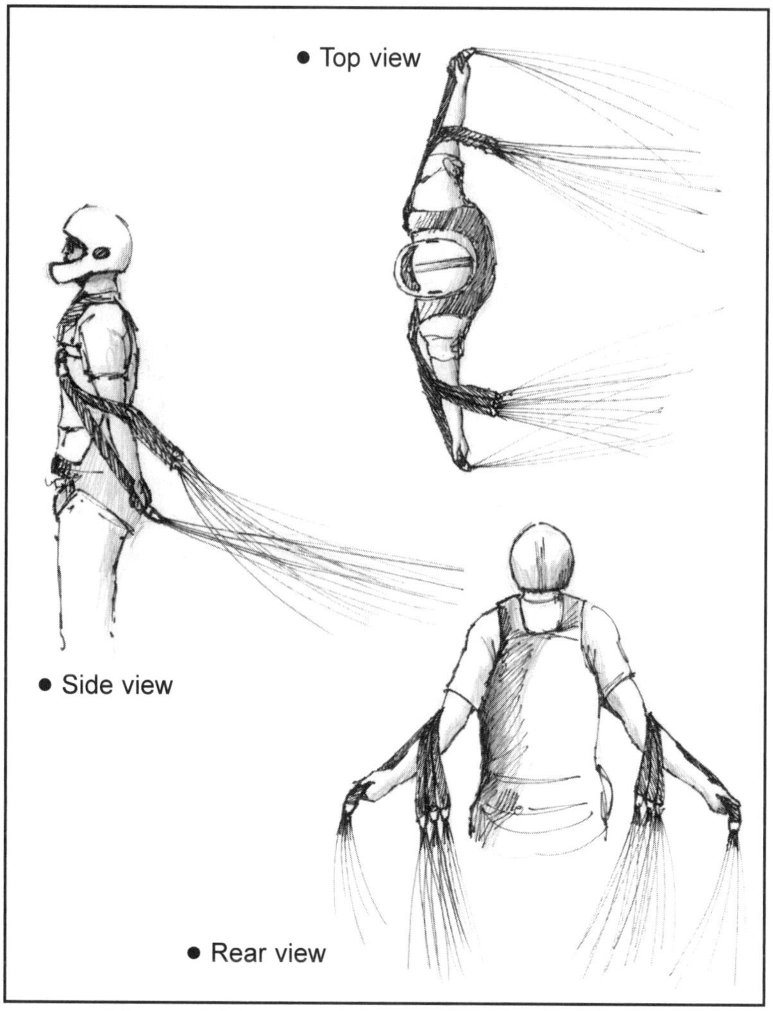

Figure 4-29: Forward Inflation Arm Position

figure with the control toggle in your hand, the A riser in your grip and all the other risers draped over your arm. Try to keep all the risers straight, but don't worry too much about the risers draped over your arm, since any twists will come out as soon as the canopy starts to lift. Repeat the process with the other arm.

3. *Arm Position* – This matter is one of the most important factors in producing problem-free inflations. Start with your arms down and back, leading with your chest. Once you begin your run, apply some upward pressure with the palms of your hands to lift the risers, but allow them to slide through your hands as you lift your arms. Your hand's motion should be at or behind the plane of your body as shown in three views in figure 4-29. This method promotes a good running stance, and applies most of the inflation forces with your legs and allows a light touch leading to a smooth transition from the risers to the controls.

4. *Smooth run* – When you have checked the wind (see pre-launch check) and your instructor gives you the OK, you will begin running. The canopy will oppose you as it rises—forces will build through the risers—but keep trying to move forward smoothly. Lean forward and keep your knees bent so you can better pull the canopy and raise yourself up if the canopy lifts you a bit. Soon the forces lighten up as the canopy gets higher and higher. Keep running!

In calm or light winds your run supplies the canopy's airspeed. Don't slow or stop during inflation or launch.

Pro Tip: *In very light winds, taking one step back to slacken the lines helps you gain momentum for a forward launch.*

This technique was especially useful for older, harder to inflate canopies. However, too much slack in the lines jerks them unnecessarily. If you can easily inflate in zero wind without line slack, do so.

5. *Release the risers* – As you continue to run, keep the forces on the A risers even, but lighten up your grip. By about halfway through the inflation you should be pulling more with your harness than pushing with your hands. At this point the backward forces in the A risers rapidly go to zero and you feel a tugging in your seat. Your hands should be open to avoid pulling down on the A risers.

Look carefully at figure 4-30. Here you see a canopy inflating. Note in the third position the pilot is still pushing forward on the A risers. They are bent at his hands so if he releases the A riser the lines will have slack and the front of the leading edge will go up which will retard the canopy. However, at this point is where the forward push should begin to ease up. The hands should be totally off the risers by the time the front of the canopy reaches directly over the pilot's head

Figure 4-30: Forward Inflation

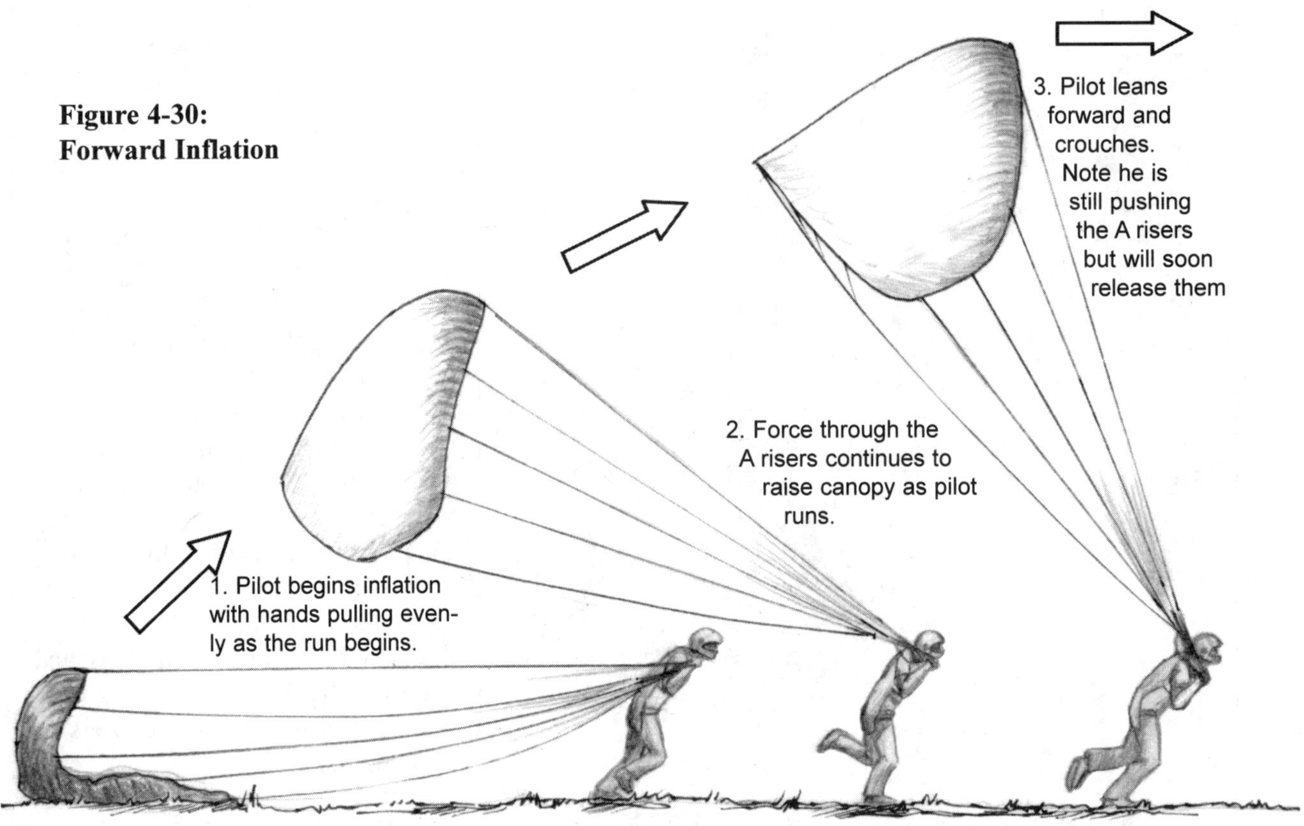

as shown by the dashed lines.

If the pilot holds on to the A risers, invariably he or she will pull down and collapse the front of the canopy. The proper technique is to *gradually release the A risers at the top of the canopy's arc and move your hands down into control position (about ear height) when the canopy is above your head. Do not stop your run.*

6. *Check the canopy* – An extremely important part of forward inflations is to look up—straight up—to check the canopy as you continue running. This matter is important, for the first time you can see the canopy on such an inflation is when it arrives over your head. If there is a malfunction, this is the time to catch it and make a correction or abort the run if necessary.

What are you looking for? A crooked canopy (one wing back), a fold (one or both tips folded in) or severe entanglement of the lines. In Reverse Inflations we discussed how to make corrections.

One of the biggest mistake beginners make when checking the canopy is to look over one shoulder. This position doesn't allow you a full view of the canopy and often twists your body so an uneven pull results. Another common mistake is to slow down the run. This tendency will allow the canopy to pass in front of the pilot which will usually cause it to collapse. You may have a natural tendency to slow down when you are not looking where you are going, but with the canopy lifting part of your weight, the consequences of a trip are not as severe as otherwise. You can practice to overcome this tendency by running carefully without a canopy and looking up.

Figure 4-30: Forward Inflation (continued)

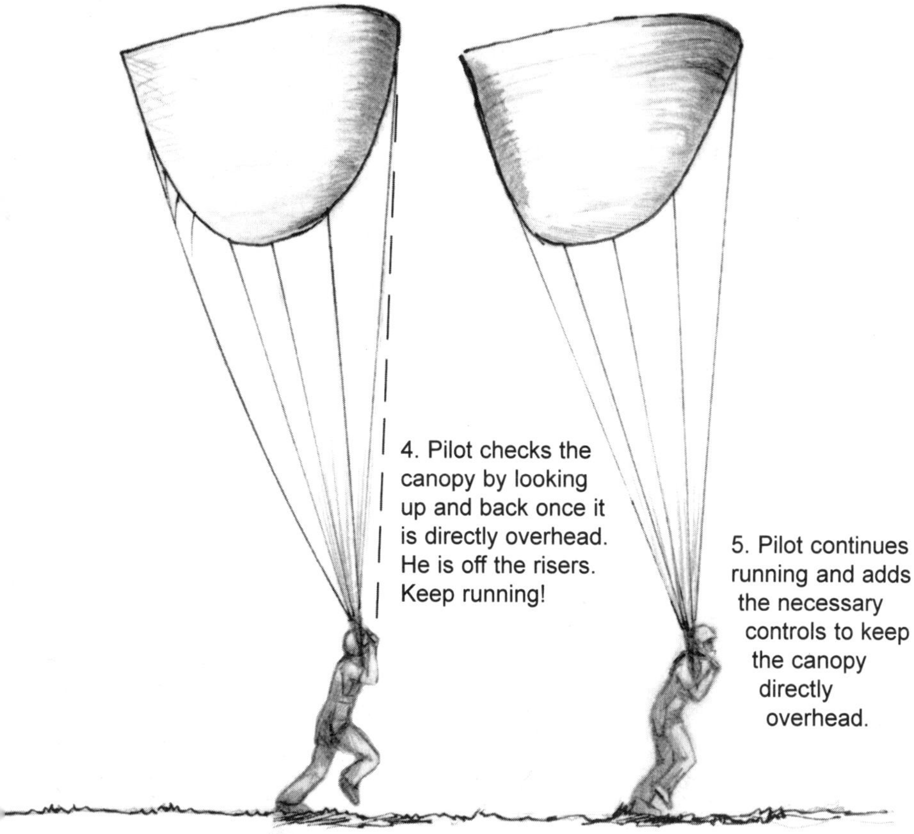

Elements of a Good Forward Inflation

- Even pull with proper hand position
- Well timed release of A risers
- Continuous run
- Canopy check

Common Mistakes

- Non-centered start
- Uneven pull or hand position
- Early released risers or pulling down risers
- Not looking straight up
- Stopping the run when checking the canopy.

Master the forward inflation as you did the reverse inflation for those few times when you will need it.

Skill Check: *To see how well you can inflate the canopy, keep a tally of how many attempts are successes. When you are scoring 9 out of 10 you are quite adept. Now work on that tenth one.*

✓ Stopping the Canopy

Just as with a reverse inflation, when you are ready to drop the canopy to the ground, we recommend you turn and face the canopy at the same time you stop it. Simply pull full down on both control lines to stop. If you are running forward the canopy should move behind you, stall and drop as shown in figure 4-31.

Figure 4-31: Dropping the Canopy

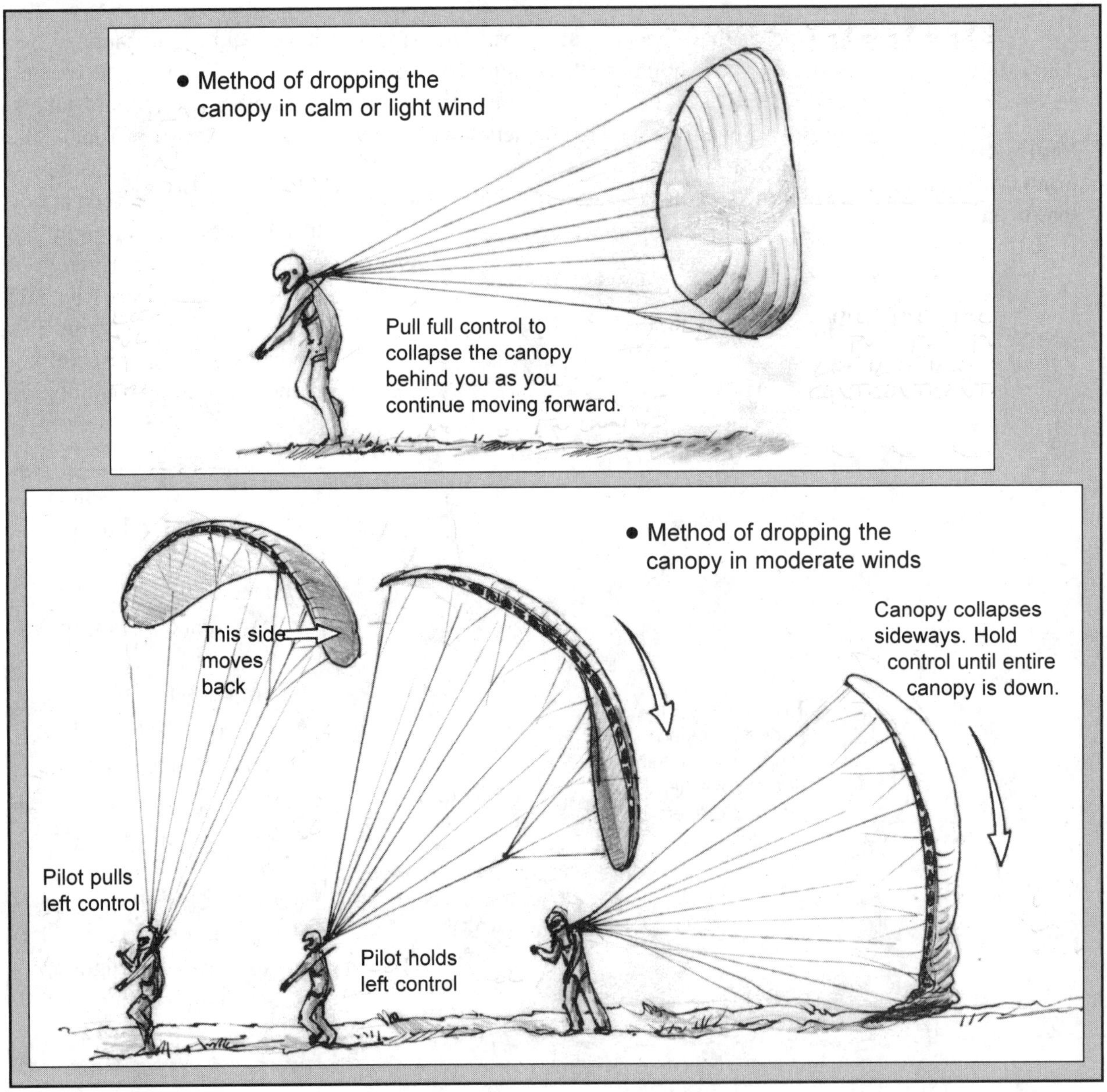

Some gliders (such as beginner gliders) are set up with long control lines to help prevent stalls. If this is the case, or if your arms are short, it may be desirable to take one or two wraps of the control lines around your hand just before you give the big pull. Do these wraps by quickly rotating your hands around the line as if you were spooling in line (which is what you are doing).

Practice this technique until you can do it quickly without looking. You will probably use it later in your flying for landing. Pulling the control lines full length to stop the canopy is known as *flaring* the canopy. A *flare* is what you will use to land gently.

Caution: *In higher winds flaring can result in you being pulled back. Strong wind procedures are described in Chapter 8. In light winds, be prepared to take a step or two back to follow the canopy. One thing you can do is pull both rear risers which will quickly drop the canopy. Alternately, you can pull just one control line fully. The canopy will then turn and drop in a nice pile as shown in figure 4-31. When you are reverse-inflating in a bit of wind you have fewer problems dropping the canopy because you can easily step towards the canopy as it drops and pulls back.*

GROUND HANDLING

Truly ground-handling includes inflation, for this topic covers all the operations and control of the glider on the ground—before stepping into the air. In fact, with success at inflation we can be satisfied that we have already accomplished a major part of ground handling.

Pilots practice kiting their glider—the best way to learn canopy control.

We can't say enough about ground handling, so we'll let the well-respected instructor Dixon White say it:

"Ground handling skill is the key to expert piloting skill."

Ground handling indeed prepares you for your entire flying career. The more adept you are at controlling your canopy on the ground, the better you will be able to launch and the more sensitive you will be in the air. Those are two marks of a good pilot. So let's learn to ground-handle.

Your Goal: *To learn how to control the glider on the ground in relation to the wind, facing both forward and backwards*

✓ Fore and Aft Control—Kiting

The act of lofting the glider and maintaining the inflation is called *kiting*. You can only ground-handle the canopy and kite it if the winds are 5 mph (8 km/h) or more. This is the minimum airflow over the canopy that allows it to hold up its own weight. You can, of course, continue running to maintain this minimum airflow which is good practice. Whether you move or are stationary, the first thing to practice is keeping the wing well inflated above your head without variations. The wind is often slightly varying so you will need to use some control inputs. Let's learn to feel their effects here.

Note: *You can (and should) practice kiting in both the forward and reverse positions. The following discussions apply to both positions except where we describe running which, of course, is much more readily accomplished when you are facing forward.*

If you are facing forward, lean forward and move forward in light winds. In stronger winds, lean forward to offset the wind force and focus on balancing the glider. The whole matter of canopy ground handling is a dance involving you and the wind. Learn finesse and involve all your senses in the experience.

When you begin, keep your hands at about ear level to maintain the canopy directly over your head. The control position varies with the glider design. Your purpose here is to learn the sweet spot that allows the glider to float in one place directly above you. Once you have accomplished this feat, try letting up on the controls and watch the canopy move forward. Don't let it go too far or it will deflate. Bring it back with a smooth, even application of both controls. Try it several times and see if you can let it go right to the point of collapse and still bring it back steadily. Practice until you can bring it back with no overshoot rearward. This act requires the right amount of control application and the right timing of release (see figure 4-32).

Figure 4-32: Fore and Aft Kiting Practice

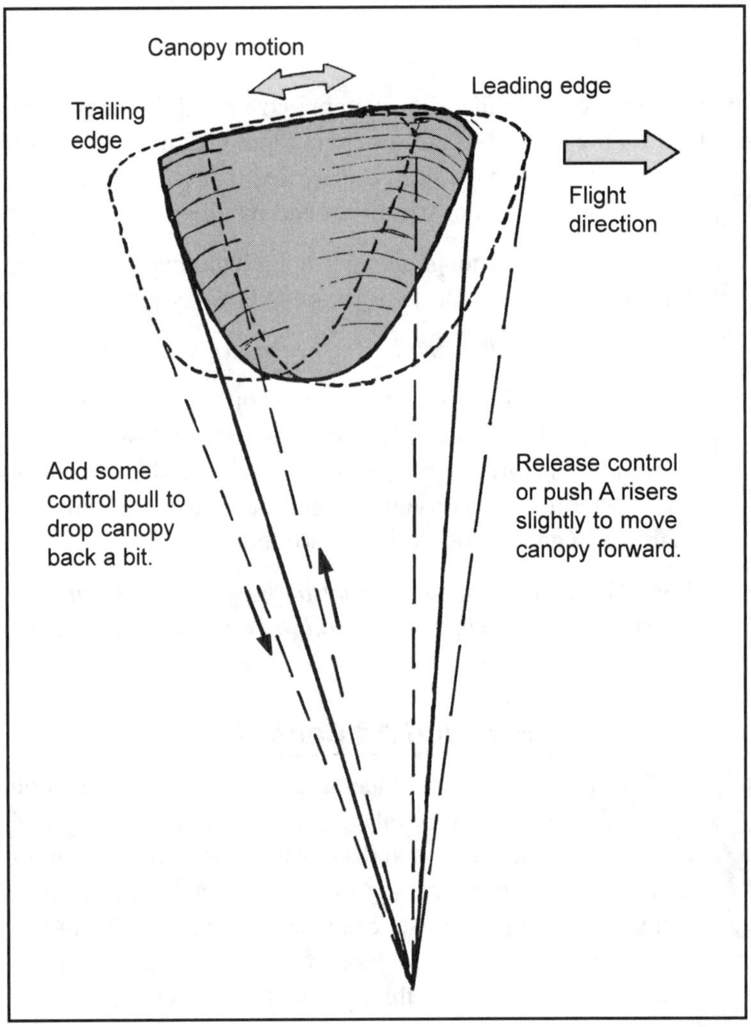

Some beginner canopies will not float forward when you release the controls. In this case, try pulling down gently on the A risers until the canopy moves forward. Release gently and control.

Did you notice the controls getting lighter as the canopy went forward? In the next practice you'll notice the control getting heavier (more force) as the canopy moves rearward. This leads us to an important technique.

Pro Tip: *To feel how your canopy is heading, sense the pressure in the controls.*

The mark of a good pilot is his or her ability to know what the canopy is doing without constantly looking at it. This is sensed through the forces in the control lines and the "pull" of the harness on your rear—true seat-of-the-pants flying.

Now apply a little even control force. Can you see the canopy move back? Don't let it go too far or it will collapse. Play with this control until you can move the canopy rearward and bring it back up smoothly without overshooting. This recovery technique involves releasing the controls then reapplying them carefully to stop the canopy's forward surge. If the canopy moves too far back and begins to collapse, you can often recover by moving forward (upwind) and pushing on the A risers. Apply controls if necessary to stop the canopy above your head.

It is normal for winds near the ground to vary in velocity or to be a bit gusty. You will notice that when the wind increases the canopy tends to move back and when it diminishes the canopy moves forward. You should practice keeping the canopy steady right above you in these conditions. To accomplish this feat let up on the controls as soon as the canopy moves back, or you feel the wind increase or you feel increased pressure in the controls. All of these items are signals which you must learn to sense. The sooner you detect a change, the less control input you have to make. With practice you will learn the exact timing and amount of control to apply for a given gust of wind. When the wind lets up suddenly, control the forward surge with a pull on the controls in a similar fashion.

Taking wraps of the control lines as you turn to face rearward will give you added control authority.

Practice these matters in both smooth and varying conditions. Perfect your technique while facing forward and reverse. These controls are the same as you'll be using in the air to maintain the position of your glider. Once you are comfortable controlling and kiting the canopy, practice turning forward and backward while maintaining a stable canopy. You should repeat this trick until you can perform it at will with control.

When you are operating in moderate winds, adopt the extreme lean position shown in figure 4-33 when facing forward. The lean helps you compensate for wind gusts and the rearward arms allow the risers to go straight up and help prevent the control pull from increasing if the canopy drops back.

Figure 4-33: Higher Wind Kiting Position

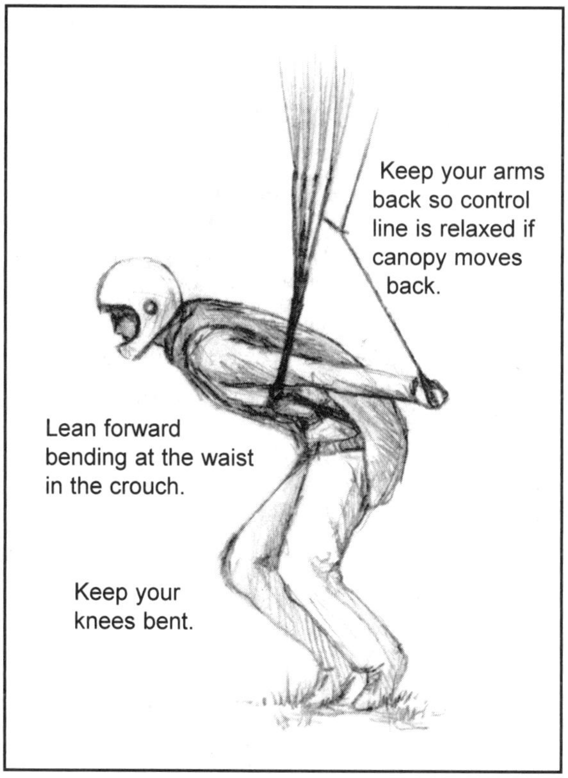

✓ Side-to-Side Controls

The next major ground-handling or kiting practice involves using the controls asymmetrically. The ability to perform these controls correctly is extremely important for handling turbulence and varying wind during launch. Also, you will be beginning to learn the proper controls for turning.

Controlling the canopy on the ground is actually more difficult than it is in the air. The reason is the situation on the ground is somewhat unstable as figure 4-34 shows on the next page. When the canopy gets blown to the side your body anchors it and it tends to go sideways even more. The only way to maintain control is to move your body. Sidestep to follow the canopy while you apply a control pull on the upper wing as shown. The idea is to get your body back under the canopy while you drop the upper wing back so the canopy so the canopy faces the crosswind. Follow these procedures when facing both forward and backward.

Figure 4-34:
Side-to-Side Control

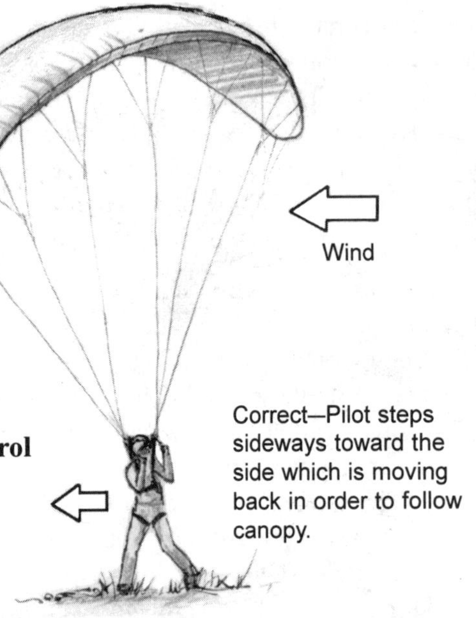

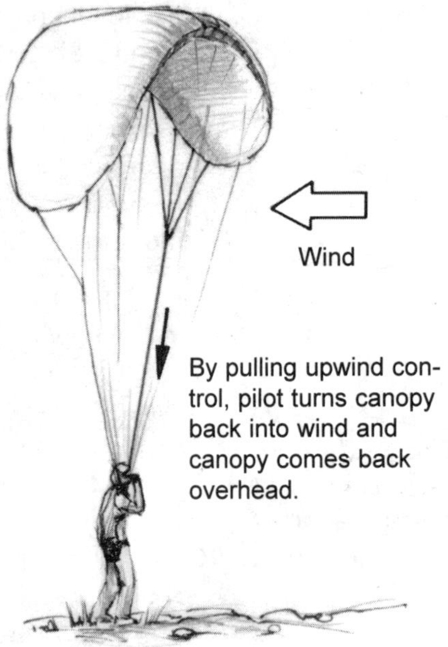

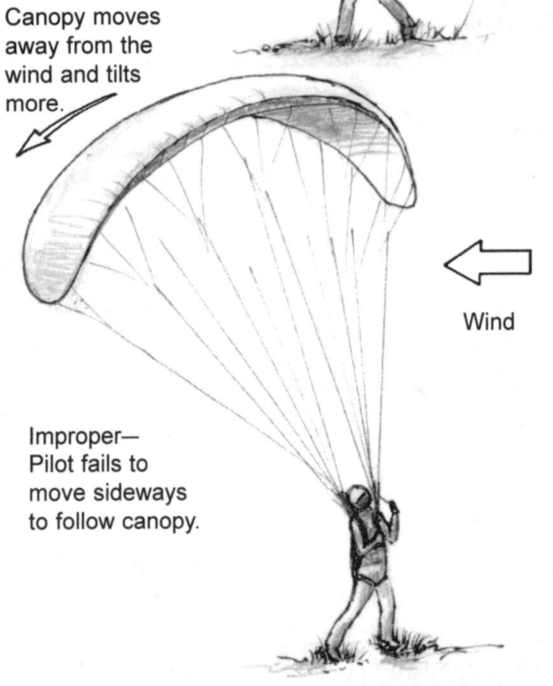

Side-to-side Control

Pull the control on the side opposite the direction to which the canopy is pulling and move towards the canopy. Remember:

- Steer left, step right
- Steer right, step left

You don't have to wait for the wind to turn your canopy. The next time you kite it, try pulling gently on one control line then releasing it. Did you notice that side of the wing falling back? Try it several times and add more control. Eventually you'll collapse that side or you'll turn the glider sideways and it will drop to the ground. Practice this dropping, for it is a useful technique for dropping a canopy in one pile (see previous figure 4-31).

A pilot reverse inflates. Note the canopy has come up crooked and the pilot is pulling the control line routing to the left to drop that side back so the other side lifts more.

If you are running with the canopy, try a slight control on one side while running to stay under the canopy. Now try going from side to side by using one control then the other. Practice all these matters while facing both forward and reverse until they become second nature. When you are ground handling in reverse, notice how an application of control on one side retards that wing so the canopy rotates but also tends to tilt away from you. This effect is the beginning of a roll which allows a turn. It is also how the canopy reacts when inflating in crossing winds. As you make your controls, try moving with the canopy to stay centered. Be sure to move a bit backwards as you do this so you don't slacken the lines. The combination of proper control application and moving your body is the technique for controlling the canopy in switching winds. Learn it well.

Listen to Lars Linde, former USHGA Safety and Training Committee Chairman:

"Everything that happens on the ground happens in the air, but it's easier to control in the air. On the ground the most important thing is body positioning while in the air your body positions automatically. The best way to become a good pilot is to spend a lot of time kiting the glider and observing the canopy. A gust changes the shape of the canopy before it turns it. Learn to feel these changes, anticipate and react before the canopy moves. Be a pilot, not a passenger."

✓ Riser Control

We have been concentrating mainly on control line inputs. Here we'll see what the risers can do. We already tried pulling down on A risers and found that the canopy moved forward. Now try pulling evenly on both rear risers. Did you see the canopy move back? What's going on? We are changing the wing's angle of attack (see chapter 3). By pulling on the front risers we essentially tilt the front of the wing down; a back riser pull tilts it up. Tilting down lowers the angle of attack which makes the wing fly faster and vice-versa.

Figure 4-35: Riser Turn Control

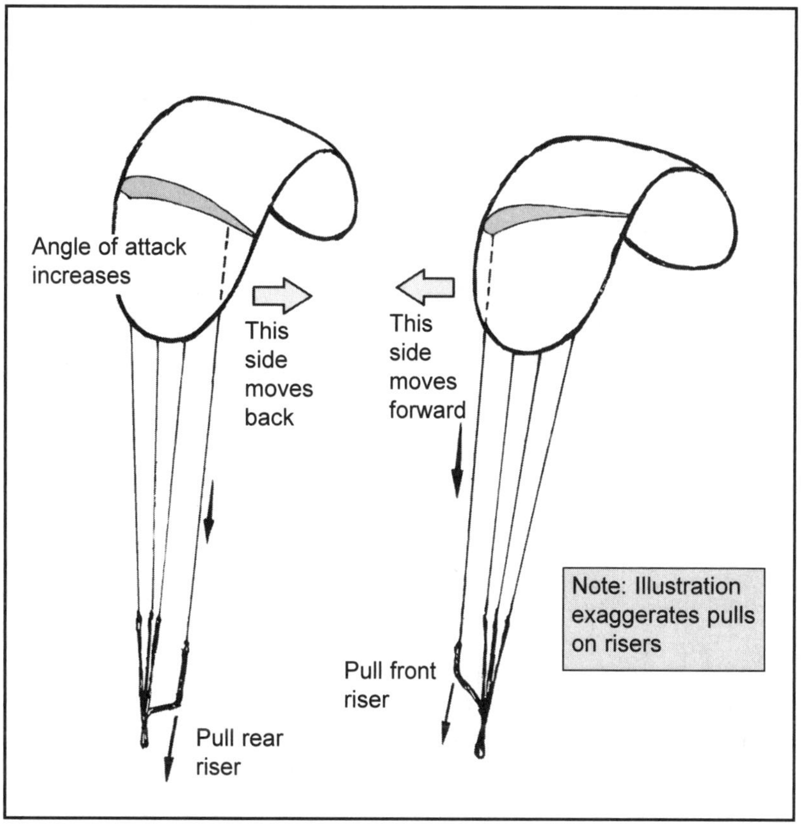

Practice this angle of attack change until you understand the limits. If you go too far with the rear riser, try bringing the canopy back up while moving upwind and hoisting the A (front) risers. Go ahead and pull the A's too much and watch the wing collapse. Now do the same with the rear risers and watch the glider drop to the ground. The latter is an important method for dropping a canopy quickly. Learn it!

Now let's try one riser at a time. If you pull gently on one rear riser you will see that side of the wing move back (naturally these changes are easier to observe if you are in reverse position). The reason is you are increasing the angle of attack and creating more drag on that side of the wing (see figure 4-35). Try the same thing with the front riser. Again, it rotates the wing, but this time in the opposite direction. That's because you are lowering the angle of attack on the pulled side.

Once the canopy is aloft, be flexible with your hands and move them readily from riser to riser to effect control. Learn automatic response through practice.

Practice this riser control until you can float the glider in varying winds purely by pulling the correct riser. If one side of the wing drops back, pull on that side's A riser. If one side moves forward, pull on that side's rear riser (we'll see later how you can control in flight with the rear risers in case of a control line problem). Finally, pull one rear riser only and see how the wing turns, then drops. This effect is often your best way to collapse a canopy in high winds.

There are other ground handling things to try, such as big ears and asymmetric tip folds, but we'll leave these subjects until Chapter 9.

SUMMARY

By now you should have a real feel for the canopy and how it obeys your every command—even if your commands are incorrect! But never fear, continued practice at inflation and ground handling will soon have you making all controls automatically, timely and precisely. Don't stop your ground handling practice at the training hill. You can continue learning all sorts of subtle things about your glider by inflating and kiting in the many varied conditions you will encounter.

No doubt you have also felt the power of the wind on the canopy by now. Perhaps it was disconcerting at first, but you soon learned that with training you could coax the canopy to combat the wind's caprices. Be aware that on the ground the entire situation is not as stable as when you are airborne. Let's experience this first hand. It's time to fly!

TEST YOURSELF (*Answers in Appendix II*)

1. When laying out a canopy the ____________ edge is furthest away from the pilot in launch position.

2. Malfunctions during inflations can result in serious problems in flight. List three such malfunctions.

3. You should use reverse inflations only in strong winds. *True or False*

4. When reverse inflating, you should change your direction of turning around according to the cross-wind component. *True or False*

5. Describe building a wall. ____

6. When inflating forward you should hold the ____ risers.

7. When in a stronger wind, how do you drop the canopy quickly? __________

8. It is permissible to unfasten your leg straps on your harness when:
   a. Climbing back up the training hill
   b. Waiting for a takeoff slot
   c. Only when the wind tails
   d. Only when removing the harness

Chapter 5

Walking on Air - First Flights

"Once you have flown you will walk the earth with your eyes turned skyward, for there you have been, and there you long to return."

— Leonardo da Vinci

Today we begin to learn to fly. This is an experience that millions of birds go through every spring, yet still it is a marvelous thing for a human being. It is a new beginning, for as the master Leonardo implied, once you have flown you will never be the same.

One prerequisite for safe and efficient progress from a fledgling to a true native of the sky is a good instructor. Such an important guide can be found by contacting your national organization. All the learning presented in this chapter is done under the assumption that you are receiving expert instruction at the time. This material is intended to give you the background understanding so you can progress more readily and devote your time at the hill to actual flying.

As Lao Tzu said, "A journey of 1,000 miles begins with the first step." Today you will take the first step on a pathway that will stretch your horizons well beyond what you can imagine.

Preparing to Be a Pilot

Your first experience of paragliding is naturally as a beginner. To achieve the USHGA Beginner level you must learn to take off, control the glider in flight and land—all with consistency and control. The way you will proceed depends on the area you are in. If you learn at a hill, you will start on level ground and learn to inflate and ground handle the glider. Then you'll proceed to run with it, still on the level. Soon you'll move to a slope and work your way up gradually until you become airborne. You will remain very close to the ground until your instructor judges your skills adequate for higher flying. The whole experience is designed to ease you gently into the air.

If you are learning in flat areas with tow training, you may begin with tandem instruction. In this case your instructor will initially perform all necessary controls and commands, then teach you to take over one thing at a time until you are piloting from lift off to landing. Again the approach is gradual so you learn comfortably.

The challenges presented by learning to paraglide are both mental and physical, so let's see how to prepare for them to maximize our fun.

Your Goal: *To prepare physically and mentally for your first lessons.*

✓ Proper Attire

For your first practice with the canopy and in the air, you should dress properly. For clothes we suggest layers that can be removed. Remember, you will be climbing hills (unless you are towing) so you'll heat up quickly once the sun comes out. You also will likely be standing around, so prepare to be cool in colder conditions. A T-shirt and long pants are fine to fly in and so is a full ski outfit. Be prepared for the local temperatures.

A new pilot kites his glider in the reverse position. Note the warm clothing for the winter months.

We mentioned footwear in Chapter 2. We'll repeat the idea that hiking boots that support the ankles are ideal while running shoes may be appropriate for your early learning. Your instructor will give you advice for the local situation. Did you forget your gloves? Not a big problem, you can learn without them, but they do help protect your hands when you are handling the lines and risers in higher winds.

✓ Physical Effort

The beginning phase of paragliding often requires the most physical exertion because you will be climbing hills and carrying the glider. Naturally the better shape you are in to begin with, the easier this practice will be. However, people of all ages (from very young to very old) and body types (from couch potatoes to body builders) have successfully learned to fly paragliders.

Since the hills won't erode away for eons, weaker individuals can simply take things more slowly for there's no rush. In fact, instructors often find that people

A pilot makes a soaring pass at Monte Cucco in Italy.

who progress more slowly often learn more thoroughly. A good school can accommodate many types of individuals and learning styles.

If you have some time (weeks) to prepare physically for your lessons, suggested training is aerobic exercises that improves stamina. However, just as useful as this training is flexibility development. Stretching exercises that limber you up are advised to allow the full use of your muscles. In fact, many instructors will have you do some stretching before you run.

You'll need some energy for your day on the hill and in the air. It is important to get a good night's rest so you are alert and fresh. Schedule your lessons so you are not taking them the day after a stressful situation like a major race or a major party.

Eat properly for exercise. This means plenty of carbohydrates (starches like potatoes, pasta or bread) the night before, and a good breakfast. Your instructor may have you pack a lunch if the lesson lasts into the afternoon. In any case, beware of "empty" calories delivered from candy bars and the like that supply quick energy that is quickly used up, leaving you fatigued. Bring along plenty of water!

We highly recommend that you begin each lesson in which you will be physically active (including inflating and ground handling) with stretching. On cool mornings a sudden acceleration can pull a cold muscle, especially if you are not used to physical exercise. Stretching is a good idea before every flight, even when you are advanced.

Early on, paragliding is about 90% physical and 10% mental effort. Soon this changes to 10% physical and 90% mental.

✓ Mental Preparation

As with most active sports some psyching up or mental control of anxiety may be required with paragliding. This process is natural as well as desirable, because without proper respect any active sport can become dangerous. The whole idea is to control anxiety by controlling risk as you gradually learn to control the glider.

If you have a fear of heights, do not despair. Fear of heights is essentially a fear of falling. You can be very high but secure and experience no fear. Conversely, you can be relatively low (on a ladder, for example) and feel insecure and therefore fearful. There are many seasoned pilots who are afraid of heights and will not get near the edge of a hill without being harnessed in their glider. When you learn to fly you build confidence gradually so the conditioned fear can be controlled and eventually eliminated.

The important point when attempting any new endeavor is to have a positive attitude. Be motivated by your own true desires to fly as freely as a bird. Come to your lessons with an eagerness to learn and you will do just fine. You, like countless others before you, *can* fly. Peter Pan understood that.

✓ At the Site

When you arrive at the training site you will likely find a gentle hill and a flat runout area*. You'll usually begin your lessons on the flat.

The hill itself will include paths to run down if ground cover exists. It will typically be from 10 to 20 degrees in steepness. This amount of slope lets you run easily to takeoff speeds.

Ideally the training site will be high enough to allow you to progress up the hill lesson after lesson. However, we don't live in an ideal world and as real estate becomes more scarce the ideal hill is becoming a rarity. As a result, most schools use several hills to cover various wind directions and various skill levels. Your instructor will direct you to the proper site for each lesson.

A pilot kites his glider straight into the wind indicated by the windsock.

✓ Watching the Wind

One of the first things you will learn in your development as a pilot is the importance of watching the wind. In Chapter 3 we noted how windspeed changes our groundspeed, so as beginners we always takeoff, fly and land heading into the wind.

For this reason we need a wind blowing directly into the hill. How do we tell the wind direction? The easiest way is to watch *streamers* and *windsocks* that are strategically placed around the site. Streamers should be placed at the landing area, at takeoff

* Much training in flatland areas takes place with towing. You may find that your hill is a tow vehicle. In that case your instructor may fly with you on a special glider built for two persons. You will learn to foot launch just as you do on a hill by running along as the towrope pulls you. At the end of this chapter we discuss what you should know as a tandem and towing beginner.

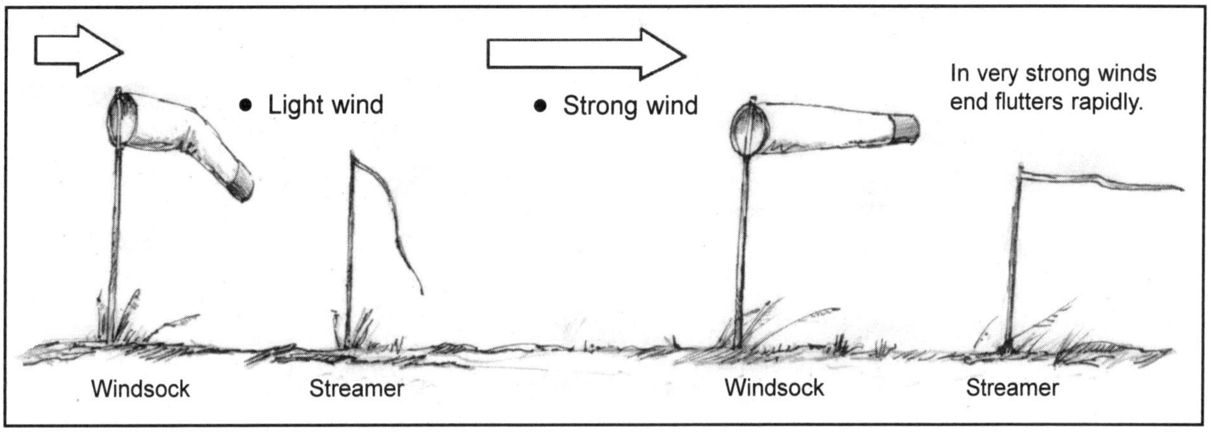

Figure 5-1: Wind Indicators in Different Conditions

and a couple points in between to really show the airflow. Don't use too many however or all the indications may be confusing. Figure 5-1 shows how windsocks and streamers indicate the wind in various conditions.

Occasionally pilots have a confusion as to the wind direction a streamer is showing. If you think of the wind as always flowing from the support stick to the tail you shouldn't have this problem. Also, try changing your position with respect to the streamer to get a better idea of the direction it is pointing.

Pro Tip: *In very light winds it is difficult to tell the exact wind direction. In grassy areas you can crumble some dry grass, let it fall and watch its drift. In dry areas kick up some dust with your shoe and watch the drift (see figure 5-2).*

Wind strength and variability are real concerns with all flying. Naturally there is a maximum safe strength for all pilots, which we discuss in later chapters. Remember, the wind force increases with the square of the velocity. For training, the ideal wind is smooth with a strength of 3 to 5 mph (5 to 8 km/h). But once again in the real world we settle for anything from no wind (calm) to as much as 8 mph (12 km/h) in very smooth conditions such as may occur near large bodies

Figure 5-2: Light Wind Indicators

of water. Also, in the real world the wind direction can be quite variable, especially on light wind days known as *light and variable* or simply L and V. Consequently we oftentimes learn to fly in slight crossing winds. Your instructor will specify the limits of direction and strength that are suitable.

Long before you arrive at the training site you can begin watching the wind. Such observation teaches you to quickly recognize conditions for your later flying and helps you assess the suitability of the day for you as a student. When your instructor gives his assessment you have a check of your own best guess. Use a wind speed (or airspeed) indicator often to check your perceptions.

Like a sailor, like a bird, we must learn to watch the wind and sense its subtle movements.

Try looking for moving trees, waving grass, smoke, flags and clothes hanging on a line to detect wind direction and strength. In the next chapter we provide a chart that helps you judge the wind.

On a typical day the wind is calm during the night and begins moving by 9 or 10 o'clock a.m. The reason for this is the sun's heating stirs things up. The wind increases to a maximum by late afternoon, then begins to die off by evening. Due to this daily cycle, most paragliding lessons begin in the morning to take advantage of the gentle conditions. In some places on some days lessons may have to be terminated by early afternoon as the air begins to churn. Don't despair, for paragliding will always be around and one thing is sure: if you wait long enough the weather will change.

Besides watching the wind and learning its moods, it is also important to learn to interpret weather signs right from the beginning of your flying experience. You should read Chapter 6 at this time (weather for paraglider pilots). In that chapter we describe the matters that a student (or pilot) should understand and use to predict flying conditions for their day in the air. Wind observations give you the immediate effects, but overall weather observation and understanding let you know what to expect.

INTO THE AIR

Here's the big moment you've been waiting for. The excitement is palpable, but you know what to expect. You've already performed plenty of controls. You've already felt the tug of the seat as your feet get lighter. You've already flown in your dreams a thousand times.

You are going to learn to step into the air with proper launch procedure which involves some safety checks and proper control inputs.

Your Goal: *To learn consistent, smooth takeoffs including good inflations and transition to flight.*

✓ Pre-Launch Check

Just before you take off, you must always perform a pre-launch check. What should you look for? Proper air conditions, proper glider setup and proper mental preparation.

When you are standing in position with the A risers in your hand, give one last look at the glider. Are you centered on the canopy and is it square to the wind? Are you holding the A risers properly with the control toggles in your hand? Are your leg straps attached? Always do a bottom to top check: *shoelaces, leg straps, chest straps, carabiners, risers and helmet strap.*

Always do a pre-launch check: shoelaces, leg straps, chest strap, carabiners, risers, helmet, traffic -- clear!

Next, look at the wind indicators around you as one last check. Remember, you need a zero wind or a straight-in smooth wind as a beginner. Later you can tolerate some variation in the wind, but for now it should be perfect. Learn to look a ways upwind to see if you can detect stronger stirring in the grass, brush or trees which would indicate a stronger gust approaching. If so, wait it out. Remember to check the variation of wind before you decide to fly. Your instructor will certainly do this for you at first, but you should learn to watch the wind from day one.

Finally prepare yourself mentally by relaxing as much as possible. From your repeated ground handling practice, you know inflation and controls will be no problem. Assume the proper position, recheck the wind, check the traffic... **CLEAR!**

✓ The Launch

We will mainly concern ourselves with reverse inflations here, but everything from turnaround onwards also applies to forward launches. In Chapter 7 we cover reverse inflations in more wind.

Normally you will have practiced inflations and kiting on level ground first, then perhaps performed them at the bottom of the slope. With a bit of wind you can kite the glider on a slight slope. Now, however, we will move far enough up the slope to actually fly. Ideally this move can be somewhat gradual so you barely get off the ground the first time. However, some training hills get steep quickly so it's best to start high enough to fly over the transition area.

A pilot prepares to reverse inflate from a ramp in the Alps. Note the clean layout of the lines.

At the chosen start point, lay out the canopy as usual (straight or horseshoe) then get into the harness in launch-ready position. When the wind looks good and your instructor gives you the go-ahead, produce a perfect inflation, check the canopy, turn around and continue running. Focus your vision on the ground about 20 feet (6 m) ahead of you. Keep running!

Safety Tip: *Remember to lean forward and crouch in your run. This posture helps pull against the backward tug of the glider, keeps your feet under you and also allows you to maintain contact with the ground longer.*

If flying is walking on air, your first flights will have you dancing on a cloud.

Now that you are on a slope you should feel the harness begin to tug more on your rear as soon as the glider inflates. Do not slow down and do not sit. Keep running until your feet are no longer touching the terrain. You're flying!

As soon as the glider is lifting half of your weight, it is useful to apply a bit of control force. Use about ¼ control (toggles about at ear level) with both hands even. This control will help the glider lift you off the ground. Make the control gradual so the canopy doesn't retard behind you. If you apply too much control, it is best to hold it when you are this low because a release may result in a forward surge of the canopy which will drop you quickly. The thing to shoot for is a nice, smooth transition into the air without canopy changes. Figure 5-3 provides an overview of an entire early flight.

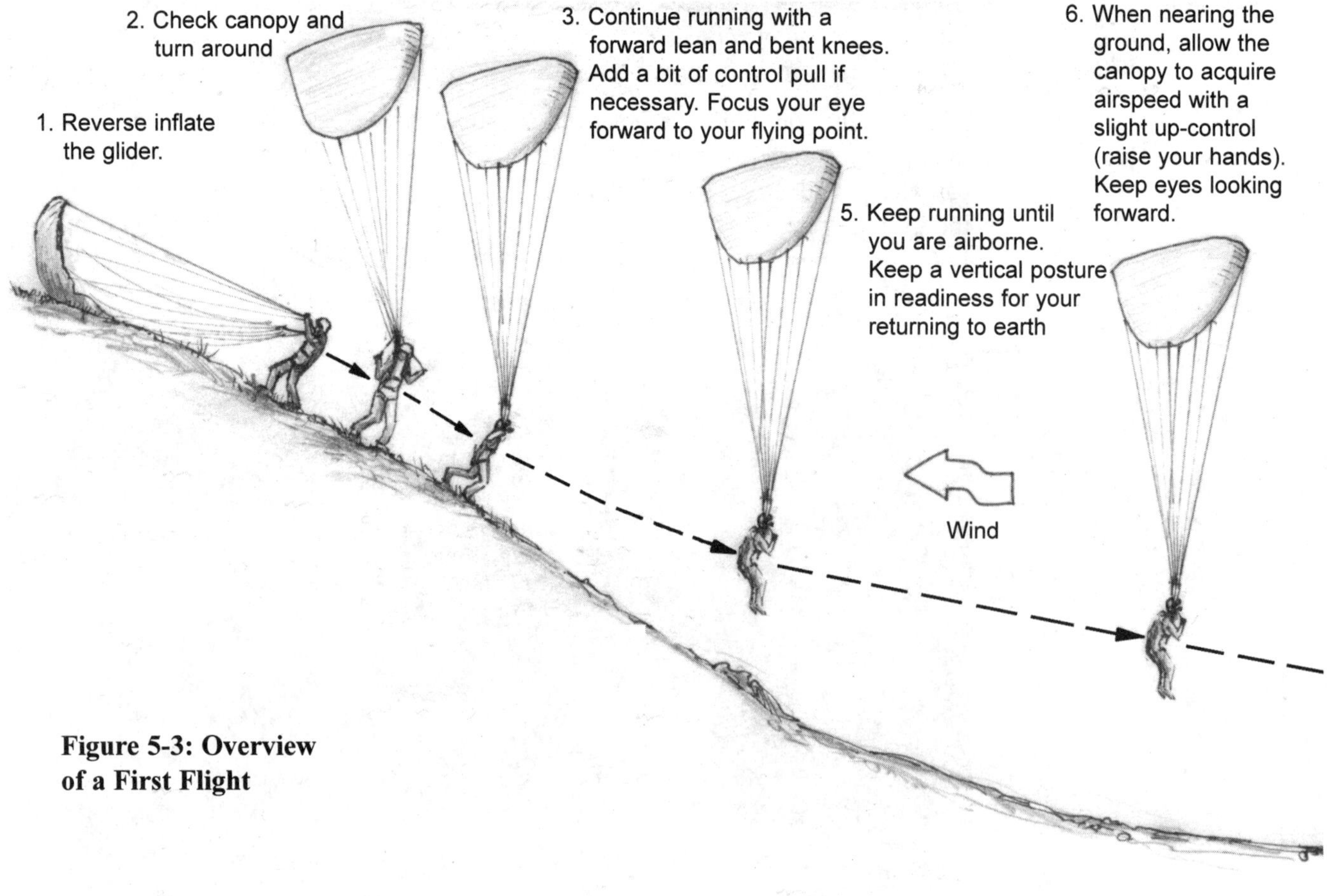

Figure 5-3: Overview of a First Flight

The proper hand position throughout the launch process is important. If you are standing still and kiting (facing forward) your hands should be near your shoulders with elbows bent. If you are running, lean through the risers with your arms back and up. This latter position may be difficult if your shoulders are inflexible, but it should be attempted as the ideal form.

Pro Tip: *To better regulate your control inputs, keep your hands following a line from the top of your head through your shoulders to your hips.*

Common Launch Mistakes and Cures

1. Slowing down the run once the glider begins lifting. Avoid this tendency by focusing on a point well ahead and visualize running to that point.
2. Sitting down too soon. Resolve to run into the air and keep your feet under you for an imminent landing.
3. Over-controlling. Make all controls slow and gradual. It's better to under-control than over-control at this point.

✓ *Imaging Practice*

Imaging is like imagining except it is carefully controlled, not a carefree fantasy. Imaging (also called visualization) has been proven over and over to be a powerful tool for learning and success in human endeavors, whether it be sports or other achievements. Most likely your instructor will have you use this technique at the training site, but you can do it at home as well.

To use imaging at this stage of your learning, simply close your eyes, relax, face the wind and play a movie in your mind of the skill you are trying to learn. As you act in this movie, try to feel the same sensations as in actual practice and go

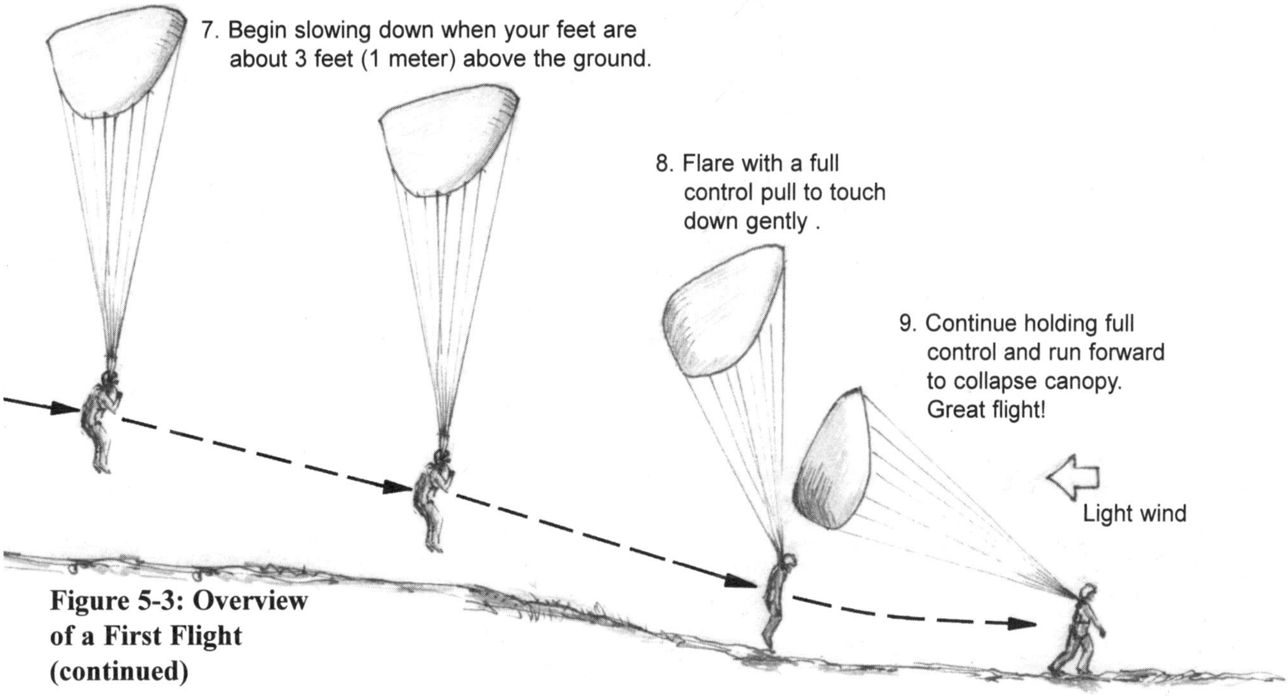

Figure 5-3: Overview of a First Flight (continued)

through the control motions. Avoid potential mishaps and concentrate on success.

The more realistic your imaging is, the more effective it is. Naturally you must know what the correct controls and results are before you can image them correctly. For that reason, instructors usually include imaging partway through the lesson rather than at the beginning.

Imagine what you want to do, then image your success.

Imaging works because it helps the mind process the sensations and proper reactions in a relaxed manner. When you are actually trying a launch, for example, things are happening fast, you're putting out physical effort and perhaps adrenaline is pumping as well. That scenario is not conducive to fast learning. Image the takeoff several times and you'll train your body's responses more readily. Use the imaging technique throughout all your flying and it will make you a better pilot.

IN FLIGHT

Your first few ventures into the air will be short but sweet. The exhilaration is beyond comparison. You may be so excited it's hard to concentrate on what you're doing. However, repetition focuses your mind so you can perform with conscious control. That's learning.

Your Goal: *To perform a smooth, straight flight without changes in airspeed from takeoff to landing, then learning careful, intentional speed control.*

A pilot sits back in the seat in easy-chair fashion.

✓ Sitting Down

Once you are airborne you may find yourself wanting to sit down. One of the primary rules for safe launches is to keep your legs and feet beneath you until you are well clear of the terrain. The reason for this rule is to allow you to continue running if you make a miscontrol or the air drops you down slightly. Develop the habit from your first flight of remaining in launch position (legs down) until you are well clear.

In the initial flights of training you are returning to earth soon after you leave it (that's necessary, unfortunately) so you should remain in launch position. In fact, any time you are within 20 feet (6 m) of the ground you should be forward in the seat in the PLF position. You may see more experienced pilots reach down to shove themselves back in their harness (which requires holding the controls in one hand). We cover this practice at the beginning of Chapter 9. Don't let go of your control toggle for any reason at the beginner level of experience.

✓ Steady Flight

Naturally, a flight doesn't end with the takeoff. You must continue controlling the glider until landing. To do this in your initial flights you normally start with the controls at a desired point (your eye level or wherever the instructor suggests) as soon as you're airborne and keep the same pressure on them until it's time to land. *Never* lock your arms, but keep them loose and floating to maintain an even pressure (from 1 to 6 lbs—.5 to 3 kg—depending on the glider and how slowly you want to fly). *Never* hold the risers or even touch them at this point. As you get higher you have more time to think about things and you can try speed control. For now, a straight, steady flight is in order.

What should you do? Begin to be aware of your sensations. Feel the amount of force on the control toggles. Feel the wind on your face and hear the sound in your ears. Do not look directly down at the ground, for it will only confuse you as it passes by. Look out ahead towards the horizon—the higher you are the further— and keep the ground in your peripheral vision. As you near the ground you can look at your intended landing point, but don't focus on your feet. Your balance and speed clues come from around you, not the ground. Don't look directly down.

Good training hill sites come in many varieties.
This training hill in Alaska shows that paragliding is a great winter sport.

Feeling the Senses

- *FEEL* – The control forces, the harness forces and the wind in your face.
- *HEAR* – The sound of the wind as you pass through the air. This sound is your best airspeed clue. Also listen to your instructor's commands.
- *LOOK* – Out ahead of you to direct your path, detect airspeed, orient yourself and keep your balance. Direct your gaze to the horizon until you are ready to land.

✓ Speed Corrections

One of the keys of proper speed control is maintaining a steady airspeed. If the glider slows too much, you will have to release some control pressure. If the glider accelerates, you will have to add control. At this point, be careful to make all corrections gradual, small (except for landing) and evenly with both hands.

If you pull the controls too rapidly, the canopy slows and your body swings forward which increases the glider's angle of attack even more. The opposite thing happens if you release the controls quickly—the canopy surges forward and leaves your body behind. At this "small controls" stage in your learning, take about 1 second to make the controls (practice them with imaging or on a simulator). Later with larger controls, do them in about 2 seconds.

Occasionally a gust or too strong control will make the canopy fall back. If you let up on the controls at the wrong time, the canopy may surge forward. Then if you hit the controls at the wrong time, you may cause it to move back rapidly. In this way you can create a pitch (fore and aft) oscillation as shown in figure 5-4. To cure or prevent such an oscillation, float the arms with about 4 pounds (2 kg) of pressure and wait until the glider naturally dampens the oscillations.

A steady canopy is the mark of an accomplished pilot.

One of the things you will learn in later flying is how much and when to apply control in order to stop a surging or retarding canopy. However, your extensive ground handling practice in the forward position prepares you well to handle canopy oscillations fore and aft (pitch). Let your muscle memory develop with practice. For now, only apply a very small amount of control correction (2 inches or 5 cm) every time your instructor tells you to or you feel yourself slowing or accelerating.

The proper controls are:

If the canopy is slowing – Float the controls upward 2 inches (5 cm).

If the canopy is accelerating – Float the controls downward 2 inches (5 cm).

The longer you delay in applying the controls the less control you should use, for you will be out of phase with the canopy.

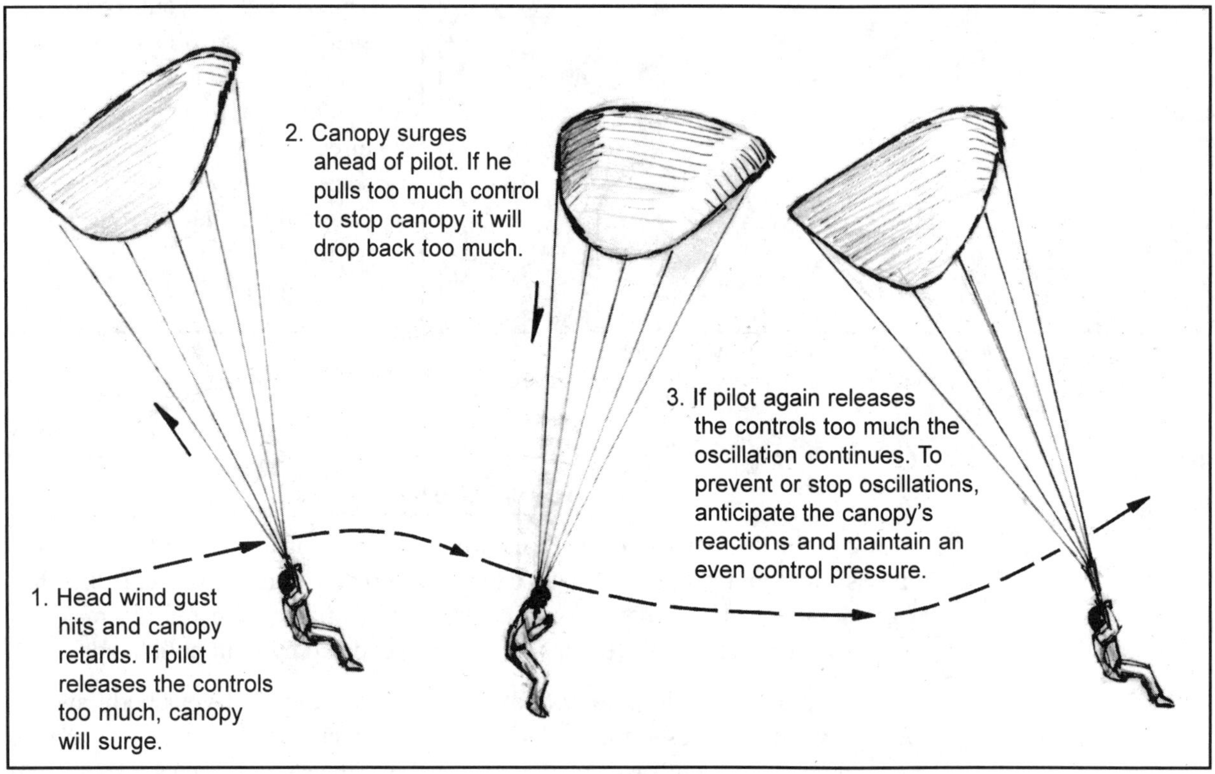

Figure 5-4: Speed Corrections

✓ Speed Control

After a few well-controlled straight flights, your instructor will probably have you start making little speed adjustments so you can feel the effects of speed control. You've probably been flying with the control toggles at your eye level. Now try them a little lower which will slow you down a bit. You'll still be in the realm of fast flight. We leave the exploration of our total speed range until you have more ground clearance. This practice is covered in Chapter 9.

Your control inputs should be in 2 inch (5 cm) increments. When your instructor requests that you speed up or slow down, let up or pull down on the controls with that amount. If he or she wants more, the command will be repeated. You should then put in another increment. If you hear the command *return!,* immediately return the controls to your normal position (usually eye level).

Your first practice will be to start flying slower. You'll accomplish this by pulling the controls down an increment (2 inches—5 cm) from your normal position. Start with one increment so even if you pull too quickly the effect isn't too dramatic (see figure 5-5). Progress in increments like this until you are eventually flying with the controls somewhere in the range of your shoulders or chest. How slow you go depends on your training hill situation. Greater height allows you to make greater control changes and still have time to focus on landing.

When you are using more control pull you will likely be instructed to let up on the controls before you are within 10 feet (3 m) of the ground in order to have energy for landing. The timing of this act is critical. Don't release the controls suddenly or the canopy will surge forward and you will dive. If you wait too long you will be accelerating as you approach the ground. It's better to hold the control and arrest the canopy with a good hard pull than to let up the controls below 10 feet (3 m).

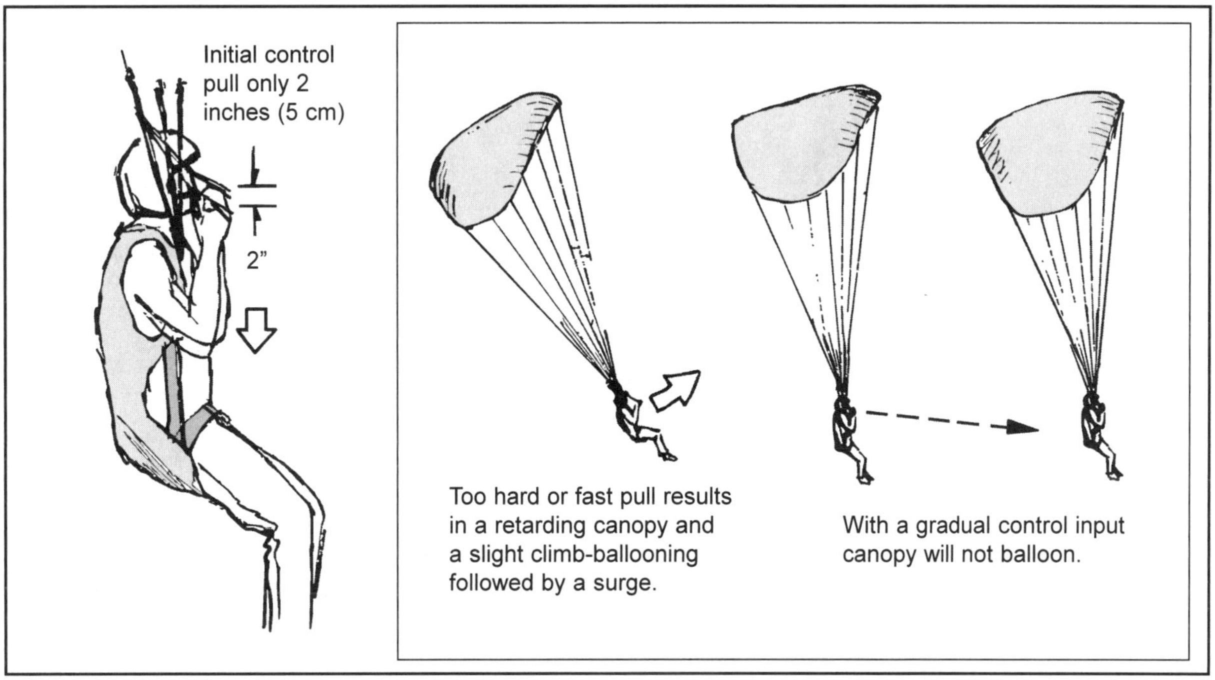

Figure 5-5: Initial Speed Control

Part of your speed control practice will be faster flight. But a training hill often doesn't allow fast (diving) flight because of clearance. Thus we leave that matter to Chapter 9 where you will learn high flight and total control.

Skill Check: *When you can vary your speed without excess diving or ballooning up, then you have learned the fundamentals of speed control.*

✓ Landing

When you get about 5 feet (1.5 m) from the ground you should drop your legs to stand more erect and approximate the PLF position (see below). When your feet are about 3 feet (1 m) off the ground, produce a full landing flare by pulling your control toggles all the way down as depicted in figure 5-6. This control action should be smooth and completed in about 1½ seconds—count Mississippi one, Mississ…

Keep your knees bent as you perform your landing for shock absorption. Also, be ready to take a few steps forward in case your flare didn't totally stop your forward speed. In wind, remember the canopy will have a tendency to pull back

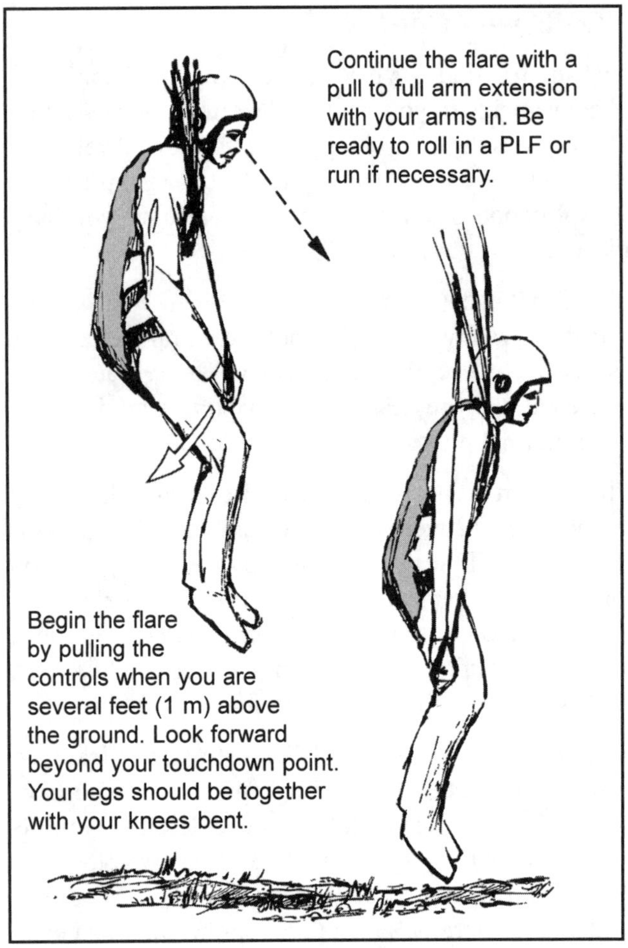

Figure 5-6: Landing Controls

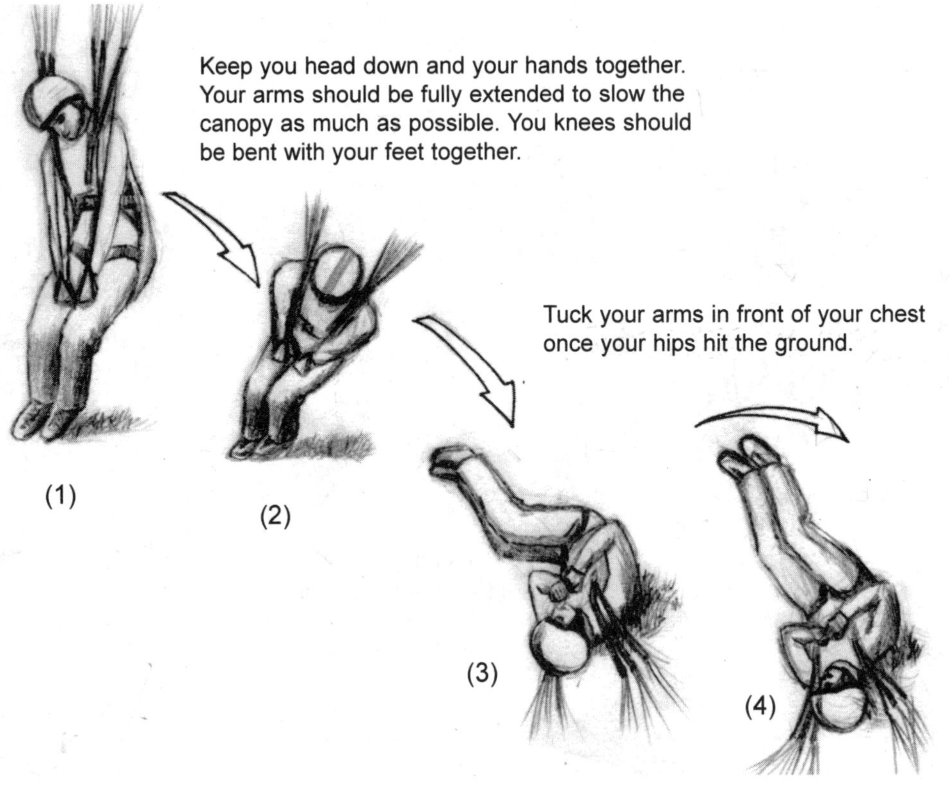

Figure 5-7: The Parachute Landing Fall

more, so keep your balance like a dancer and follow the canopy if necessary. Later, in Chapter 8, you will learn how to handle higher wind on landing.

The action described in figure 5-6 is a simple form of landing appropriate for your beginning trials. Practice it over and over along with your entire launch and flight skills. The main determinant of a good landing is a well-timed flare. The timing comes with practice and the practice is fun. In Chapter 8 you'll learn more landing finesse.

Skill Check: *Fill out a log book at the end of each training session with a comment on every single flight. Do you notice the progress? When you can inflate and launch in one smooth motion, fly straight ahead without significant climbs or dives and land gently, you have achieved the first phase of flight. When you can repeat this process consistently, you are ready for more challenges.*

✓ Performing a PLF

A PLF is an important part of every paraglider pilot's safety skills. You learn it here to help out with landing mistakes. The acronym PLF stands for "parachute landing fall" and as you can guess it originated with parachutists. It is a method of hitting the ground feet first, then rolling in a manner designed to dissipate the downward kinetic energy more or less sideways in order to avoid breaking bones in a hard landing.

A PLF is an emergency procedure —have it ready for emergencies.

The PLF roll sequence is shown in figure 5-7 (from 1 to 7). The pilot has applied full controls and is otherwise relaxing his body. The limbs are inward and together. The pilot collapses to one side to land on his thigh and rear, then roll over on his back to his other side.

Here are some details of a PLF:

1. Be ready to do a PLF on all landing approaches.
2. In an impending hard landing situation, assume the PLF position first, then worry about corrections.
3. Keep your ankles and knees together for your two legs will act like splints supporting each other. Your knees should be flexed with your feet angled about 45° downward.
4. Keep your elbows at your sides and your hands in. Your arms will be straight down to pull full control. Your head should be down with your chin in. Your shoulders should be rounded to assure that you flex forward upon impact (see drawings).

Figure 5-7: The Parachute Landing Fall (continued)

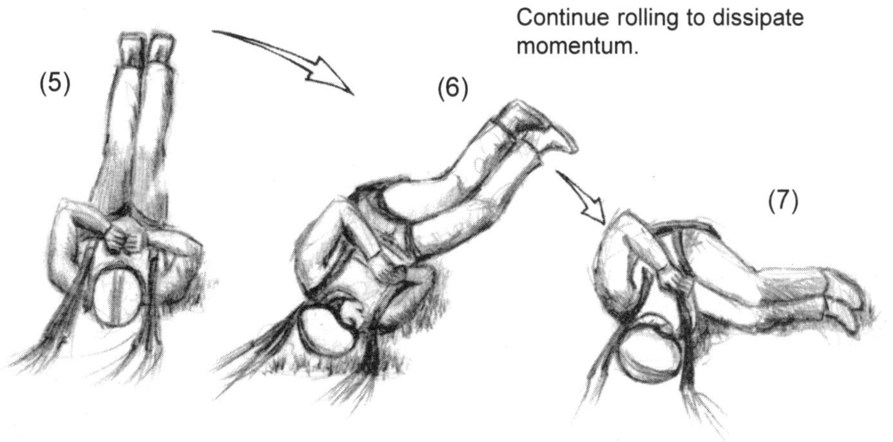

5. Yell to expel air just before touching down in a PLF. This action reduces the risk of internal damage.
6. Practice the PLF periodically. The few minutes spent on it in a training session are not enough to assure its proper performance in an emergency. Practice on your own.
7. Don't expect to always do a stand-up landing. Be ready to exhibit your skillfully performed PLF if your landing proves unskillful.

You should practice this maneuver by dropping to the ground and going through the roll, first from a standing position, then from a chair, then from a three foot (1 m) stand. Do this practice on soft ground, although you can't necessarily choose the ground cover when you need a PLF. Don't neglect this important skill, for it can save a broken bone.

✓ *Finishing the Flight*

Once you have landed, use the methods described in the last chapter to drop the canopy (pages 63 and 74). As you get more comfortable with landings, you should start controlling the canopy with more care after landing. To do this control, bring your hands back up after the flare to let the canopy rise back over your head as you run forward (see figure 5-8). If you raise your hands too quickly the canopy will surge forward, too slowly and it will drop back. Practice until you get it right.

The idea is to control the canopy and kite it off the landing field out of the way of others. Always check to make sure that you aren't going to obstruct landing

Figure 5-8: Kiting a Canopy After Landing

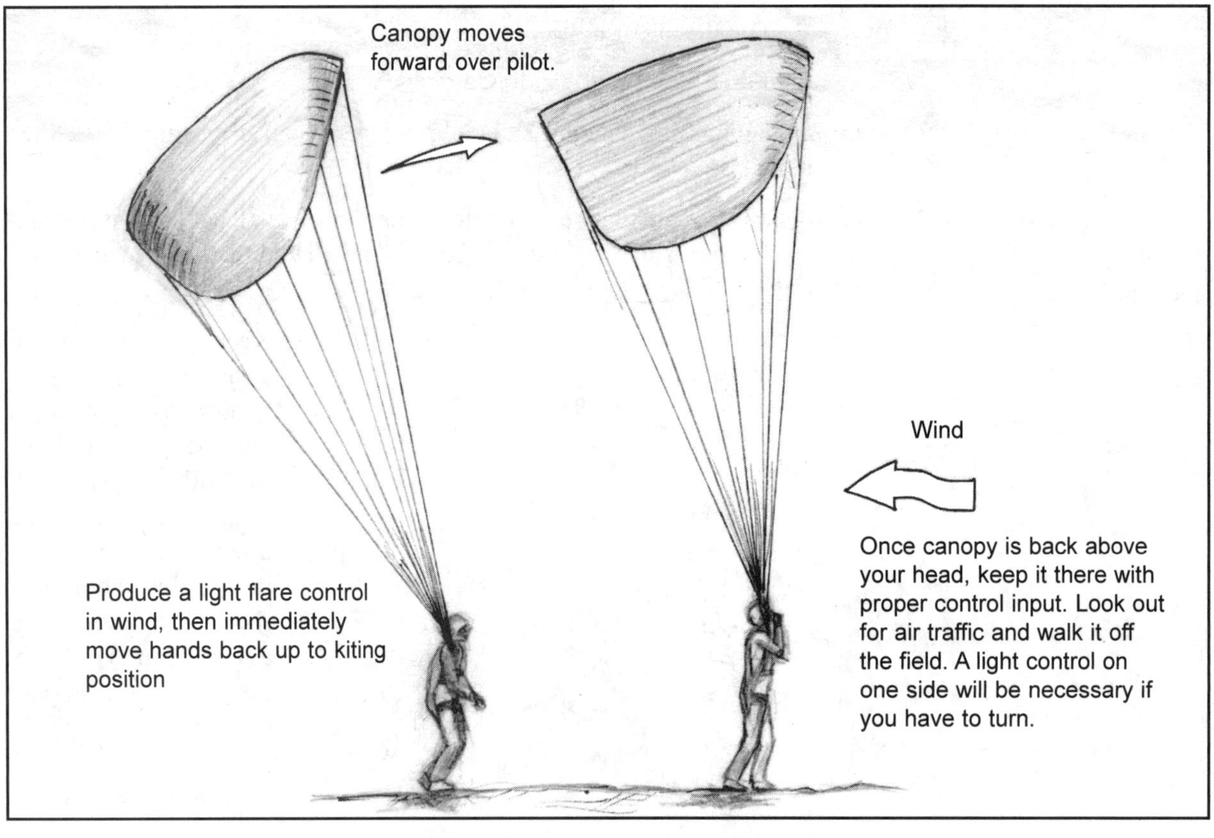

gliders when you do this kiting, both as a beginner and when you start flying sites other than the training hill.

Naturally, you shouldn't be flying in strong winds at your level of learning, but we'll provide you with an emergency procedure here for you to practice now so you are ready if you ever need it.

> **High Wind or Gusty Wind Emergency**
>
> In strong winds or gusts, quickly turn and face the canopy and pull on one control line. Keep pulling in this line hand over hand until you reach the canopy. Hold onto the canopy and it will no longer produce enough drag to pull you dangerously.

If you are facing the canopy and it is on the ground when a gust hits, pull on the rear risers or both control lines hard as illustrated before in figure 4-19.

✓ More Handling Tricks

Sooner or later, your canopy will be the victim of an errant little perverse wind that will flip it over after you land or when you are getting ready to launch. This behavior is shown in figure 5-9. Sometimes the problem is operating in cross-winds or not pulling the control lines evenly, but at any rate the mess must be corrected. If you have at least a 5 mph (8 km/h) wind you can perform a slick trick. Face the canopy, grasp the control line on the side that did the swinging over and tease the trailing edge into the airflow. The idea is to get that side to lift and swing back over to its correct (untwisted) configuration as shown in the figure.

If the wind isn't accommodating, you'll have to walk to the canopy, lift the side that swung over and toss it to the correct side. It helps to have an assistant arrange the canopy in this case, for in very light winds you need a good canopy layout.

Figure 5-9: Righting a Flipped Canopy

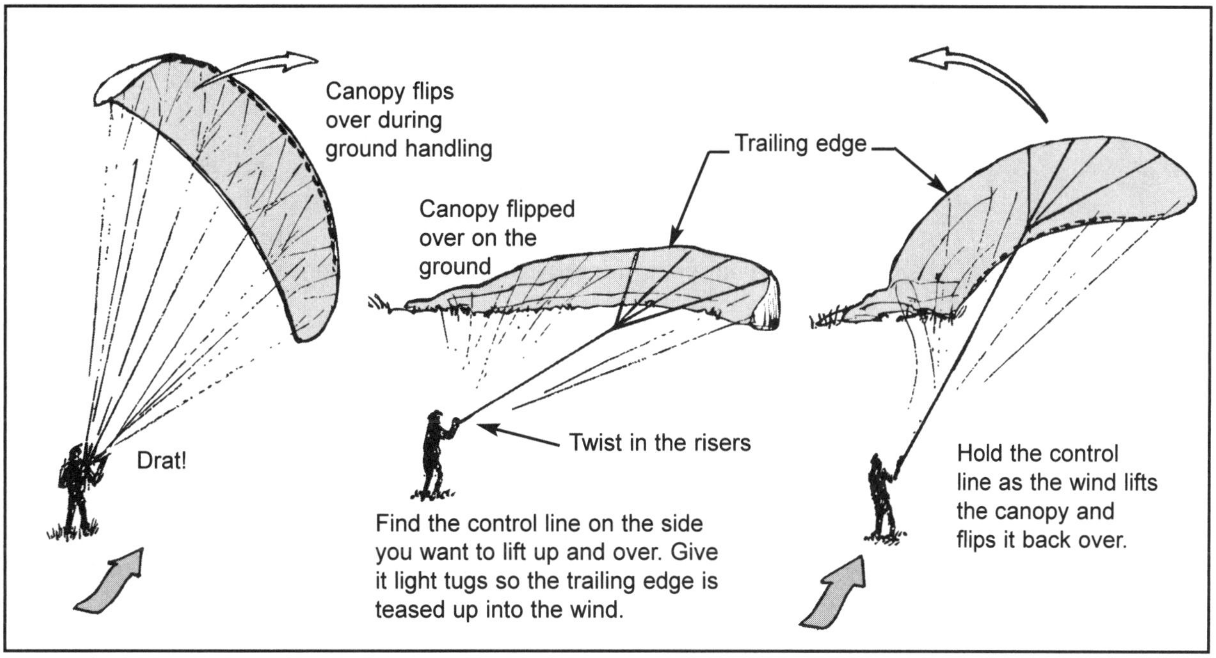

Kiting the glider back up the hill for another go is another useful trick which can be performed in smooth winds greater than about 8 mph (13 km/h). To perform this practice, you must land well and keep the glider kiting over your head as we just described above. Now, while keeping proper canopy position above you, quickly turn around. Add a little more control to make the canopy back up and walk with it. Keep it balanced. With a bit more wind and good canopy control it will help pull you up the hill—what a pleasant thought!

✓ Turn Corrections and Controls

Once you are producing consistent flights with steady speed control you can learn slight turn corrections and control. Begin with corrections. If you find yourself aiming in the wrong direction or you feel the glider lift to one side, you should make a smooth, gentle turn correction. Do this correction by pulling down on the control on the side you want to point towards. If you want to aim more right, pull right as shown in figure 5-10. Similarly, if you feel the canopy tilt so that the right wing moves up, pull on the right control to drop it back and down. Failure to correct a tilting (or rolling) canopy in a timely fashion will result in a turn toward the lower side. That's why you correct by pulling the high side control.

Learning to turn frees you in three dimensions.

By now you should be feeling the difference in control pressure to help guide you as to which control to pull. The control on the lifted wing will develop more pressure and that's your signal to pull it down. Your ground handling practice helps you here. In fact, you can learn to do turn corrections while running along kiting the glider on the ground.

How much control input do you need? Initially limit your correction controls to about 4 inches (10 cm) and hold it about 2 seconds. Later your instructor may have you making larger controls, but soon you'll be doing intentional turns and will quickly learn how much control is needed for a desired turn response. After you are comfortable correcting with one control toggle, move on to correcting with two. That is, raise the opposite hand while you lower the proper correcting hand. This practice is exactly like that in ground handling. Of course, in order for the up control to have an effect, you must have been flying with some control pressure to begin with.

Using a two-hand, see-saw motion in the manner described is a more effective turn control mechanism, so begin this practice with about 2 inches (5 cm) of control. You can always add more control momentarily but remember, there is a bit of lag between your control input and the glider response.

✓ Learning Turns

Now we progress to intentional turns. Since you will start out from takeoff flying into the wind, any turn you produce will result in your flying slightly crosswind. For that reason we suggest beginning this practice in very light or zero winds. The training hill usually doesn't allow you to make large turns because of height restrictions. However, you can learn the proper controls, the proper timing and your glider's reactions.

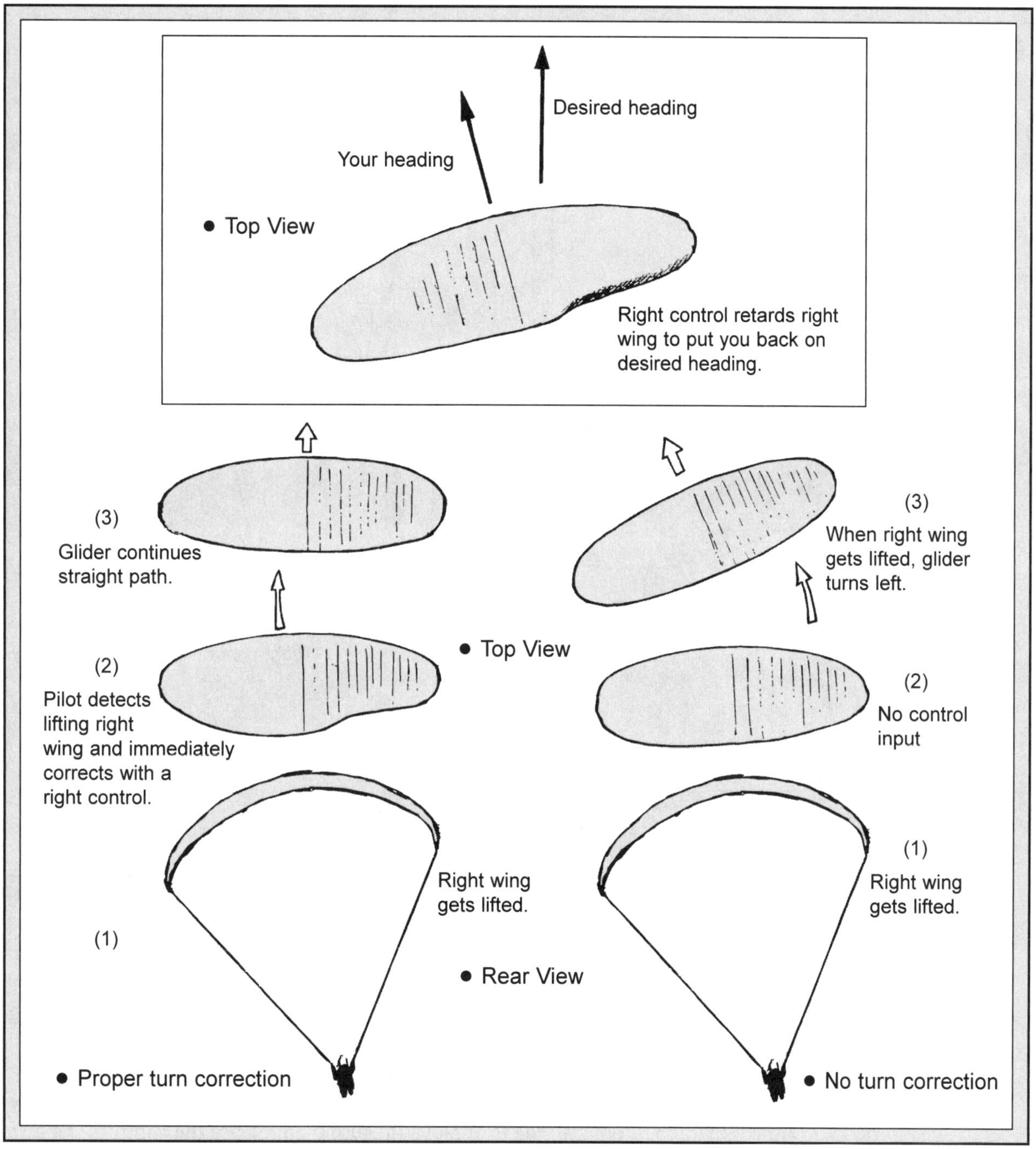

Figure 5-10: Turn Corrections

Let's start with very small controls and heading changes. Once you have established straight flights with ground clearance of at least 10 feet (3 m), turn your head dramatically to the side (let's assume left) where you want to go (in order to broadcast your intentions), then gently pull about 2 inches (5 cm) of control with your left hand. Hold this for a couple of seconds or until you feel the glider react and change heading. Then return the control to its normal flying position.

Not bad, huh? Repeat this process first left then right until you can achieve your desired heading through the proper control timing. Follow this practice with

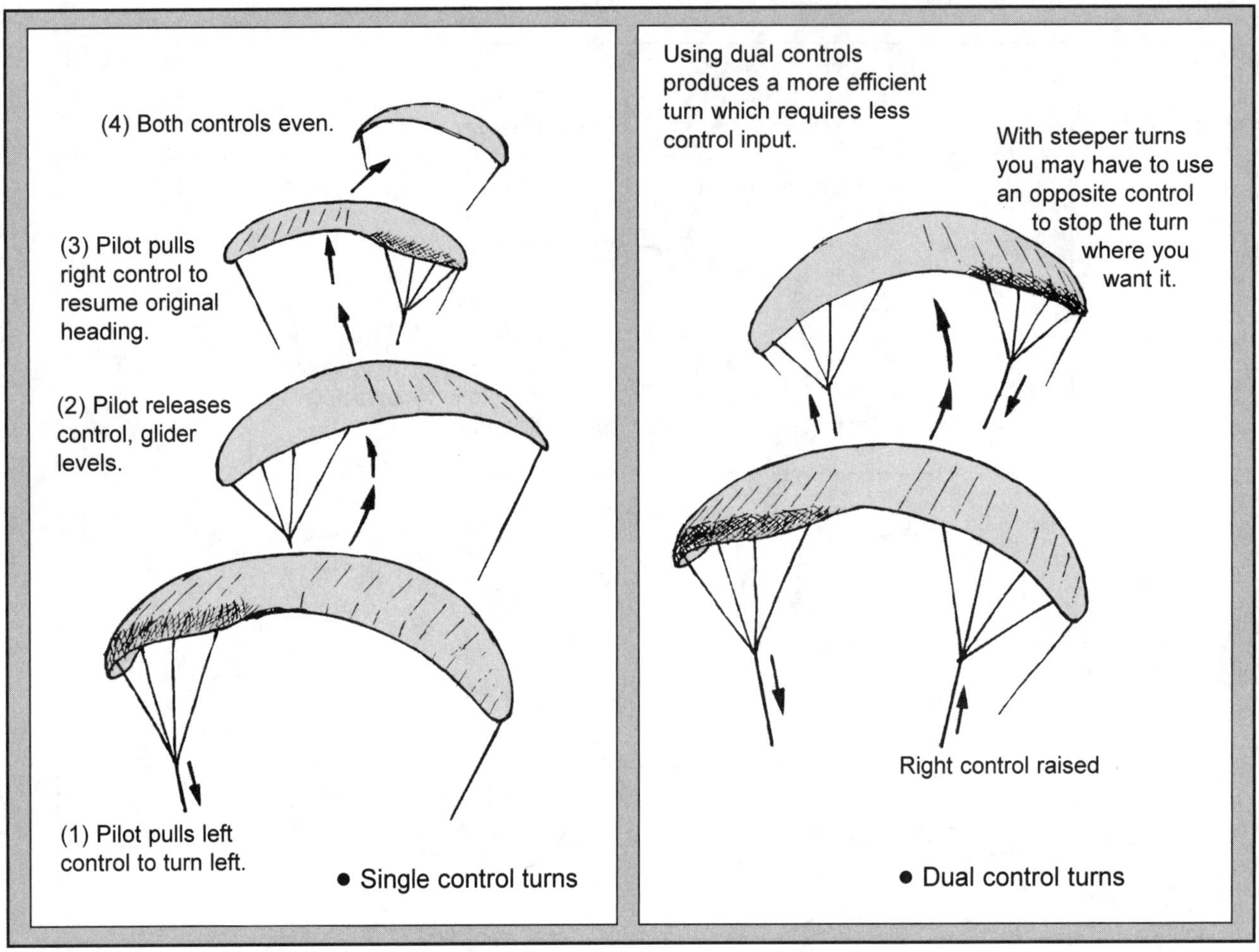

Figure 5-11: Turn Control

turns of more heading change. Your instructor will guide you as to how much of a turn you have clearance for.

When you learn turns in this manner, you will be producing very shallow turns. Thus each time you stop your control input the glider will likely level out and fly straight ahead. Later, when you are making steeper turns, you will also be making a speed adjustment and you will have to stop the turn with an outside control. We'll leave the speed adjustment for Chapter 9 when we discuss efficient turning, but for now let's concentrate on a smooth turn entry and exit. Figure 5-11 shows the whole process. You pull on one side to achieve the turn then release the control. The turn may continue so you pull on the other side about an equal amount to stop the turn. Your current focus should be on producing the correct amount of turn your instructor requested and coming out of the turn exactly where you wanted.

Follow this practice with steeper turns produced by applying more control pull—up to 6 inches (15 cm). Finally, progress to shallow linked turns—a left turn followed by a right turn and vice-versa (see figure 5-12). Remember, there is a limit as to how much you can turn at the training hill. Be conservative and save your turn enthusiasm for higher flights. Later, you'll learn more intricate and efficient turn controls.

Turn Practice Summary

You began learning turns with just a single control (one side pull). The reason is to let you feel how much pull is required and to allow slower turn response so you don't get surprised. Next you graduate to using both controls in our familiar see-saw fashion. The glider responds more effectively to these dual controls, so start out their use with smaller control movements and small directional changes (45° or less). Gradually progress to more control inputs, more direction change and linked turns just as before. All these turns should be shallow because you are close to the ground and do not have room for error. However, be prepared to exit the turn with opposite control as shown in figure 5-11. Soon you'll be flying higher and turning with abandon.

Your instructor may have you begin experimenting with weight-shift at this point. The addition of weight-shift to your turns adds a bit of complication and requires less control input. We cover this matter in detail in Chapter 9.

One good turn deserves another.

Skill Check: *The true test of your beginning turn skills is to see if you can turn at will, initiate smoothly and come out of your turn heading exactly where you intended. Using the same criteria for linked turns is a further test.*

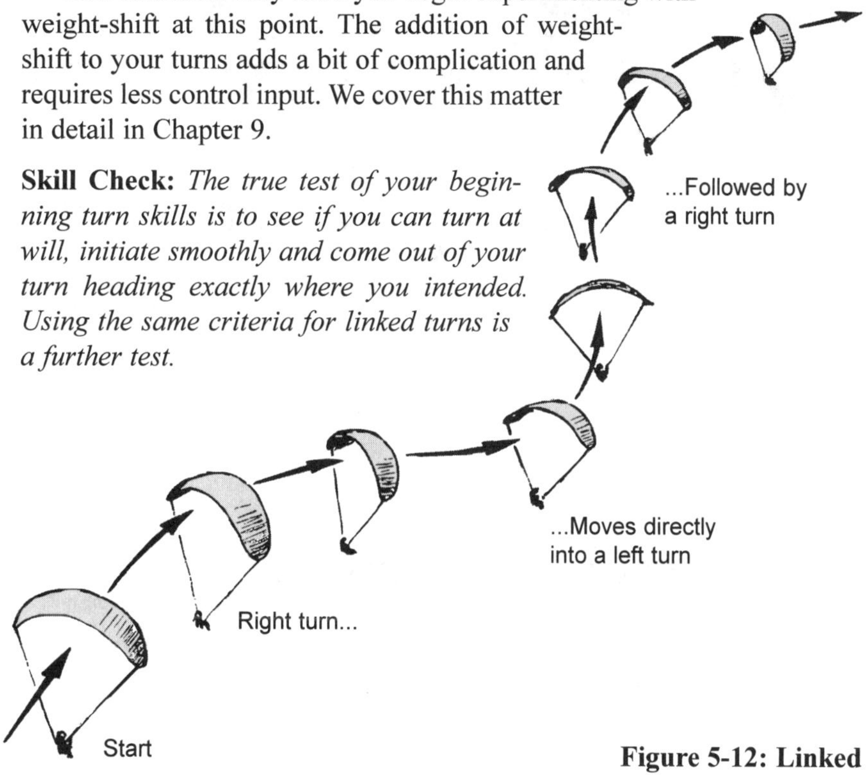

Figure 5-12: Linked Turns

Tandem Flying as a Student

Many schools use tandem flying as part of their training program. Tandem flying involves flying a large glider designed to hold two people—you and the instructor. With tandem flying you can quickly gain airtime and acquire a feel for the controls and perspective of flight. Often schools incorporate tandem flying into the lessons at several different points so you can get a fresh perspective of what you are trying to learn.

The launch process on your part is somewhat different from when you are flying alone. Normally on a tandem flight you will be in front of your instructor and a little below him or her once in flight. You will not have the risers over your arms or in your hands (the instructor will). Naturally you will not have the controls either unless you are given them in flight. Sometimes the instructor will reverse inflate, but you'll remain facing forward while he or she turns as much as possible within the available harness slack. There are only two things for you to do, but they are important:

Close view of a tandem flight just after takeoff at Millau, France.

1. *Focus on running to a point your instructor indicates when he or she says go!*
2. *Continue holding on to your shoulder straps or where your instructor indicates.*

In flight you should remain motionless as much as possible unless your instructor says otherwise (you may be asked to tilt your body to help weight-shift). During landing perform exactly as you would when alone, but without the controls. Reach a foot for the ground and be prepared to take a few steps if necessary.

Towing as a Student

Flatland schools usually use towing as a substitute for a training hill. Normally this begins with tandem training. All the matters outlined in the tandem discussion apply. You will be running on the flat ground and may have to run for a number of steps, but you will keep getting lighter.

Towing can take several forms: behind a truck with a payout reel or from a ground based winch or car. In any case, your duty is about the same: run, then remain still in front of your instructor until you are high enough to take over the controls. Whatever you do, don't sit down until you are at least 10 feet (3 m) above the ground.

Eventually you will be the pilot in command under tow. You will learn launch procedures and towline control. You will practice all the skills we discussed in this chapter—launching, corrections and landing. The differences may be higher control forces because of the greater load on the glider. Also, the canopy position will be different because the towline will pull you forward in relation to the canopy. But once you release from the line you are back to flying at normal trim position.

At some point you will be ready to fly solo on tow. The practice may begin with the tow vehicle "trolling" you along while you practice control on the line. You'll learn to release from higher and higher for some real freedom. For a com-

A pilot lifts off on a towing flight.

plete guide to towing paragliders, see the book *Towing Aloft* by Dennis Pagen and Bill Bryden (available at the address given on the title page of this manual).

Summary

You have come a long way, from a mere spectator watching birds and longing for the grace to be borne on silent wings, to a fledgling pilot actually finding a new home in the sky. You have covered the most difficult part—perfecting takeoff and landing skills.

This is the first step, the basics. However, this is arguably the most important learning, for only by perfecting skills at this level can you have a solid foundation to build upon in the future. Every item in this chapter is chosen to develop these basics, so patient practice is of great importance. Don't rush things, for flying is fun at all levels. Hone your beginner skills to a keen edge and you'll be a much better pilot later.

Once you have solid control techniques, you can build on them gradually with new challenges and the exploration of new heights. The sky's the limit.

TEST YOURSELF (*Answers in Appendix II.)*

1. Select all the correct answers concerning reverse inflations and forward launches:
   a) Forward launches are more complicated than reverse.
   b) Reverse inflations may be performed in less than 5 mph wind.
   c) Pilots should learn both launch methods.
   d) Forward launches can be performed in zero wind.
   e) Reverse inflations require you to let go of the controls.
2. When forward launching, holding on to the A risers too long will result in ______________ ____________________________.

TEST YOURSELF (continued)

3. To drop a canopy to the ground when kiting you should pull the _____ risers.

4. In flight your control inputs should be ____________.

5. In order to land you should reach for the ground and move the controls all the way up.
   True or False

6. Sitting down in flight too early may result in ____

7. When kiting your glider in the landing field, first check for _____________.

8. Describe how to correct for a left turn: ____
____

USHGA Beginning Rating Checklist

General Description - A Beginner pilot has the knowledge and basic skills necessary to fly and practice under direct instructor supervision and within significant operating limitations. The pilot understands the USHGA paragliding rating systems and recommended operating limitations.

Beginner Rating - Foot Launch - Required Witnessed Tasks

A. Attends and completes a basic ground school.

B. Layout and preflight of canopy and harness

C. Demonstrates canopy handling skills sufficient to launch from a training hill under control.

D. With each flight demonstrates method(s) of establishing that pilot is properly connected to the canopy, with cleared lines and risers, just prior to inflation.

E. Launch unassisted showing:
   1) Aggressive inflation and run
   2) Pendulum control during launch
   3) Directional control
   4) Smooth transition from running to flying, during launch

F. Airspeed recognition and control
   1) Two flights., predetermined to show:
      a) Constant airspeed
      b) Smooth straight flight
      c) Safe, smooth landing, on feet, into wind
   2) Two flights, predetermined to show:
      a) Confident, slight variation in airspeed showing awareness of control inputs and pendulum control
      b) Smoothly increasing airspeed, and smoothly, slowing airspeed showing good control
      c) Safe, smooth landing on feet, into wind

G. Shows the ability to recognize and understand how different wind conditions at this site will affect flights.
   1) Wind direction
   2) Wind velocity
   3) Terrain shape
   4) Obstructions

H. On each flight, demonstrates proper post-landing procedure to include, but not limited to:
1) Canopy deflation
2) Canopy immobilization
3) Checking traffic
4) Removal of canopy from landing area
5) Disconnection from the canopy.

I. Demonstrates of understanding of the importance of proper packing, storage and care of the canopy.

J. Has read and understands USHGA statements and definitions of good judgement and maturity.

K. Must pass the USHGA Beginner Paragliding written exam.

L. Must agree to all the provisions of the USHGA standard waiver and assumption of risk agreement for the Beginner rating and deliver an original signed copy to the USHGA office.

Beginner Rating - Tow Launch - Required Witnessed Tasks

A. Must demonstrate the above-mentioned Beginner paraglider rating foot-launching tasks except C, E.

B. Must demonstrate system setup and preflight, including a complete discussion of all those factors which are particular to the specific tow system used and those factors which are relevant to towing in general. Must demonstrate complete understanding of both normal and emergency procedures, including checklists for normal procedures and the indications of an impending emergency and convince the Instructor of his ability to recognize and execute emergency procedures.

C. Demonstrates successful, confident, controlled launches and flight under tow to release at altitude, with a smooth transition to flying, with proper directional and pitch control resulting in proper tracking of the towline and appropriate maintenance of proper towline tension and airspeed. Such demonstrations may be made in ideal wind conditions.

D. Demonstrates understanding of all Beginner-level Paragliding Tow Discussion Topics.

E. Must agree to all the provisions of the USHGA standard waiver and assumption of risk agreement for the Beginner rating and deliver an original signed copy to the USHGA office.

Recommended Operating Limitations for Beginner Pilots:

A. Should exceed these limitations only after demonstrating complete mastery of the required Beginner paragliding tasks (above), and only after acquiring a full understanding of the potential problems and dangerous situations which may arise from exceeding these limitations.

B. All flights should be made under the direct supervision of a USHGA Certified Basic or Advanced Paragliding Instructor.

C. Should fly only in steady winds of 12 mph (19 km/h) or less.

D. Should foot launch only on slopes of 3:1 to 4:1, where wind is within 15° of being straight up the slope.

E. Should launch only when there are no obstructions within 60° to either side of the intended flight path.

F. Should fly appropriate sites for this skill level.

G. Should fly a canopy recommended by the manufacturer as suitable for Beginner or Novice pilots.

DATE	GLIDER	SITE		WIND - WEATHER				REMARKS	AIR-TIME	FLT #
(M/D/Y)	SITE, TYPE	LOCATION	HGT	DIR	SPEED	CLD COVER	TYPE AIR			

Use this log book to fill in your first flights and you will always have a record of this memorable experience.

A pilot climbs high above the clouds. He is flying legally only if his cloud clearance is ample.

Chapter 6

THE NATURE OF THE SKY WEATHER FOR PARAGLIDING

PART II - Novice Flight

*" **Clouds are the memories and dreams of pilots—they leave them behind when they come back to earth, but every time they go back into the sky, they can relive them.**"*
Jimmy Buffet (from *Where is Joe Merchant*)

Early pilots had little or no weather information. They flew on a wing and a prayer. What weather knowledge they had was acquired in the Trial and Error School of Higher Learning. Those who flew regular mail routes, like Charles Lindberg, either became very adept at predicting weather by the natural signs, or they didn't last. But observation from one point of view isn't really sufficient to cover sudden changes and other possibilities. Even Lucky Lindy had to parachute to safety one night when his airplane was engulphed in a thunderstorm.

The hardest part of paragliding is learning to understand the conditions. You can learn to fly from point A to point B in a few days, but learning the conditions takes a lifetime.
—Lars Linde

Today we have global satellites, weather stations and computer networks giving us twenty-four hour details about the weather. What the public receives, however, is what the majority wants: the daily temperature, sunshine and precipitation. This is not enough for pilots, be they of the 747 or paraglider variety.

We need to know more about the general and specific behavior of the atmosphere before we fling our bodies into its midst. In this chapter we are going to explore the personality and moods of the sky so we can better interpret the available information. We'll also see how to get this information.

We begin our exploration of the weather with effects that work locally, then expand our vision to the larger scale. The emphasis is on practical knowledge that you can use in your everyday flying. We'll endeavor to keep it light and airy.

A Note to the Beginner

If you are just beginning to fly you may be overwhelmed by the amount of new ideas presented in relation to the weather. Do not worry about learning all the details at once. Concentrate first on the matters affecting your immediate flying such as the wind direction, strength and possible turbulence. Ask your instructor questions and the answers you receive will be illustrated with what you can see. Then begin to notice clouds and other effects and you will add to your overall understanding. Ultimately we recommend our book *Understanding the Sky* to complete your learning of weather lore for sport flying.

✓ Limits

Part of the reason we learn about the sky is for safety, and part is to enhance our performance. As a beginner, you should concentrate on the safety aspect and learn the limits for safe flying. Throughout this manual we provide suggested limits of wind factors that have been carefully developed by instructors after years of experience. If you combine these guidelines with observation of the conditions and your instructor's advice you will safely remain within the limits of your ability. With this mature attitude, you will find the sky is not a strange, mysterious void but a true aerial playground.

The Wind—Breath of the Sky

Simply put, the cause of wind is pressure differences in the air. The air flows to equalize these pressure differences, and we feel this flow as wind. The pressure differences are caused by uneven heating of the earth's surface, so we can ultimately say that wind is caused by the sun. A general rule is stated in the box.

> **Wind and Circulation Rule**
> The air flows from cooler to warmer areas.

This rule works very well on the local scale. By local we mean effects that take place over a small area, say 10 to 20 miles (15 to 30 km) or less. We call such effects *local effects* and the wind they cause *local wind.* On the larger scale the rule also is valid, but the circulation is complicated by other factors.

The reason this rule works is simple: As the sun warms the ground, the overlying air heats and expands. It becomes less dense, so the nearby cooler air (which is more dense) flows in to displace it.

Another important rule we can establish is that the local winds which are given birth by the sun are thus dependent on the daily sun cycle. So we can generalize:

> **Daily Circulation Rule**
> Local winds tend to follow a daily cycle whereby they arise during the day, peak in the early afternoon and die or reverse by nightfall.

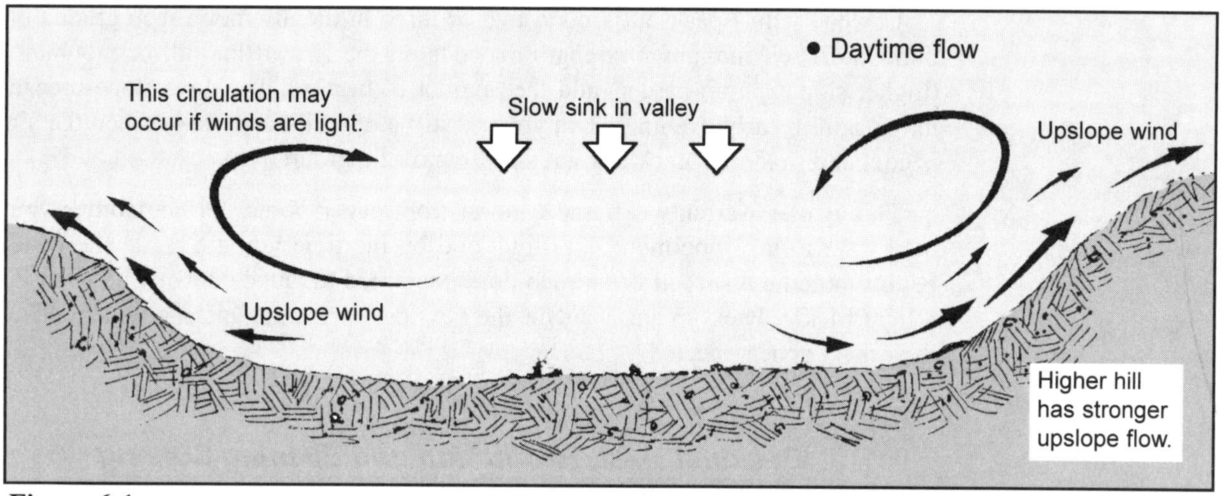

Figure 6-1: Upslope Winds

✓ Upslope and Downslope Winds

Any sunny day will send a wind up a hill as long as the general wind isn't from the back of the hill.

One of the most important things paraglider pilots learn about the wind is how it changes on a slope during the day. We pointed out in the previous chapter that it is desirable to have the wind blowing into our training hill in order to fly. The same general rule applies to all flying sites. The flow of the overall or widespread wind greatly determines the wind direction and strength. However, local effects can and do alter this flow by adding to or subtracting from it. If there is no overall wind, the local effects will be the only wind you see, and its behavior is important.

Near a slope, the sun's heating makes the air less dense, so cooler air in the valley moves in to replace it. This sets up a circulation up the slope and down in the valley as shown in figure 6-1. This breeze is called an *upslope wind* (anabatic flow is the scientific term).

Upslope winds increase with the sun's daily heating and die out when the sun fades. As the ground cools the opposite effect happens. The air next to the slope becomes cooler and denser than the general air and it begins sliding downhill as shown in figure 6-2. Thus a reverse flow, a *downslope breeze* (catabatic flow is the scientific term) begins in the evening and lasts all night until the morning sun reverses it. It can be quite strong near high mountains.

Figure 6-2: Downslope Winds

Upslope winds peak out around two or three in the afternoon then gradually diminish. Their maximum strength depends on the size of the hill or mountain (higher slopes pump harder) and the amount of heating. The downslope breeze may begin in early evening when you are still at the hill, especially when the air is cool and shadows block the sun's heating of the ground.

This matter warrants extreme caution from eager pilots, for sometimes you may want to take one more late flight and by the time you are ready the wind begins to come downhill or tail. Do not be tempted to launch downwind even in a light trickle. Pack up and wait for the next day when the upslope winds make their daily appearance.

✓ *Residual Heat, Restitution and Evening Soaring*

A special condition can occur due to a combination of upslope and downslope flows. Sometimes on a day of good heating a valley will be filled with warm, buoyant air. When the downslope winds set in they may slide under this air, making it well up as in figure 6-2. This action can provide gentle lift that lets you float around without sinking. The valley seems to be breathing one long-lasting evening sigh.

When restitution winds are smooth, pilots call them "wonder winds" or "magic air".

A more common occurrence is when an eastern facing slope becomes shaded first so only one side of the valley exhibits downsloping winds. Then the other side gets added lift which often is very smooth and widespread as shown in figure 6-3. Such evening soaring conditions may be wonderful sources of lift which are ideal for pilots learning to soar.

But a word of caution: sometimes these residual heat conditions can get quite strong or exhibit surprising turbulence. Usually strong conditions happen when the general wind adds to the restitution wind. The turbulence happens at times in high mountains when cooling in the canyons or valleys sends strong rivers of cool air below the warm rising air to form shear turbulence (see below) where the cool and warm layers meet as shown in figure 6-3.

Figure 6-3: The Cause of Restitution Winds

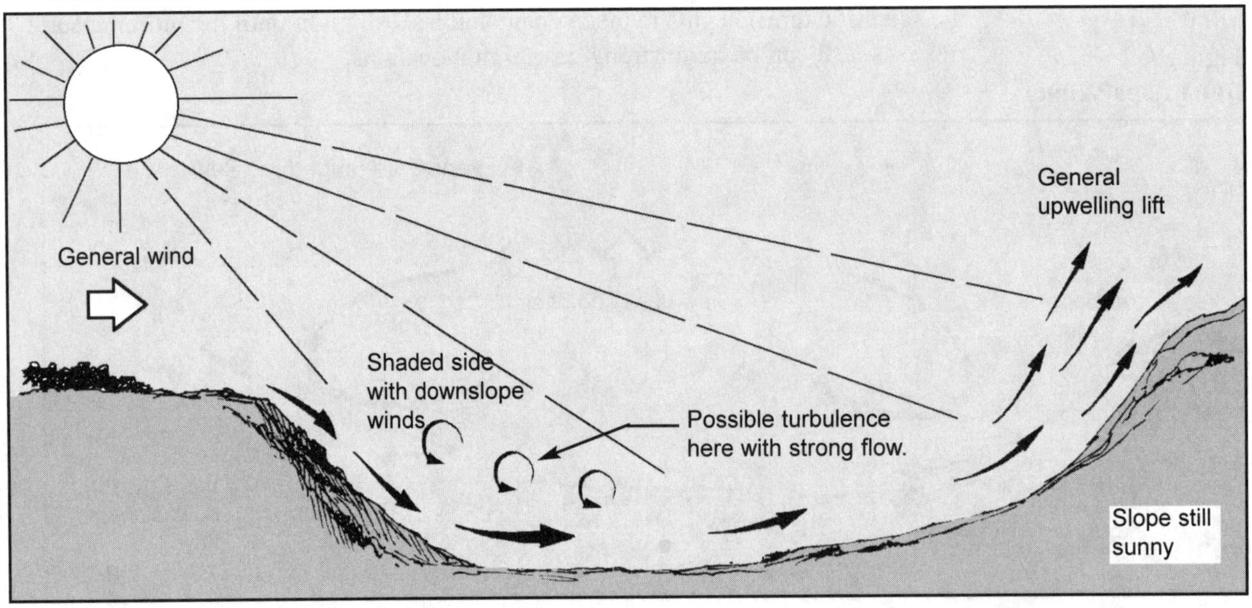

✓ The Sea Breeze

Another common circulation takes place near shorelines. This is called the *sea breeze* and it again is due to different amounts of heating on the earth's surface. During the day the sun hits the ground which in turn heats the air, making it less dense. The water's surface does not get heated as much because the sun's rays penetrate the water, so the air above the water stays cool and dense. As a result, a flow from water to land begins. This is the sea breeze, as shown in figure 6-4. Anyone living close to an ocean, sea or lake will recognize the daily cooling breeze flowing inland from the water.

At night the opposite effect occurs because the land cools faster than the water. A land breeze sets in and this continues until the morning sun reverses the flow.

The sea breeze may reach tens of miles inland near the ocean, but is often confined to an area closer to the shore. The strength of this wind peaks in the afternoon and can be very strong where desert areas border the sea. Generally, though, the sea breeze consists of gentle and smooth flows that are ideal for training or simple soaring. This feature is a prime reason why so many paragliding schools are located near shorelines.

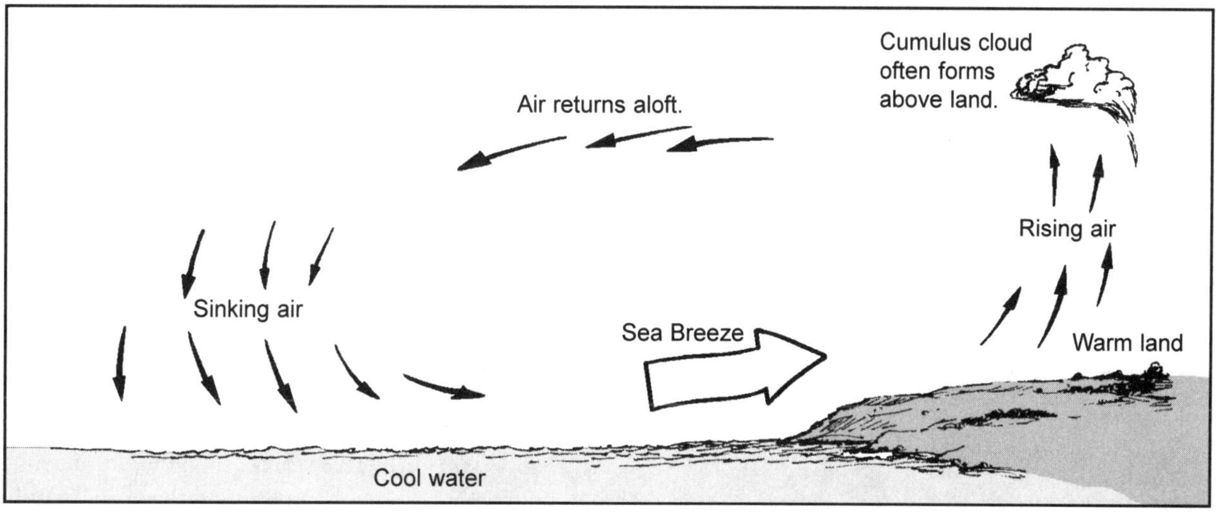

Figure 6-4: Sea Breeze Formation

✓ Other Circulations

Here we briefly mention other forms of local circulation. These can be between forest and field areas, sunny and shaded areas or near snowfields. A large forest will be much cooler than the nearby fields, as you can tell anytime you take a walk in the woods on a warm day. When a large cloud bank blocks the sun, a small circulation may be set up between the shaded and sunny areas. Sometimes the shading of a hill by a large passing cloud can make the wind temporarily turn to downslope. Also, when a large snowy area lies next to bare ground, a light circulation can occur.

To summarize what we have learned about local winds; here's a convenient listing:

Local Wind Factors

- They alter the behavior of the general wind.
- They flow from cool to warm areas.
- They are subject to the sun's heating effects.
- They peak during the maximum heating or maximum cooling.
- They are generally quite mild except in desert or high mountain conditions.
- The wind strength, direction and gustiness are all affected by the daily heating.

JUDGING CONDITIONS

When we speak of conditions we are mainly referring to the wind itself. True, the presence of rain, vertical air movement and the possibility of changes are also a concern, but for our early flying we are mostly interested in "what's happening now."

We are looking at three aspects of the wind:
1) What is its strength?
2) From which direction is it coming?
3) How smooth or turbulent is it?

When you learned to judge these factors you can say you understand the nature of the wind.

✓ *Wind Strength*

We have all felt the wind on our face, our body. We know that the stronger it blows—the faster it moves—the more force we feel. In fact, the force of the wind varies with the square of the speed. What that means is that a wind of twice the speed has four times the effective force. If you are accustomed to an 8 mph breeze, a 16 mph one will be four times as forceful. At 24 mph the force is nine times stronger. We should point out that this increase in force is felt mainly on the ground (while launching and after landing), for once the glider is flying it moves in relation to the airflow and the control forces are normal. However, it should be clear that wind strength is an important factor in the safety of flight, for a more energetic wind can create more violent turbulence and hinder our penetration to a landing field.

A wind streamer keeps a pilot informed of the airflow near the ground.

We tell the wind speed, and thus strength, by using a windspeed meter (or windmeter), feeling the wind and observing its effects on the environment. We mentioned in Chapter 2 how you can use a windspeed meter to develop a feeling for wind strength. It is also important to learn to interpret the effects of wind on the surroundings to better judge its strength. Those ripples on a pond, those rustling leaves and those whipping flags all tell a story.

The sound of the wind in the trees and brush is a good indicator of wind strength. However, be aware of the sudden increase in volume in the spring when the leaves first appear.

Here is a chart that provides a guideline to wind strength effects. Refer to this chart as you observe wind in nature and gradually you'll learn accurate wind strength judgement. Try judging the wind out your window right now.

Wind Table

Wind Velocity		**Effects on Environment**
calm	—	Smoke straight up. No movement in vegetation
0-3 mph (0-5 km/h)	—	Smoke straight up. Leaves begin to rustle.
3-5 mph (5-8 km/h)	—	Smoke leans, twigs move.
5-10 mph (8-16 km/h)	—	Smoke leans about 45°. Small branches and grass begin to move. Clothes move on a line.
10-18 mph (16-29 km/h)	—	Smoke lies about 30° up from horizontal. Whole branches begin moving. Grass waves. Clothes wave on a line.
18-25 mph (29-40 km/h)	—	Smoke lies flat. Large branches wave and grass ripples. Dust swirls begin.
25-35 mph (40-56 km/h)	—	Large limbs and medium trunks sway. Clothes flap. Dust and snow blow readily.
35 and over (56 km/h)	—	Larger trees sway, cars rock. Difficulty walking into wind.

One of the best wind indicator is your face. Learn to feel wind strength just as you do airspeed—after all, they are the same feeling.

The important thing to judge is how much wind is acceptable for your skill level. You can soon learn to do this by carefully observing the conditions each day you fly and noting what you felt in the air. Your instructor and other experienced pilots will help you in your judgement. With time you will be able to accurately assess the wind strength. If in doubt, be sure to err on the side of safety.

✓ *Wind Direction*

In many cases, the wind direction due to the general and local effects changes very slowly, so it is not a safety concern. However, due to turbulence, passing fronts or storms (topics we discuss later), we often find the wind's direction changing. So we must again be observant to be safe pilots.

To help recall the wind convention, remember: a New Yorker comes from New York. A north wind comes from the north!

In the English speaking world, we identify the wind direction as the compass direction that the wind is coming *from.* Thus, a west wind is blowing from west to east.

To detect wind direction, we use streamers, windsocks, smoke, blowing dust, flags, leaves on trees, wind lines on water, etc. In brief, anything that can be affected by the wind can give us direction clues. We mentioned some of these in Chapter 4. Here we'll add some explanations.

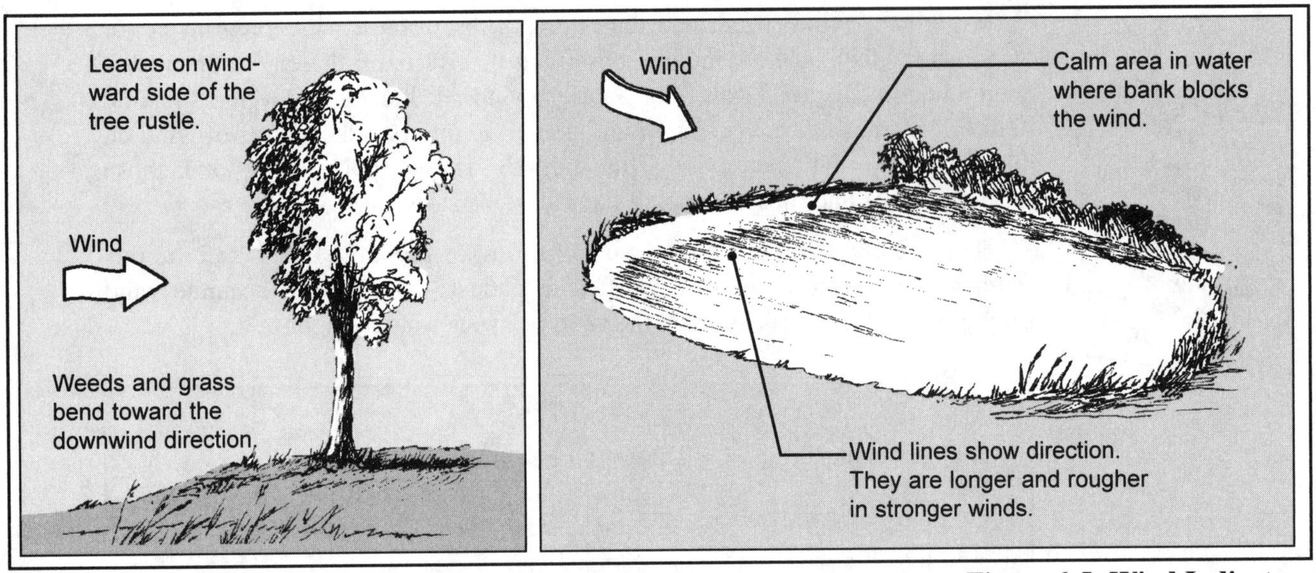

Figure 6-5: Wind Indicators

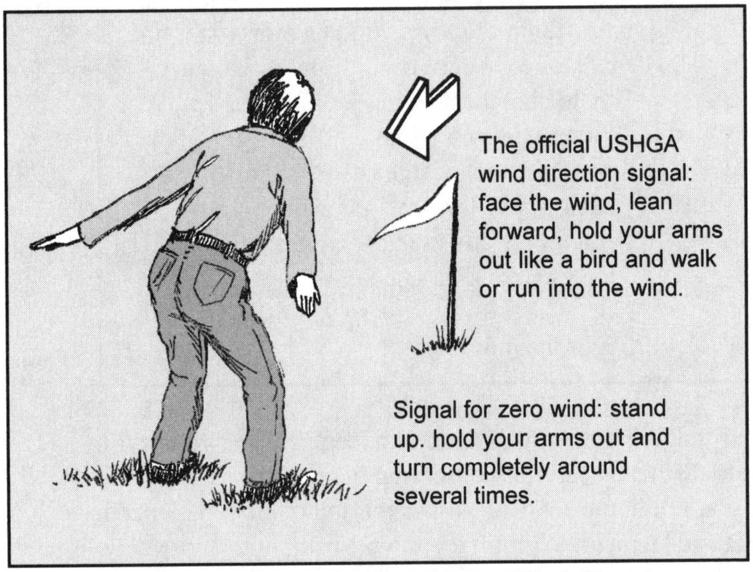

Figure 6-6: The Official USHGA Wind Signal

First, to use trees, brush and weeds as wind indicators, note how they bend and which side is most affected by the wind (see figure 6-5). In light winds, the side of a tree the wind strikes first and thus ripples most is the side the wind comes from. Watch the leaves on trees closely and see if you can tell which way the wind is blowing. In stronger winds, the trees will lean. Tall grass and weeds lean with the wind and show ripples flowing along in the wind direction.

In very light winds, water develops ripples that spread out from the shore. There is normally a dead air spot right next to the windward shore where the wind is blocked; then lines form across the water giving the wind direction as shown in the previous figure.

Keep an eye out for all the clues to wind direction as you walk, ride or fly around and you'll soon be judging the wind with a sixth sense. Note that the most reliable wind indicators are streamers, flags and socks. Be cautious of relying on the signals from pilots on the ground for they can give an ambiguous signal. Are they pointing the way the wind is blowing to or from? Figure 6-6 shows the official USHGA wind signal which avoids this ambiguity.

✓ *Wind and Turbulence*

One factor that can turn the wind from friend to foe is *turbulence*. Turbulence is a swirling of the air similar to the eddies and swirls in a fast moving river. We feel these swirls as changes in wind speed and direction.

The problem turbulence introduces to our flying (and to all of aviation) is that it causes changing forces on our wings so we must make constant corrections to keep on an even keel. In fact at some level of turbulence we can no longer maintain control, so an important part of judging the wind is making an assessment of the intensity of the turbulence.

There are three factors to judge concerning turbulence. They are:

1) *Gust differential* — This is how much difference there is between the maximum and the minimum wind strength. A wind changing from 5 to 15 mph is more gusty and more difficult to fly in than one changing from 10 to 15 mph.

2) *Gust frequency* — Gusts that come more often warrant more concern because our controls will have to be made more frequently. It should be clear that if gusts are too frequent we will have trouble keeping up with them.

The stronger the wind the stronger the expected turbulence.

3) *Gust sharpness* — Gusts that come more suddenly are more of a problem than gusts that come and go more slowly. For example, if a gust reaches full strength in half a second it will be more severe than one that builds and dies in six seconds. In figure 6-7 we show a trace of the wind speed on two different days. Day I shows higher differential, higher frequency and greater sharpness of gusts. Day II is much less turbulent and a much safer day on which to fly if all else is equal.

A beginner pilot may have difficulty judging such changes, so all flights should be performed in smooth conditions. You will probably notice that your lessons take place in such "whipped cream" air.

With practice and time you will learn to judge the gust strength of the wind. Just like with judging the general wind strength, feeling the changes and watching the wind's effect on dust, flags, smoke and trees is the best way to tell the

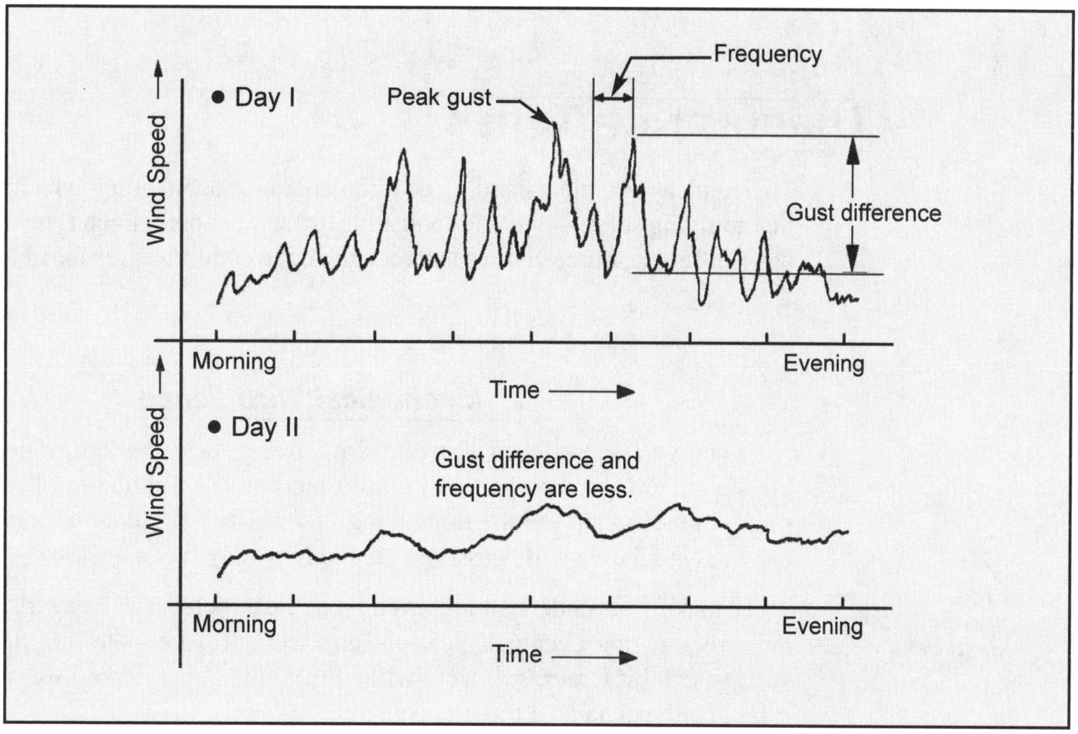

Figure 6-7: Turbulence Factors

Figure 6-8: Wind's Effect on Smoke and Flags

wind's gustiness (see figure 6-8). We have included operating guidelines at the end of each part of this manual that include a gust factor maximum.

Safety Tip: *As a rule of thumb, avoid flying in conditions where the strength changes more than 5 mph (8 km/h) or the direction changes more than 30° in less than 5 minutes. Fast changes can fold a canopy.*

TURBULENCE TYPES

To complete our understanding of turbulence—understanding which is so important to flying safety—we will look at its different causes. Each type of turbulence has a different source and we can recognize the conditions responsible for each of them.

✓ Mechanical Turbulence

Turbulence comes in all strengths:
- *Ripples*
- *Bumps*
- *Slams*
- *and Dunks.*

If you've ever watched a stream flow over a rocky bottom or around strewn boulders and logs, you have witnessed mechanical turbulence. The eddies and swirls in the stream are the same thing that we feel in the air as turbulence. You can imagine these swirls passing you and how they feel as gusts.

To make your understanding even more clear, note how the swirls caused by a large rock or log eventually die out downstream. Also note that faster flowing streams exhibit larger, stronger swirls. From this object lesson we make general rules about mechanical turbulence:

Mechanical Turbulence
• Occurs when the wind flows around solid objects. • Is stronger in a stronger airflow. • Exists downwind from the object and gradually dies out downstream. • May be a standing (non-moving) eddy or random swirls.

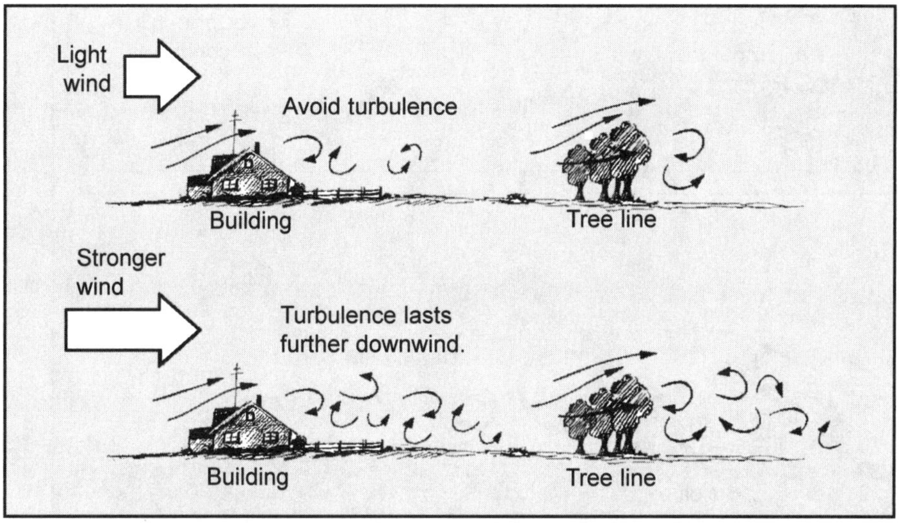

Figure 6-9: Mechanical Turbulence and Wind Strength

From this summary we can draw the following conclusions: The stronger the wind, the more likely we are to encounter strong mechanical turbulence. The closer we fly behind obstructions, the more likely we are to encounter turbulence and the stronger it will be.

What are obstructions? Houses, trees, hills and any other solid object qualifies, as figure 6-9 illustrates. Note how the turbulence extends further downwind in stronger wind. The broader a front the obstruction presents to the wind, the more disruptive it is to the airflow, so the more turbulence it produces. A long row of buildings or trees or a long ridge is conducive to mechanical turbulence because the wind can't readily flow around them, as figure 6-10 makes clear.

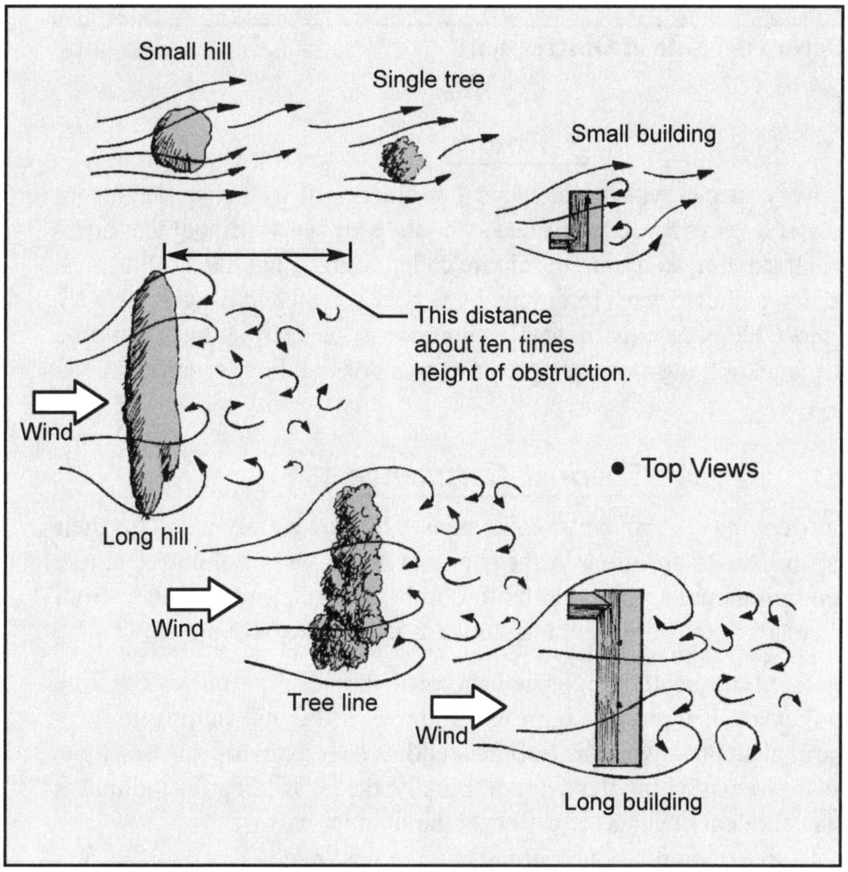

Figure 6-10: Turbulence Behind Broad Obstructions

To avoid mechanical turbulence, *fly in light winds if you are inexperienced and avoid flying too close to the downwind side of solids at all times* (note: *Understanding the Sky* provides more detailed guidelines to this matter).

Safety Tip: *As a general rule, expect turbulence to extend 10 times the height of an obstruction downwind from its top as shown in figure 6-10. This rule applies to mountains, ridges, hills, buildings, tree lines, walls, etc. In strong winds the distance is further, in light winds less.*

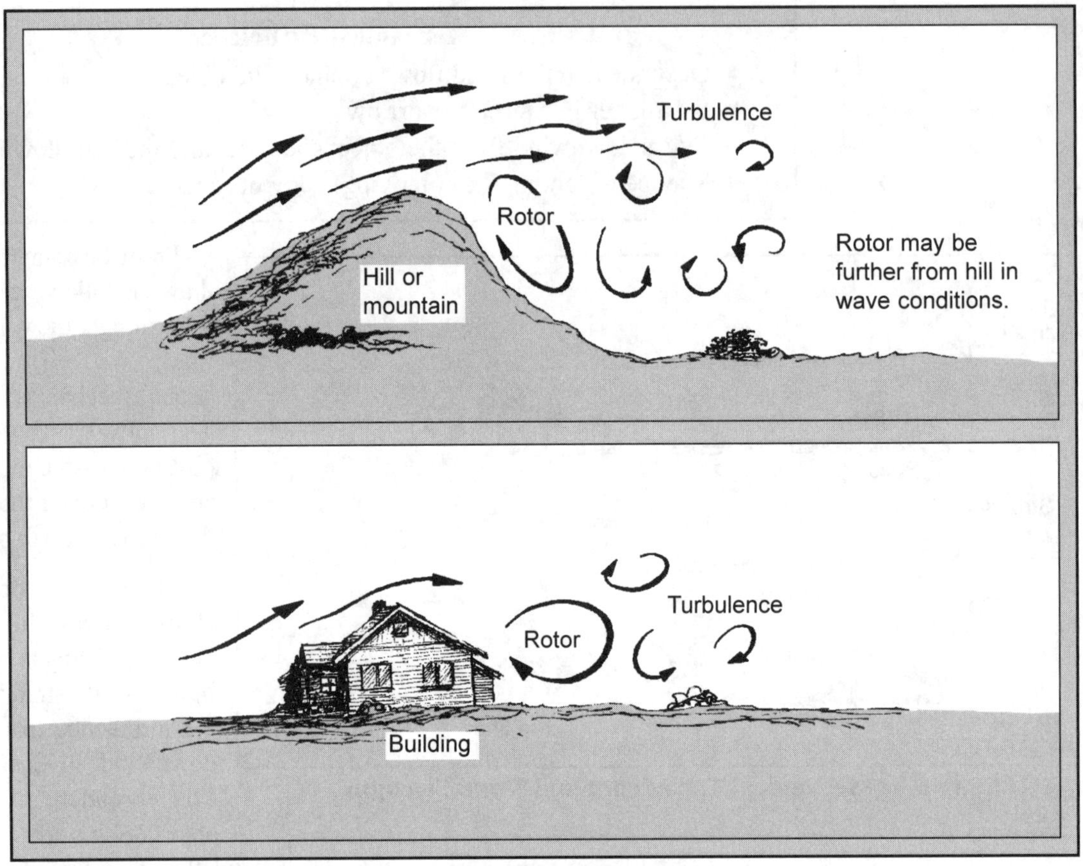

Figure 6-11: Rotor on Downwind Side of Obstructions

✓ Rotors

A rotor is a very special type of mechanical turbulence. If we revisit our stream and look behind a large rock we may see a standing eddy—a whirlpool that doesn't move. Similar standing eddies in the air are called rotors. Figure 6-11 illustrates a rotor behind a large hill. Such a rotor may be very powerful and dangerous to fly in. They are more likely to exist in stable conditions (see below), but should be expected at all times. Flying close behind a mountain or building is very risky.

✓ Thermal Turbulence

Thermals are common inhabitants of our skies. They are simply warm (as their name implies) masses of air, often in the shape of a bubble or column, that rise upward. Skilled pilots use these updrafts to climb skyward, and even hop from one to the other as if they were stepping stones to travel cross-country.

The reason thermals create turbulence is because they disrupt the smooth flow of the air as they push upward. Figure 6-12 shows a thermal bulling its way through the normal airflow. Note the turbulent eddies near its borders. Also, some unsteady flow may exist within the thermal. Finally, the air feeding the thermal at the bottom may suddenly change direction as the thermal passes.

We list some observations about thermals on the next page.

Thermal Lore

- They are warm or humid masses of air that develop over a warm area of ground.
- They rise at various rates, depending on size and their excess heat, and thus vary in buoyancy.
- They often create cumulus clouds as they rise to the condensation level.
- Their activity is known as convection.
- They usually vary with the time of day, so that the maximum heating produces the maximum frequency and strength of thermals.
- They usually appear above easily heated areas such as plowed fields, dry weeds, dry crops, pavement and quarries. They are least likely to appear above water, damp ground and cool forests.

From these generalizations we can make some conclusions regarding thermal turbulence. First, it is most likely to be encountered on sunny days above areas that heat most readily and to be strongest during the midday peak heating. Also, desert areas that heat up most readily will exhibit the strongest thermals and associated turbulence. Overcast skies often prevent thermals from developing.

Thermals are like bubbles reaching for the top of the sky.

Since thermals often rise to cloud base their turbulence is likely to be encountered at all heights, unlike mechanical turbulence that is found near ground obstructions or mountain tops. Up in the air, light thermals feel like upward bumps that smooth out once you are inside them. Stronger thermals may produce swirls that push you around vigorously. On the ground the turbulence is felt as strong gusts that come in cycles and change the wind direction, sometimes as much as 180°.

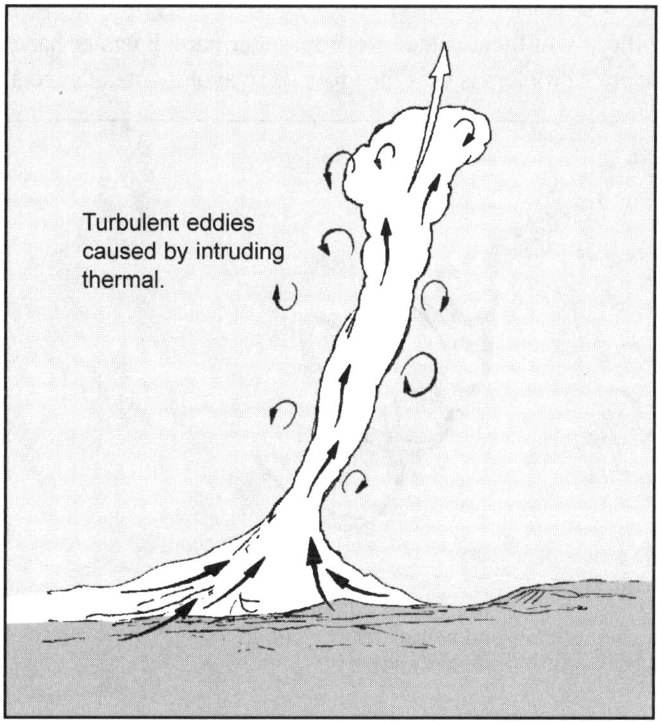

Figure 6-12: Thermal Turbulence

What goes up must come down, so normally areas of sink are present when thermals are bubbling. This sink may be widespread, so expect it anytime you expect thermals. Allow yourself extra altitude to reach your landing area in thermal conditions.

It should be clear that if thermals can exist without wind, so can thermal turbulence. One complication with no-wind thermals is that they can vary the wind in the landing field radically and rapidly. Beginners should not fly when thermals are prevalent. For that reason training often takes place in the morning or evening, when the sun's heating is mild. More experienced pilots fly in thermals and enjoy them, but they must be treated like any form of turbulence with the limits of strength, direction and gust factor carefully obeyed for safety's sake. It should be noted that thermals added to general wind can make the turbulence from both sources worse.

✓ Shear Turbulence

A third form of turbulence is occasionally encountered. This is shear turbulence, caused by layers of air moving against each other. This effect is similar to that of wind blowing over water and forming ripples or waves. In this case the water is replaced by a layer of cool, dense air.

We are most likely to meet up with shear turbulence in the evening when an inversion layer develops (see below) or after a downslope breeze begins. Shear turbulence is not too much of a problem because it is rare and not overly strong as long as the winds that cause it are reasonably light.

When crossing wakes, maintain good control pressure. Do not follow directly behind another glider.

✓ Wing Tip Vortices

There is a fourth form of turbulence known as tip vortices (a vortex is simply a swirl or eddy) or wake turbulence. All aircraft produce these vortices and the larger, heavier craft produce much larger and stronger swirls.

Figure 6-13 shows how vortices are produced by a wing and how they behave. Note that they gradually sink behind the wing as they expand. In the air, the vortices from another paraglider merely feel like a sharp bump and are not a problem as long as you maintain good canopy pressure. However, vortices from a larger aircraft can fold your wing into a tangled mess. The air changes faster than you can correct for its effect in this case.

To avoid vortices, avoid flying directly behind another aircraft (including a paraglider or a hang glider) or flying through its path for several minutes. The swirls from a large transport airplane can last up to 5 minutes on a still, stable day.

Most likely the only vortices you'll encounter are from other paragliders or hang gliders. The only time these are a problem is near the ground. To avoid vortex dangers

Figure 6-13: Wing Tip Vortices

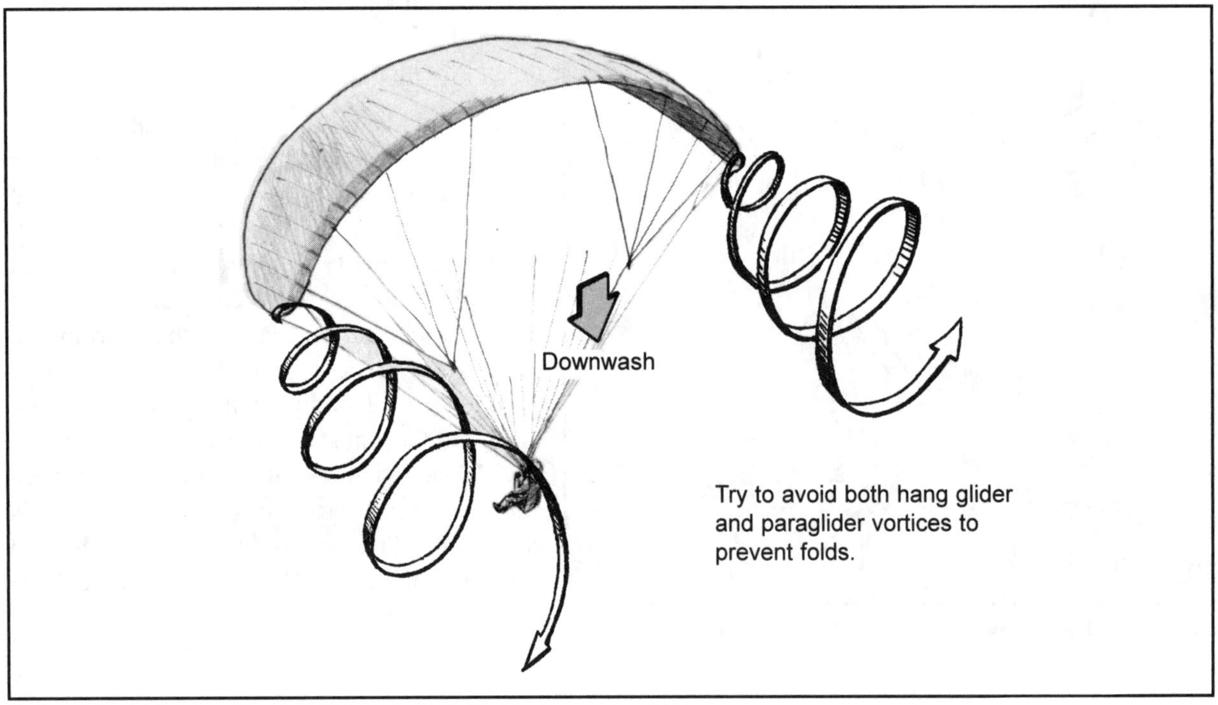

do not take off behind another glider *for at least fifteen seconds.* Landing is a more serious time to encounter vortices for you can't always choose when you land. We discuss avoiding vortices in this situation in Chapter 8.

✓ Watching the Wind

We have outlined many characteristics of the wind. Most of our understanding requires some imagination, for we can't actually see the wind—we can only see and feel its effects. However, we can see the motion of water and in many ways flowing streams and slower moving rivers will directly mimic the action of the air.

We learn to "see" the wind by using observation and imagination.

A primary part of our development as pilots is learning to read the conditions. It is an ongoing process whereby we watch the wind, feel the pressures, ask the opinion of others, experience the affects in flight and correlate what we felt with what we observed. Repeat this process with every flight and soon you will be seeing the air's flow in your mind's eye and your judgements will be sound.

Pro Tip: *Watch the birds. Our feathered friends are masters of the air. The soaring birds have light wing loadings so they judge when to fly and when not to. If they are flying around with their wings tucked in, chances are the wind is too strong for safe flying. If they are soaring on outstretched wings, conditions are probably fine. Watch the birds' drift as they circle to get a read on the wind direction and strength.*

Lifting Air

One of the joys of our powerless mode of flight is to gain altitude and float on high through the benefit of invisible air currents. This act is known as soaring—staying aloft on the air's energy. To perform such a feat we must understand a bit about the air's behavior. Essentially there are four main sources of lift that are used by paraglider pilots. We review these sources here and discuss the techniques of soaring in Chapter 12.

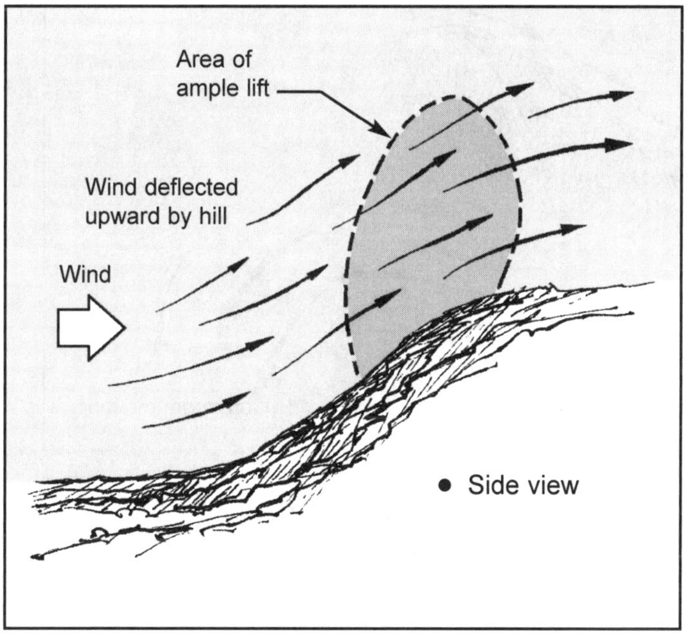

Figure 6-14: Lift Created by a Slope

✓ Slope or Ridge Lift

Lift is air moving with some upward velocity. The most commonly encountered and easily exploited source of lift is *slope lift*, also known as *ridge lift*, deflected lift or orographic lift. This lift occurs when wind strikes a hill, ridge, cliff or other obstruction and gets deflected upward, as shown in figure 6-14. There are many aspects of slope lift that should be understood, but we'll leave them for later when you learn to practice ridge soaring.

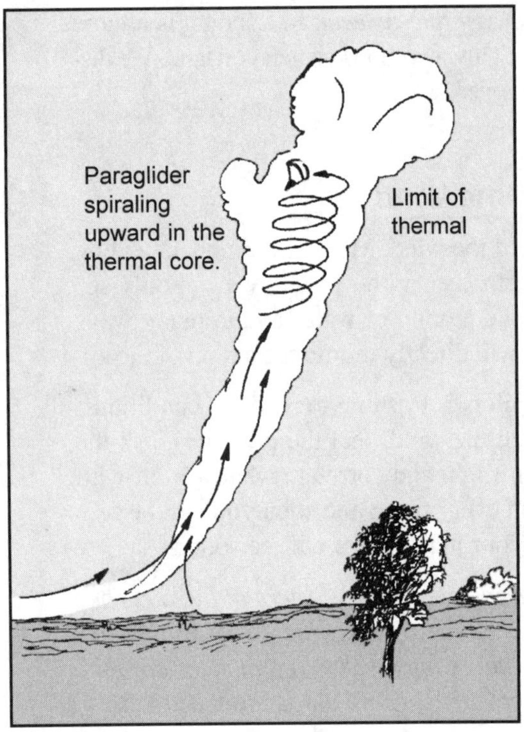

Figure 6-15: Using Thermal Lift

✓ *Thermal Lift*

We have mentioned thermals earlier in relation to the turbulence they produce. They also produce great lift which more advanced pilots enjoy. They can carry us to cloud base and allow long flights.

To use thermal lift, we detect a thermal through the feel of the disturbance around the border and the feel of an upward surge of lift. Another way is to use an instrument called a variometer that measures pressure drop as we climb and serves as a sensitive rate of climb indicator. Once inside the thermal we must circle to stay in its confines (see figure 6-15). That's why you have witnessed soaring birds turning endless circles. They were climbing in a thermal. Go to any advanced site and you are sure to see paraglider pilots wheeling in an upward spiral as well. They're thermaling!

✓ *Convergence Lift*

Often the air flows together from different directions. This effect can occur when downslope breezes meet in the middle of a valley or when flows meet behind a large mountain. As figure 6-16 shows, the flows coming together well up and create lift. The coming together is known as convergence. Convergence lift is often light, but can provide an enjoyable ride on buoyant, magic air.

Figure 6-16: Convergence Lift

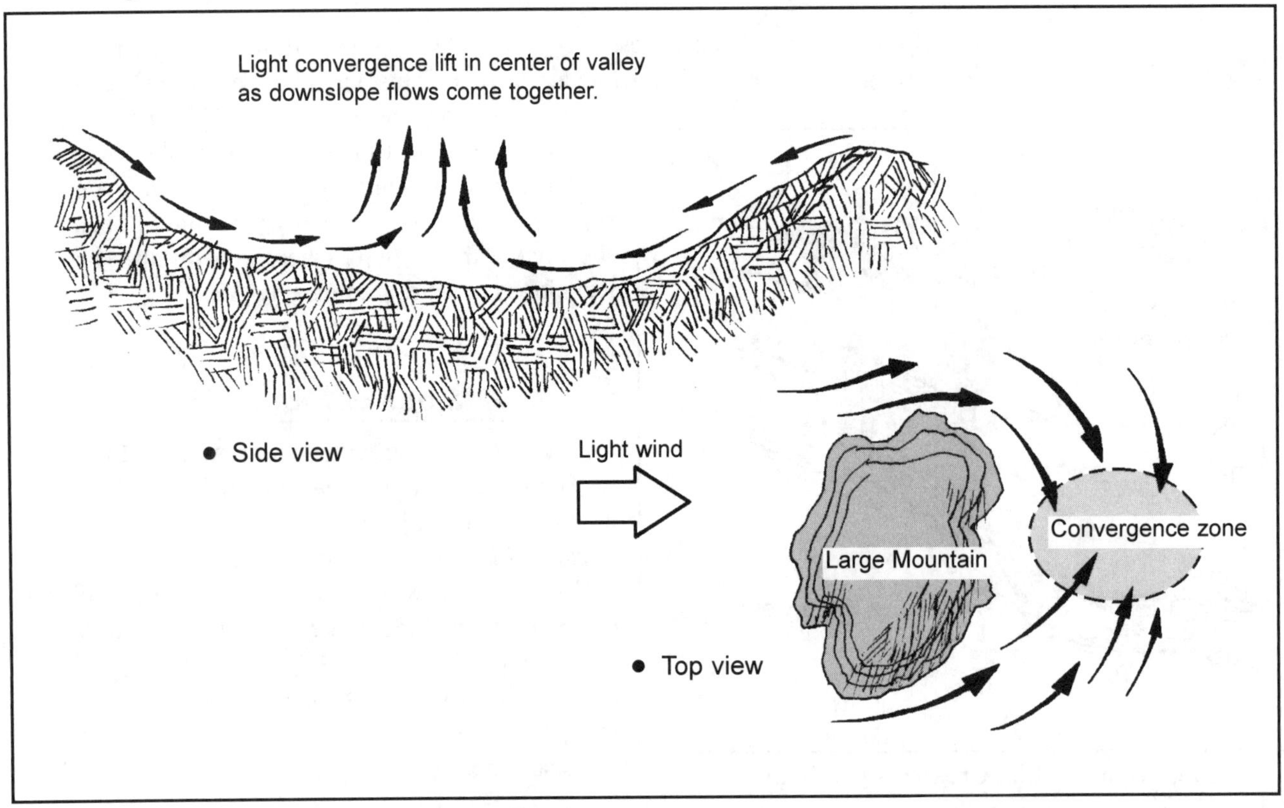

✓ *Wave Lift*

The air, like any fluid, can exhibit waves. Very specific conditions are needed to form waves, but these conditions occur frequently enough when the air rebounds up and down after passing over a mountain. In the right conditions waves can be very smooth and delightful, although they can harbor severe turbulence in high winds. Paraglider pilots have ridden waves for hours at times.

Finding lifting air is one of the great skills of a good pilot. At the beginning level, you aren't concerned with this feat. However, within a few months you can begin to learn the basics of soaring, and then all the waking hours you can spare will be spent thinking of where to find lift and how to use it.

THE AIR YOU FLY IN

At this point we turn our attention to the weather on the large scale and the way the atmosphere behaves. We will start with the nature of the atmosphere. You'll find that most of this makes sense, for after all you've been living and breathing in that atmosphere for many years now.

✓ *The Density Effect*

From science class at school you perhaps know that the earth's atmosphere is made up of gases— mostly nitrogen and oxygen—that have mass and are thus held to the earth by gravity. This gives us a clue to a feature very important to pilots: the air closer to the earth is more dense because the lower layers are compressed by the air above them.

Another factor affecting density is the temperature of the air. As we explained when discussing circulation, warm air expands and becomes less dense. Conversely, cool air is more dense. In a similar manner, humid air is less dense. You may think that adding water vapor to the air would make it heavier or more dense, but actually the opposite is true because water vapor is lighter than dry air, since water molecules are lighter than either nitrogen or oxygen. Figure 6-17 shows the general structure of the air.

The final determinant of density is the pressure systems in the area. A high pressure system means that more air is piled up aloft, so the pressure we measure below is higher. The effect is denser air at the surface because it is compressed.

Air Density

Air is less dense:

- The higher the altitude.
- The warmer the air.
- The higher the humidity.
- The lower the surface pressure.

What all this means to pilots is important, since our wings fly by deflecting the airflow. The less dense the air the faster we have to fly to create lift. Consequently, on a hot, humid day at altitude we fly faster at any given angle of attack or control position. You won't be able to detect this, for you will feel the same forces and the same wind flow on your face because the faster speed makes up for the lower density. Just fly the same as always and everything will be fine.

However, during takeoff and landing in these conditions you will notice a difference. Besides being shorter of breath at high altitude you *will have to run faster* to take off and your landing will be a bit faster. As a rule of thumb, we add about 2% to all our flying speeds for every 1,000 feet (300 m) increase in altitude. So if you take off at 5,000 feet you would add 10% to your takeoff speed compared to sea level. That's only about 1.5 mph (2.4 km/h), but an extra step or two may be required. At higher altitudes, even more speed is needed. Remember, a good take-off technique will work well at any altitude.

✓ Pressure Effect

The pressure of the air is simply a measure of how much air is above with its weight pressing down. Naturally, if we are higher we feel less pressure since there is less air above us to press down. As we mentioned before, pressure combined with temperature and humidity at a given altitude determine the air's density.

Pressure, like density, is greatest at the surface and becomes gradually less with altitude (the change with altitude is nearly linear in the lower 10,000 feet or 3,000 m) as shown in figure 6-17. This property gives us a convenient way to measure altitude. We can use an instrument like a barometer designed to measure the air's pressure, mark it off in feet or meters and it will read out our altitude as we go up or down. We call such an instrument an *altimeter* since it measures altitude. If we perform a few tricks on the instrument we can even get it to display the rate of change of the pressure which is our rate of change of altitude. Such an instrument is known as a *variometer*, and it measures our rate of climb or sink. More advanced pilots use sensitive variometers a great deal when searching for lift.

Figure 6-17: General Air Structure

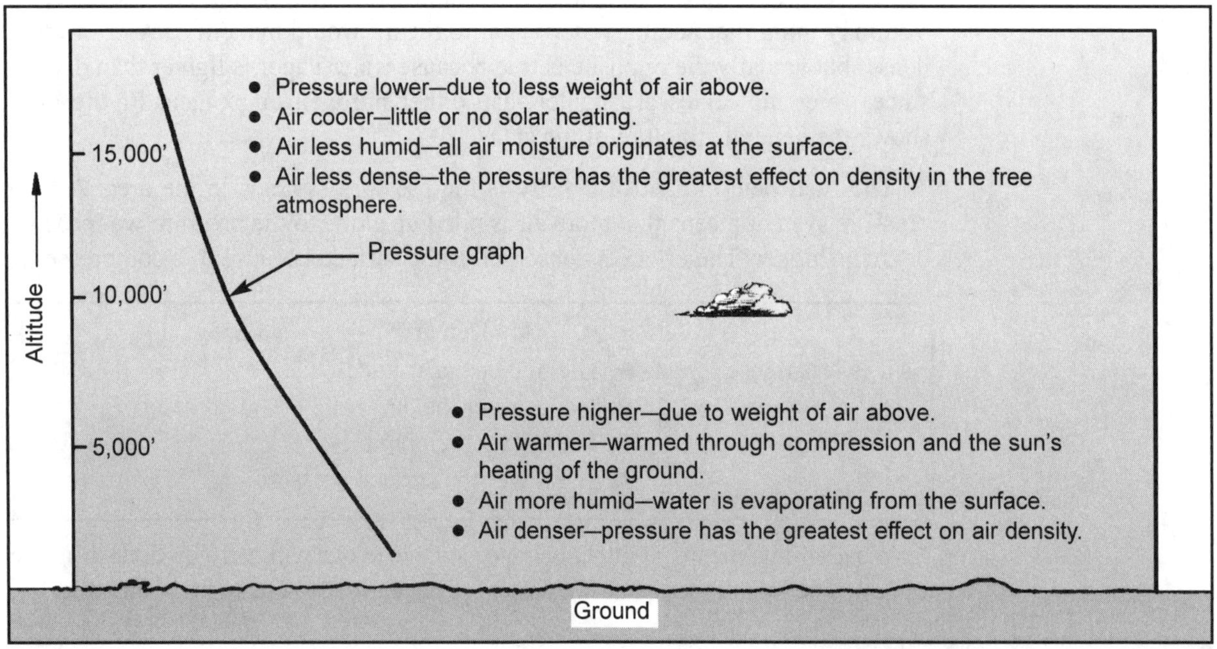

✓ Temperature Effect

Let's look a little closer at air temperature effects on our flying. At a given pressure, cooler air will be more dense than warmer air since the molecules are packed more tightly together. Denser air will create greater forces on our wings for a given true airspeed. We often find a site soarable in the cold winter air with wind speeds noticeably less than those necessary in the summer. But unfortunately, reduced heating from the sun, more clouds and snow cover typical of winter diminish the development of thermals, so they are less common in the cold months.

✓ Stability

A very important aspect of the air's temperature is that it generally cools off with increasing altitude, as we scc in figure 6-18. Anyone living in mountainous country has experienced the drop in temperature as the mountains are climbed. That's why high peaks often bear snow year around. A typical temperature drop is 5.5° F per 1,000 feet (1° C per 100 m).

This feature of the air is important because it determines the *stability* of the air. The stability of the air is defined as its tendency to resist vertical motion. Stable air reduces or suppresses thermals, while unstable air promotes them. To see how this works, imagine a parcel of air warmed by the surface below so its temperature is higher and its density lower than the surrounding air. This parcel or mass will begin to rise since it is lighter and more buoyant than its surroundings. But as it rises it expands due to lowering pressure, and it cools. If the surrounding air cools faster than or at an equal rate to the expanding parcel, the parcel will continue to rise and the conditions are said to be unstable, as illustrated in figure 6-19. Such a rising parcel is called a thermal.

Figure 6-18: Cooling of the Air with Altitude

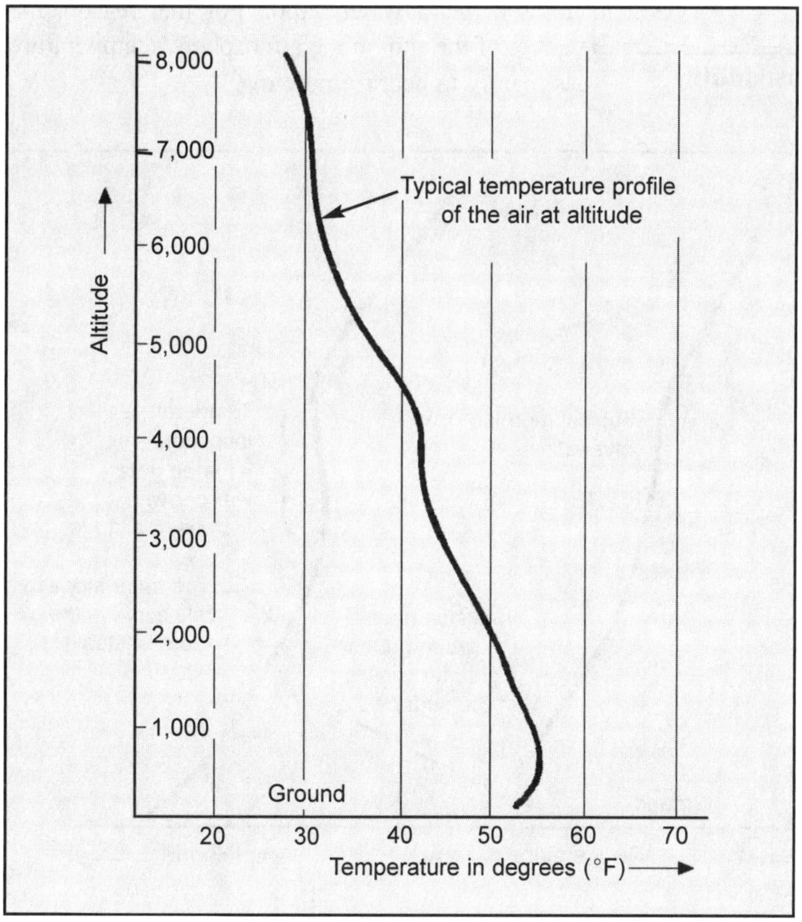

Conversely, if the temperature of the surrounding air doesn't cool fast enough with altitude, the parcel will stop its climb and the air is said to be stable. The measure of the air's temperature with altitude is called *lapse rate* or *temperature profile.* When you hear other pilots speaking of a good lapse rate they are saying the air's temperature cools fast enough with altitude so that conditions are unstable and thermals develop—soaring in them is possible. If you hear more advanced pilots glumly mumbling about a poor

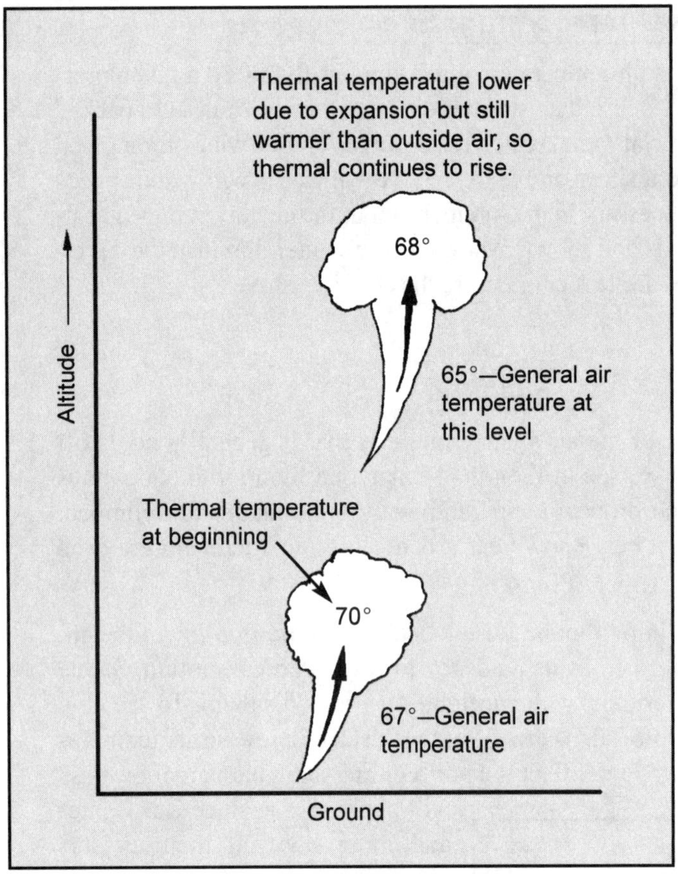

Figure 6-19: The Meaning of Instability

lapse rate they are meaning that the temperature profile is not so good for promoting thermals and the day is stable. Figure 6-20 shows typical lapse rates or temperature profiles. You can get the best idea of the stability of the air by getting the *thermal index* reading from your weather briefer.

In the figure we see a bump on each graph line. This bump is a *temperature inversion.* With an inversion the air's temperature actually increases with altitude. Such an inversion is common in the air, especially in greener, moister areas. A thermal will often rise to an inversion layer and bump up against it like a helium balloon against a ceiling. The thermal will only penetrate the inversion if the thermal is very strong or the inversion layer is thin.

✓ Thermal Production

We have discussed how the sun heats the ground and causes thermals. What we didn't mention is that the sun's rays pass through the air without heating it much at all, except when it contains a lot of water vapor. For that reason, the effect of the sun on the atmosphere's temperature is mainly to heat from below.

Stable air suppresses vertical motion in the air. Unstable air promotes vertical movement.

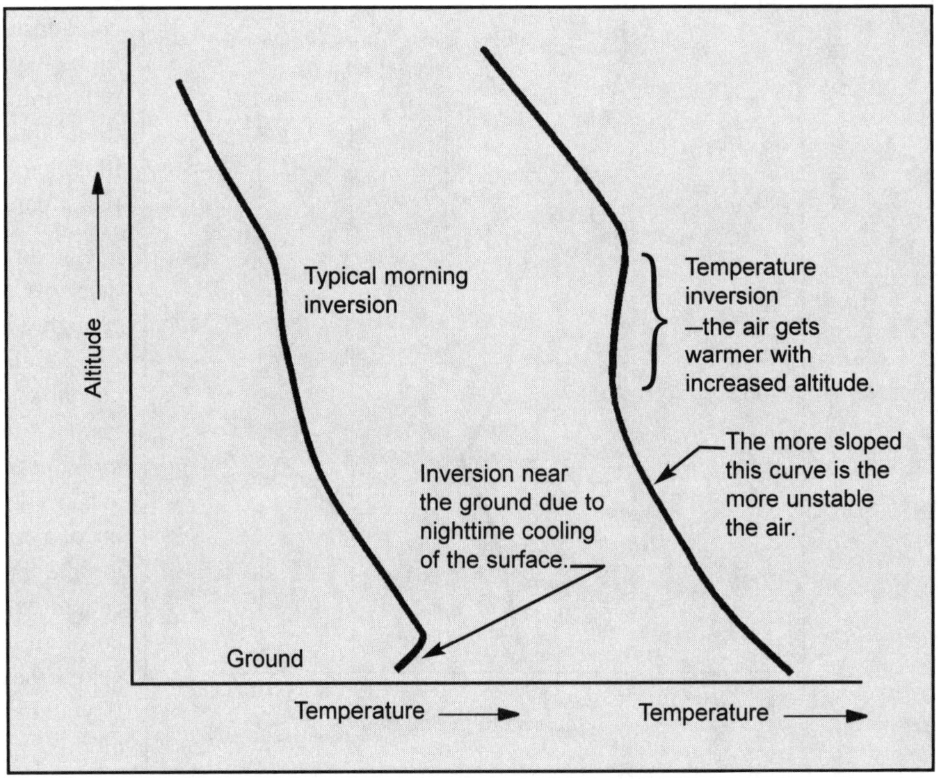

Figure 6-20: Typical Lapse Rates

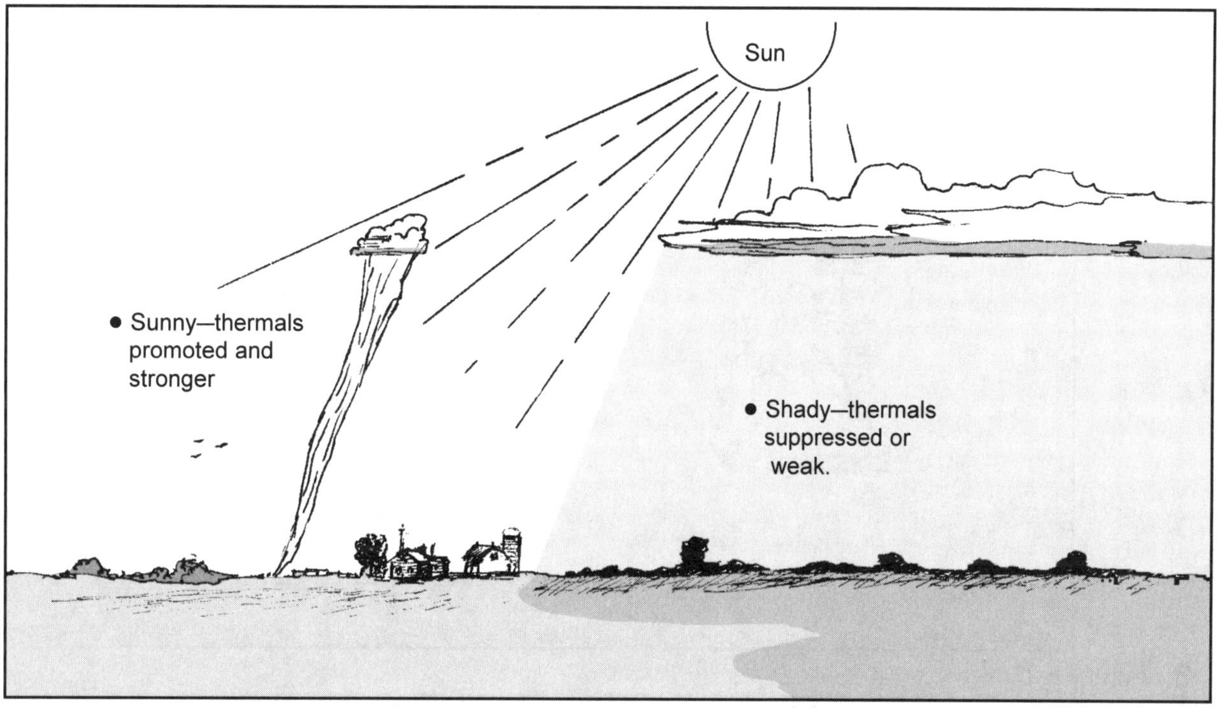

Figure 6-21: The Sun's Effect on Thermal Production

As we saw above, heating the lower layer of the air leads to instability and thermal production. So a general rule is that sunny days exhibit the most thermals and cloudy days the least, as shown in figure 6-21. Also shown in the figure is how thermals rise to form cumulus clouds, which we investigate next.

Clouds

Everyone knows clouds as those wisps, puff balls or layers in the air that seem to float along with a life of their own. But pilots need to become a little more knowledgeable than the average person about clouds for they reveal a great deal about conditions.

✓ Cloud Causes

All clouds are formed by air that is rising. You learned that a rising thermal expands and cools. This is true of any air layer or parcel. When the air cools to the point where the water vapor condenses—like the droplets on the glass of an iced drink—a cloud is formed.

Clouds are the off-spring of lifted air.

✓ Cloud Types

There are two main types of clouds: puffy or tumbled clouds called *cumulus*, and flat, layered clouds called *stratus* (see figure 6-22). Cumulus clouds are caused by rising thermals or updrafts as mentioned. Their tumbled, cauliflower

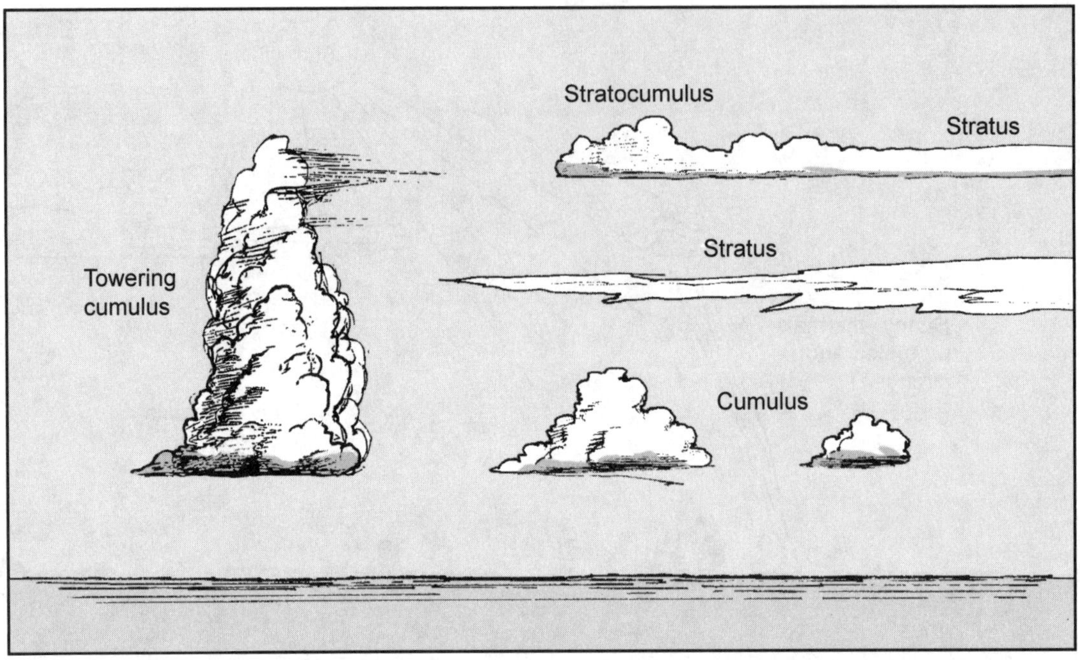

Figure 6-22: The Two Main Cloud types

appearance is due to the fact that the uprising air arrives in surges and rolls over. If you watch a fast-growing cumulus cloud you can see the edges bulging and tumbling about.

Stratus clouds occur when a whole widespread layer of air is lifted. The underside appears totally flat because the entire layer reaches the condensation level at the same altitude. We can have a combination of these clouds as shown in figure 6-22.

The suffix "nimbus" indicates rain.

Other names associated with clouds that you may hear are *alto* and *cirrus* as in *altostratus*, *altocumulus*, *cirrostratus* and *cirrocumulus*. These designations simply refer to the cloud's height with alto being medium-high clouds and cirro being very high clouds.

Cumulus clouds promise thermal lift.

A very important cloud for paraglider pilots to recognize is the *cumulonimbus* cloud, which is a thunderstorm. As shown in figure 6-23, a thunderstorm is just an overgrown thermal cloud that keeps pumping in humid conditions. Thunderstorms are dangerous because they can exhibit very high winds (including tornados), severe turbulence, lightning, strong precipitation and cloud suck (lift too strong to get away from). *Understanding the Sky* and *Performance Paragliding* (the companion book to this manual) tell you a lot more about how to avoid and survive thunderstorms.

Figure 6-23: Thunderstorm Cumulonimbus

Two other important clouds you will see are *cap clouds* and *lenticular* clouds. Cap clouds, as shown in figure 6-24, are formed from upslope breezes along the flanks of a large mountain that rise high enough to reach the condensation level. You will often see them as clouds that sit above a mountain all afternoon without moving. Their significance to paraglider pilots is that they can cover high mountain flying sites on moist days. Tropical islands with mountains often suffer from cloudiness due to this factor.

Lenticular clouds are lens-shaped in cross section and are created as atmospheric waves undulate up and down (see figure 6-25). They are formed in the uprising portion of the air and die out in the downward moving part. There can be many repeated "lennis" as they are nicknamed or they may be stacked one upon the other as shown. Lenticular clouds are common in the evening after a cold front passes, and they signify the presence of waves in the air similar to waves on water.

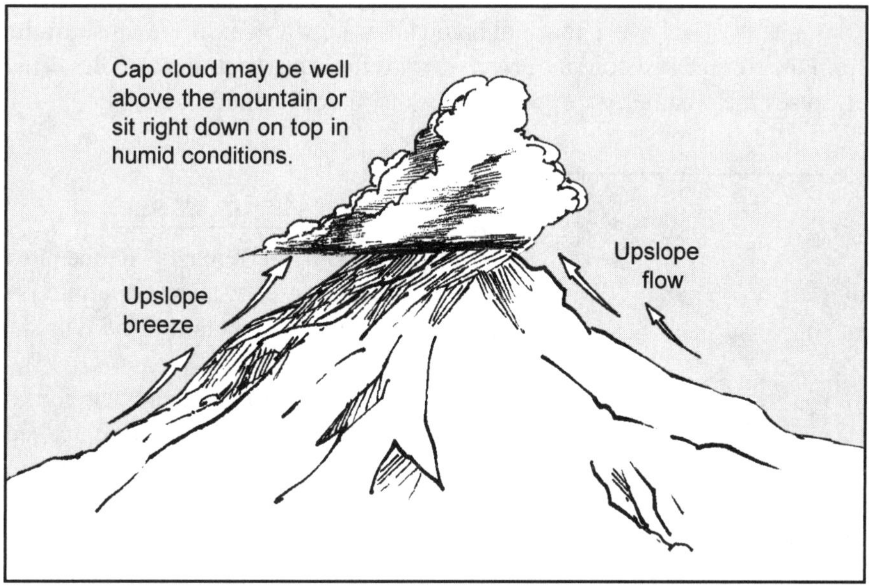

Figure 6-24: A Cap Cloud

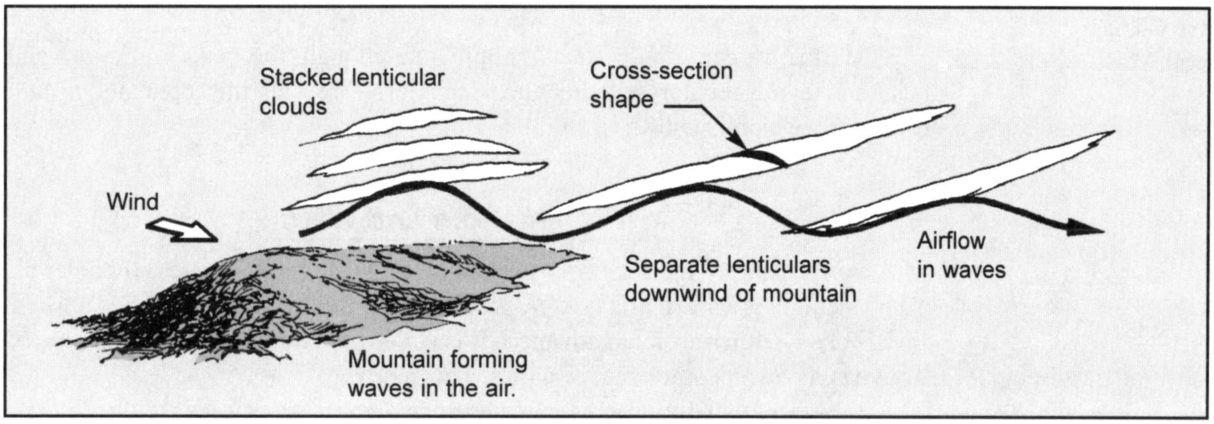

Figure 6-25: Lenticular Clouds

General Weather

What the weatherman means when he says the word "weather" is the large-scale movement of the air that covers half a continent. This aspect of the atmosphere is important because it affects our flying for days to come. We can't possibly delve into this broad subject in detail here, but we'll point out the highlights so you can begin to learn more as you gain experience.

✓ *Circulation*

The air circulates around our globe due to the sun's heating at the equator and tropics. This circulation would be a simple matter except that the greatly different surface areas (water, jungle, snow, grasslands, desert, etc.) result in uneven heating. In addition, the rotation of the earth creates an effect known as *Coriolis force* that tends to turn the flowing air to the right in the northern hemisphere and to the left in the southern hemisphere.

The result of this complex circulation is a buildup of air at the poles and in the tropics at around 30° latitude. This buildup causes high pressure areas on the surface. Consequently we have a constant battle between warm air driven toward the poles from the tropics and cold air driven toward the equator from the poles. This conflict causes the weather we experience in the middle latitudes.

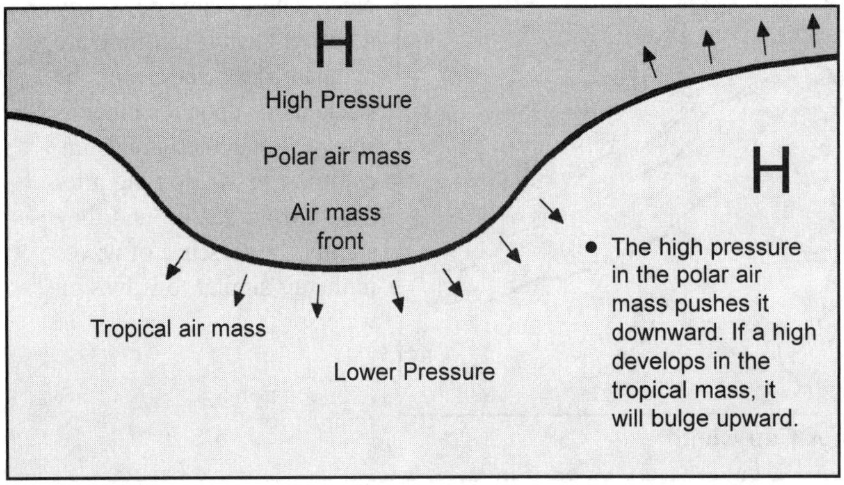

Figure 6-26: Frontal Movement

✓ *Air Masses*

When air remains in one area long enough, it takes on a uniform quality of temperature and humidity and is known as an *air mass*. The air masses that gather their forces at the poles are naturally called polar, while those at the tropics are called tropical. These air masses may be moist (designated maritime) or dry (designated continental) depending on whether they've been sitting over water or land.

When the pressure in an air mass builds up enough, the mass breaks out and pushes into the territory of the opposing mass. We call this breakout *frontal movement*, which is shown in figure 6-26.

✓ *Fronts—Cold and Warm*

A front is simply the boundary between two different air masses. If cooler air is advancing it is known as a *cold front*. If warmer air advances it is called a *warm front*. When neither air mass advances it is called a stationary front. Each type of front has its own characteristics, as we shall see.

A cold front usually progresses faster than a warm front because the cold air

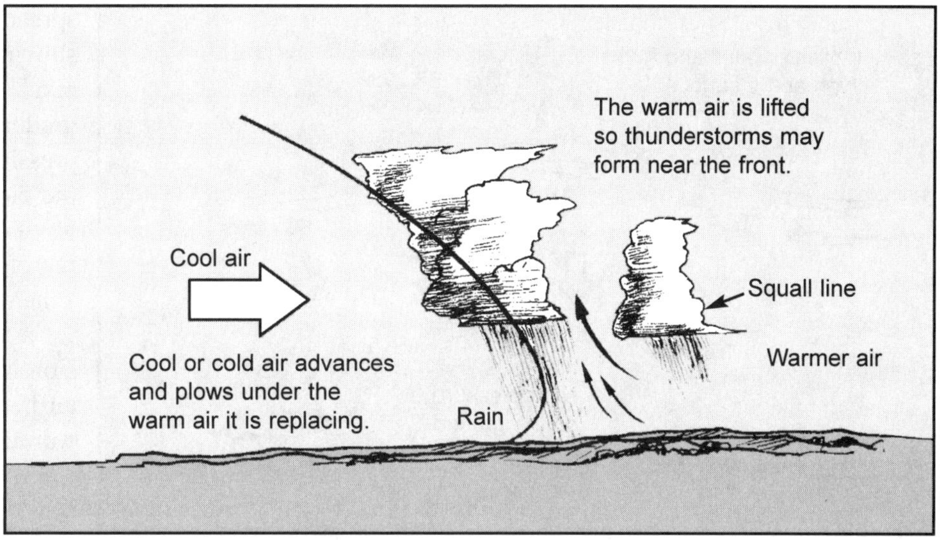

Figure 6-27: The Nature of a Cold Front

tends to be more vigorous. Figure 6-27 shows how the cooler air is denser, so it plows under the warm air that it is displacing, lifts it and forms clouds and often thunderstorms. Once the front passes, the skies generally clear, the air is cool and dry, and good flying weather usually results.

A warm front tends to be slower moving. Even though it is pushing the cold air away, it rides up over the cool air and thus forms widespread layers of clouds. Figure 6-28 shows how the advancing front displays wispy cirrus clouds then alto clouds followed by stratus type clouds that get lower and lower. Widespread rain accompanies warm fronts—often for days—and the air after the front passes is usually warm, humid and frequently stable.

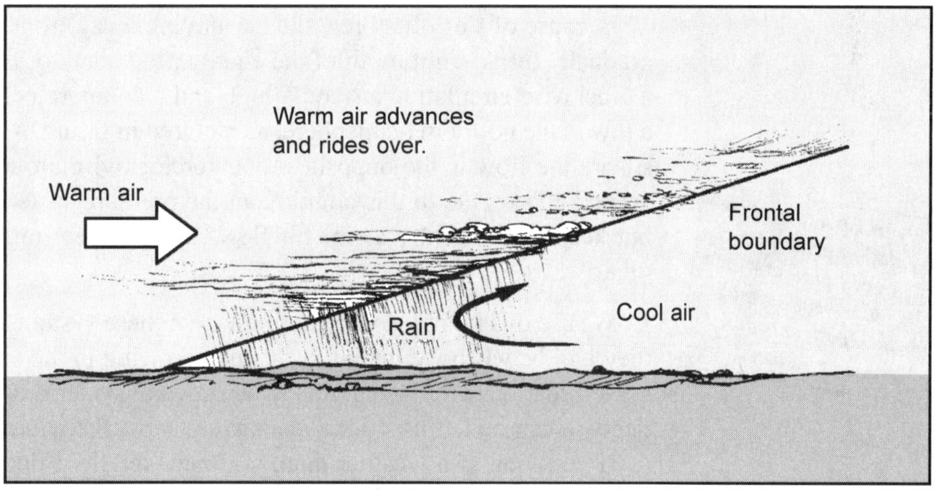

Figure 6-28: The Nature of a Warm Front

✓ *Pressure Systems*

The last matter of general weather we'll investigate is pressure systems. You have already learned a bit about *high pressure systems* or highs. These are formed when air is piled aloft due to circulation effects. The air slowly sinks in a high and

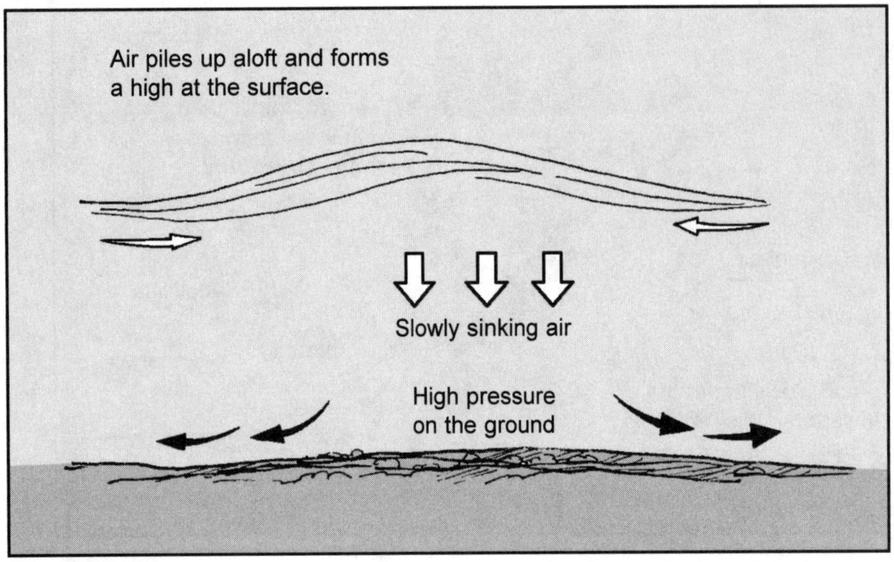

Figure 6-29:
High Pressure System

spreads out at the surface as shown in figure 6-29. The result of the sinking air is a gradual warming (due to compression) and a drying out of the clouds. Highs usually follow cold fronts, which is one reason for the clearing skies. When highs have been around for a while, they cause inversion layers and stability of the air because of the sinking and warming effect.

A *low pressure system* is a secondary effect that is also a result of global circulation (see *Understanding the Sky* to grasp the more complex matters of lows). Lows have rising air in their midst and the air flows inward at the surface. Because of the rising air, an abundance of cloud is formed around a low in moist areas, and these systems create the gully-washing, frog-choking rains known to much of the temperate region.

✓ Circulation Around Pressure Systems

An important concept for all pilots to understand is how the wind flows around pressure systems. If we know this we can forecast our own winds given weather maps which display highs and lows.

Because of Coriolis force, the air moving away from a high and toward a low gradually turns (right north of the equator, left south of the equator). The result is a clockwise circulation around a high and a counterclockwise circulation around a low in the northern hemisphere, as pictured in figure 6-30. In the southern hemisphere the flow is the opposite—counterclockwise around a high and clockwise around a low. Due to this circulation, the pressure systems don't equalize readily but keep their identity as the air flows around them rather than from one to the other.

By understanding circulation principles you can predict winds from a weather map showing fronts and pressure systems.

We use this knowledge by noting where these systems are on the map or where they will be when we intend to fly, then drawing in the expected flows. Of course some situations aren't so simple when several systems are present, but with practice you can be letting your friends know what the winds will be tomorrow.

If you look at a weather map, you may see light lines drawn on it which are known as *isobars*. An isobar is a line connecting points of equal pressure. You will note that isobars are often drawn as circles or distorted ovals around highs and lows (see figure 6-30). In addition, isobars pass through fronts and usually are kinked at the front boundary.

Now here's an interesting fact you can use. The winds away from the surface always follow the isobars. If you have a surface map, you can figure that the winds generally follow the isobars but cross them from 30° to 45° (depending on the terrain roughness—30° over water and 45° over mountains). Which way do the winds flow and cross? Above we explained that winds flow clockwise around a high and vice versa around a low (northern hemisphere). Trace such a path along the isobars and you'll get the wind flow. Also note that the flow is away from high pressure and towards a low and you'll see the direction of the angle of flow with respect to the isobars.

One final bit of isobar lore: the closer the isobars are portrayed on the weather map, the stronger the winds. The further apart they are, the lighter the winds. Now armed with a weather map showing isobars you can predict wind direction and strength.

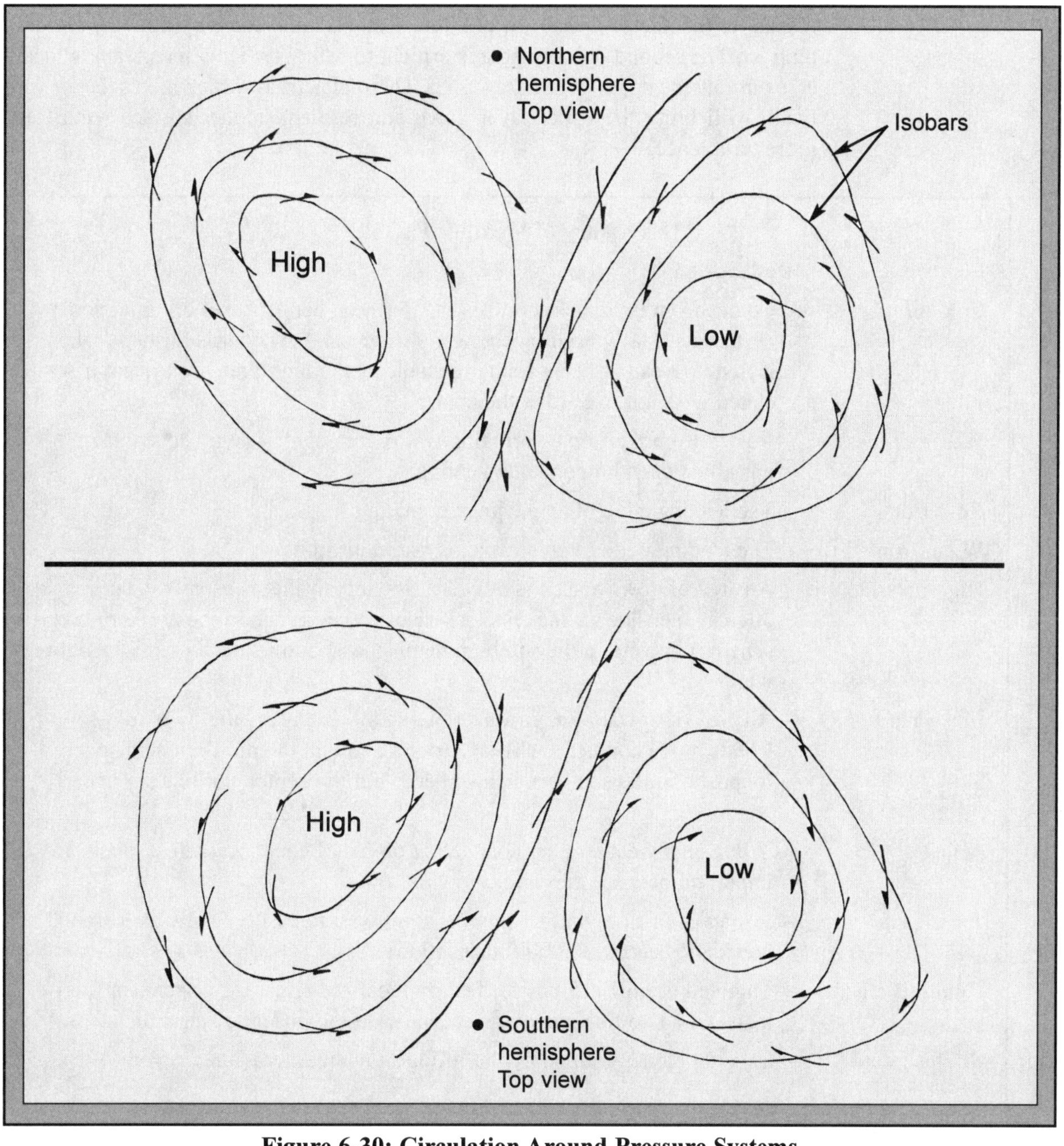

Figure 6-30: Circulation Around Pressure Systems

✓ Jet Stream

High aloft—20,000 feet (6,000 m) or more—flows the *jet stream*. This rapid river of wind circumnavigates the globe with winds up to 100 mph (160 km/h) or more. The jet stream is caused by the imbalances at the interface between large cool and warm air masses. In fact, there are two such streams in the northern hemisphere. The polar jet stream accompanies cold fronts barreling down from the poles. It tends to be stronger and more complete. The sub-tropical jet is more intermittent and lies around the 30th parallel.

Both of these jet streams tend to undulate north and south as upper-level pressure systems vary. Their significance to you as a pilot is mainly two-fold. When a jet stream dips down, it usually is a sign of poor weather within 24 hours. The reason for this factor is that low pressure systems are created by a southward dipping jet. The second point to note is critical to safety. When a jet stream is located over you, the upper winds are strong. Thermal activity which mixes the air vertically will bring the upper layer down and result in higher surface winds and more turbulence.

Summary of Weather Factors

Feature	Description
Cold air mass	—Comes from the poles (north in the northern hemisphere) bringing cool air that is usually dry (except near west coasts). The air is unstable when it first arrives but becomes more stable as the high pressure system associated with it moves over the area.
Warm air mass	—Comes from the tropics bringing warm, humid air. It may be stable or unstable depending on solar heating.
Cold front	—The leading edge of an advancing cold air mass.
Warm front	—The leading edge of an advancing warm air mass.
High pressure	—An area of downward and outward flowing air that is usually clear and often stable. The air moves clockwise around a high in the northern hemisphere (opposite in the southern hemisphere) and its center exhibits light winds.
Low pressure	—An area of upward and inward flowing air that is invariably cloudy. Winds move counterclockwise around a low in the northern hemisphere (opposite in the southern hemisphere) and the center exhibits stronger winds.
Isobar	—A line on a weather map connecting points of equal pressure. Isobars are important because they indicate wind flow.
Jet stream	—A river of fast moving air flowing from west to east typically over frontal boundaries above 30,000 ft (10,000 m).
Cumulus clouds	—Tumbled or puffy clouds formed by localized rising air currents or thermals. Good vertical development indicates an unstable lapse rate.
Stratus clouds	—Layered clouds formed by the lifting of a widespread area of air.

Obtaining Weather Information

In order to predict or judge the weather you must have a source of information. The weather sources available to pilots vary from country to country. However, here are a few suggestions that should prove fruitful.

▲ *General media* — Newspapers and the like provide weather maps of varying degree of usefulness. Some provide depictions of fronts and pressure systems which help you figure out winds. Others merely list the temperature and expected precipitation. Find a newspaper that shows surface details and you can predict winds.

▲ *Weather TV* — In the U.S. at least, a cable channel provides non-stop weather information which you can use to interpret conditions. This information is generally accurate but does not provide a lot of pressure system detail.

Aviation weather — In many countries paraglider pilots are considered part of the aviation community and are allowed to use the aviation weather services. Ask more experienced pilots or your instructor how to do this. In the U.S. this service is available by telephone at 1-800-WX-BRIEF. If you state you are a paragliding student or pilot you will get whatever information you request (see below).

Stratocumulus clouds showing here are often a sign of higher winds aloft and an upcoming weather change.

Internet web sites — The Internet has several weather providers. You can get frequently updated surface map information plus forecasts from these sources. Here is a list of useful links:

General weather information:
www.paraglide.com

Aviation forecast: http://adds.awc-kc.noaa.gov/

National Weather Service:
http://iwin.nws.noaa.gov/iwin/graphicsversion/bigmain.html

Soaring forecast:
http://csrp.tamu.edu/soar/for.html

Weather maps:
http://www.weather.com/weather/maps/
http://www.intellicast.com/
http://weather.unisys.com/

▲ *Weather radio* — In the U.S., the National Oceanic and Atmospheric Administration (NOAA) broadcasts 24 hour weather reports to most places in the country. You can pick these broadcasts up on a ham radio or a special radio available at electronic supply stores. This report provides winds and significant weather information 12 to 24 hours in advance.

Once you are familiar with the sources of weather available to pilots in your area, learn to use this information by remembering what was predicted compared to what you actually experienced. Then temper your own or the weatherman's predictions to get more realistic forecasts. For example, if you are often finding more wind than what was predicted, you know how to judge things for the future. With practice you may become as accurate as a professional weatherman—as many pilots are—for you have the advantage of first-hand experience in the actual conditions with your body as an instrument.

Cumulus clouds form just above launch at a soaring site in the Italian Alps.

Weather understanding is an important part of every pilot's development. Here's what Dixon White suggests:

"By day two of training we request every student show up with weather info about the jet stream, highs and lows, isobars, upper level winds, temperatures, the thermal index and general forecasts. We show them how to access this information and use it. With such hands-on experience pilots soon learn to predict flying conditions very well.

My students who do a great deal of kiting find that certain models of weather are more or less appropriate for ideal experience. It's interesting how there are similarities between those ideal kiting conditions models and those models that are best for some types of flying. Study the weather every day and quiz your local pilots—who did get to fly when you couldn't—about that day's conditions so you can start to improve your ability to anticipate the weather conditions you prefer."

Finally, we again direct your attention to *Understanding the Sky*, the book about weather for sport pilots. You will learn more about the sky we fly in and gain more understanding about how all the factors discussed in this chapter interconnect.

RISK MANAGEMENT

All our weather knowledge does us little good if we can't put it to practical use. As your experience grows you'll learn how to analyze conditions you observe for ways to improve your performance—soaring. However, from the very beginning it is important to learn to judge conditions in regards to safety. We call this risk management.

The weather and local conditions can be readily judged using the information provided in this chapter with a little practice. Your instructor will help you "see" the air and make safety decisions based on what you observe. Later in your flying experience you should ask more advanced pilots for their assessment of conditions as a way to rate your own conclusions.

When we judge conditions we are looking for things that could affect safety at the present time, and for signs of imminent change. For example, wind strength, gustiness and direction are primary factors relating to current safety. The temperature, lapse rate, presence of clouds and time of day are also current safety factors. The position of the jet stream and closeness of isobars allow you to assess the expected winds for the day. Other matters such as a falling barometer, the rapid buildup of clouds, an approaching front or cloud bands are signs of change.

The best way to learn how to evaluate these matters is to use the Weather to Fly™ system. This system is a package including the Weather to Fly™ video, CD, log book and the weather book *Understanding the Sky.* It has been put together by Paul Hamilton with the expertise of Dixon White. The address for obtaining this system appears in Appendix I.

SUMMARY

We have distilled the whole workings of the atmosphere of the entire planet into a meager few pages. Our goal is to provide a general idea of how the weather works at this stage. It will take more time and experience before you can really understand the details that help make you a master of the air.

But don't fret, because before long you will be seeing things in nature that you never noticed before. The light bulb will go off in your head and you'll think, "Oh yeah, I read about that." Then you'll gain new insight. Each little bit of understanding adds up to a whole that provides you with the overall picture. That hawk with tucked wings, that bit of fluff floating by, that drifting cloud—all are clues that speak a clear message once you know the language. Weather lore is fascinating simply because there are so many aspects to it and because it now has such practical use to us pilots. As if flying through a pristine sky from cloud to cloud for the pure joy of it had any practical use other than to keep us sane and happy!

TEST YOURSELF (*Answers in Appendix II*)

1. Name three aspects of gusts that affect the safety of flying: __________, __________, __________.
2. Turbulence should be most expected at what period of day? ______________
3. The higher the wind strength, the lighter the expected mechanical turbulence.
   True or false
4. Normally we expect (*upslope, downslope*) winds to set in as the sun goes down.
5. Which cloud types are most likely to suggest turbulent conditions in light or calm winds: cumulus or stratus?
6. Name three types of turbulence we may encounter in the air: ____________, ____________, ____________.
7. Where would you expect a rotor to lurk on a windy day?
8. Which pressure system is most likely to produce widespread clouds and rain?

A pilot running on a light wind launch. Note the forward lean and hand position. The higher hand position lets the canopy accelerate.

Pilots kiting and launching. Note the protective air bag harness and lower pilot hand position.

Chapter 7

PERFECTING TAKEOFF SKILLS

"Simple style is like white light. It is complex but in its complexity, is not obvious."

— Anatole France

At the training hill you learn to inflate and take off in carefully controlled conditions. The slope is gentle, the wind is right, your instructor is all but holding your hand. Gradually you will learn to handle a bit more wind and even keep your composure in the midst of mild wind variations. Perhaps you accomplish some launches from steeper or flatter slopes than those you have been initially trained on. Your progress is apparent and appealing. But now it's time to take the next step.

That step is to add versatility to your takeoffs while at the same time perfecting your technique. Versatility means being able to safely handle slopes of all types of steepness, ramps and rough ground, as well as crossing and more variable wind. Perfecting technique means doing all this with absolute control of the canopy, including inflation, kiting and launch airspeed.

Truly you will develop your launch skills along with your flying and landing skills. However, we have placed all these matters into their own chapters so you can better focus on their particular demands. It is important to note that not all the takeoff situations and techniques included in this chapter are suitable for novice pilots. However, we cover them here for your future learning and to make you aware of what complications are involved with more challenging situations. You must always proceed to learn more advanced skills gradually. For example, launch in winds only slightly varying in direction for many times before working your way up to gustier conditions. The same applies to steeper slopes—gradually increase the steepness from the familiar training hill.

We begin with the perfect takeoff.

Takeoff Perfection

Before we jump into takeoff complication, we should first determine what is a proper takeoff. We'll do a review of what we learned before. Such a periodic review helps solidifying good technique. If you recall when you first learned launch skills, you were overwhelmed with the new sensations. Eventually you settled down and were well in control. Now that you are more or less relaxed in the process and have a bit of experience under your canopy, you can better relate to our discussion of perfection.

To quote Dixon White:

"Knowing you can perform a good launch allows you to relax and examine the best air to launch into. Solid launching skills, and knowing you have solid launching skills, has a very powerful effect on the whole flight. Pilots who are worried about launching are already engineering their attitudes and experience

Figure 7-1: Launch Elements

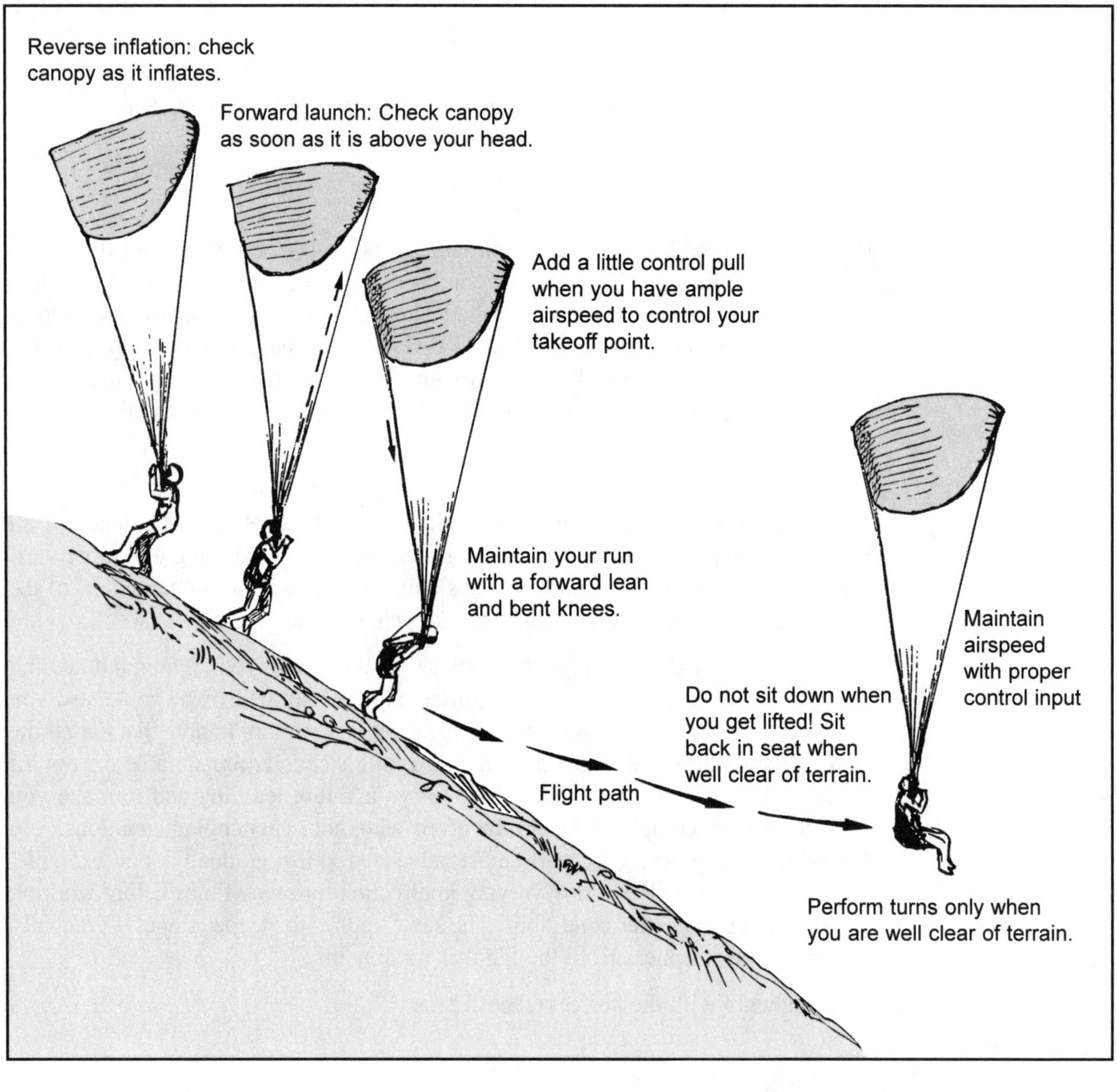

in a negative light. Feeling good about your launch, that you won't hurt yourself or your gear, that you won't embarrass yourself, has a powerful effect on the enjoyment of the whole experience. In the process of becoming great at launches you will develop a much more solid integration with the glider, which means you'll manage roll and pitch better. It can also mean you'll be better attuned to information the glider can tell you about the lifting parcels of air. In the course of practicing launches you'll also fine-tune your understanding of the weather."

✓ Takeoff Review

When you first learn to fly you are often heeding the words of an instructor. Soon, however, you gain an ability to think in the air and start making conscious decisions and controls on our own. Eventually you gain enough experience and body sense (sometimes called kinesthetic awareness or muscle memory) that many of your operations will become automatic. It is our job here to make sure your automatic responses develop correctly. Your instructor and other skilled observers can provide necessary feedback to check your performance until you are self-aware enough to check yourself.

Perfect your basic takeoff skills so you can add slight variations to handle all situations.

The complete takeoff for novice through advanced pilots consists of the same technique. It includes a clean inflation, canopy check, an accelerating run, a stable canopy over your head and a smooth transition into the air as shown in figure 7-1. Here are the important elements:

▲ *Preflight and pre-launch checks* – These checks are standard and should precede every launch. The last thing you check is air traffic and the suitability of the wind.

▲ *Eye focus and canopy check* – With a forward launch look ahead to the horizon or a little below. The run path should be in your peripheral vision. As soon as the canopy is up, look up and back from a proper inflation. This point is important and is a required skill. With a reverse inflation you look at the canopy during inflation, of course, then look ahead as you turn around.

▲ *Control position* – Put the controls in the proper takeoff position as soon as you come off the front risers. For most designs beyond beginner gliders, a bit of control input as the canopy reaches the point above your head helps keep it from getting ahead of you. However, if you are forward launching, your run keeps your body up with your canopy and less control input is needed. The amount of required control depends on the glider, but ear/eye level is normal.

▲ *Maintain the run* – Once you start your run, keep it going. Naturally you may be slowed as the canopy lifts, but keep up your effort. Do not slow for the canopy check or when you start getting lifted. Do not assume the sitting position to clear the ground—keep running. A forward-leaning, knees-bent posture allows you to extend and remain in contact with the ground if you get lifted unexpectedly.

▲ *Control the takeoff point* – With good acceleration of the canopy and your body, the glider has ample flying speed and will take off automatically. However, on flat slopes or in other situations where a fast run is needed, you may use a little less control pull so the canopy doesn't get behind you, then, when at top speed, a smooth application of a slight control pull will increase the wing's angle of attack and lift you into the air.

One of the signs of a good pilot is the smooth transition to flying. If you hit too much control at this point the canopy will retard—you'll lift off—then it will surge and you'll drop. Your goal should be a smooth transition into the air with no speed bump in the flight path as the figure depicts.

▲ *Maintain good airspeed* – As you fly away from the ground you should have plenty of control airspeed. This airspeed should be faster than minimum sink rate at least until you have established your path and gained ample clearance from other gliders, trees and anything you wouldn't want to hit. Ideally your control for liftoff will be the correct position for this safe post-launch airspeed.

Figure 7-2: Inflation Problems

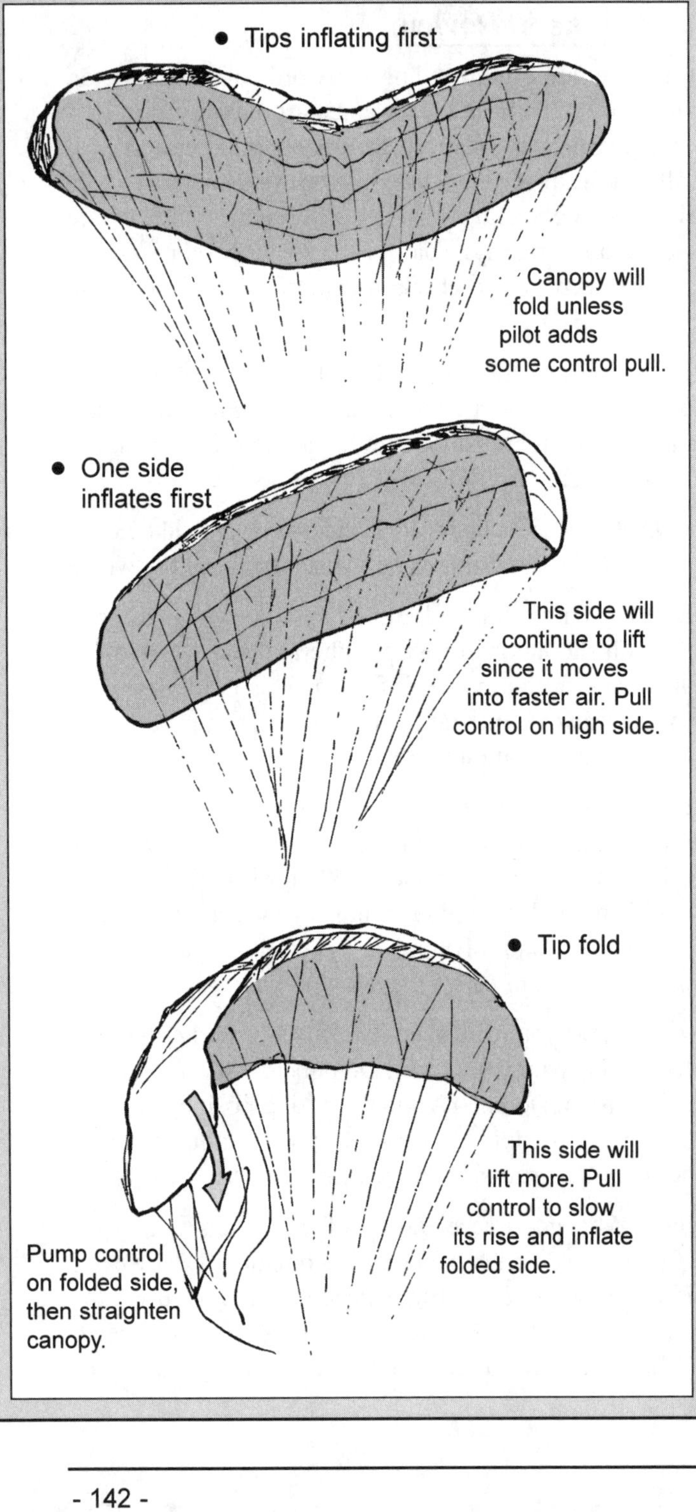

▲ *Sit down when definitely clear* – Only attempt to sit back in your harness when you have achieved at least 20 feet (6 m) or more clearance and have established a "good" airspeed.

▲ *Turn only when definitely clear* – A turn to catch lift parallel to a hill should only be initiated when you have established your steady airspeed and are at least 20 feet (6 m) from all obstructions.

These elements should be checked frequently. As mentioned, when you are developing your skills your awareness may not be total, so a trained observer should watch your takeoffs and give you feedback. As you gain more flying time, don't become complacent. Just because you haven't aborted or blown a launch doesn't mean your skills are impeccable. Remember, an ounce of prevention…

✓ *Solving Launch Problems*

Now let's look at common launch errors and correct them.

▲ *Poor inflations* – We can group poor inflations into two categories: uneven or tip first inflations or a wing fold (see figure 7-2). In the first case there are several causes. If the wind is crossed to the canopy, the tip furthest downwind will inflate first and lift. If the canopy is laid out unevenly or you are not standing on the centerline, one side will inflate first. Occasionally, a twig or weed will catch a line of the canopy and retard one side. The way to stop all these situations from occurring is to

A pilot performs a light wind launch. Note the forward lean and control position aiding liftoff.

take more care in your canopy layout with regard to the wind and ground cover, as well as your launch initiation position relative to the canopy.

Wing tip folds are a different sort of beast although their cause is often the same as the above problems. If the wind is cross to the canopy it can blow the upwind wing in. The problem with tip folds is that they may persist and be hard to remove if you are kiting the glider and the canopy isn't fully pressurized. To remove folds, keep the glider facing directly into the wind and give a few medium (about 1 foot—30 cm) pulls on the control line routing to the folded side. Better layouts and positioning can help prevent inflation tip folds except in gusty conditions.

▲ *Overshooting canopy* – A canopy that seriously overshoots during inflation will usually suffer a frontal fold. In this case the angle of attack gets so low that the airfoil form and pressurization are not maintained. The result tends to be a tucking under of the leading edge and a total loss of wing shape. If you are trying to launch at the time, you may go tumbling down the hill dragging your canopy.

The surging or overshooting canopy problem should be taken care of in your ground handling practice. Inflate the canopy multiple times and get it to stop right above your head with the proper control amount and timing. Usually this problem shows in forward launches in light winds, so this is what you should practice over and over. Lift the canopy, loft it above your head, then drop it to try again. On a steeper slope it is easier to run and put forward momentum into the canopy. The result is that the canopy may accelerate faster and have more tendency to overshoot. The requirement here is for a bit more control line input. Once you have mastered flat ground surge control, practice it repeatedly on the slope and you should solve the problem.

▲ *Popping off* – Another common mistake is pulling on the controls too much once flying speed is achieved (and sometimes before) and popping into the air. Usually the error is realized and the pilot lets off the controls and dives as described earlier. The dangers of this practice are flying too slowly near the ground after the initial pop, then diving close to the ground.

The only way to cure this problem is practice on the training hill. Make your controls smooth and gradual. Diminish your control input if necessary. A smooth arcing path as we saw in figure 7-1 is the goal.

▲ *Sitting down prematurely* – You may see pilots run then lift their feet so their seat barely clears the ground (or doesn't). This practice can eventually lead to a sore butt or worse. By adopting a forward lean in your run you can reduce this tendency. Also imagine you have to run twice as far as you would expect—keep your feet under you.

▲ *Turning around* – Another common early problem is stopping the downhill progress when you turn around in a reverse inflation. In this case the canopy may overshoot you and have a fold. You can cure the problem by practicing the turn around first without the glider. Back up, then keep moving as you turn around. Keep your hands and body in actual takeoff positions and form an image of the proper motions as you practice. Next try the procedure with your glider in a steady wind. Inflate and kite it in the reverse position, then turn forward as you keep it aloft. Continue moving into the wind and reverse yourself again while maintaining the canopy and your movement into the wind. When you can perform this act several times in a row you should have solved your launch problems. Remember, if you are reverse inflating and immediately turning around to run, the canopy may have momentum which you must control with control lines.

With all launch problems (as well as many of the moving skills in paragliding) one of the best exercises is to form an image of the complete task (imaging). Do this by relaxing, closing your eyes, assuming the right position and going through the launch in your mind. Follow the actual timing and imagine the feel of the wind, the control pressures, glider tug, etc. This type of exercise is a powerful learning tool for all new flying skills.

✓ Aborting a Launch

Sometimes a launch must be stopped. When a tip fold is present or a major stick or tangle in the lines appears it is best to stop the launch. Normally the matter is only critical or a problem with a forward launch, for you shouldn't turn around to start running without a perfect canopy in a reverse inflation.

The most automatic way to abort launch is to pull both controls full on as if performing a landing. However, if you have already gained some speed, it's likely that this control will lift you off the ground, especially on a steep hill. The return to earth may not be gentle.

Figure 7-3: The Danger of Rotors

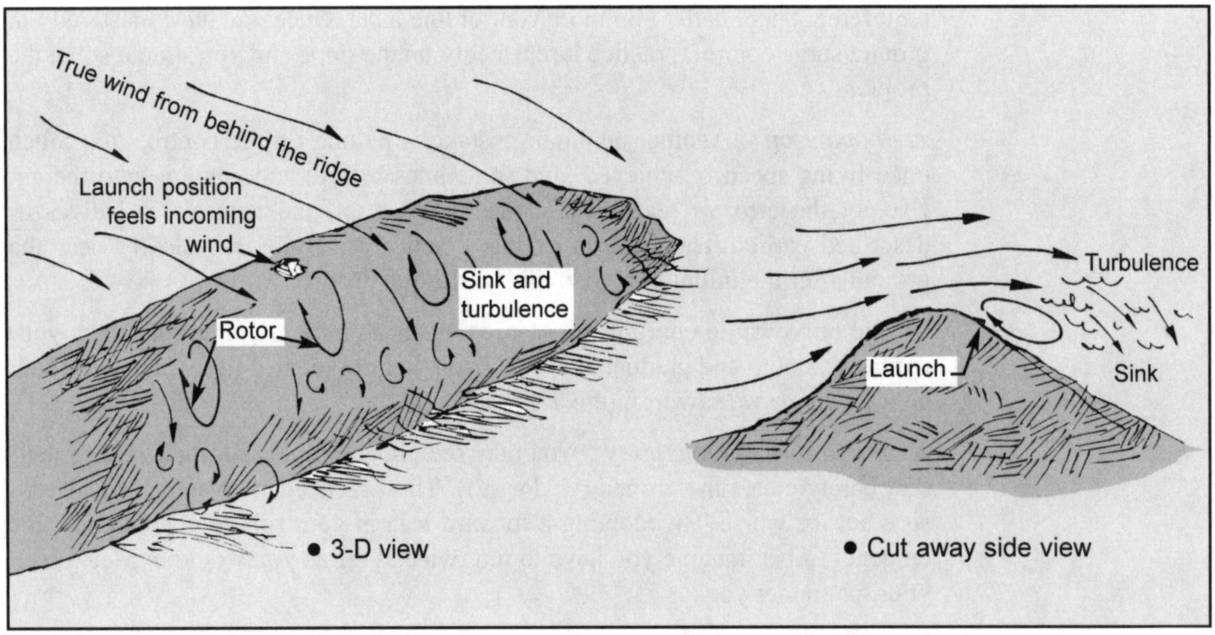

It is far better to pull both rear risers vigorously to abort a launch. The glider will stop which helps you slow down and it will fall behind you to act as a brake. It is very desirable to pull both risers simultaneously so that so that the glider doesn't pull to one side forcefully. In order to act efficiently in the excitement of the situation, it is important to practice launch terminations at the training hill. Do it first on flat ground, then try it on the slope. The knowledge that you can readily stop the canopy if necessary helps you make an easier decision to abort launch and thus makes you a safer pilot.

Safety Tip: *One more advanced gliders the rear risers may be partially attached to the other risers. Practice finding and pulling these risers frequently at the training hill long before you need them in a launch situation.*

Launch Conditions Variations

Now you can apply all your hard work (or hard fun?) perfecting your takeoff skills. It's time to move beyond the well-groomed slopes and predictable gentle winds of the training hill. You'll start stretching your skills and exercising your judgement by learning to launch in various wind conditions.

Safety Check: *When you start flying higher, in wind and with less supervision, it is extremely important that you avoid mistaking a rotor for a nice wind coming in at launch. Figure 7-3 shows how this happens. Often you can launch just fine in this condition, but once you are out from the slope a bit, you may hit severe turbulence and sink. Canopy folds are possible.*

Former female world champion Judy Leden and her husband, Chris Dawes, were in Argentina for a film project. They planned to fly from a mountain ridge several thousand feet high (900 m), and set out climbing in the morning. The wind appeared to be gently wafting up the slope and they anticipated a great thermal flight. Judy climbed nearly to the top and figured she was high enough. Chris climbed the rest of the way and began hollering at Judy just as she took off. Judy didn't get far before everything went to pieces. Her canopy thrashed and became uncontrollable. She was a puppet along for the ride. Three times in the "flight" her canopy was a ball below her. Each time she pulled it out, but the severe turbulence didn't stop until she was nearly at the ground—a feature she kissed, incidentally. What was Chris yelling? "Judy, don't launch, it's blowing 45 (mph) out of the back here!"

As with all new skills, add launch variations gradually. Start with slightly crossing winds or slight direction changes or slightly steeper slopes.

To avoid Judy's fate, look at cloud drift, birds drifting in circles or the general wind far from the mountain to get an idea of the true wind. Weather reports are also a good source of information as to the true wind. Unfortunately none of these items were available to Judy and Chris in the remote part of Argentina. It's always advisable to check the wind at the top of the mountain before launching. Also, using radios to constantly keep all members in a group in touch is a good safety measure.

Your Goal: *To eventually be able to launch safely in all conditions acceptable in the Recommended Operating Limitations for your skill level.*

A pilot at Millau, France turns to face forward after reverse inflating.

✓ Calm Winds, Tail Winds, High Altitude

What do all these conditions have in common? They all present inflation difficulties. Let's look at each in turn.

▲ *Calm winds* – By now you probably have experienced calm wind launches. The main challenge in this case is to keep moving *and* to check the canopy as you run. On a steep slope with a forward launch in light or zero wind, you do not have a lot of margin for error. You should be 100% confident in your inflations and have 100% inflation success in calm winds before you attempt such a feat. We consider a steep slope to be over 30º. A very careful layout of the canopy, including a reasonable horseshoe and symmetrical tip positions is mandatory in this situation.

▲ *Tail winds* – You may hear talk of experienced (in skill, not necessarily in judgement) pilots performing tail wind takeoffs. Some of these stories may be exaggerated and some may be of the "barely survived" type. At any rate we highly discourage attempting a tail wind launch, even in a slight trickle. One problem may be that a slight trickle at ground level may be stronger 20 feet (6 m) up where your canopy will be. You may feel a steep slope will help you get off the hill, but as we mentioned above, you have much less ability to control an aborted launch on a steep slope. Don't be tempted in tail winds.

> *At a meet in Kössen, Austria in 1990, we witnessed an incident that fortunately ended up laughable since no injuries resulted. Many pilots were laid out ready to fly, but the wind was a downhill trickle and the air was humid. Once in a while a pilot would launch and barely clear the road that cut across the slope below takeoff. One unfortunate pilot ran and ran and just managed to get off only to impact on the road and roll in a ball down the other side. He took a few minutes to gather his wits and scattered equipment then started back up the hill. Just as he crested the road, he encountered a female pilot who was following his flight path with a little more success, but now he was in her way. With no time to alter course, she headed right at him with her feet held straight out to clear the road. Not wanting to plant a pair of boots on his face, she spread her legs and wedged his head. Our hapless pilot was carried backwards as the pair went rolling together in a dual ball this time. It was never reported whether or not they became friends. There are better ways to meet people than launching in tail winds.*

Safety Tip: *In very light or calm winds, take great care that the wind doesn't trickle down the hill at times. This warning especially applies at the end of the day.*

▲ *High Altitudes* – The higher you go, the less dense the air so you have to move your canopy faster to achieve lift-off. In general, this matter isn't a problem except at the extremes. If calm or occasional tail winds are present, high altitude launches require careful judgement of the conditions and perfect launch performance.

The advice we give for calm winds applies doubly in the high altitude situation.

The launch at Telluride, Colorado is above 12,000 feet (3,660) MSL. The soaring and scenery are spectacular, but the altitude combined with variable winds due to thermals lifting from different quarters makes this launch tricky at times. We have witnessed wings suffering asymmetrical or frontal folds during takeoff when less wary pilots do not time the cycles to get a nice straight-in wind. Judgement and skill are required at sites like this.

✓ Crosswind Technique

Crossing winds present a special problem for paragliders, for until the pilot is in the air, the glider must be pointed into the wind. Here's how we proceed.

▲ *Crosswind launches* – When the wind is cross on launch, lay the canopy out so it is facing the wind as shown in figure 7-4. Position yourself on the centerline as shown. Inflate as normal, turn around (if you are reversed) and begin your run into the wind then curve your path to aim straight down the hill as shown. During the run you must add a bit extra control on the downslope wing (right in our drawing) to turn the glider to follow your path. If you do not, the glider will wander off to the side as shown on the right figure labeled "Wrong!".

This procedure is not difficult but does require a bit of practice. Learn crosswind launch controls on flat ground or a very shallow slope. Start into the wind, inflate, run and turn the glider to follow a curving path within three steps. Repeat until you have perfect timing, perfect control.

Figure 7-4: Crosswind Launches on Shallow Slopes

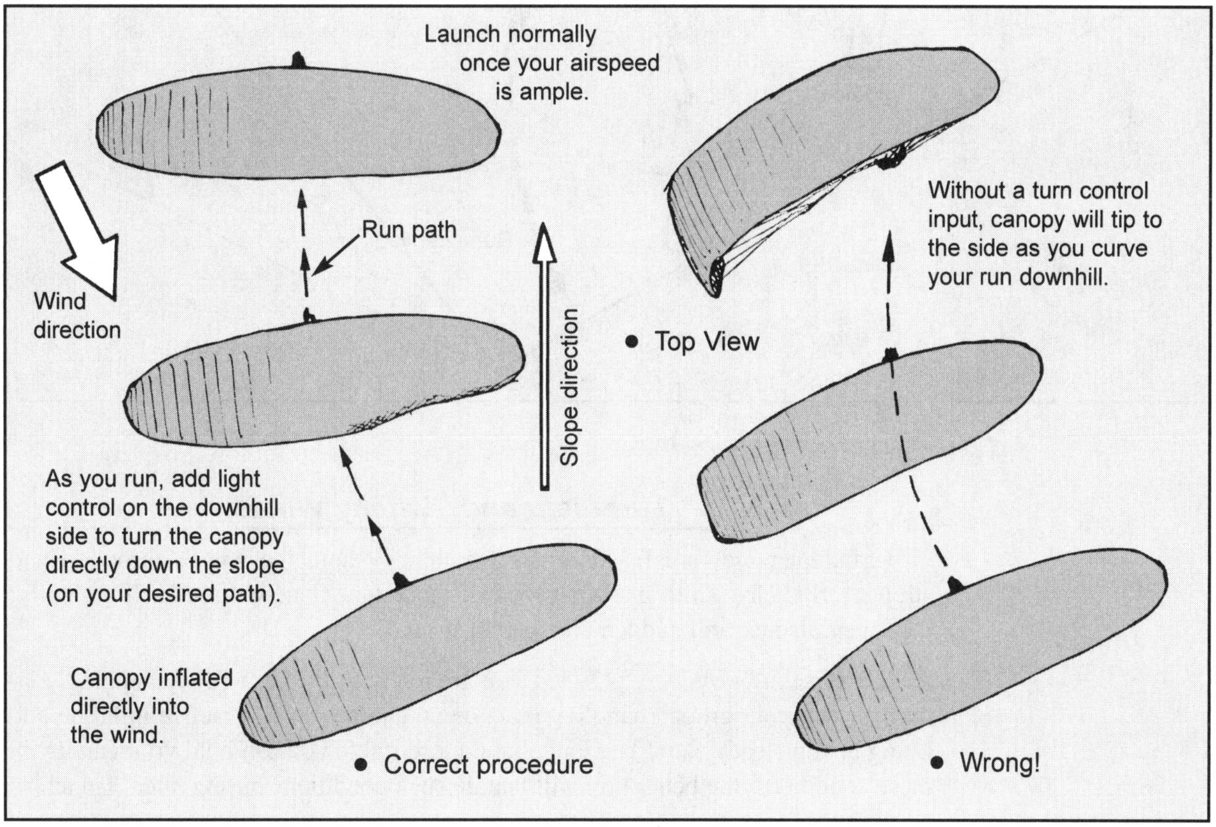

On a steep slope a crosswind launch is different. The canopy is tilted so when it inflates it is already in a bank—it wants to turn as figure 7-5 shows. Its turn direction is down the hill so you may not need any control input to make the glider follow your curving path. Thus, the picture that arises is a variation of technique according to the steepness of the hill. The guiding principle in all crosswind launches is to move under the canopy and control the canopy to keep it centered (side-to-side) over your head. Note that the steeper a launch hill, the more a crossing wind gets deflected sideways, so steep hills make a crosswind situation worse.

Pro Tip: *Good canopy awareness and control is required for crosswind launches. Move your body under the center of the canopy and apply controls to keep it heading exactly where you are going.*

Figure 7-5: Steep Slope Crosswind Technique

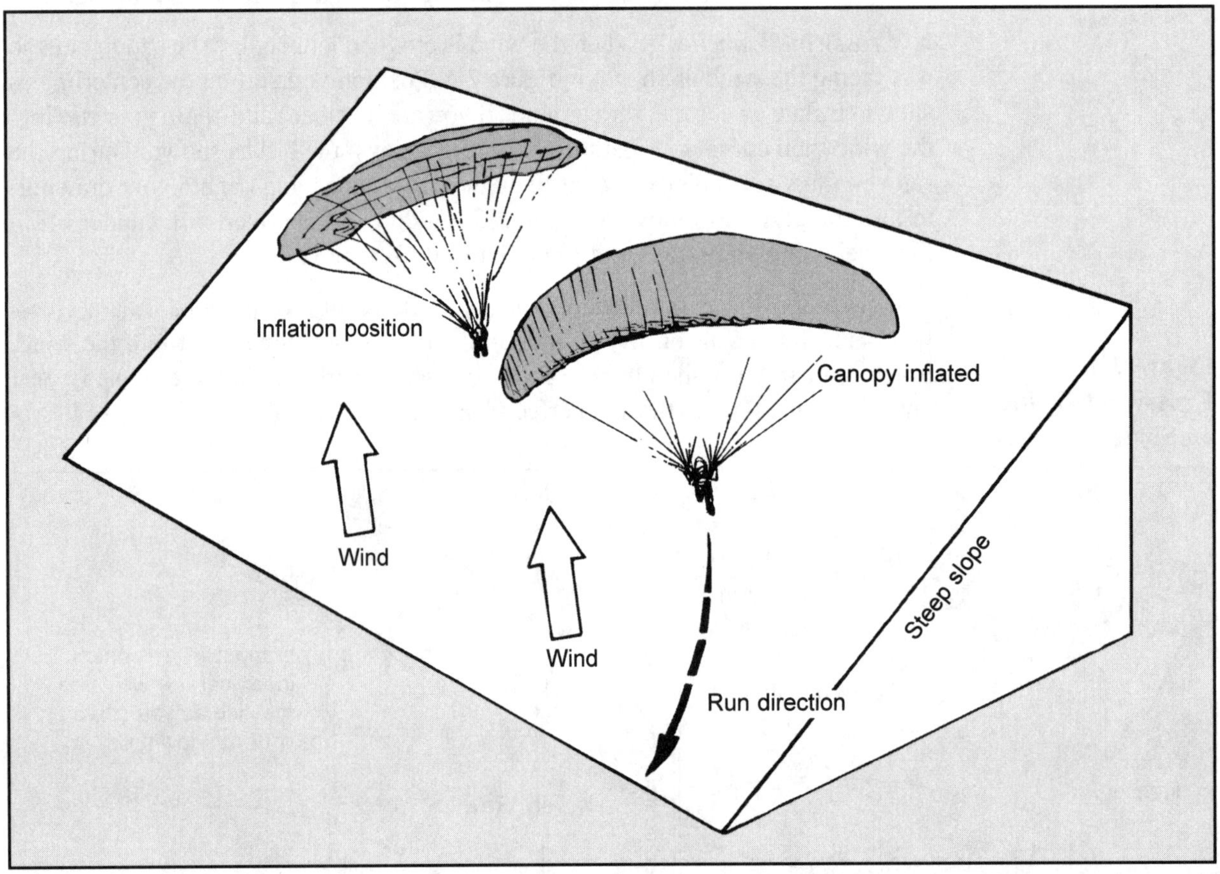

✓ Turbulent and Switchy Winds

Turbulent winds are felt as gusts which vary both the wind's direction and strength. Switchy winds are what we call those that change direction frequently; they may also exhibit sudden changes in strength.

▲ *Launches in Turbulence* – When you are ready to fly in more challenging conditions, step your ground handling practice up another level. Practice inflating and kiting in non-steady flow. The longer and more stable you can hold your canopy in these conditions, the better you will handle such conditions during a real launch.

Remember: The maximum safe wind strengths are lower the more turbulent it is.

The technique for handling turbulent launch winds is to inflate in the reverse position, stabilize the canopy then turn and go without hesitation when the canopy is relatively steady. Such a feat usually demands constant input on the controls, and possibly the risers. Remember all that ground handling riser practice? Here's where it pays off. You may find you have to move around a bit to stay under the canopy. At times, you may lose control and have to re-inflate. That's par for the course, especially when you are inexperienced. Tip folds are fairly common in gusty conditions. Work them out as described earlier in the Solving Launch Problems section.

One of the most important points when launching in turbulent conditions is to time the cycles before you attempt to launch. Usually turbulent cycles are fairly regular, especially if they are caused by thermals. Often you can go in a lull between gusts—if the duration is long enough. Of course, if turbulent changes are so frequent you can never stabilize the canopy for more than a couple seconds, perhaps conditions are too unreliable for a safe launch. We have seen several gliders draped in the trees because the pilot didn't heed such a guideline.

Crossing winds at a launch which is a slot in the trees can present some real turbulence problems. As figure 7-6 shows, the trees turbulate the air in the launch slot. Be aware of the true wind direction when you are launching in these conditions, for wind is often directed up the slope by the slot. Look down to see which way the trees are bending and check to see which side of the slot is getting more wind in order to determine the wind direction. If winds are too cross the turbulence may be more than you can control. In general, in a crossing wind your first turn should be into the wind.

Figure 7-6: Launching in a Slot

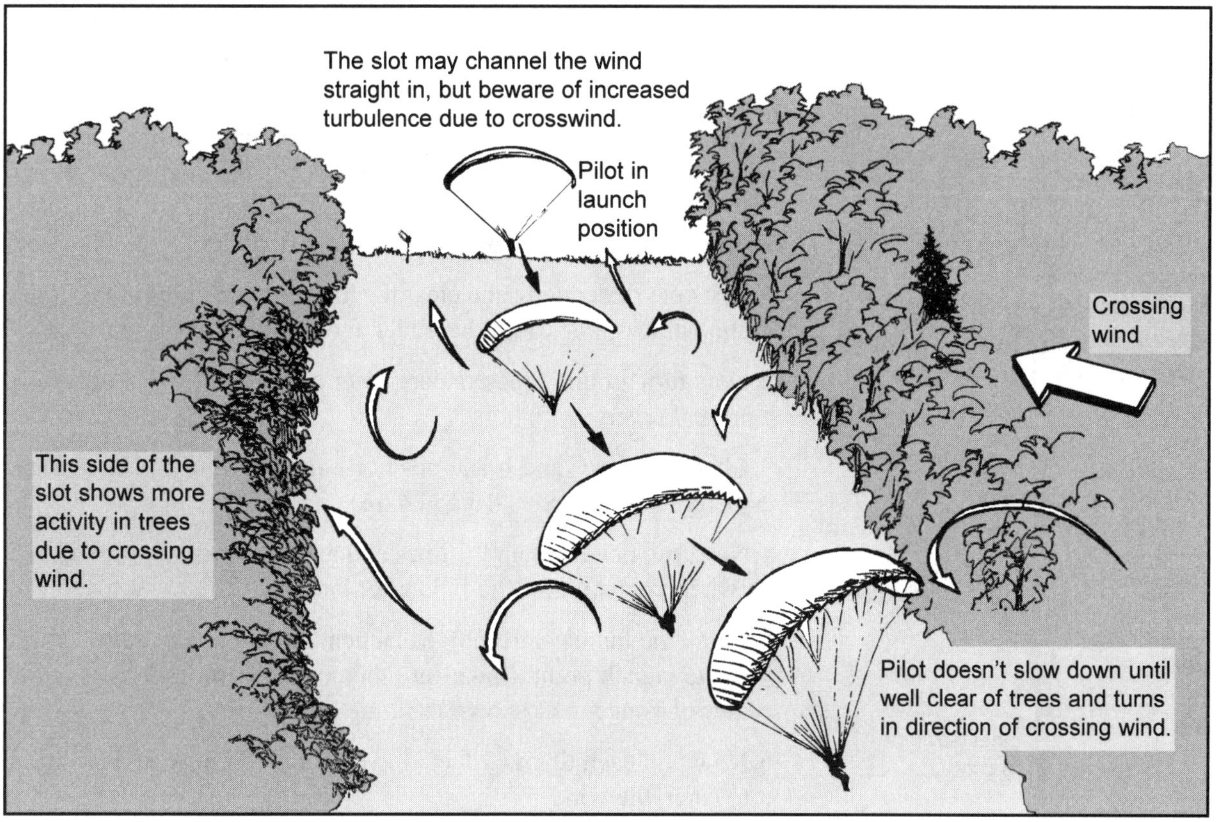

▲ *Switchy winds* – When the wind direction varies on launch, you have two options: wait for an in cycle (if possible) or treat it like a crosswind and use the appropriate procedure. The option you take depends on your patience (and the patience of those waiting behind you) and the frequency of the cycles. For example, if the wind blows in occasionally but only for a few seconds, it's better to take the crosswind option. Winds are often switchy in thermaling conditions, so learning to time cycles, launch in crossing winds and launch efficiently (without undue delays) will be a great asset when you progress to soaring.

Safety Tip: *Do not let other pilots push you to launch before you are ready.*

Everyone wants to get off the hill in reasonable conditions, so if pilots are anxious behind you and you don't feel you are ready, move aside. With novice skills you need to be relaxed, not hurried on launch. It's much easier to move than it is to pick yourself out of the bushes after a hurried and unsuccessful launch.

✓ Stronger Winds

Probably you have encountered slightly stronger winds in your training experience. By the time you acquire all your novice level skills, you should be comfortable in winds that average about 12 mph (20 km/h) with brief maximums up to 15 mph (24 km/h). These limits are set because more wind brings more complications as we shall see in Chapter 9.

The complications presented on launch consist of possibly more turbulence, faster inflations, more inflation force and possibly getting dragged. The way to solve these problems is to always use a reverse inflation in windy conditions with special control techniques. The recommended reverse inflation is a new one which we describe below, but first let's see how to prepare for such a launch.

Figure 7-7: Laying the Canopy in a Rosette or Bunch for High Winds

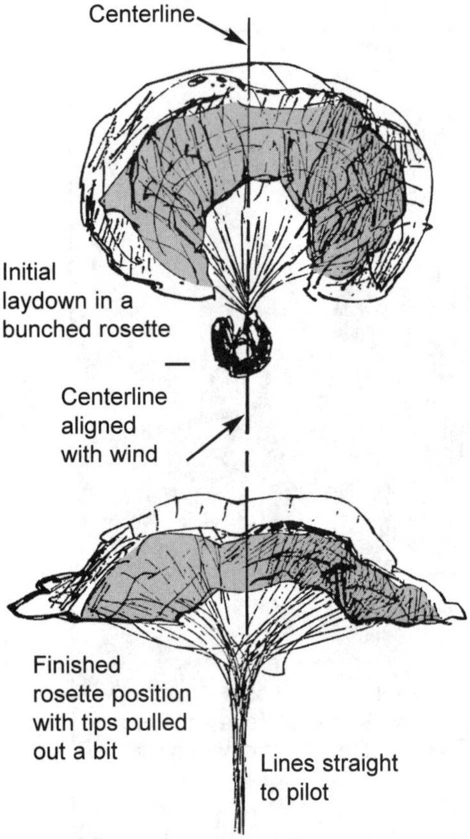

When conditions are windy right at the launch point, it may be impossible to lay out the canopy without it blowing away unless many assistants are helping. This situation is unreliable, so an alternate method is the best policy. Here's what you do:

1. Lay out your canopy and clear the lines in a sheltered area. Climb in the harness, hook into the glider and preflight as normal.
2. Now turn around (duck under a riser group) as if you were doing a normal reverse inflation.
3. Check the riser and brake positions and take the brakes in hand as before (see figures 4-15 and 4-16).
4. Now, run one hand up the lines and gather the canopy just as you do when you walk off the field after landing.
5. Carry the bundle carefully to launch when it's your turn and set it at the launch point. Make sure the center of the canopy is at the center of your run as shown in figure 7-7.
6. Now pull each tip out 3 feet (1 m)—not much for you don't want it to catch the wind.

7. Finally, carefully back up while you feed out the lines from your hand holding the loops.

When the lines are fully extended you are ready to launch. You should use the technique described below.

✓ High Wind Reverse Inflation

This method is the third reverse method we alluded to in Chapter 4. Why a new method? The reason is, higher winds can tend to drag the canopy (and your bruisable body) downwind. Going for a drag is not the most enjoyable paragliding exploit. You learned before that pulling fully on the control lines can drop or disable a canopy, but in higher winds it may pick you off your feet and carry you downwind before it can be disabled, then it may stay inflated near the ground as we go for a bouncing tour of the countryside. You also learned that a good pull on the rear risers disables the canopy quicker (and more definitely) than the control lines. So our refined reverse inflation is controlled with the front risers and rear (or C) risers exclusively.

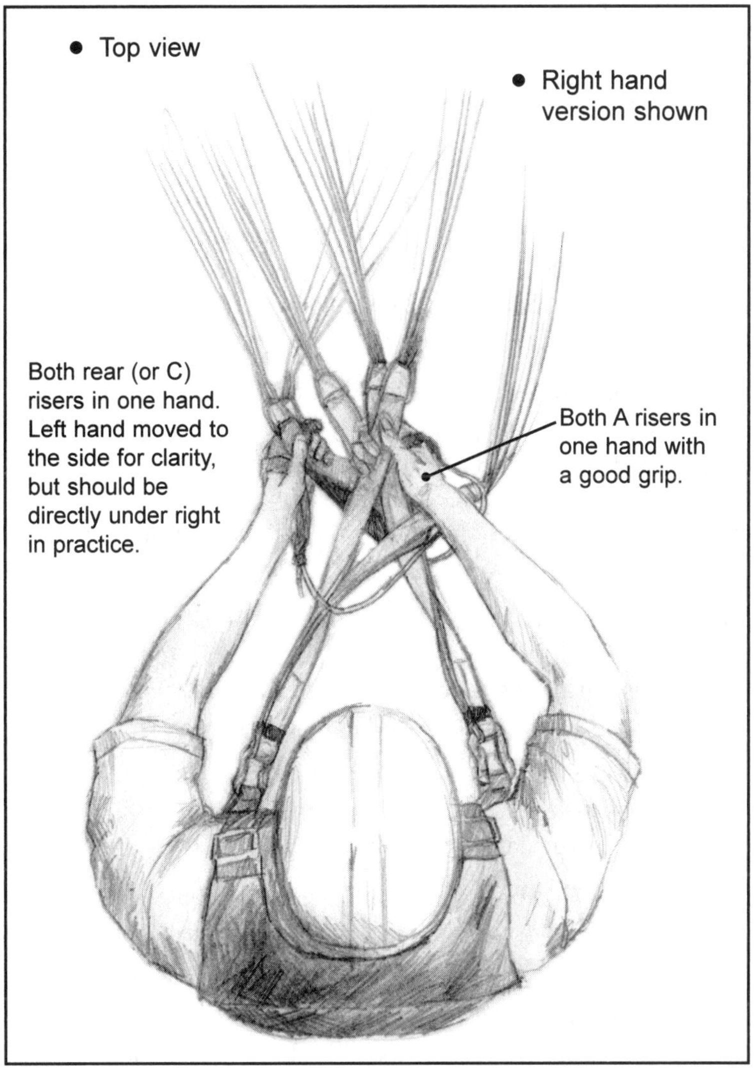

Figure 7-8: Higher Wind riser Arrangement

Here's the technique: With the control toggles in the proper hand (and routing), grasp *both* A risers with your preferred hand (right hand for right handers usually), and *both* rear (or C) risers with the other hand. The A riser hand is on top of all risers and the rear riser hand under all risers as shown in figure 7-8. To make sure you don't twist risers, move the A riser and the in-between risers of the lower group out of the way as shown in the figure so you can grasp that rear riser with the rear riser of the upper group. Holding both risers where they naturally cross is best.

Note: Using the C risers rather than the rear risers may be a more effective way to control the inflation on some gliders. Some pilots use a variation whereby they hold both A risers in one hand and *both* control toggles in the other. There are two drawbacks to this method: First, the control lines are not as effective in bringing down the canopy and may result in you getting lifted. Second, you have to transfer one of the controls to your free hand (the one pulling on the A risers) once the canopy is up. We do not recommend this method.

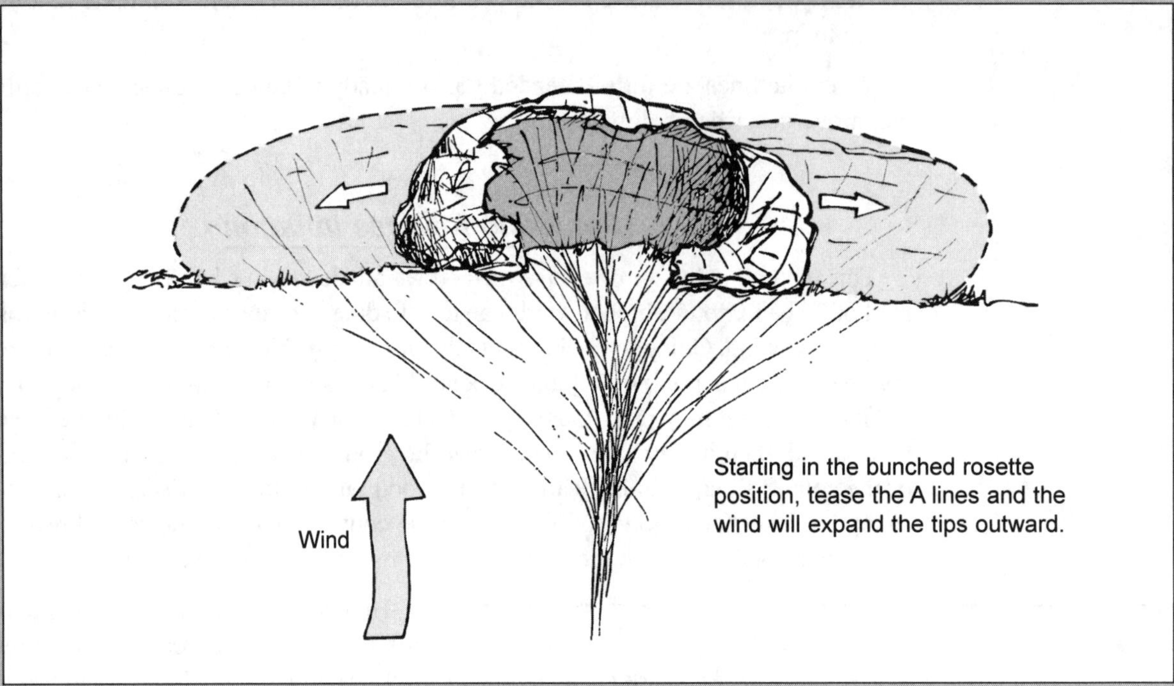

Figure 7-9: High Wind Inflation

Next, tease the canopy by jostling the A risers upward slightly. If you do this carefully, you'll build a short wall that will expand and automatically spread the wings out as shown in figure 7-9. Continue this tease until the wing is about ¾ expanded (less in higher winds). Don't go all the way or it may inflate itself. Now do your traffic check and wind check on launch.

One of the benefits of a rosette layout is the lines aren't spread as much so you have less chance of catching debris.

The launch is fairly normal: lift the A risers with your upper arm and watch the canopy fill. Correct for any wing turning by moving the pairs of risers to the left or right. For example, if the wing is moving to your left (right wing down), pull the A risers right and the rear (or C) ones left as shown in figure 7-10. The effect is to pull more on the right side A lines and less on the left to bring the right side up. This method of control is very powerful.

In general, your inflation should be quick in order to minimize the time the wing is in the low-to-the-ground, high-drag position. If you feel you are getting tugged too much, take steps toward the canopy as needed. If things get out of hand you can always quickly drop the canopy by pulling the rear or C risers.

The higher the wind, the more important it is to have a perfect canopy inflation before turning to run.

Once the canopy is nearly aloft, let go of the risers, use your controls, turn around and launch. It is not wise to spend a lot of time kiting a glider in high winds on the slope of launch, especially if trees or other obstructions are around.

This entire launch procedure is a bit more complicated than what you have previously experienced, but not overly so. As usual, you should perfect it by practicing first on flat ground. Ground handling is our tried and true training program.

Skill Check: *Your windy condition launch skills are well-developed when you can handle stronger winds, turbulence and crosswinds (within your operating limits, of course). Launches should be controlled, not barely managed affairs.*

✓ Turn Around Emergency Procedure

It is possible to have the misfortune (or misjudgment) to turn the wrong way when reverse inflating. It happens occasionally. If you find yourself in this situation, don't panic. Fly the glider away from the hill to get plenty of clearance. The glider will be plenty stable because the risers are so close together. Once you are clear, keep the controls in your hands and grasp the risers to force them apart. This action should start you turning and straighten you out. You will have to do a complete 360 with your body in this case.

It has also happened that a pilot gets pulled off while facing backwards and flies away. If this happens, don't panic but control the canopy as before until you are well clear and again, spread the risers to help rotate your body around. Keep the glider pointed into the wind and it will fly away from the hill. Then go back to the training hill and repractice the turnaround until it's ingrained in your brain.

✓ Windy Assisted Launches

When you first started launching in a bit of wind, perhaps you performed a forward launch with your instructor pulling on your harness straps. This practice helped prevent you being pulled backwards. However, soon you learned to reverse inflate so you no longer needed a forward harness pull.

Sometimes pilots will have assistants hold back on their harness in windy conditions. We recommend against this practice for several reasons. First, this assis-

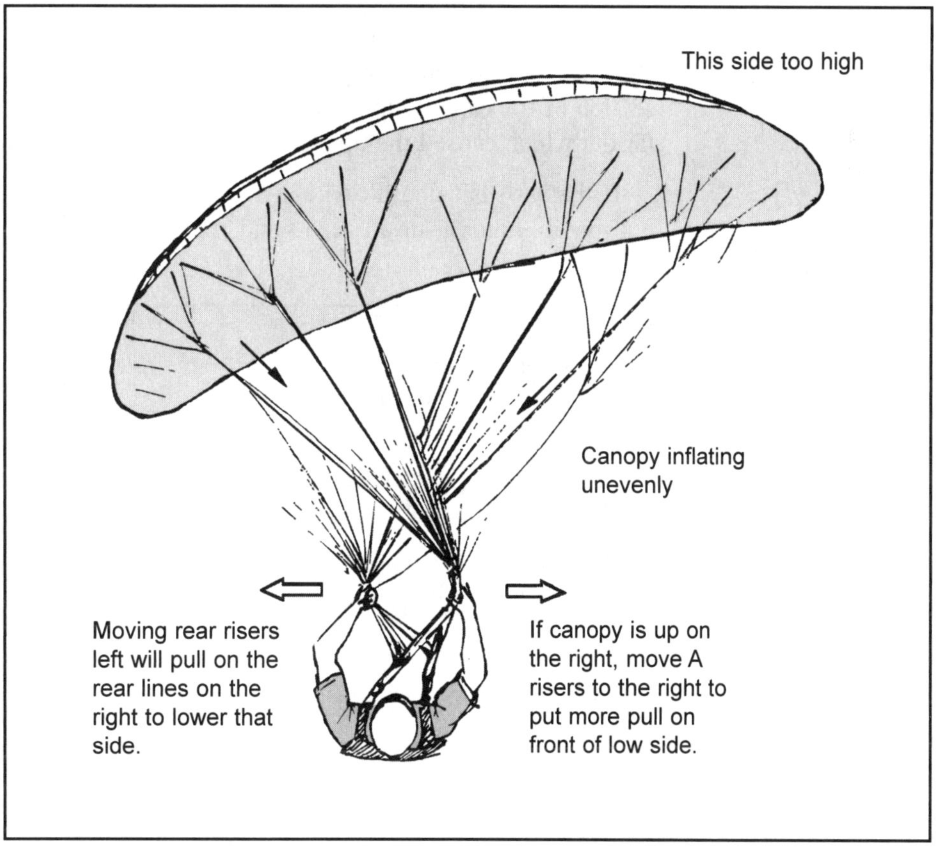

Figure 7-10: Uneven Inflation Control

tance may allow you to inflate in winds that are far too strong, resulting in getting blown back or encountering strong turbulence. Secondly, the assistant may complicate matters and increases the danger if he or she grabs the wrong thing such as your parachute handle. If the assistant doesn't turn with you, your turnaround may be hindered or your path blocked. Also the assistant may let go at the wrong time or even hang on, which can have serious consequences in high winds.

If you have learned proper reverse inflation techniques with good canopy control, you should be able to inflate in any wind strength that is safe for flying. Remember, wind complicates all flying. The complications go up with the strength of the wind, so set your own safe limits and launch in only what you can handle unassisted.

✓ Light Wind Assistance

In light or calm winds, when you don't have a long runway or perhaps when the canopy is laid out in a depression below your launch point, it is preferable to have assistance. As figure 7-11 shows, lifting the leading edge can help it catch the air and inflate the canopy quicker.

It is important for the assistant to do the right thing. He (or she) should grasp the canopy with the thumb or forefinger on the upper surface with a little pinch of cloth as shown in figure 7-12. He should raise his arms as the canopy rises and let it pull out of his hands. If the assistant holds onto the leading edge, the canopy may be retarded. An alternative holding method is with your palms up, fingers curled as shown in the figure.

Only a single assistant should be used carefully centered on the canopy as shown. Two assistants simply complicate matters. The idea is to raise the center section of the canopy up so that area fills first. We have seen assistants hold too far out which caused the tips to inflate and a fold to form in the center.

If the assistant doesn't actually lift the canopy when the pilot starts moving, but merely holds it up in zero wind, sometimes the canopy just moves forward

Figure 7-11: Canopy Lifting Assistance

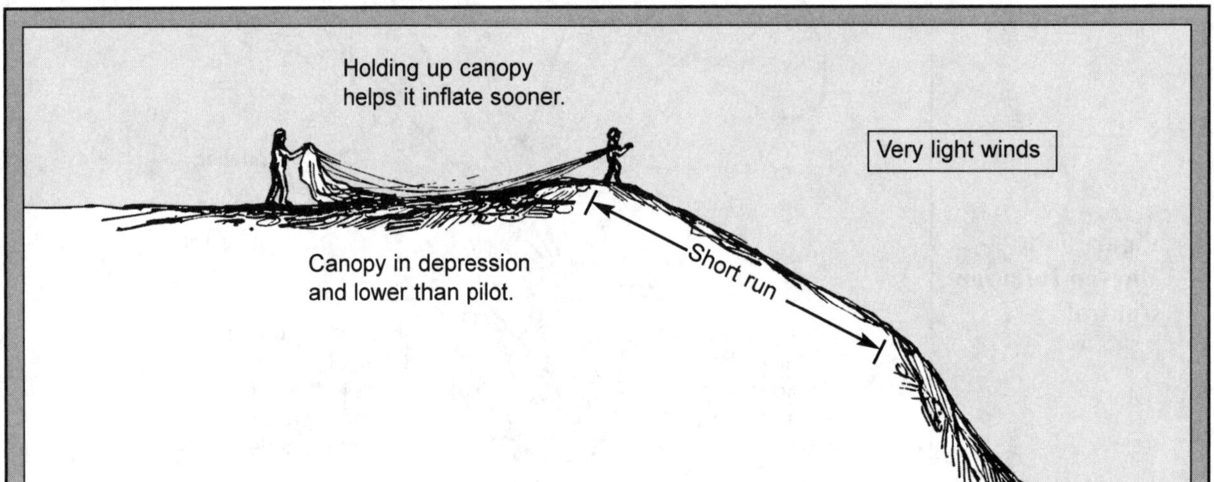

without upward momentum and fails to inflate. Proper assistance is an upward motion coordinated with the pilot's initial motion pulling the canopy.

LAUNCH SLOPE VARIATIONS

Here we turn our attention to launch complications which result from different slope shapes and ground cover of launch areas. Certainly many launches are groomed grass or carpet covered and ideal. However, as your flying experience expands, you'll encounter many sites with merely the terrain Mother Nature provided.

Your Goal: *To learn to assess the suitability of a launch for your skill level and to launch safely from a variety of launch shapes.*

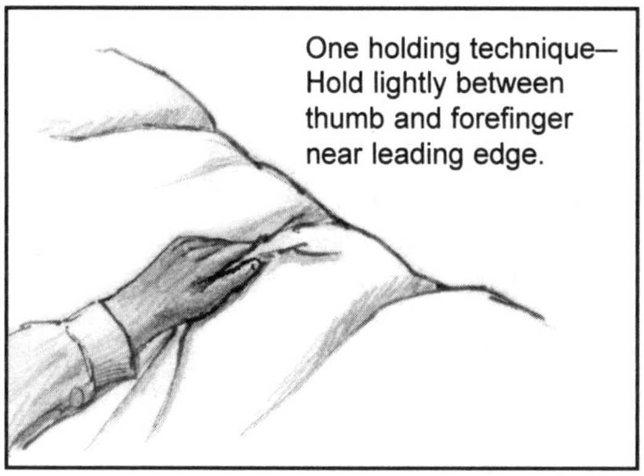

Figure 7-12: Assisting with Canopy Inflation

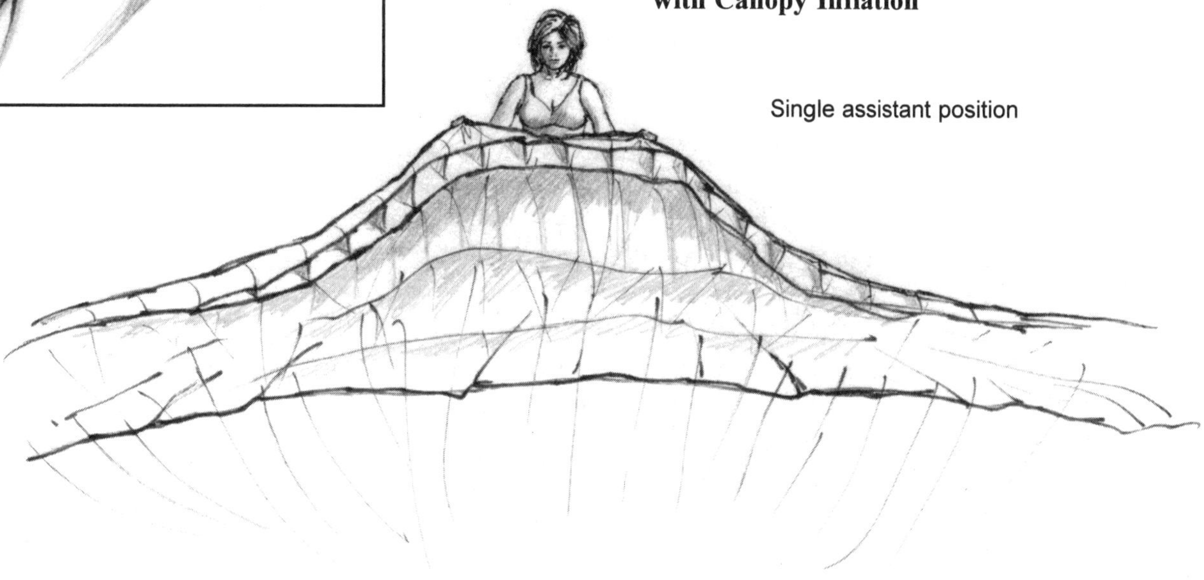

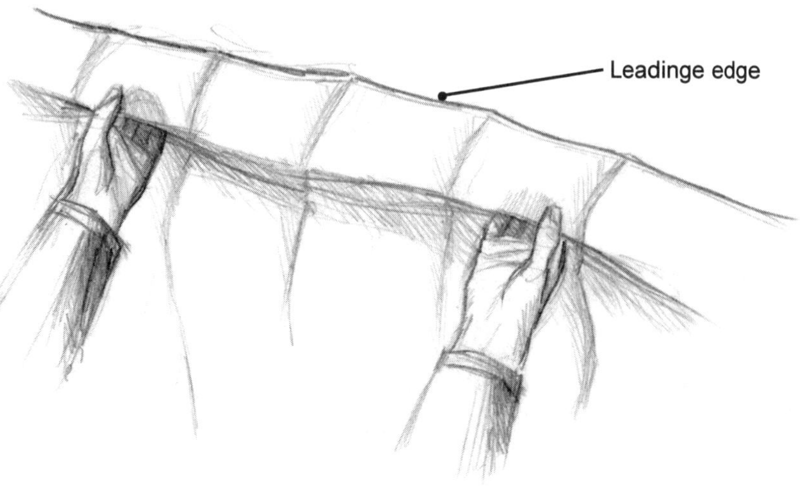

✓ Flat Slope Launches

Flat slopes (20° or less to the horizon) are generally not too much of a problem, especially if there is a little wind present. Of course, if the slope is flatter than your glider's ability to outglide it, there must be a steepening or drop off eventually or you won't be able to get airborne. Perhaps it's better to start at the steeper portion.

Safety Tip: *At this point in your flying experience you should not be a site pioneer. Fly sites that other paraglider pilots have used on a regular basis and have proven safe. Use the same launch point that others have used successfully, for multiple flights.*

Keep your feet under you as you lift off to avoid bouncing your butt if you sink a bit.

On a flat slope, inflate as you normally do and produce a good run. You should apply *less* control input than in a steeper launch because it's important to keep the canopy moving and you must avoid entering the air too soon. Hitting the controls and popping off may have you returning to earth as you reduce your input to gain airspeed. The secret to a good flat slope launch is to apply just enough control to keep the canopy right above your head and keep running until the terrain falls away or you outglide the slope. *Don't sit down* until you are well clear.

✓ Steep Slope Launches

Steep slopes are very easy to launch from, but are very unforgiving of a mistake or malfunction. When forward launching it is difficult to check the canopy for proper inflation because launch happens quickly. In addition, if you see a problem, it is difficult to abort without tumbling down the hill. For that reason, we highly recommend only using reverse inflation if a slope is steeper than 35 or 40 degrees.

A pilot gets assistance on launch.

Safety Tip: *Use reverse inflation techniques on very steep slopes in order to check the canopy.*

The above policy generally means you need some wind to launch from a steep slope. Fortunately, steep slopes tend to be conducive to producing upslope breezes. In addition, it is possible to perform a reverse inflation on a steep slope in almost zero wind. The steep slope helps you back up to inflate the canopy and you don't have to move far because you are dropping down more which accelerates the canopy forward. The tricky part is checking the canopy and quickly turning around to run in very short order. In more wind it is easy. This technique is for more experienced pilots and should be learned at the training hill by perfecting the inflate-inspect-rotate-run with repeated practice.

Skill Check: *When launching on a slope more shallow or steep than normal, your launch technique is most likely fine if you can complete the launch without varying the attitude or canopy position until liftoff.*

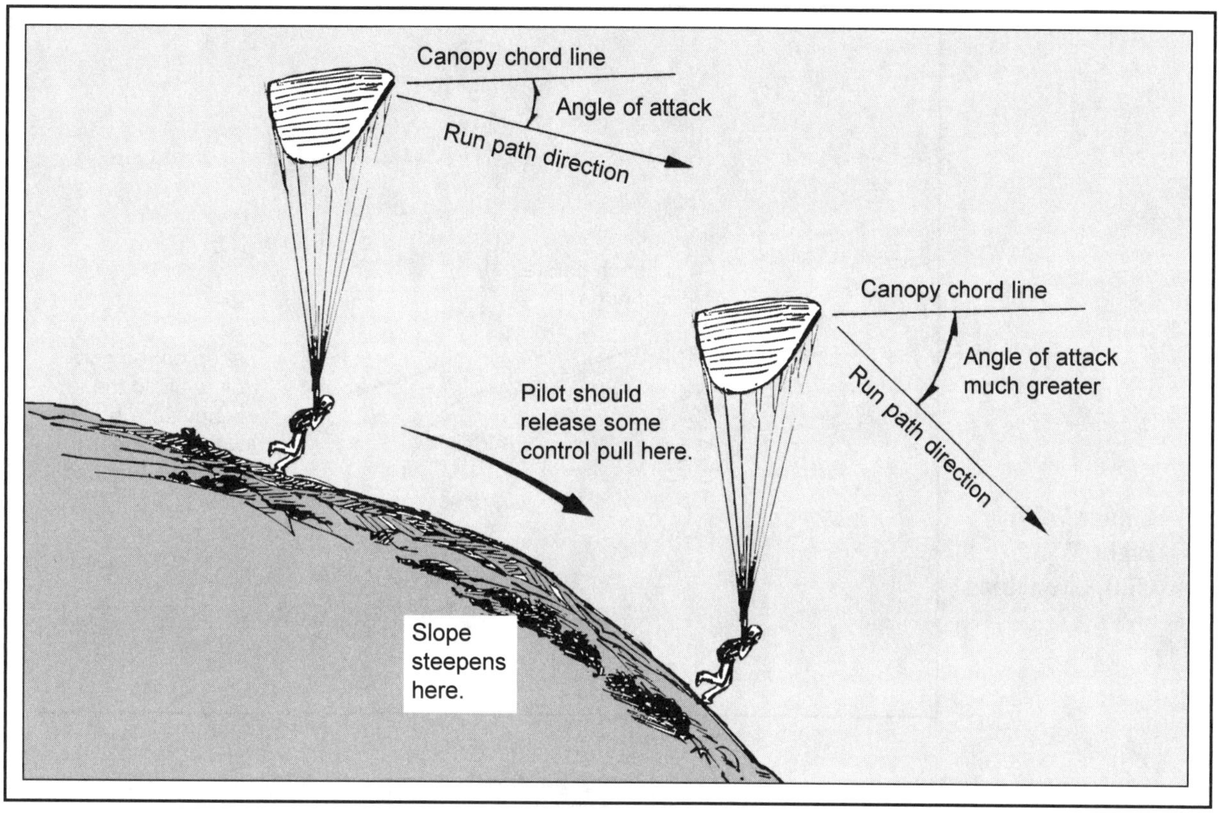

Figure 7-13: Launching from a Steepening Slope

✓ Launches from Flat to Steeper

Most flat slopes eventually drop off. There is no problem as long as we have enough room to get the glider flying and lifting our weight first. But what about a slope that allows only a few steps before it drops? What about a slope that gets flatter in the middle of the launch run?

To handle a slope that gets steeper midway through your launch run, be aware of the problem that as you start running down the steeper part, you increase the wing's angle of attack as shown in figure 7-13. So the proper technique is to monitor the control pressure. If the angle of attack increases where you transition to the steeper run, the pressure will build, so you should raise the controls to maintain the normal takeoff pressure. If you don't do this you may enter the air with too little airspeed. Once again, the key to performing this launch variation perfectly is to feel the controls and keep the canopy directly above your head.

Canopy control means always maintaining the proper angle of attack and positioning by maintaining the proper control pressure.

Many cliff sites (such as famous St Hilaire du Touvet in France) actually have slopes that lead to the cliff. They should be treated like slope launches with limited running room. If you ever run out of room on a cliff run and can't stop, let up on the controls to trim speed for the canopy will have to dive to gain ample airspeed to fly away. These situations are depicted in figure 7-14. Note that a cliff launch is a more demanding launch—it could prove disastrous if you get it wrong—and not for novice pilots without guidance.

Ramps have often been constructed at many hang gliding sites. Sometimes they are short and sometimes they drop off after a few steps. Use your short run technique (inflation assistance) in this case. Remember, some ramps are not suitable for paragliders in calm wind, so do not overexert your guardian angel.

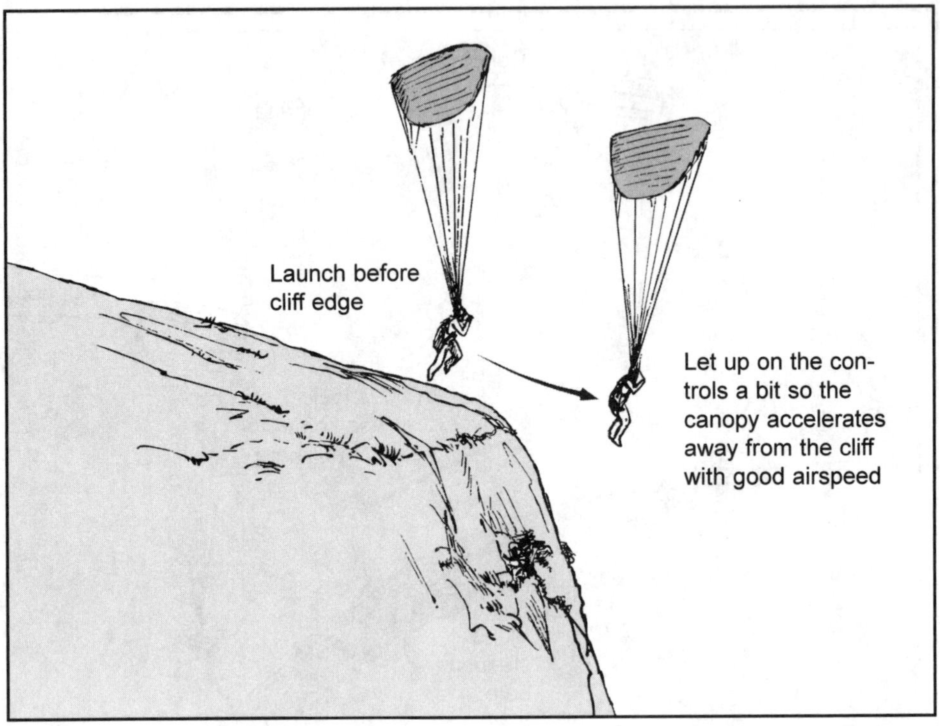

Figure 7-14: Light Wind Cliff Launching

✓ Launches from Steep to Flatter

A more serious situation with a varying launch slope—and one you are likely to encounter at some point—is when the slope gets shallower during your run. The problem occurs when you transition to the flatter slope and off-load the canopy. Without the lines pulling it, the canopy will lose inflation and may drop behind you or accelerate ahead if you can't keep up.

> *I experienced this problem at Ellenville, NY in 1992. I was using a small borrowed canopy. It took me so long to get off in the morning still air that I couldn't clear the slope lower down where it flattened out. I continued running when I hit the ground, but the canopy collapsed and I tumbled through the bushes.*

In the situation of a flattening run you must apply some control pull at the transition point as drawn in figure 7-15. Note that this is the opposite control necessary for the steepening launch situation. Again, this technique is not for very inexperienced pilots. However, many of us encounter this situation at the training hill when we start at the very bottom and transition to the flat. That is the ideal situation to practice the flattening slope technique. As for so many of the skills in paragliding, the safest approach is to practice new skills in a controlled manner first.

Be aware that some slopes change so much that attempting a launch is unsafe. Can you start lower down on the flatter slope? Even if you fly successfully from such a varying slope in wind, do not assume you can do so in a calm.

✓ Windy Cliff Launches

Launches on most windy cliffs will be similar to the situation shown before in figure 7-14. In most case you can inflate and launch well before the cliff edge. However, if the running room does not allow it, you will have to inflate closer to

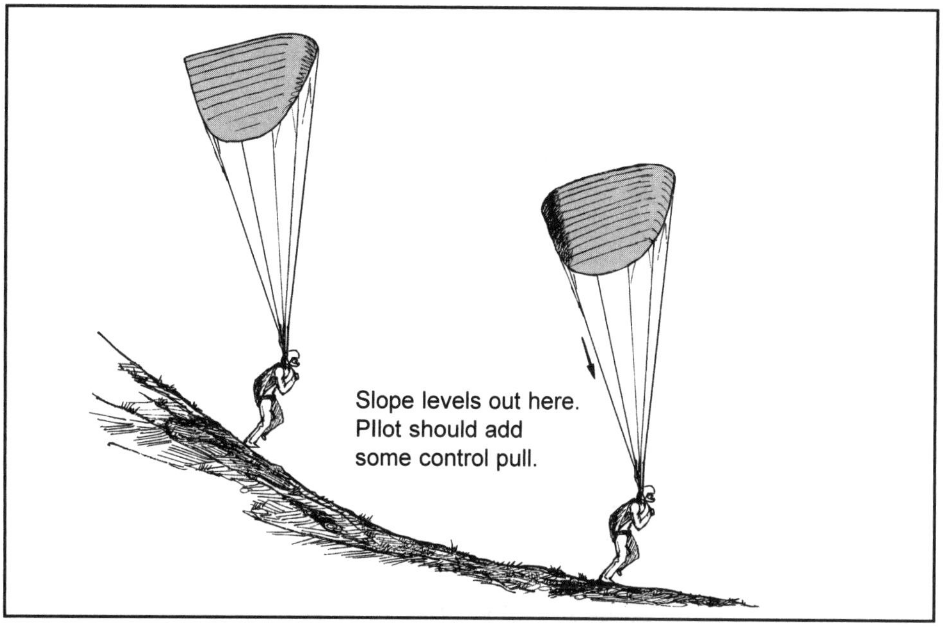

Figure 7-15: Launching on a Flattening Slope

the edge. The problem here is that rotors, eddies or swirls can be present near the cliff edge as we see in figure 7-16. This factor is more true the sharper the cliff edge. An undercut cliff is the worst case.

As a general rule of thumb, you should have at least 20 feet (6 m) of clearance from the edge with the glider laid out before you inflate. This buffer is to give you a chance in case you trip. When you inflate, kite the canopy for a few moments. If you can't maintain a steady canopy for 10 seconds or so, the situation is too unreliable. Sure, you may turn around and get away, or you may end up as a piece of impromptu artwork somewhere on the cliff side. We do not live on luck in this sport, we live on good judgement.

Figure 7-16: Windy Cliff Hazards

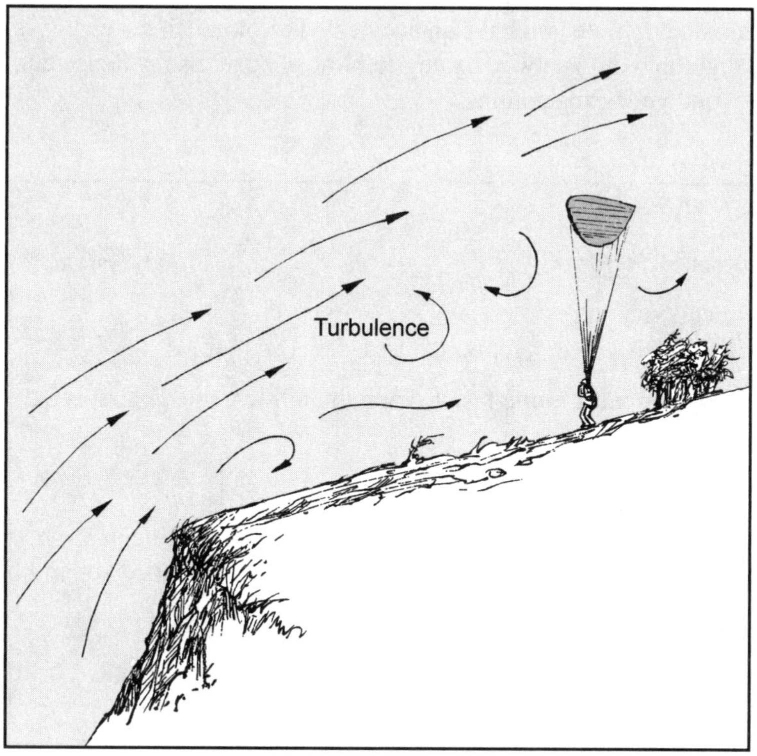

✓ *Rough Terrain*

The final launch complication we'll address is when the ground cover is uneven or consists of brush, weeds or other nasties. We'll note at the beginning that we know a number of pilots who purchased large plastic tarps to spread out when they encounter such a situation. Other solutions include laying out the canopy in a bunch in a small clear area and letting it open as it climbs. Multiple assistants can help hold the canopy up to clear brush or snagging weeds. One of the problems with chopping down brush and weeds is that you produce many sharp spikes that may catch the canopy or more commonly puncture it if you deflate on launch. In general, it's better to let weeds live unless you can pull them out by the roots or push them flat. Try to lay the canopy out on top of them so that the leading edge is held up.

If you have rough ground to run over, avoid tripping by stepping high with your feet. You can practice this run on the flat, but don't let anyone see you or you'll see a few heads shake. In some cases it is necessary to watch where you place your feet—perhaps there are rocks, ruts or holes to avoid. In this situation you must drop your gaze lower so you can better direct your feet. Uneven terrain in light winds may call for a forward inflation, but certainly it's better to have some wind so you can reverse inflate and check the canopy without having to look up while you are running.

SUMMARY

Every good flight begins with a good launch. Perfecting your techniques requires not only practice, but also an understanding of what a good launch entails. Further understanding is necessary in order to safely handle all the many possible launch variations you'll encounter, from gusty, windy or cross conditions (or any combination thereof) to steep, shallow or precipitous slopes. Each situation demands specific controls and considerations. You must learn the different techniques appropriate for the different launch challenges one at a time.

An important critical element in all these launch matters is judgement. Sometimes the launch factors—wind or terrain—are too challenging for your current skill level. Learn to choose not to launch if you aren't 100% confident that you can handle the situation. Did you base that confidence on actual preparation such as practicing ground handling in turbulence? If not, then your judgement is suspect.

We're not here to dampen your enthusiasm. But remember, the great pleasure of paragliding ceases the minute you have an accident. The mountains are always going to be there and the wind is always going to blow. So exercise patience and you'll progress beyond your expectations.

TEST YOURSELF (*Answers in Appendix II*)

1. When running for takeoff you should look ____________________.
2. Checking the canopy after inflation is necessary in:
   a. forward launch only   b. reverse launch only   c. both a and b   d. neither a or b
3. Tail wind launches are dangerous because you must move faster down the hill to achieve airspeed.
   True or False
4. In a crossing wind, how do you orient the canopy layout?
   a. Pointing down the hill's slope   b. Pointing into the wind
5. The possibility of collapses and wing folds on launch is (*increased* or *decreased*) in turbulent conditions.
6. Besides the possibility of turbulence, the main danger of strong wind launches is _______________.
7. In general, it is safer to launch from a slope that is:
   a. Shallow and then gets steeper   b. Steep and then gets shallower

Chapter 8

Perfecting Landing Judgement

"Takeoffs are optional, landings are mandatory."
— Anon

Coming back to earth is always a bittersweet affair, for it means the flight is over. But it also means we can share our aerial adventures with our eager flying buddies or whomever else will listen. Landing is not without its challenges, however, for while we can choose when we take off, our landing time is largely determined by the whims of the wind gods.

Landings often take a while to perfect because they require so much judgement. You'll begin with the basics early on in your novice training, but must continue building experience until landing setups are nearly automatic. The better you become at landing judgement, the more varied sites you can safely fly.

We have organized this chapter to begin with the last act of landing—the final approach and flare. This part of landing is used early on, even at the training hill. Later, we add landing setup for flying higher. Finally, we explain about landing variations so you can handle all sorts of situations.

We mentioned that landing can be a challenge, yet it can also be very rewarding, for nothing polishes off a great flight like a perfect landing. So without delay we begin your training in the art of alighting lightly.

From Final to Flare

Transitioning from flight to the ground can be separated into the landing setup and the final approach. The landing setup consists of starting near the landing field and maneuvering to position ourselves to enter the final approach. The final approach, or "final", is a straight-ahead fast glide from an altitude of 20 to 80 feet (6 to 24 m) down to the ground in a vertical or standing up posture. If any turns are made at all they are minor and only serve to accommodate a wind shift or avoid an obstruction. The final consists of several parts which we shall examine here.

Be aware of each part of your approach as it happens.

Your Goal: *To learn the separate parts of a final approach and be able to master each part.*

✓ The Four Phase Final

We break the final into four parts to better understand and perform it. As figure 8-1 shows, these four parts are 1) *The steep glide*, 2) *the roundout*, 3) *the near ground glide* and 4) *the flare*. Each part has a specific purpose and technique, so let's look at them individually.

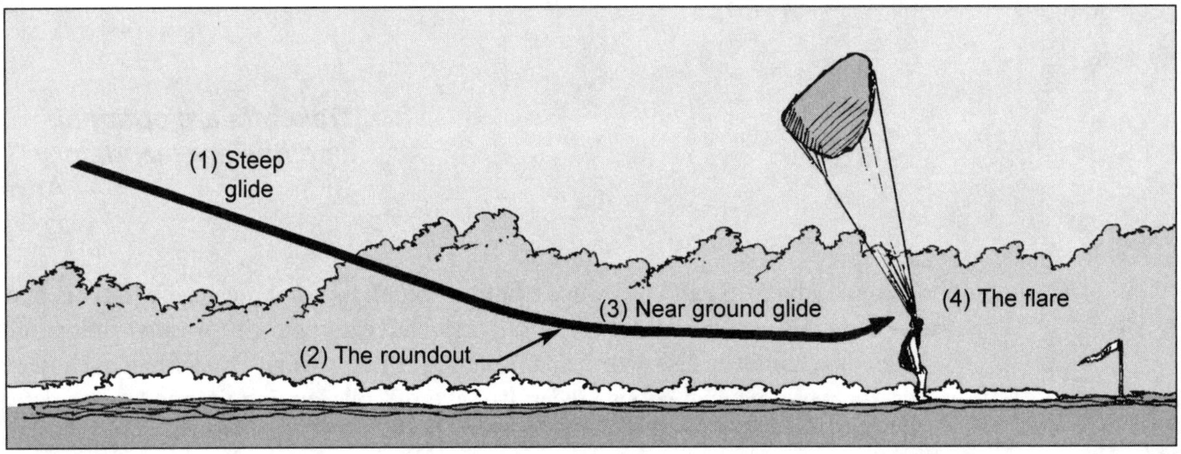

Figure 8-1: The Four Phase Final

1) *The steep glide* - The entire landing approach must be accomplished at a good maneuvering speed to combat control problems associated with turbulence near the ground. The final can not only introduce you to turbulence but also to *wind gradient*. Wind gradient is a change in wind close to the ground and we'll cover it in detail below. For now, note that wind gradient can cause surges or stalls close to the ground and is handled simply by flying with ample speed.

Final Glide Speeds

In smooth conditions make all landing approaches close to trim speed. It is good policy to hold some control pressure in order to feel the wing and to be able to let up if the canopy retards. In turbulent conditions hold a bit more control input—to help pressurize the canopy.

In conditions where a wind shadow exists (see below) or in stable conditions with lots of wind, it is wise to apply even more speed on final. Note that most inexperienced pilots come in for landing too slowly rather than too fast.

The whole point of the fast glide portion of the final approach is to get you down quickly through the turbulent layer and the gradient, afford you good control, then set you up for the next step.

2) *The roundout* - The roundout is a simple control pull input that slows the glider and lets it float above the ground a bit. This control reduces the vertical descent and changes it to a more horizontal glide so that only the forward motion must be stopped.

The control input in the roundout should be smooth and last about one second. Do not make a jerky or abrupt control or the glider will balloon upwards and lose too much airspeed. This control should be initiated when your feet are about 10 feet (3 m) off the ground and should be complete when they are 2 to 3 feet (0.6 to 1 m) up. A slow, smooth pitch control produces these results. In figure 8-2 we show the fast glide and the roundout.

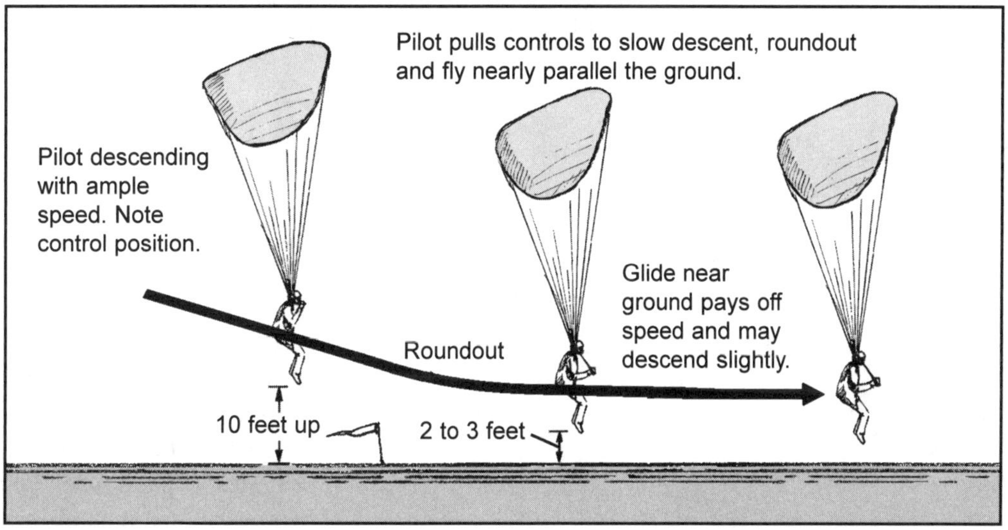

Figure 8-2: Roundout and Near Ground Glide

3) *Near ground glide* - This part of the final approach is very important because here you bleed off speed, stabilize your canopy, and descend to flare height. Your speed will continue to slow, you'll lose a bit more height and in a couple of seconds it will be time to flare

The whole idea of the near ground glide is to bleed speed off gradually so the glider essentially parallels the ground. You pay off speed to earn a nearly level flight path, as in figure 8-2. Such a practice allows you time to sense your airspeed, position and path.

Watch the experts perform a full flare. You can recognize them by their feathers.

4) *The flare* - The act of flaring is a matter of timing and control pull speed. We discussed flare procedure and technique in Chapter 5 and this material should be reviewed. We'll repeat that your legs should be down with one foot leading. Be ready to do a PLF and give a full-on control input to stop the glider when your foot is about 2 feet (0.6 m) off the ground (see figure 8-3). Note that in higher winds you shouldn't give a full flare control as we will explain later in this chapter.

A summary of the four phase final and the elements of a good landing follows on the next page.

Landing Techniques Summary

Landing sequence

- Begin with a straight controlled fast glide through the gradient (fast glide phase).
- Slow down the amount necessary to begin gliding almost parallel to the ground (roundout phase).
- Continue gliding near the ground while slowing to just above stall (near ground glide phase).
- When the glider begins to lose energy fast, pull down smoothly on the controls (flare phase).

Landing tips

- Use a long final
- Maintain heading and wings level
- Keep a level gaze
- Keep the controls between your chin and the top of your head
- Be sensitive to your glider's input
- Keep an erect posture
- Keep your legs under you and be ready to run with slightly bent knees.
- Flare smoothly
- Produce a gentle touchdown

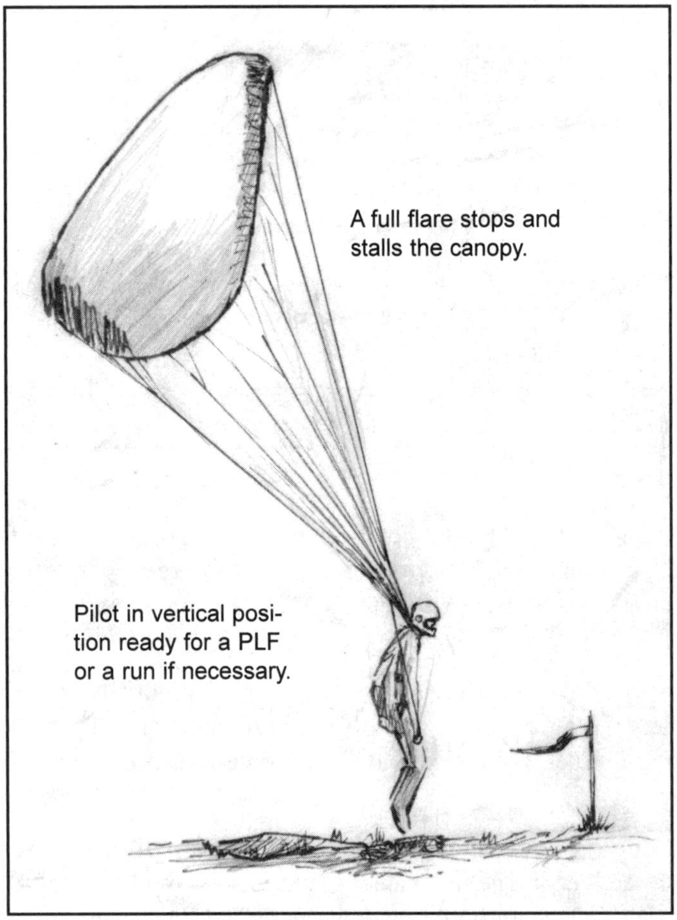

Figure 8-3: Flare and Touchdown.

Safety Tip: *Remember, your legs are your landing gear. Put your gear down (stand vertically) well before you are near the ground.*

Some pilots use what is essentially a two-phase final. They perform the approach and produce a flare when they are about 3 feet (1 m) above the ground. This technique is perhaps simpler than the four-phase final described above, but often results in more forward speed or harder impacts when you touch down. The two-phase final requires better timing of when to start the flare and how fast to make it. The four-phase final requires timing and finesse as well, but seems in most cases to be easier once you get the feel for how much control to apply for the roundout. Listen to the advice of your instructor as to which method is preferred in your typical conditions and be aware of both methods.

✓ *Troubleshooting Landings*

Here we review some common landing problems and offer some tried and true solutions. Note that we are generally considering non-windy and smooth conditions. We deal with wind, wind gradient and turbulence later in this chapter.

▲ *Coming in too high* – On your early high flights you may misjudge your turn to final approach. If you are too high and your field is narrow, lift the controls to trim position (full up) to dive down. *Do not* use the foot stirrup (speed bar) close to the ground as glider stability is reduced (we cover foot stirrup use in the next chapter). Be sure to start slowing down once you are within about 10 feet (3 m) of the ground in this case. Clearly you should allow plenty of overshoot room and plan to land in the first half of the field until you have perfected the accuracy of landings.

Every good flight ends with a good landing.

If the field is wide and you are too high, you may perform S turns to burn some altitude. Turning increases your sink rate and your progress along the field is shortened. Do not make slow turns and do not get close to trees, buildings, poles or power lines. Allow yourself plenty of margin for error. Stop your turns about 20 feet (6 m) above the terrain for a safe, straight final. Resolve to work on your landing setups for the future.

▲ *Oscillating on final* – Sometimes your control inputs are overreactions. Too abrupt a control input can retard a canopy which then produces a surge then a swing back. The best thing to do in this situation is to relax with your hands at approach position and let the oscillations dampen out. If you try catching them your timing may be off and you could make the situation worse. Such an event is dangerous near the ground.

To prevent oscillations in the first place, avoid abrupt control movements. Make every input smooth and gradual to allow the wing to adjust its speed to the angle of attack you have applied. Be gentle and use finesse rather than aggression.

▲ *Flaring too high or too early* – If you flare too soon you may end up suspended well above the ground. The best thing to do in this situation is to hold the controls full down to slow your descent speed and prepare for a hard landing (see figure 8-4). Do a PLF which we described in Chapter 5. By all means, *do not* release or let up on the controls or the canopy may surge and dive you to the ground. If you catch the error early—as soon as you begin the flare input—you may move the controls up to the ¾ position, then reapply them hard once you are within 3 feet (1 m) of the ground. But always beware of allowing a surge close to the ground.

Side view of an imminent landing. Note the pilot's hand position which allows good speed for a roundout and flare.

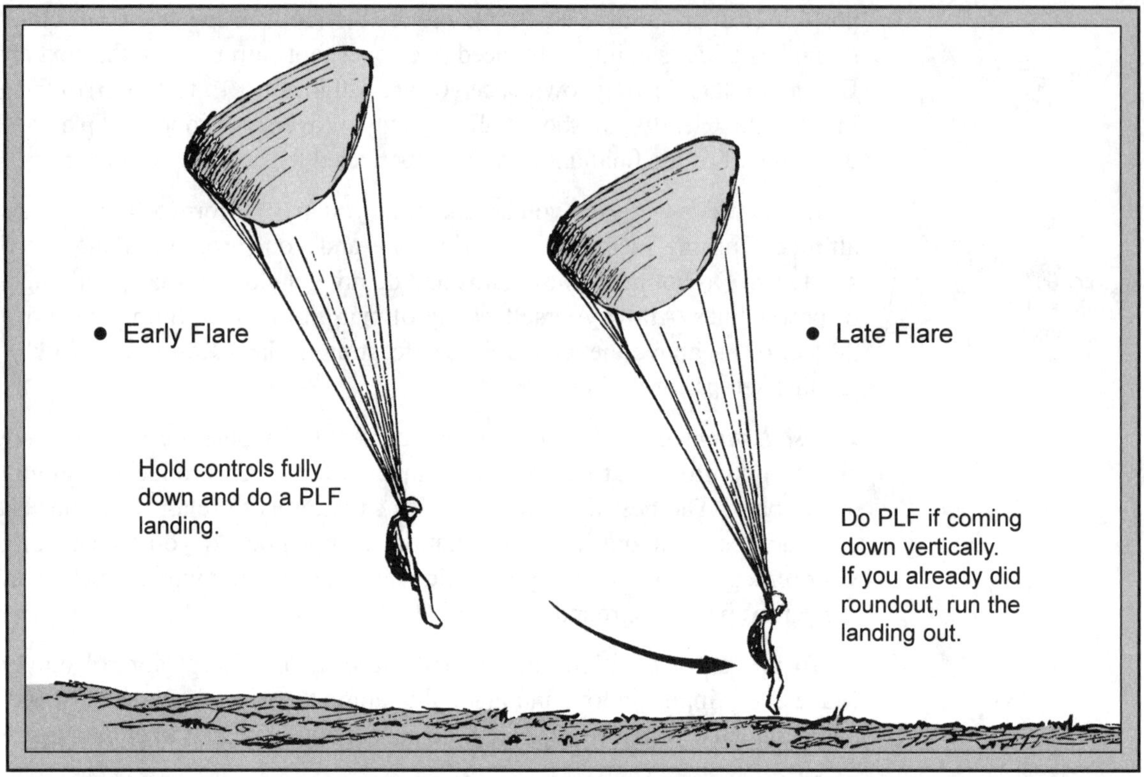

Figure 8-4: Flaring Early or Late

▲ *Flaring late or weakly* – A late or weak flare will result in a harder touchdown than necessary. Be prepared to do a PLF if this is the case. If you use the four-phase final, a late flare usually only results in the requirement to run rather than a PLF.

Weak flares may be the result of not enough control line pull available. You can remedy this problem by taking one or two wraps on the control lines with your hands prior to the flare. This process was described in Chapter 4. You should perform this act when you are about 15 feet (4.5 m) above the ground. Don't wait too late and don't do it too soon for you may have less than full control motion with the wraps.

The opposite of a weak flare is flaring too quickly and ballooning up. In this situation, the glider may stall and drop you hard, possibly on your back. Avoid such an outcome by making your flare gradual and smooth.

All landing control problems—ballooning, too early or too late a flare—should be taken care of by practicing timing on a simulator (a glider harness and controls hung in a frame) and imaging (playing the scene in your mind). Concentrate on the right timing and amount of controls. Timing includes how fast you produce the required control pull. For example, if you are producing a full flare from the shoulder position, the control should span about a second.

The second thing to do is pay attention to your visual field. That means make a conscious effort to look ahead of you at least 50 feet (15 m) as you near the ground and lower your eyes gradually to be looking about 10 feet (3 m) ahead as you arrive at flare height (see figure 8-5). This process is natural if you are looking ahead at your expected landing point. Do not look directly down until you have produced the flare, for the effect of ground rushing by is confusing.

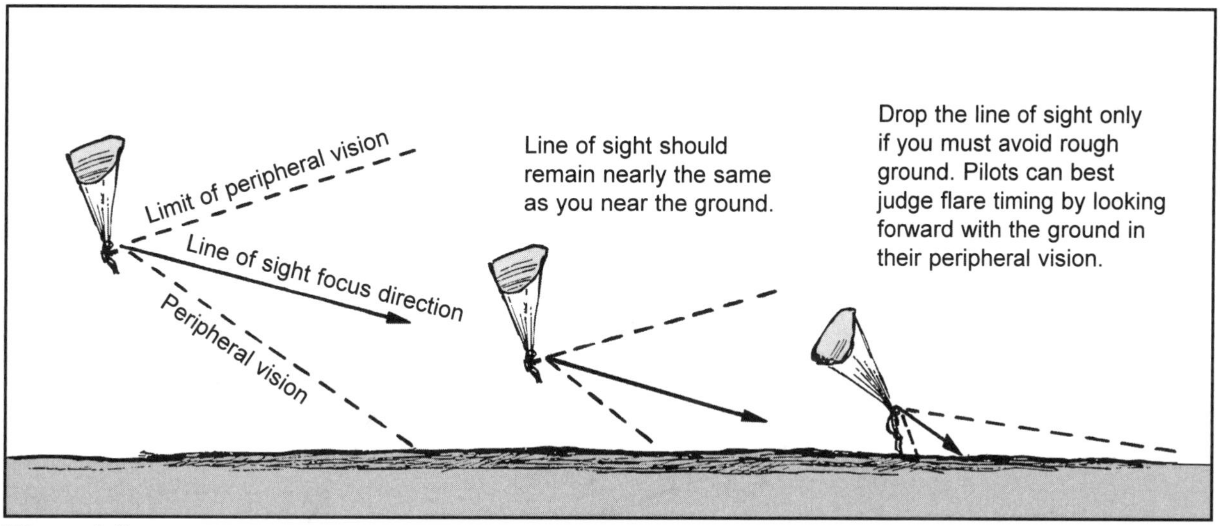

Figure 8-5: Eye Focus During Landing

Pro Tip: *Keep your eye focus ahead a bit past your landing point so you are looking about 10 feet (3 m) beyond it at flare time.*

A perfect landing is a gentle no-stepper (in all but calm winds) with the canopy dropping behind you or kiting under control. Make landings part of the fun of flying by pulling them off with flair. Settle for no less than perfection.

Skill Check: *When you are automatically producing no-step gentle landings consistently, you have mastered the last act of landing. In order to tell if you have perfected landing skills, log each landing and grade it on a scale of one to ten. If nine out of ten landings are seven or better then you have achieved proper landing skills.*

✓ Wind Gradient

A change of some factor over a distance is called a gradient. The change of wind speed as we near the ground is known as a wind gradient. The drag and friction of the surface including grass, bushes, trees, buildings, mountains, etc tends to slow the wind the lower you go. A wind gradient is present at some level whenever the wind blows. The problem this presents on landing is shown in figure 8-6. Here we see a pilot descending through the gradient. Initially, he (or she) may have ample airspeed. However, as he descends he encounters a rapidly diminishing head wind, so that the glider may surge forward and drop the pilot, possibly followed by a pendulum action whereby the pilot swings forward and stalls the canopy.

Entering a wind gradient is a non-event if you carry ample speed.

Let's make this clear. Ground speed has nothing to do with this problem and the ground only serves to cause the wind gradient (of course the ground will hurt if the pilot hits it). The problem is caused by the glider's inertia. A glider will produce the airspeed appropriate for the angle of attack the pilot sets. However, it takes the glider some moments to adjust to changes in airflow due to its inertia. If you imagine flying into a strong head wind that suddenly stops, you can see how a glider will be left with no airspeed and then surge forward. You can experience this with your body by leaning into a strong, gusty wind. When the wind diminishes for a moment, you fall forward.

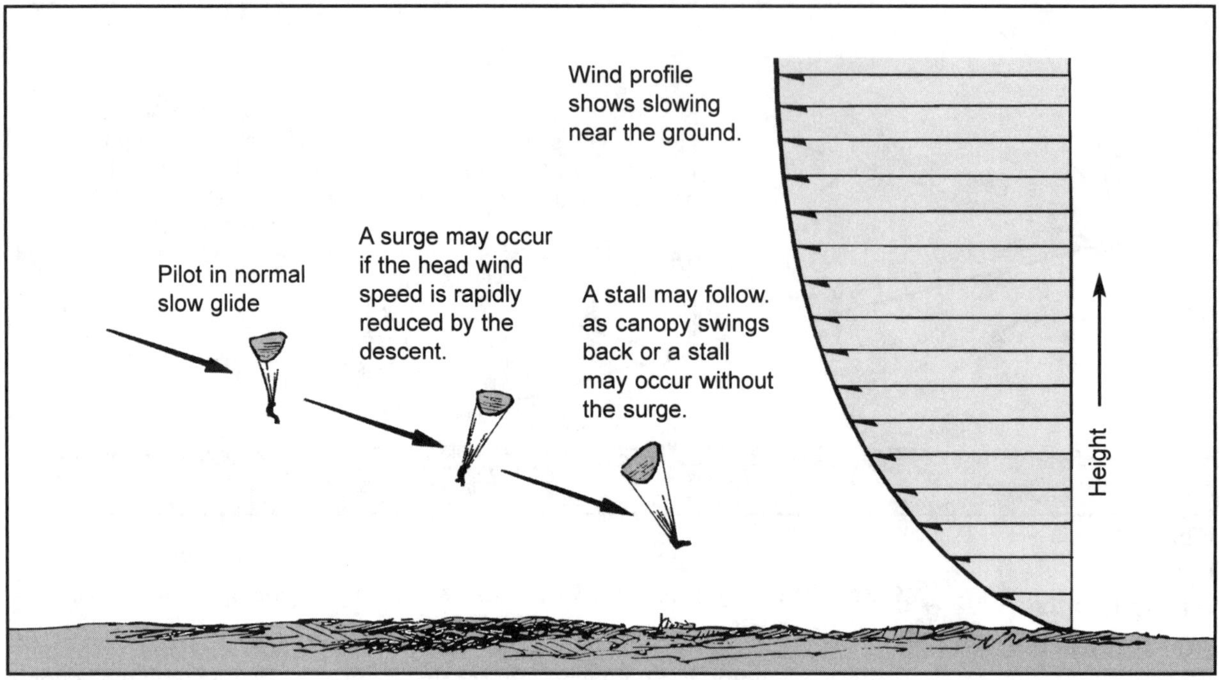

Figure 8-6: Possible Effects of Landing in a Wind Gradient.

In a gradient, the continuously lowering head wind will lower your airspeed before your glider can react. The more severe the gradient, the more airspeed is lost and a glider can surge ahead as it naturally tries to regain airspeed. A severe gradient can occur at the edge of an obstruction and is known as a *wind shadow* as shown in figure 8-7. You are most likely to encounter such a gradient when landing behind a tree line or buildings.

The problem with surges or stalls in a wind gradient is that they normally occur close to the ground with too little altitude to recover. Because the head wind continues to diminish lower down, a stall recovery takes more altitude than usual. If you do surge or stall in a wind gradient, you must monitor the canopy to apply

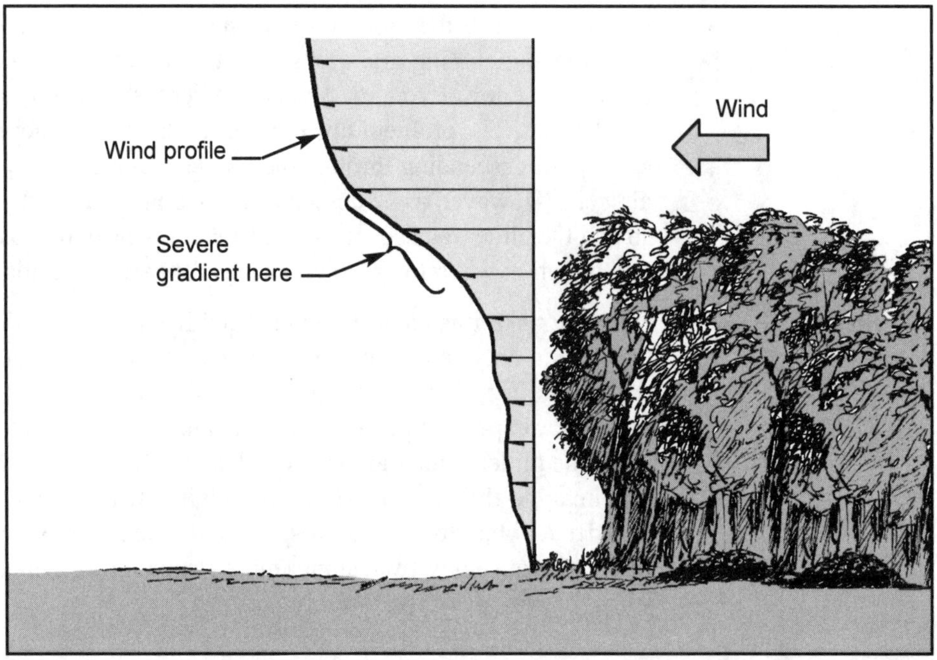

Figure 8-7: Wind Shadow Effect

controls to either dampen the surge or relieve the dropping back of the canopy. In general, if a stall occurs close to the ground, you will not be able to regain normal flying mode. Your control inputs may also make things worse. In any case, be ready to perform a PLF and a rapid, vigorous full flare just before impact. Usually this action will suffice to prevent excess damage to you and your confidence.

It is obviously of great importance to avoid surging and stalling in a wind gradient. The way to do this is to acquire additional airspeed *before* you encounter the gradient. Remember, wind gradient can be worse on stable days because there is little vertical mixing of the air to spread out the wind speeds. Also, in a severe wind shadow or gradient you should expect some turbulence.

Avoiding Wind Gradient Problems

- Carry ample speed on your final approach
- Do not land close behind barriers to the wind.
- Add extra speed on strong wind, stable days and whenever you suspect a wind shadow (Note, *do not* use a speed system below 300 ft—100 m—since these systems reduce inflation reliability).

LANDING SETUPS

All of the above discussion focused on the final approach and flare. But how do we get to the point of final in the first place? This is where our *landing setup* takes the stage. The landing setup is the part of the flight that takes place around the landing field where we lose altitude and position ourselves properly to come in for a landing. Since we don't have an engine, we have a great need to do it right the first time.

Before we proceed, here are some ground rules:

- When you first start flying high, it is important to *reach your landing field with at least 300 feet (100 m) of altitude.* This altitude gives you plenty of time to observe conditions in the field, collect your thoughts, plan a landing setup and watch for other gliders.
- Another rule is to *begin focusing your attention on the landing field* as soon as you are 300 feet (90 m) or lower. Don't allow yourself to drift beyond an easy glide to the field. This caution is especially important in wind.
- The final rule is to *clear all turns*, even in the landing pattern (see Chapter 9), and observe the right-of-way conventions (see Chapter 10).

The two most common landing setups used in paragliding are described below.

Your Goal: *To learn the judgement and control skill to land near a pre-selected spot with both the figure eight and the aircraft landing approach.*

✓ The Figure Eight Setup

The first landing method is known as the *figure eight* because that's the pattern we describe in the air. This method is often used in higher winds because it

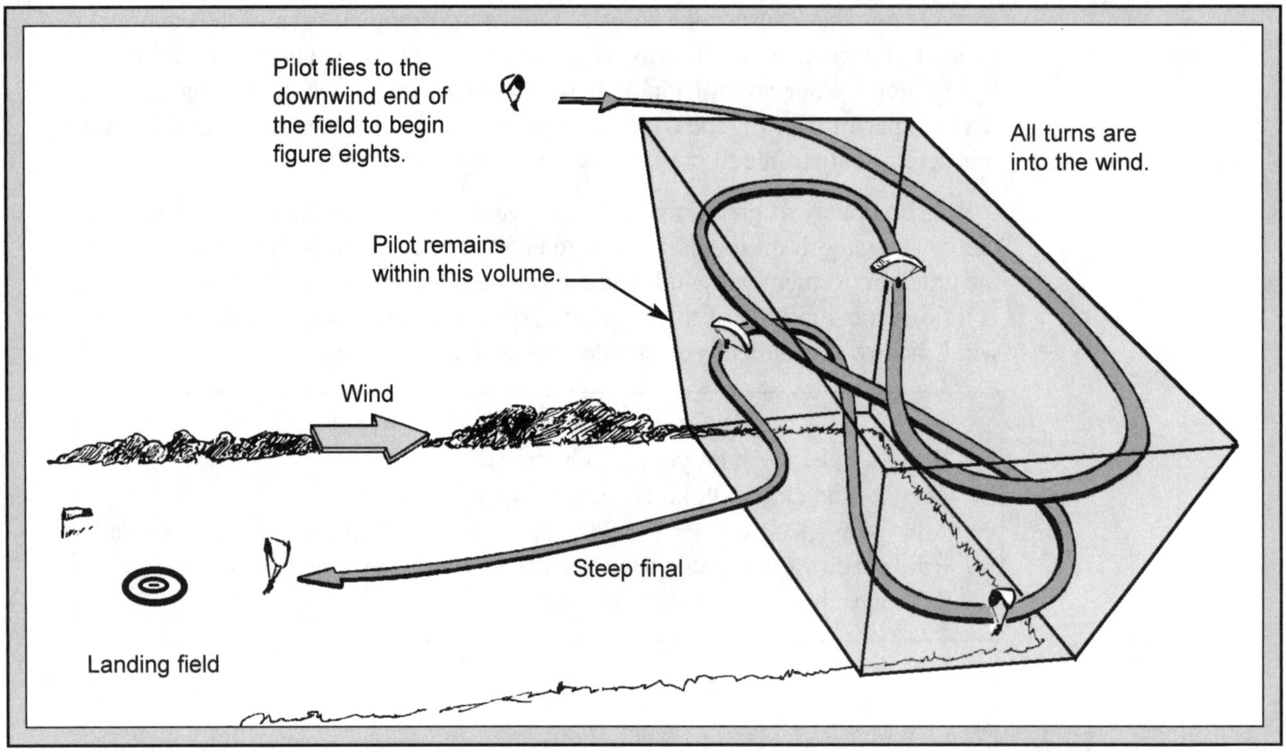

Figure 8-8: The Figure Eight Setup Pattern

requires no downwind flying. It is also *recommended for your first high flights* because of the absence of downwind flying.

Figure 8-8 shows the features of a figure eight setup. The pilot approaches the landing field at its downwind end. There he or she remains, performing a figure eight flight path while descending to the proper height to enter the final.

The reason we must perform figure eights rather than 180° turns is because 180° turns will step us up the field as shown in figure 8-9. The result will be running out of field and some serious decision making. In zero wind the figure 8 pattern

Figure 8-9: Figure Eight Problems

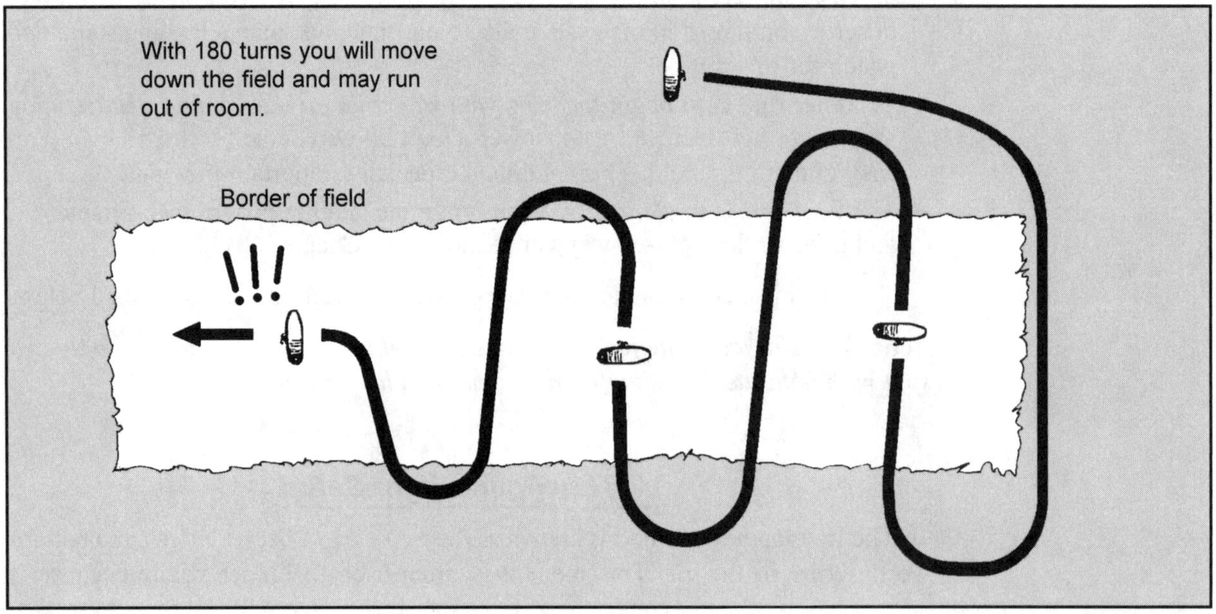

requires turns greater than 180°, so it's best to learn the landing setup in a bit of smooth wind at turn altitude and to land in a very large field until you have perfected your turns.

Placing yourself properly in 3-dimensional space is critical to landing.

Follow the figure eight track and you will see that alternating left and right turns are required. A glider has inertia and other sources of resistance to reversing turn direction, so the legs of the figure eight must be long enough to allow you to turn with ease. You should practice alternating turns of lesser heading change while you are at the training hill.

Another feature to notice in your setup is that the legs of the figure eight get a bit shorter as you get lower. This is necessary to prevent getting too far away from the confines of the landing field. For this same reason you must not be beyond the downwind edge of the field, at least in your early experience. Figure 8-10 shows a ground observer's view of the figure eight setup. Study this drawing and figure 8-8 until you clearly understand the necessary path.

Figure 8-10: The Figure Eight Setup from the Ground

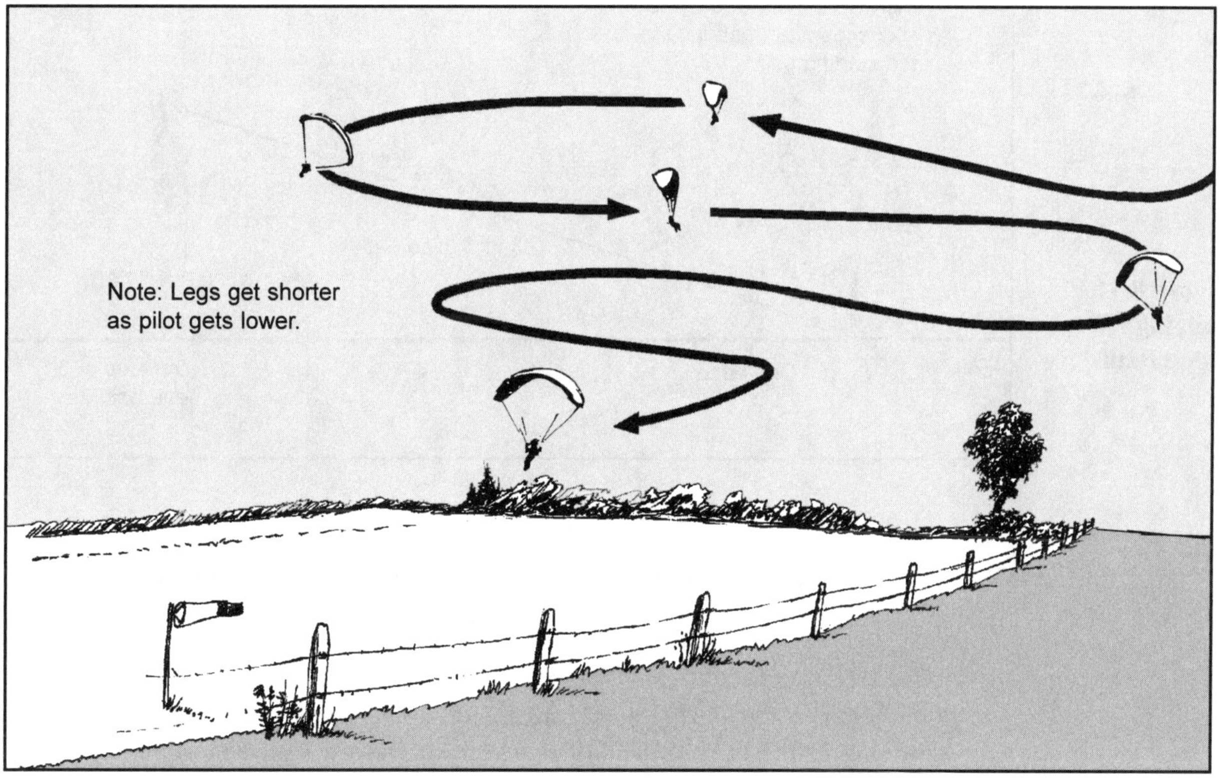

✓ Judging Final Height

One of the important matters in the figure eight setup is knowing when to stop doing figures and turn onto final. The answer is when you are about 3 to 1 glide to your chosen landing point. (A glide is a descending path. A 3 to 1 glide is a path that goes forward 3 feet for every one foot it drops. The matter of glide ratio and glide path is explained in Chapter 9.) A more experienced pilot can choose to land almost anywhere in a field, but you should pick a point about a third of the way along to allow you plenty of runout.

The real question is, how do you determine a 3 to 1 path to your landing point? The secret is practice before flying. Most people's eyes are located 2 paces above the ground. That means if you pace off 6 paces and look where your feet were at your standing point, you will be looking at a 3 to 1 glide path, as shown in figure 8-11.

Look along this path and practice judging the *angle* below the horizon. Any time you are above the ground—on a balcony or hill—try estimating the 3 to 1 angle. With a little practice you will become surprisingly accurate.

The reason we use 3 to 1 is because that's a good path to follow in our fast approach sitting-up position. Our gliders will certainly glide better than 3 to 1 if we want them to, and this is a safety valve in sink or a head wind. Also, if we use a greater glide path than 3 to 1 we must continue the figure eight pattern too low to the ground if the field is not very long.

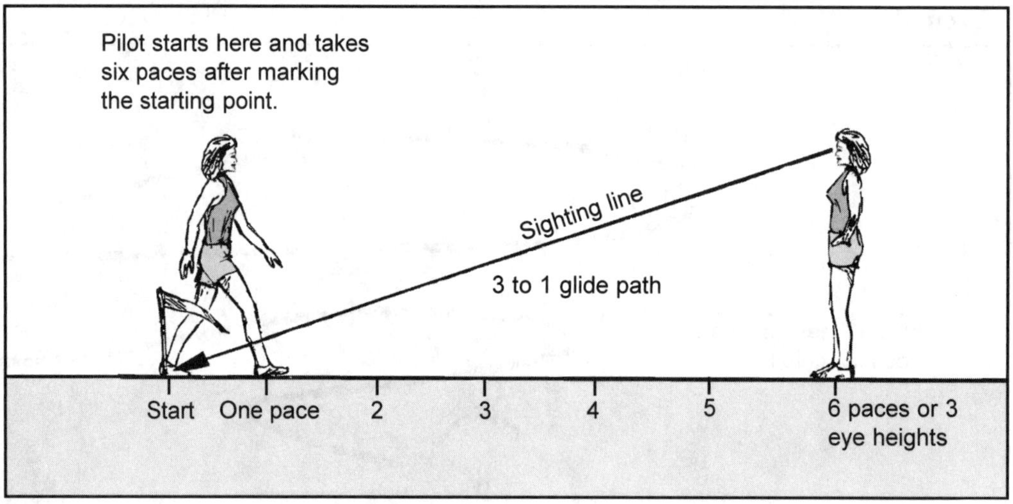

Figure 8-11: Judging a 3 to 1 Glide Path

✓ *Preventing Setup and Approach Problems*

There are common problems to the figure eight approach, but we can easily prevent them with a little thought.

▲ *Turning S-turns* – The most common problem is that already mentioned—performing 180° S-turns instead of figure eights. As we saw, this eats up a landing field and you may end up at the upwind end still high and dry. Practice figure eights up in the air by using a road or an imaginary line perpendicular to the wind and remaining above it while going back and forth.

Once you have started your approach, never turn away from the field.

▲ *Turning too often* – When you are high let your figure eight legs extend long enough so you can turn leisurely. As you get lower, you must not stray too far to the side of the field, especially in wind, so you don't get caught short. Of course, with a long landing field you can enter the final while you are still high.

▲ *Turning too slowly* – Your flight in the figure eight pattern should be at a good maneuvering speed (see Chapter 3 and chapter 9). You should carry speed through your turns. They should be fast enough to provide good glider response.

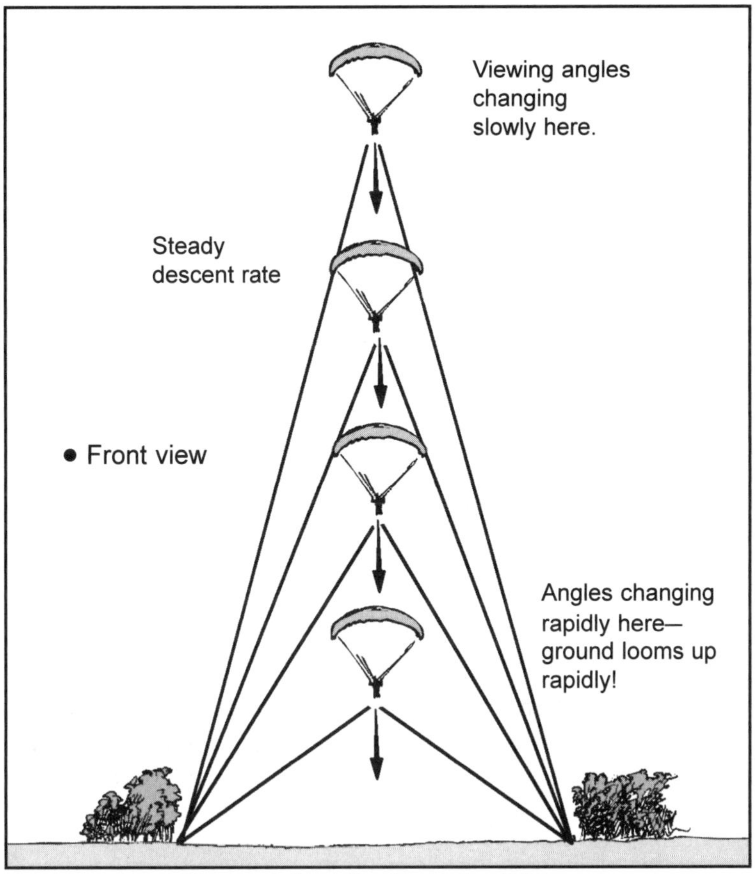

Figure 8-12: The Looming Effect

▲ *Waiting too long to turn on final* – When you are losing altitude starting above 300 feet (100 m) for instance, the scenery changes slowly. It doesn't seem like you're descending very fast. However, as you get closer to the ground things change rapidly and you can get caught turning late onto final. This perception experience is called *looming* and is due to the angle that our eyes perceive, as shown in figure 8-12. If you descend at a constant rate, the visual angle to a point on the ground changes very slowly when you are high and far away. When you get closer, this angle changes rapidly and you can get caught by surprise if you aren't aware of this fact.

To avoid looming problems, do not look down at the ground to judge height, but look ahead to your landing point and judge the 3 to 1 glide.

▲ *Not centering the final approach* — If you do not control your last turn or time it properly you will not end up on the centerline of your field, and you may have to perform additional corrections to line up. These corrections can be a problem in turbulence, or can cause you to lose more altitude than desired. In a wide field, this may not be a problem, but narrow strips are common and it is very important to be able to land along a chosen line. Practice this with every landing and you will learn proper final turn timing.

Skill Check: *Your figure eight setup skills are well-developed when you can enter the pattern from any height and lose altitude on the downwind side of a field, then enter a final that sets you down between a third and half way down the field. Your pattern turns should be well-controlled and no "upwind creep" should occur.*

✓ *The Aircraft Pattern*

Our second landing approach is the one you should develop well and use in almost all your landings. This setup is known as the *aircraft type approach* or the downwind, base and final (DBF) approach. The reason for these names is because airplanes use this method exclusively and it consists of three legs. This is the preferred approach for all capable pilots in winds up to 8 mph (13 km/h).

Look at figure 8-13 and you will see a bird's eye view of the DBF approach. Here we have shown a *downwind leg* flown directly parallel to the landing field, as the ground track shows. Next we have a *base leg* or crosswind leg flown perpendicular to the field. Finally, we have a *final leg* into the wind.

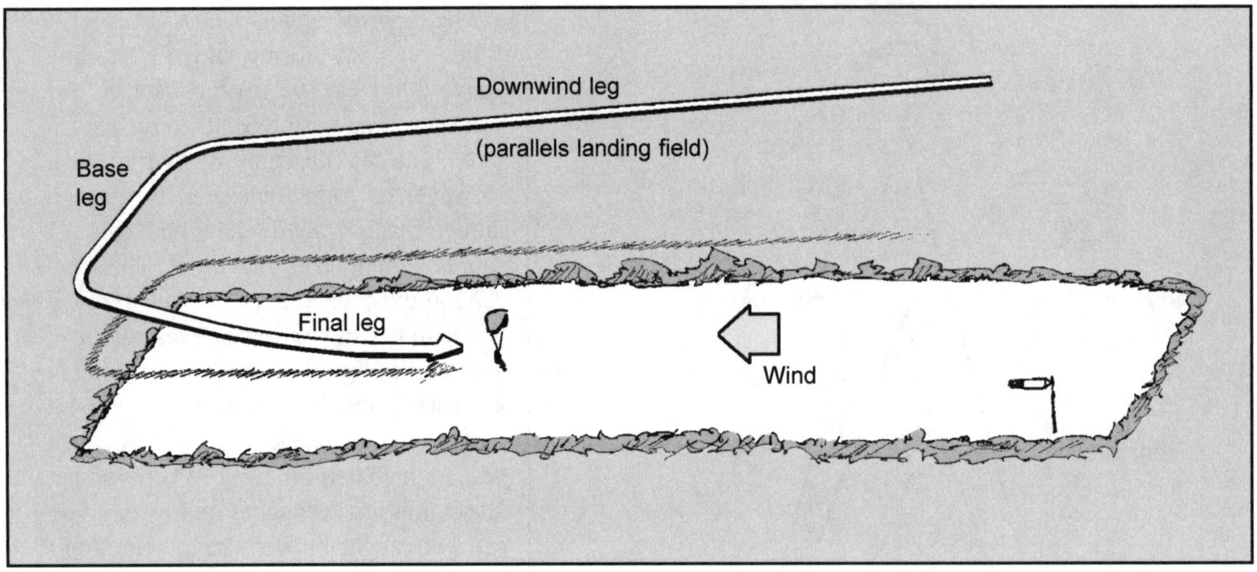

Figure 8-13: The Downwind, Base and Final Approach

Now let's look at the mechanics of a DBF approach. As the top view in figure 8-14 shows, we begin at a point upwind of our desired landing point called:

▲ *The initiation point* — The position of the initiation point is not too critical but should be closer to the desired landing point for smaller fields. The height at the initiation point should be at least 75 feet (25 m) for pilots new to the DBF method. However, we do not judge altitudes, we judge *angles.*

In figure 8-15 we see the angle rules for our position. We always begin within a 45° angle (1 to 1 glide) to the edge of our field. This is a safety rule so we don't get caught short. Note that if the field is surrounded by high obstructions the 45° rule is applied from the top of the obstructions, as shown in figure 8-15. Of course, a pilot learning the DBF approach should not be doing so at challenging landing fields.

Figure 8-14: The Mechanics of a DBF Approach

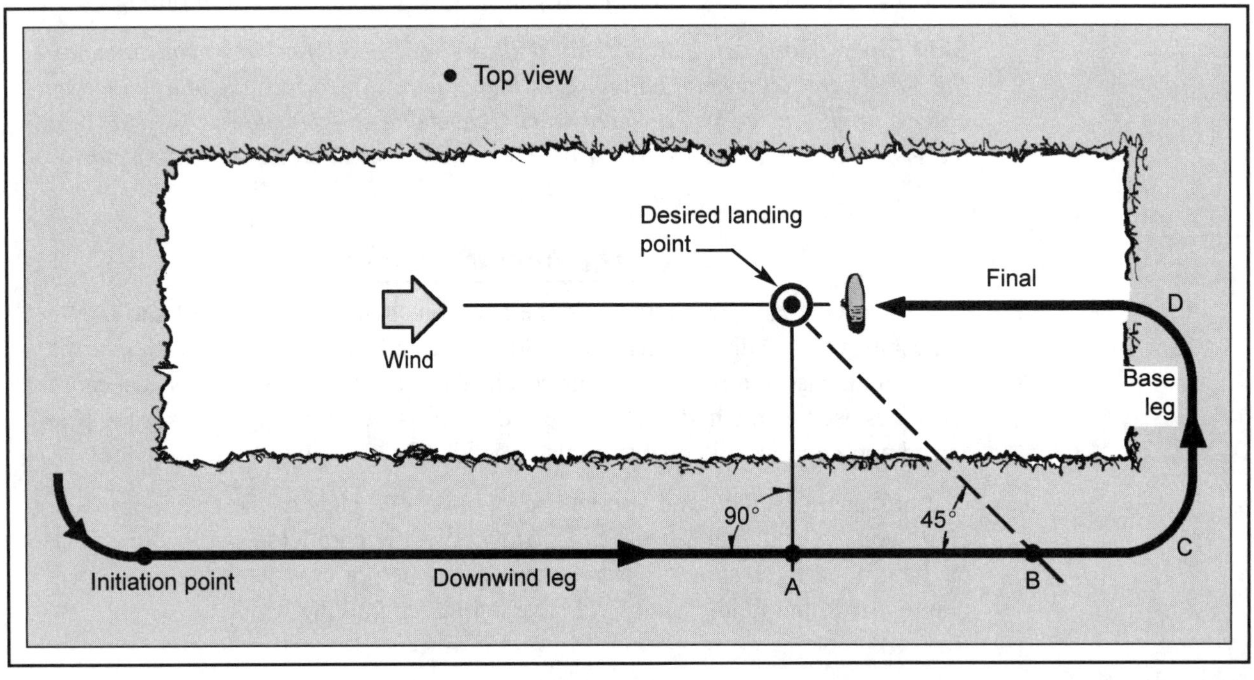

▲ *The downwind leg* — Once you have positioned yourself properly at the initiation point, you begin your downwind leg at a good maneuvering speed. This speed should be a normal approach speed with the controls from chin to the top of your head. The downwind leg is parallel to the landing line. There are three important points on the downwind leg (refer to figure 8-14). Point A is directly across from your desired landing point. There you should focus on your landing point, recheck your angle to it, and judge the wind strength at your DBF pattern height. Point A is an important checkpoint that gives you the first good clue to your progress. The idea is to have a 2 to 1 glide or a 30° angle to the landing point at this position.

The DBF pattern is versatile—take it with you wherever you fly.

Point B is another important point for it simply establishes the minimum position of the first turn. Point B is at a 45° angle from the landing point, as shown in figure 8-14. In normal conditions the first turn at C would not be made before point B. Only in strong winds or sink would we alter this rule.

▲ *Turn to base leg* — The initiation of the turn at point C is judged with our familiar 3 to 1 rule to the landing point (see Judging Final Height on page 171). The idea is to end up with a 3 to 1 glide by the time you reach point D.

Figure 8-15: Altering Positioning With Obstructions

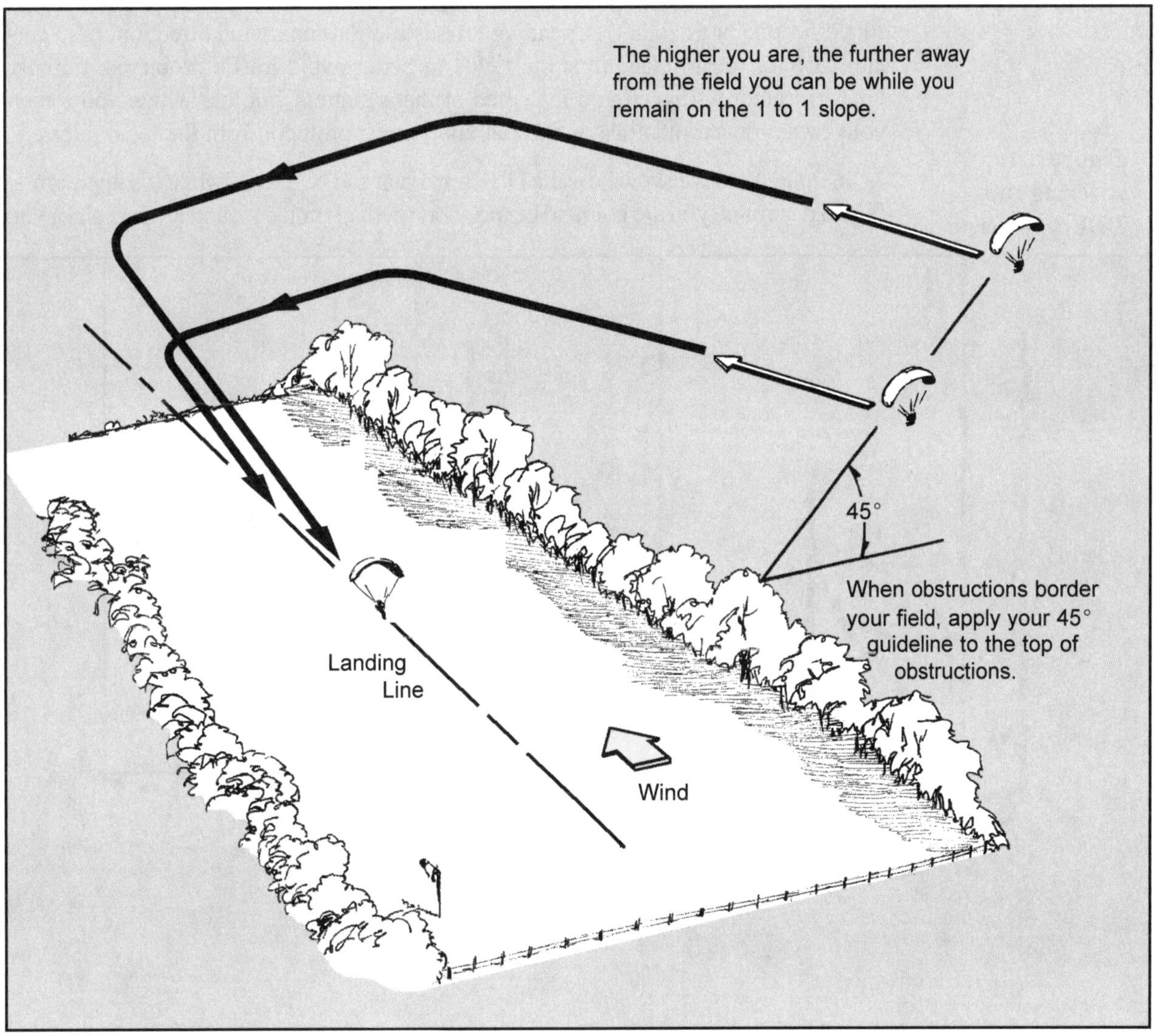

Both turns on your approach should be well-controlled turns to control altitude and maintain good maneuvering airspeed. Do not try to perform maximum efficiency turns on the approach for these turns are too close to stall.

▲ *The base leg* — The base leg is the crosswind leg and is flown in a crabbed heading if necessary so that the ground track is perpendicular to the landing line (see chapter 9 for a discussion of crabbing).

▲ *Turn to final leg* — The last turn at D is initiated so that it leaves you tracking directly into the wind along your chosen *landing line*. The landing line is the desired flight path through the chosen landing point. The line is usually in the center of the field. The last turn is a fast turn that establishes the additional airspeed needed for the final leg.

▲ *The final leg* — This leg is our familiar four-phase final and should be performed as such. Review this material at the beginning of this chapter.

The beauty of the DBF approach is that it can be transported to almost any landing field and altered to suit conditions. Once you learn to judge the important angles you do not have to learn a new setup with every new site. By all means you should practice the pattern using left hand and right hand turns, for many fields are better suited for one or the other, depending on obstructions and wind direction. Also certain customs are in place in some fields to better avoid traffic problems. Initially your instructor will keep you informed of these matters, but later when you are on your own, you should always find out about local protocol from the local pilots.

Figure 8-16: Altering the DBF Approach

In figure 8-16 we show how the DBF approach can be altered while it's in progress. If you determine you are going to be too short (perhaps you encounter strong sink) you

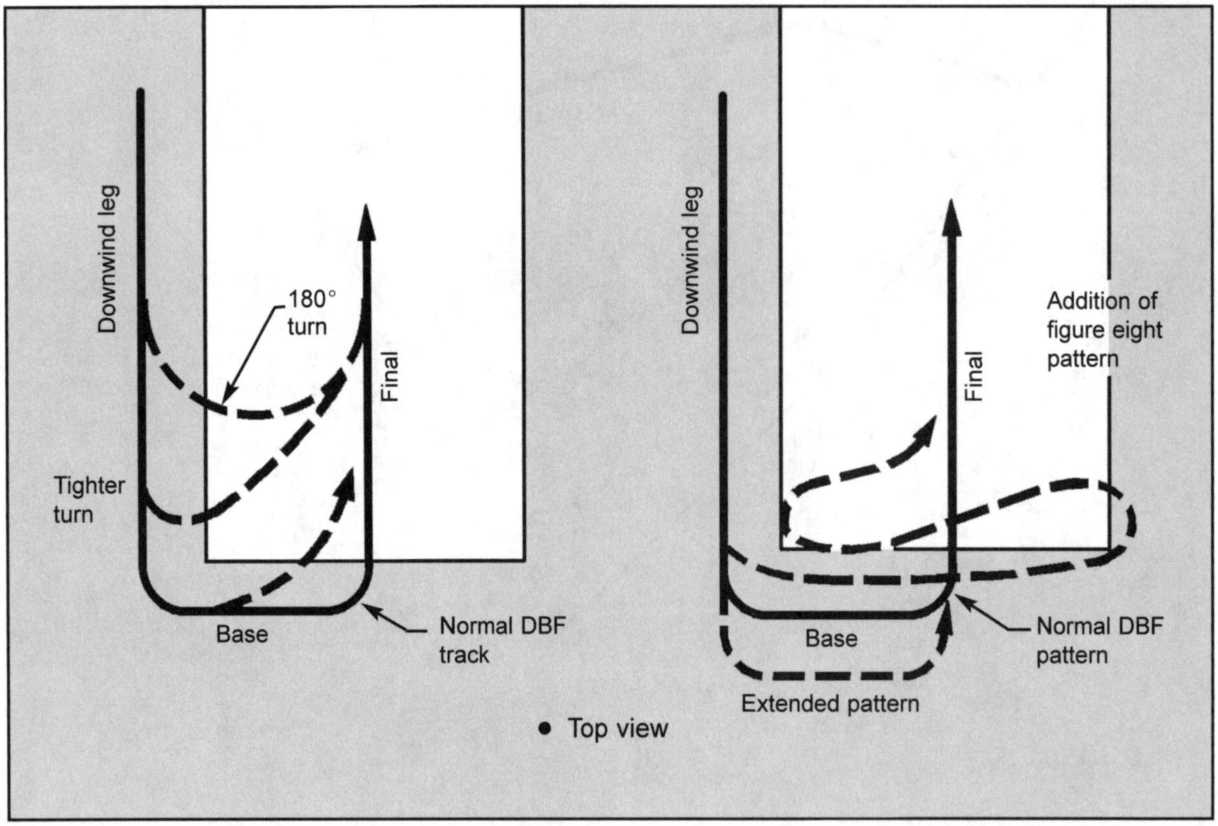

can turn sooner on your downwind leg or even drift closer to the field and make a 180° turn to final as shown. Your base leg can also be altered to angle toward the field.

If you appear to be too high at your checkpoint A or B you can extend your downwind leg as shown. In higher wind you may not want to do this for fear of not reaching the field once you turn back into the wind. In that case you may wish to make a longer base leg or even add a figure eight as indicated. This is the combination type of setup. Finally, we should mention that the turns at C and D can be used to control altitude by performing diving turns once you are more experienced with these maneuvers.

Caution*: Diving turns can create lots of airspeed, which can result in control problems when you level out (oscillations). Be sure to practice these turns with ample ground clearance before you use them in a landing setup.*

✓ *Entering the DBF Pattern—Staging Areas*

With the figure eight pattern you do not worry too much about entering it, for you can come in at any height and begin your figure eights. With the DBF pattern, however, you need to figure out a way to place yourself at the initiation point with a reasonable altitude.

The easiest way for a pilot to learn this method is to develop the concept of *staging areas*. These are areas near the landing field where you will lose altitude. As figure 8-17 shows in a bird's eye view, there are four standard staging areas around any field. These areas are off the end of the field and to either side. This is where you should lose your altitude to arrive at your initiation point.

Here are the steps for choosing the best of the four areas:

1) Arrive at the landing field with at least 300 feet (100 m) of altitude so you have plenty of time to judge conditions and position yourself. Be sure to constantly clear your turns as you set up your landing.

Figure 8-17: The Four Standard Staging Areas

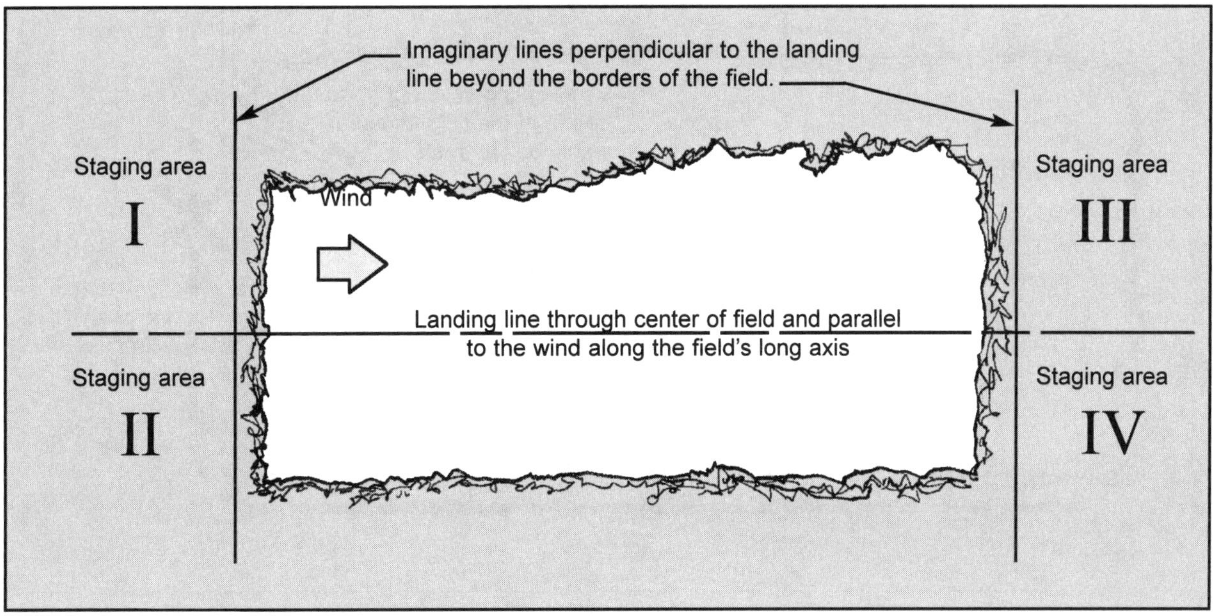

2) Now *check the wind direction* to decide the proper landing direction. In light or variable winds this decision should be made with careful observation. In stronger winds you will probably know the proper direction from your drift aloft. If the wind is perpendicular to the long axis of the field (and the field is narrow), plan to land crosswind, but favor the direction the wind shifts toward most often. Draw an imaginary landing line along the long axis through the center of the field.

3) Next fly to the *upwind* end of the field. You have now narrowed your staging area down to two possibilities which are I and II in figure 8-17. Draw another imaginary line off the end of the field to further describe the staging areas.

4) Now, choose on which side of the field you will perform your DBF approach. This choice depends on the size of obstructions along either side, the shape of the field and the actual wind direction. If all else is equal, choose your downwind leg and thus your staging area on the side with the least or lowest obstructions (trees, power lines, buildings or hills). Also be aware that if the wind is angling across a long, narrow field your downwind leg, and thus the staging area, should be on the downwind side as shown in figure 8-18. In your early practice, your instructor may indicate which side of the landing field to use before you fly.

5) You have narrowed the staging areas down to one, so enter it and begin losing altitude. The preferred way to lose altitude is with wide, gentle 360° turns. There are orientation problems associated with multiple 360's, especially if they are tight, so keep your turns shallow (we cover all the aspects of 360's in Chapter 13). If you have taken tandem flights, you will have experienced many 360's and such landing setups. If you haven't, your instructor will give you the preferred pattern to lose altitude and arrive at the initiation point.

Figure 8-18: Approach for an Angling Wind

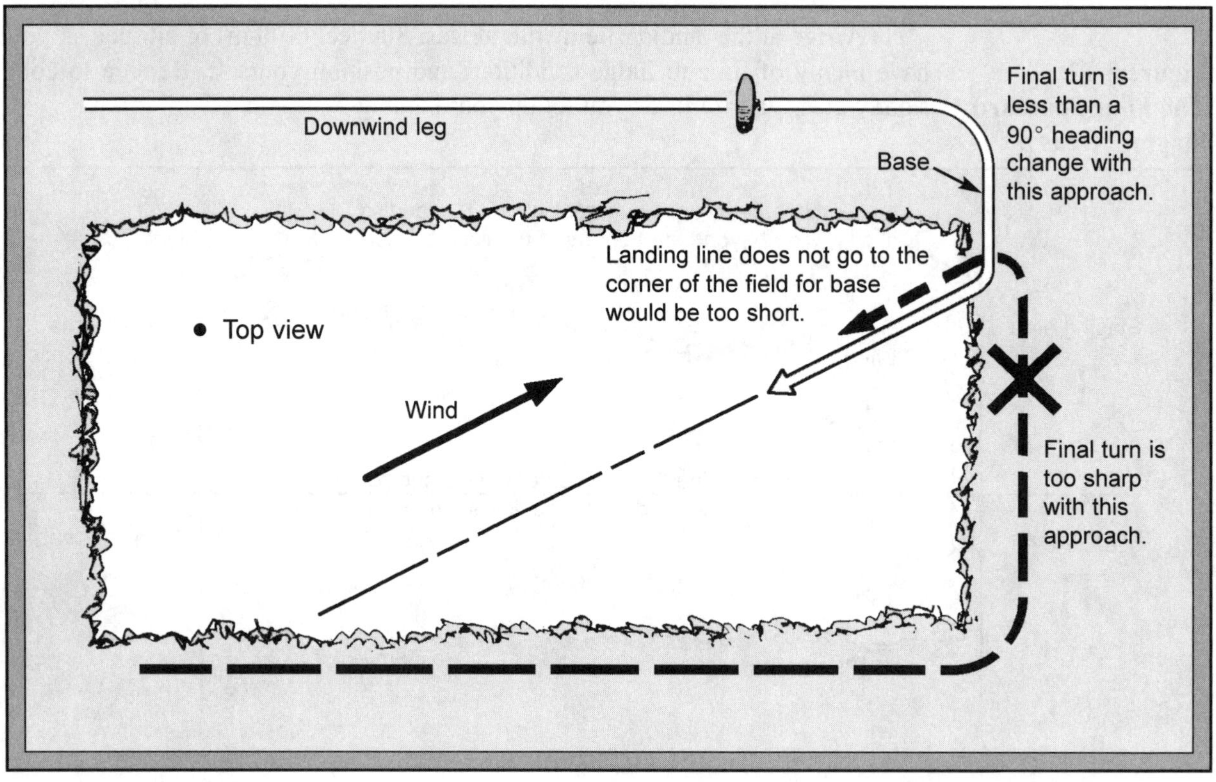

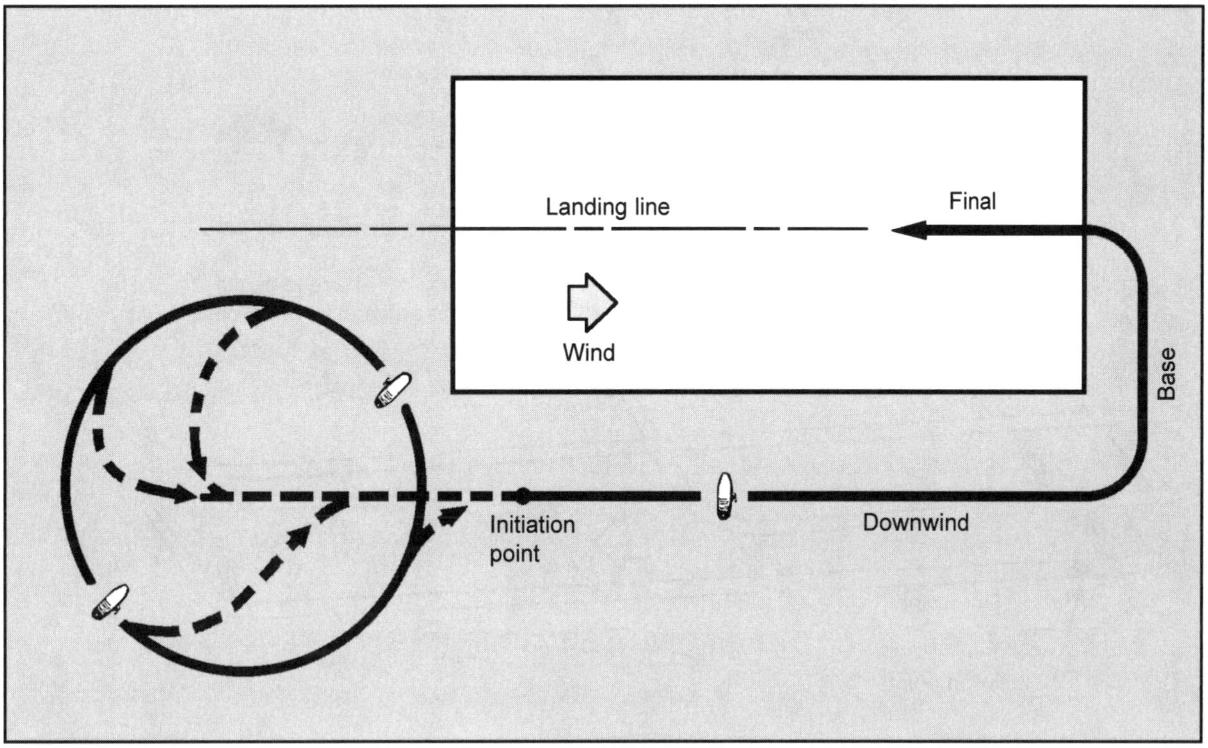

Figure 8-19: Leaving the 360 for the DBF Approach

When you perform your 360's, keep your eye on the field at least 50% of the time and make a conscious effort to judge your initiation point. Also, have your DBF pattern clearly in mind. You should continually monitor the wind by detecting your drift and watching wind indicators. Alter your DBF plan if necessary to accommodate wind changes, but be sure to allow yourself plenty of room. It is better to land off the direct into-the-wind line than to risk hitting an obstruction.

As a final note, look at the 360 pattern in figure 8-19. Here you see a number of possibilities for leaving the 360 to reach the initiation point at the desirable height. Where you leave depends on your height and the amount of wind. You can readily see how this setup allows plenty of options.

Arriving at the perfect initiation point requires lots of practice, but don't worry. The DBF approach allows plenty of leeway because it is designed to help control your altitude. Remember, it is better to be too high than too low.

Pro Tip: Never *fly directly over your landing point for you lose eye contact with it and thus your angle reference. It will take additional time to reestablish this reference and by then you may be too low.*

✓ Boxing the Field

Another method of losing altitude in a landing setup to enter the DBF is called "boxing the field." In this method we fly around the entire landing field using 90° turns at the corners. This method keeps us closer to the field at all times. However, it often requires that we cut across the field at some point to reach our initiation point, especially if the field is large. Figure 8-20 illustrates this concept.

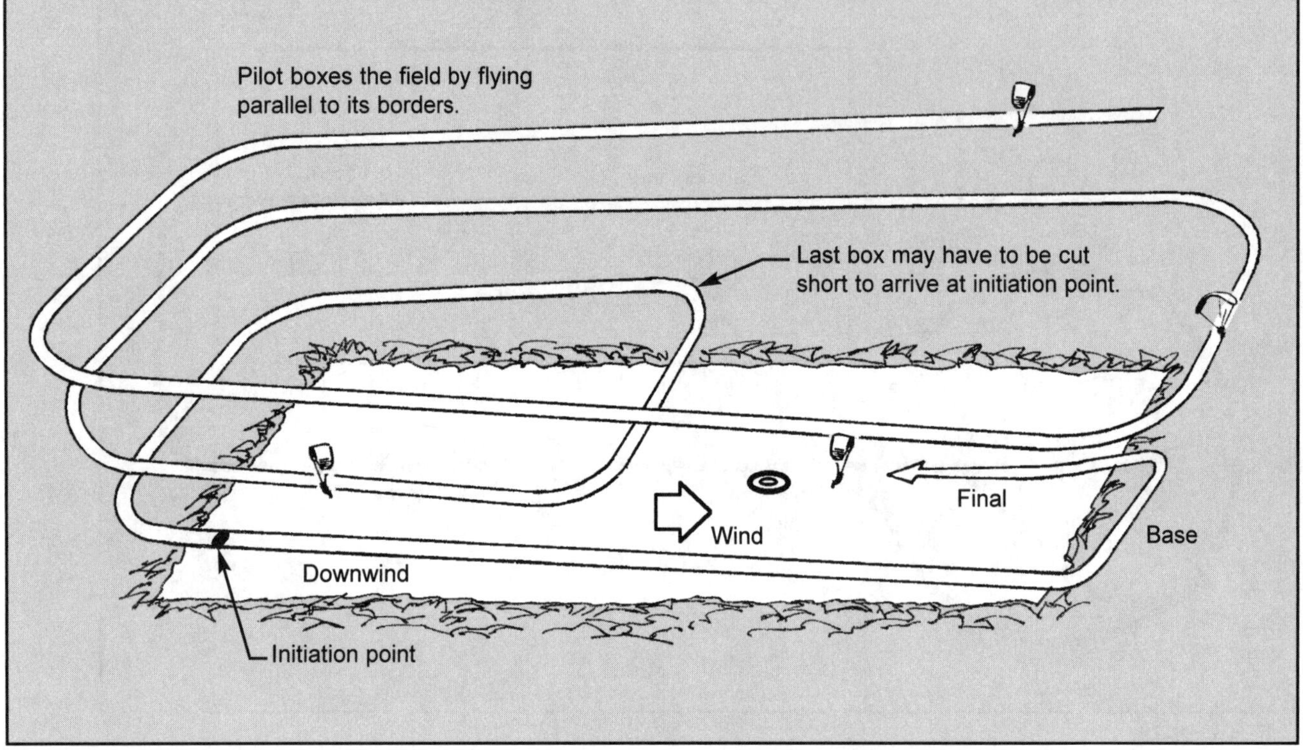

Figure 8-20: Boxing the Field Setup

When using the boxing method, you should arrive at the field with the standard minimum height of 300 feet (100 m). Remain within a 45° glide of the field as shown. In stronger winds you should stay closer to the field and in winds over 8 mph (13 km/h), do not go past the downwind end of the field. This policy ensures you don't get caught short trying to penetrate a head wind.

All your turns during the process of boxing the field should be in the same direction as the turns in your DBF approach. As figure 8-20 shows, the boxing process flows into the DBF. Just as with the 360 method, you should constantly monitor the winds and your position in relation to the initiation point. Remember to use angles to make this judgement. With each lap of the field you should recognize how much lower you are getting. Chances are you'll have to cut the last lap short or make it a bit longer to arrive at the initiation point, but remember, the DBF has a lot of room for adjustment.

One big advantage of boxing the field is that you avoid using 360's in your early high flights. One disadvantage is you sometimes have to make the last box very short which almost the same as doing a 360. However, in calm winds boxing the field is ideal, for you can turn in at any point as long as your field is large enough. Your instructor will tell you the best method to use according to the location and the training situation.

✓ Learning the DBF

We must first point out how important it is to learn the DBF at a large landing field in calm or very light, smooth winds. But even in strong conditions you can begin your learning by starting on the ground.

Your instructor may lay out a small-scale imaginary field on the ground with markers. Your job is to select a landing point (a clump of grass or rock will do) and walk through the entire setup while vocally noting each important point and position. The height of your eyes determines the angles you should be judging (see figure 8-21).

Remember your angles as you dangle in the breeze.

Find the initiation point (within 45° to the field border and 30° from the landing line, whichever is closer), walk the downwind leg and identify points A and B. Then judge the 3 to 1 path to the landing point and turn at C. Finally, walk the base leg and turn again at D with a 3 to 1 path to the landing point.

Your instructor will check your judgement at the critical points. If you are alone, mark them and then go back to pace them off as a check (a 45° angle is 2 paces, a 3 to 1 glide path is 6 paces). Practicing in this manner in both directions with a variety of imaginary landing fields is the best way to perfect your aim.

Once you have practiced enough on the ground it's time to perform in the air. Many schools add a tandem flight here so you can see the judgement from the air. On your own you should remember to relax, and focus on your landing point and your airspeed. With a large field your base leg can (and should) be long so that you have time to fly level and make your judgement for the turn to final. The main thing to note is that in calm winds you are essentially doing nothing new. The downwind leg is not downwind at all—you are merely flying past your landing point and then linking two 90° turns to land. You've done that many times at the training hill, haven't you?

Note that in all our previous discussion we chose our landing point about 1/3 of the way along the field (when heading upwind). If the field is small or obstructed on the downwind side, this is too short a glide path and the halfway point is a

Figure 8-21: Practicing Landing Judgement

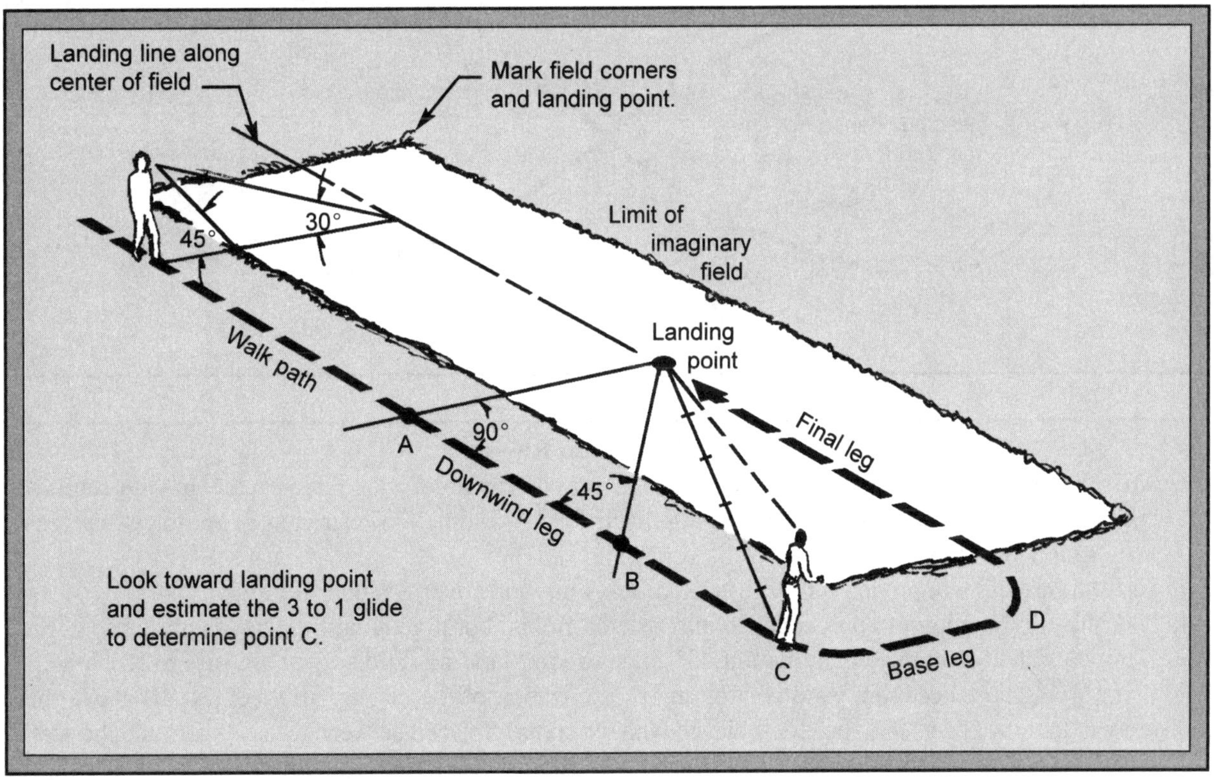

better choice. However, inexperienced pilots should only be landing in big, clear, comfortable fields. Leave the more challenging landings for later when you can control your position and path precisely.

The most common mistake inexperienced pilots make with the DBF approach is to perform the legs and the turns too slowly or to make the base leg too short. Remember to always monitor your airspeed and practice spacing your turns properly.

Skill Check: *Your downwind, base and final approach landing skills are perfect when you can confidently enter the pattern, fly straight legs that rarely require alteration, produce fast, controlled turns and enter a final with proper speed, right on course.*

✓ Landing Setup Rules

Here we summarize the rules for using our two setup systems. Note that only in higher winds do we suggest the figure eight method once you are equally able to perform both. The reason for this is that in higher winds it is harder with the DBF to judge where to turn to the base leg since the downwind leg occurs so fast. Also, the figure eight method gets easier in higher wind because the turns require less heading change. This is due to the crab angle, as shown in figure 8-22.

Figure 8-22: Figure Eight Setup in More Wind

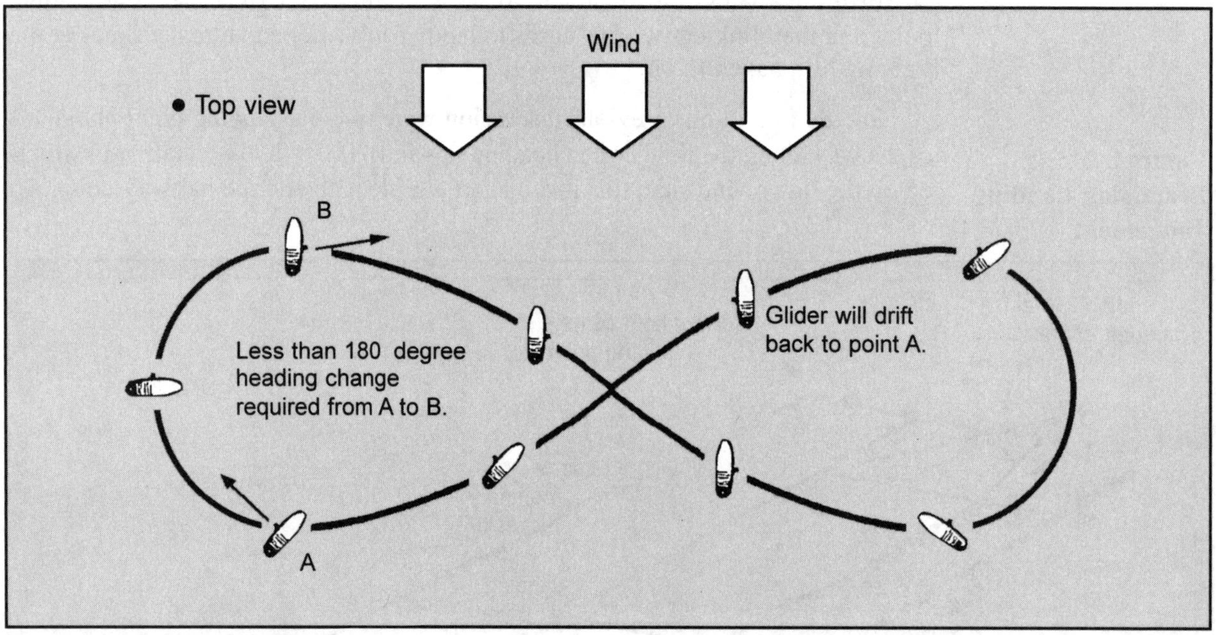

Landing Setup Rules

1. Arrive at your landing field with 300 feet (100 m) so you can assess conditions and plan your setup.
2. Use a downwind, base and final (DBF) approach for all landings in restricted fields for winds less than 8 mph (13 km/h).
3. Use a figure eight approach for all restricted fields in winds over 8 mph (13 km/h).
4. Use any combination of approaches in unrestricted fields (but practicing the DBF is preferred).
5. Below 300 feet (100 m) remain within 45° angle to your landing field so you are sure to reach it.
6. Use a flying speed offering good control and pressurization of the canopy 100 feet (30 m) above the ground. In most gliders this speed is achieved with the controls placed between the shoulders and the ears.

Landing Variations

All of our above discussion assumed perfect winds and a perfect field, if not perfect skills. However, in our imperfect world we will encounter many landing situations that are not so ideal. Here you will learn how to land in all types of terrain and conditions.

Your Goal: *To be able to handle all manner of wind conditions and terrain during landing which are within the recommended limits for your skill level.*

✓ *Wind Factors*

In all our discussion to this point we have been assuming a pleasant breeze and landing directly into the wind. Yet in the natural world we encounter all manner of variation from this ideal. Here we show how to handle many of the conditions we typically encounter.

Wind can make a landing easy—unless it is unsteady.

▲ *Calm winds landings* — Perhaps you have experienced calm winds at the training hill. If so, you will have learned how important a well-timed flare is and the importance of being prepared to run a couple steps. In addition to these matters, be especially cautious of the possibility of a gradient at some point above the field (if the winds are also light at takeoff this probably doesn't exist). Furthermore, be aware that heated layers build more readily in calm conditions, so expect the possibility of some lift or even turbulence on your final approach during a midday landing. In any case you should expect to glide farther on final in a calm than with a head wind.

▲ *Strong wind landings* — It should be clear that at some wind strength the turbulence along the ground will be too strong for control. This factor largely depends on what the surface cover is. On a shore or an open plain you can tolerate more wind in landing than you can in hilly or tree-covered areas.

One important caution in stronger wind is to be careful not to be caught downwind of the field. Do this by getting to the landing field high and avoiding the area on the downwind side. The volume of safe airspace gets shifted, as shown in figure 8-23.

Figure 8-23: Shifting Setup Positions Upwind

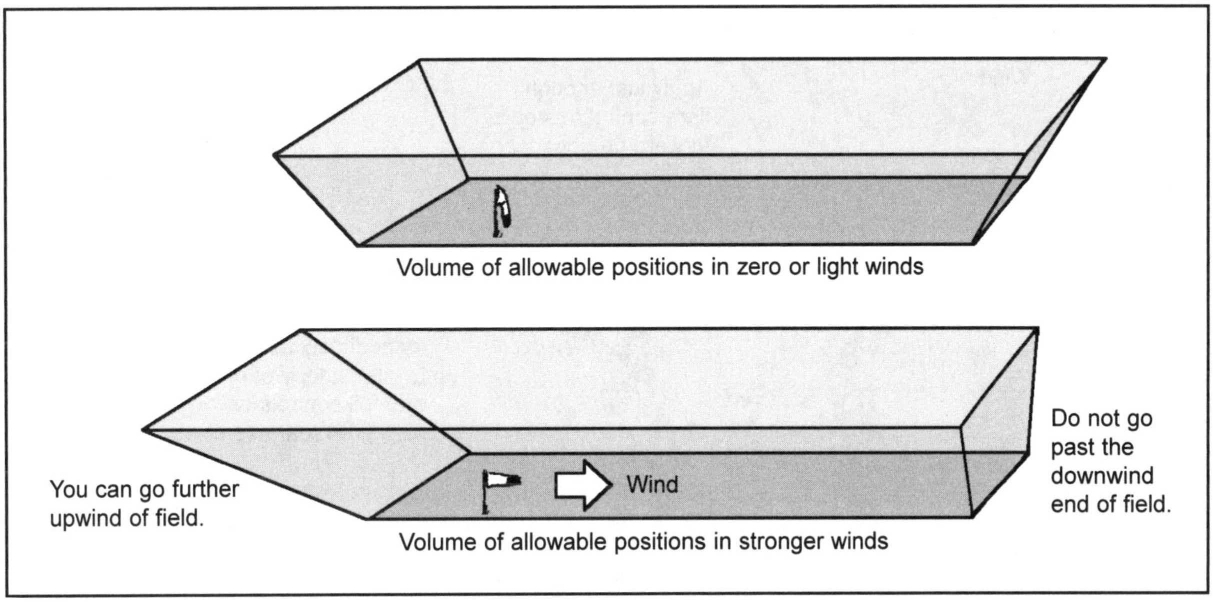

In very strong winds your descent on final may be almost vertical, so don't get caught short and use plenty of airspeed. Keep your glider flying straight with precise controls. Your flare should be a very short and smooth pull down on the control lines—just enough to stop your vertical descent. In fact, the flare and roundout become one and the same if you can't penetrate upwind. If you produce a full flare, chances are you will be pulled backwards.

As soon as your feet touch, collapse the glider by pulling down abruptly on the rear risers. Better yet, you should immediately spin around to face backwards as shown in figure 8-24. In this position you can kite the canopy or collapse it much more safely. If a stronger gust blows through you can follow the canopy. Collapse it with the rear risers or kite it off the field if conditions aren't too turbulent and more gliders are not landing.

Getting dragged after landing in very strong winds is a real possibility. Naturally you shouldn't be intentionally flying in such conditions, but sometimes pilots fail to heed signs of stronger weather. We deal with the matter of getting dragged in Emergency Procedures in Chapter 10.

▲ *Switchy wind landings* — Often a windsock will show variable winds, usually due to thermal conditions. Sometimes the variations can be from side to side in the field or even up and down the field. The best way to handle these conditions is to come in on final along a landing line that splits the difference of the sock

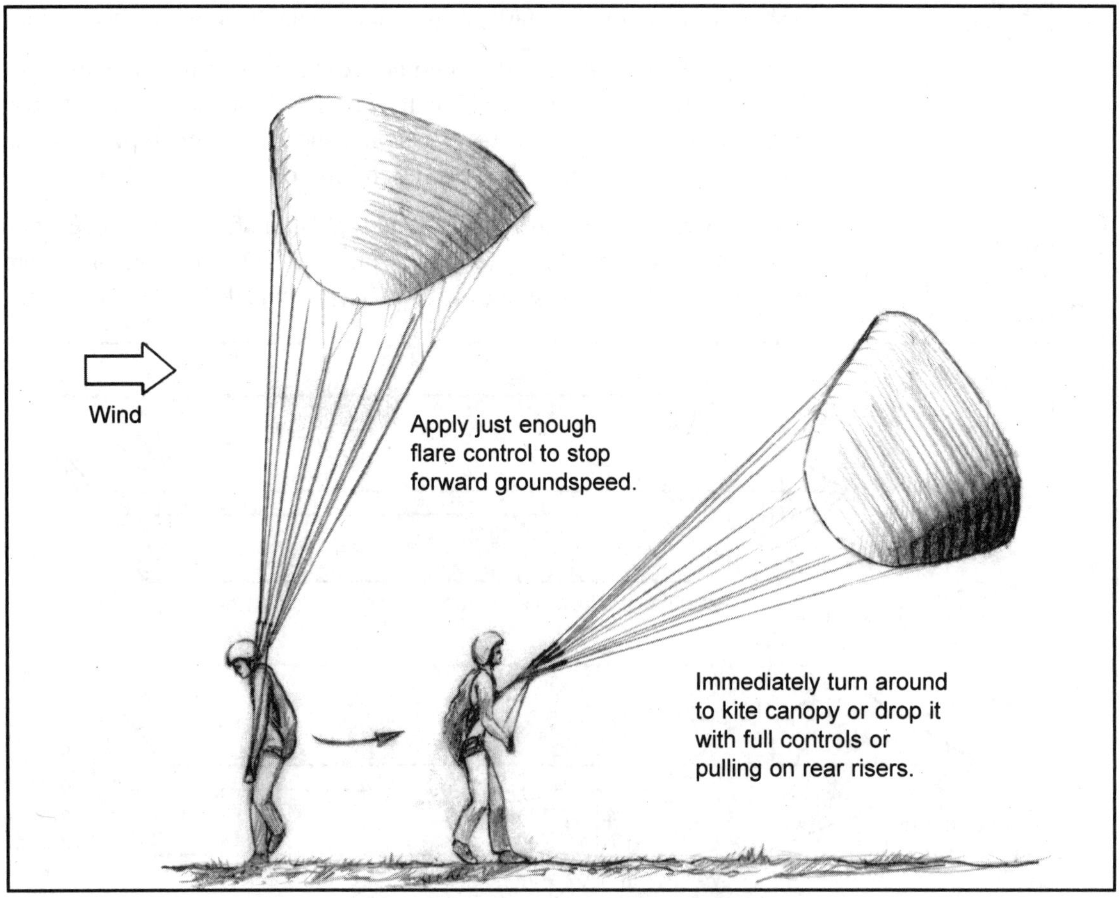

Figure 8-24: Stronger Wind Landing

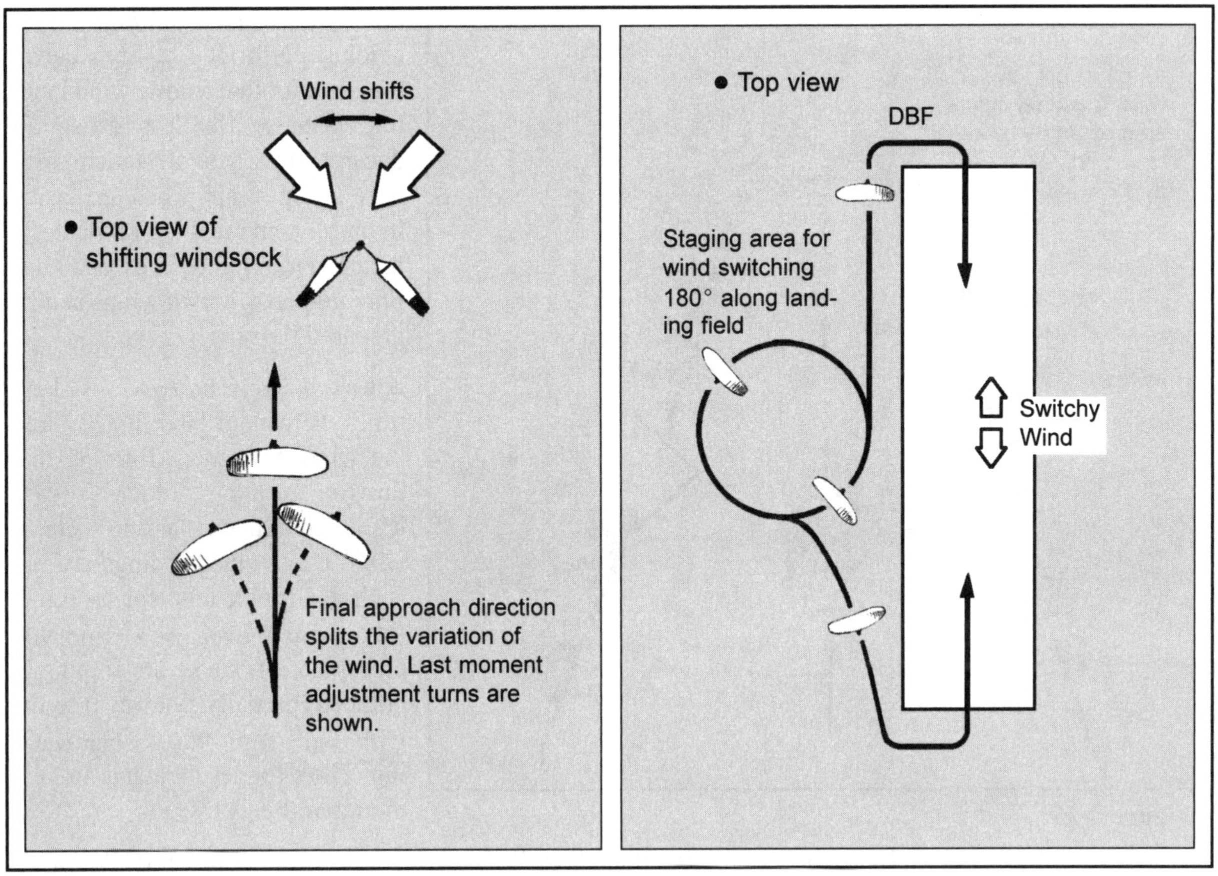

Figure 8-25: Handling Switchy Winds on Landing

indications, as shown in figure 8-25. Then you will only have to do a mild cross-wind landing (see below) at worse. Some more experienced pilots will add a slight turn at the last moment, but if not performed in an expert manner this can cause more problems than it solves.

If the wind swaps 180° on a low final you simply must land downwind, unless the field is wide open and you have enough altitude to turn. If the wind appears to be switching 180° when you arrive high at the field, the best procedure is to use a staging area midway along the field, as the previous figure indicates. Losing altitude in this staging area allows you the most options and you can perform a DBF in either direction according to the current wind direction when you reach the proper height. This procedure requires more skill if your field is not long, since the down-wind leg will be shortened and you have less time to judge your position. Pilots new to high flying are cautioned to only fly in consistent conditions.

▲ *Downwind landings* — Downwind landings are not always pleasant, but they can be performed safely in light winds. Sometimes it's difficult to tell if you are in a light downwind if you do not have a streamer in front of you. In any case, perfect landing technique will rescue you because a very strong flare (take wraps to insure this) produced with your feet right at ground level is required. You must be ready to run (as always) and do so the instant you touch down. Run means sprint. Note that world-class sprinters top out a little above 20 mph (32 km/h). With your harness on you'll probably do half of that. Clearly the upper limit of tail wind we can safely land in is below 5 mph (8 km/h).

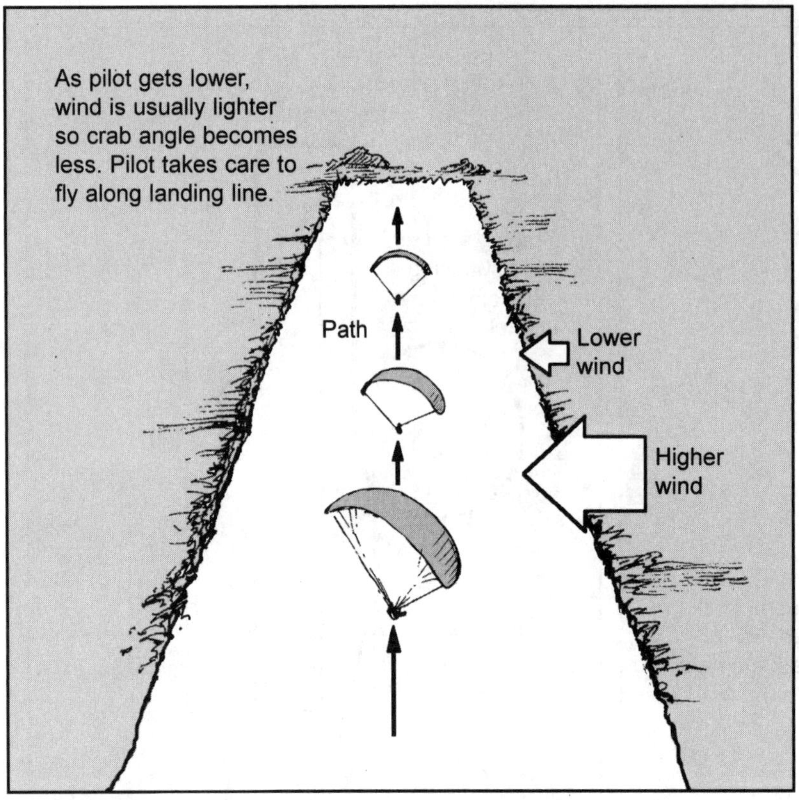

Figure 8-26: Landing in a Crosswind

▲ *High altitude landings* — A landing performed at high elevations is a lot like a downwind landing, because the less dense air doesn't slow you as much upon flare and true airspeeds are faster. In ample wind this isn't much of a factor, but in light winds you must flare vigorously with wraps and be prepared to run.

▲ *Crosswind landings* — Many times we cannot land directly into the wind without performing low turns or gyrations. In such a case it's much better to land in a crosswind. Crosswind landings are not such a difficult feat, for as figure 8-26 shows, even in a wind 90° cross to our field we are coming in crabbed (partially sideways) so the wind isn't truly 90° to our heading. The stronger the wind, in fact, the more this is true.

To perform a crosswind landing, fly directly along your landing line. When it's time to flare—at the right height and airspeed—do so gently and run a few steps so your angle of attack isn't too high, just as you would in stronger wind. This is a matter of judgement, for in light crosswinds there is little effect and you should flare normally. Expect your canopy to drop to the downwind side of you so be ready to step in that direction.

Sometimes turbulence will try to turn you downwind (to the left in our figure) as it rolls over the ground. Keep your maneuvering speed to the end and maintain positive roll control.

Pro Tip: *In crossing winds, land with ample speed, a roundout and a nearly level glide so that your glider can adjust itself to the crosswind component with little input from you. Some pilots develop a phobia for landing in crosswinds, but with the proper technique they're easy!*

▲ *Turbulent landings* — We have already mentioned the necessity to maintain good control speed on approaches. In turbulence you must maintain good control speed and good pressurization to prevent tip folds. As you will see in Chapter 9 and 12, these two requirements are not necessarily compatible. For the most part, you should keep your controls between your shoulders and ears during turbulent approaches. Make all turns and corrections coordinated and shallow. Weight-shift for efficiency unless conditions are too turbulent. In that case it's better to try to remain centered and just use the control lines.

One matter of concern is landing in the wake turbulence, or *tip vortices,* from another paraglider or worse, a hang glider. This is a particular problem on calm or stable days for the wake can linger for up to 30 seconds. As shown previously in figure 5-13, our wake consists of two organized vortices or swirls originating at each wing tip. The more slowly we fly, the more energetic they are. Away from the terrain they are no problem—they feel like a strong bump—but during landing they can cause a tip fold.

To avoid vortices, simply land beyond another pilot's touchdown point since vortices stop being produced the instant the wing stops flying. Also, be sure to fly above the preceding glider's path (see figure 8-27), since vortices sink. Alternatively, avoid landing within 20 seconds of another glider. Finally, if all else fails, try to land by the side of the preceding glider's track and carefully control your path because you may experience a tip fold. If the wind is crossing, land on the upwind side of a preceding pilot since vortices drift with the wind.

Figure 8-27: Avoiding Vortices During Landing

✓ *Landing on Slopes*

Just as sometimes we must accept less than ideal winds, we occasionally settle for less than perfectly flat fields. Here's how we handle these situations.

▲ *Uphill landings* – Uphill landings are dangerous on a paraglider except for the most gentle rise. The reason for this statement is that a paraglider doesn't have a lot of speed retention that can be used to glide up parallel to the slope. Avoid uphill landings, and use a side slope landing (crosswind if necessary) instead.

Slopes are very hard to detect from the air without lots of practice. Know your landing field well.

▲ *Downhill landings* - Landing downhill is like landing downwind: you can do it if your landing techniques are perfected and the downward slope isn't too severe. The way to do it is to float along as low as you can safely manage. You will travel much farther than normally—have plenty of runout room—so don't get impatient. The most important thing is to carefully sense airspeed to avoid a stall and produce a strong flare just before a stall occurs. Be ready to do a PLF for you may end up higher than you expect. The most common mistake in downhill landings is flaring too early. Note that there is an illusion that makes it appear that you are approaching too low.

▲ *Side slope landings* – Sometimes you may encounter a field that is sloped sideways to the wind. In a lightly sloped field the procedure is a no-brainer. Just land normally. For a steep slope use the techniques given next.

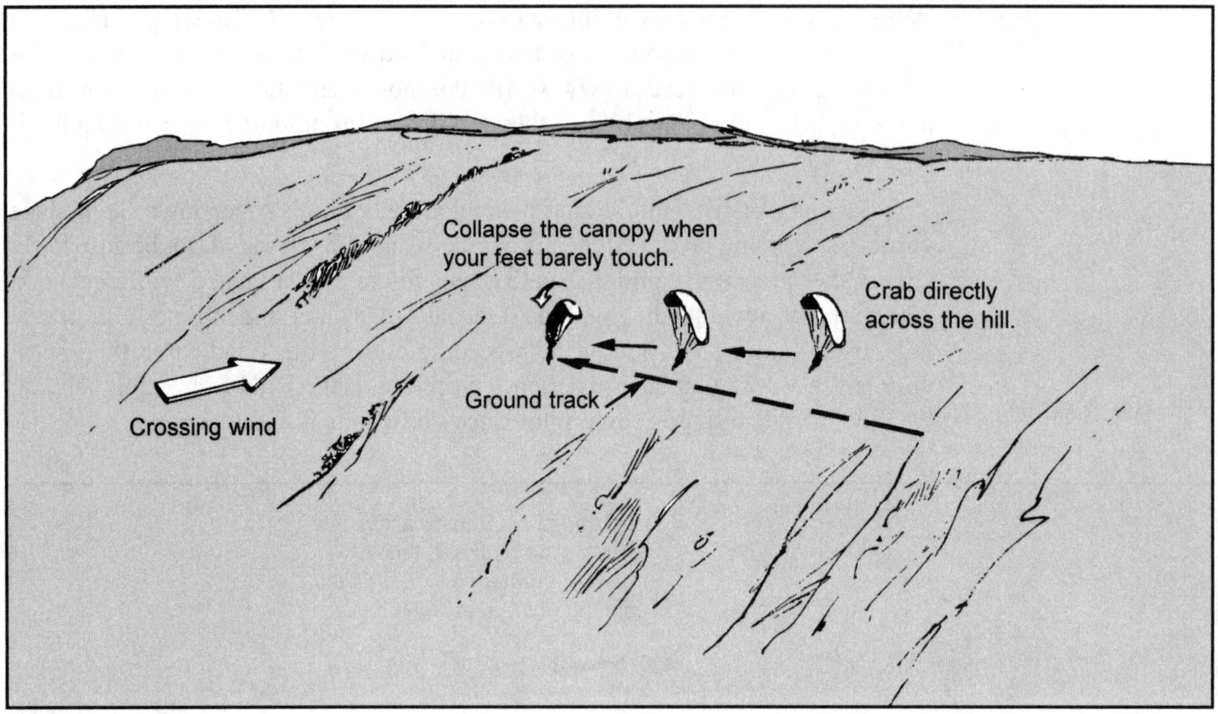

Figure 8-28: Landing on a Side Slope

At times pilots land on the side of soaring slopes to ease the climb back up to the top. The famous south side slope at Point of the Mountain, Utah is ideal for this practice. The main problem to avoid with this technique is flaring too high. Set yourself up for such a landing by flying into any quartering wind to reduce your groundspeed as shown in figure 8-28. If there is no wind or a straight-in wind, you have the option of landing in either direction across the slope.

With a wind you will be crabbing a bit (your canopy turned somewhat away from the slope as you fly directly across the hill). Maintain your path directly across the slope until you descend to where you can touch the ground. Keep your feet under you and prepare to run. Once you are low enough and your airspeed has reduced to landing speed, flare gently with the *uphill* control line getting about twice the pull of the downhill one. The idea is to drop the glider sideways on the hill. Be careful to avoid flaring too much, for you may rise up and come down too hard on the sloping terrain. It is best to run a few steps on a side slope.

Be sure to use more flare on the upslope wing to prevent the canopy from turning in the downslope direction and pulling you away from the hill. Finally, you may choose to barely flare at all, then continue kiting the canopy for your walk back up the hill (see Chapter 5). This procedure requires more skill and should not be tried until you have experienced multiple side slope landings. It is important to avoid reentering flight with a partially folded canopy.

✓ *Landing in Varied Ground Cover*

Legs are wonderful inventions. "Two-feet drive" can go many places where 4-wheel drive can only sit and spin. Consequently we can land in places an airplane could not. Here are a few situations you may encounter.

▲ *Rough ground landing* — Because rough ground is not easy to run in it is important to perform a non-running landing. This means producing a good flare to stop your forward groundspeed. You should practice your landings at an easy field to perfect no-step touchdowns before attempting to alight in rough fields.

▲ *Landing in high weeds or crops* — Sometimes, the weeds in our landing fields grow fast in the course of a couple of months and we are caught with a surprise challenge. Treat such cover just like rough ground and flare well to come down vertically. Running is out of the question in very high weeds.

Be aware that high weeds may grab your legs and foot stirrup while you are still flying. This is not a problem as long as you are aware of the possibility and flare before the weeds trip you. Usually you can let your legs drag in the weeds top, *but not your foot stirrup.*

Crops are handled similarly, although you should do everything in your power to avoid landing in crops (if you do, go to the farmer, apologize, and offer to pay for the damage. Sites have been lost by careless pilots landing in crops). If you land in crops carry your glider out as soon as you can gather it.

> *Once I landed in weeds taller than my head. I flared high, came down vertically and rolled upon landing. The only problem was finding my way out of the field and wrestling my glider through all the botany.*

▲ *Emergency landings* — Other hostile landing environments include tree landings, water landings and power lines. We will take up these subjects under *Emergency Procedures* in Chapter 9.

A pilot with full arm extension during the landing flare. Note that the canopy is still over the pilot's head. Taking wraps on the controls will help drop the canopy behind.

✓ Landing in Traffic

Even though you may be careful to avoid flying near other gliders at this stage in your aerial development, conditions and fate may conspire to have you nearing the landing field at the same altitude as another pilot. This can be a common occurrence at popular sites when lift dies in the evening. Here are a few procedures to carry you through such a situation without a conflict.

To begin, always look carefully for other air traffic as you approach the landing field and enter the landing phase. If you see another glider, don't stop there—keep looking. Clear all your turns. We have been in landing mode along with as many as five other gliders on occasion. It is important to track the course of all other gliders near your level and below you, so you do not suddenly get surprised by a looming white wing.

Next, assume the other pilot or pilots do not see you. That means you must stay out of their way (no matter who has the right-of-way—see Chapter 10). They may be less observant and concentrating on their landing setup. Position yourself on the opposite side of the landing field from their position, if possible, to lose your altitude. Try to perform a regular pattern with no sudden changes of direction. Broadcast your intentions (see Flying in Traffic in Chapter 13).

The most important matter is the general rule which all pilots should heed: *in a landing conflict situation, the lowest pilot should lose altitude as fast as possible*. This act provides safe separation between gliders. If more than two pilots are involved, the next lowest will follow the first one down after a few moments and so forth.

How do you lose altitude fast? Use your fast flight and diving turn skills. Once you are able to perform steep multiple 360s (see Chapter 13) you can use these maneuvers as well. If several pilots begin losing altitude at the same time, watch who is performing the steepest turns—he or she will probably lose altitude the fastest. If it isn't you, stop your dive to allow the separation to take place. Remember, there's plenty of time for several gliders to set up a landing if they arrive with a minimum of 300 feet (100 m) of altitude. Be sure to avoid any preceding pilot's vortices by remaining above or upwind of his path as shown previously in figure 8-27.

✓ Accuracy Landings

From the very beginning of our sport, pilots have been fascinated with hitting a spot. This enthrallment has caused no shortage of entertainment in landing fields worldwide, for more often than not pilots throw caution and all their well-versed skills to the wind in their eagerness to land on the spot, whether it is in good form or in a heap. Your goal is to develop judgement that allows you to both arrive where you wish in a landing and avoid endangering yourself in trying to hit a spot. Landing where you chose is an important part of safe and precise flying and should be the goal of all pilots.

A pilot lands on a boat ramp during water tow operations.

▲ *Running a line* — The first part of an accurate landing is becoming skillful at flying along a precise ground track. Since landing normally requires some turns for setup, this means you must be able to make exacting turns which begin and end on a perfect heading. The place to practice such skills is at altitude. Use a ground reference and perform 90° turns to specific headings. Two intersecting roads are perfect for this practice.

To describe the proper ground track, try to end your turn to final directly downwind of your landing spot (see figure 8-29). This allows you to level out and concentrate only on speed and glide control. If you have to make additional directional adjustments to head to the line, you will not be able to control your distance and perhaps not even achieve the line.

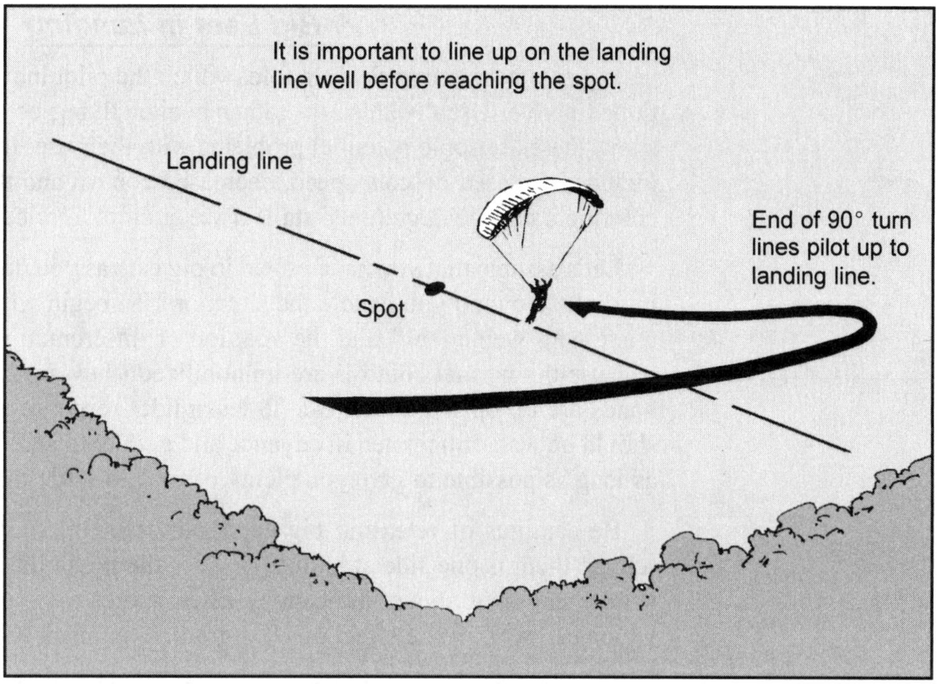

Figure 8-29: Accuracy Landings

So the first rule of accuracy landings is: *The landing begins in the setup. Follow the proper ground track and turn precisely to end the last turn directly on the landing line.*

▲ *Reaching a point* — The second part of accuracy landings is gliding the right distance. This too begins in the landing setup, for here we determine the height of our final and the glide ratio to the landing point. This is why we stress flying to a standard such as a 3 to 1 glide. By repeating the same judgement time after time the process becomes automatic like swinging a tennis racket or throwing a dart.

Spot landings help you practice your landing setup judgement. Don't change your flare timing for a missed spot or you will be compounding bad judgement.

Once the standard is engraved in your brain, you can consciously alter it for such factors as head wind or sink. This becomes a matter of more experience, but is not beyond the abilities of a novice pilot. By far the best thing you can do to develop your landing accuracy is *to attempt to do an accurate landing every time you land.* If your usual flying site doesn't have a course laid out, you can simply spread an old towel for the center point and use a line parallel to the wind through this center point as the landing line.

The best way to alter the steepness of your glide path on final approach is to vary your speed. Be very careful not to fly too slowly or to dive radically at the ground. Do not use your foot stirrup close to the ground. You may try sitting more upright or standing to increase your drag if you are too high. On the other hand, fly at best glide over the ground (see Chapter 9 and 12) if you are low. At the novice level, develop landing accuracy in the air before you turn on final. By all means, flare when it's time to, not when the spot says "you just passed me."

Perhaps you have seen or heard of pilots using "big ears" (intentional tip folds) to increase the steepness of their glide path on final. This practice is dangerous for several reasons which we will cover in the next chapter. We'll discuss landing with big ears below, in case you pulled them in while at a high altitude.

✓ Big Ears in Landing

Big ears are symmetrical tip folds which the pilot induces in his wing as mentioned above. Usually this is a rather benign flying configuration, but as we'll show, there are some potential problems with their use. In landings, the problems include increased descent speed, decreased control and the (small) possibility of entering a *parachutage* (deep stall) if we attempt to release them when low.

Let's assume that you have pulled in big ears as you descended over your landing field. How do you handle the situation? To begin with your controls must be made with weight shift and the addition of differential big ears (see Chapter 9) because the normal controls are immobilized (they are partially folded and your hands are up on the lines). Due to less glider response in this mode, your turns should be well-anticipated in advance and gentle. In addition your final should be as long as possible to give you plenty of time to settle things down.

Be cautious of releasing big ears below 100 feet (30 m). A benign way to release them is one side at a time. Besides the possibility of a parachutage (deep stall) mentioned above, the canopy often surges a bit upon release of big ears which can worsen the effect of wind gradient or shadow.

Landing with big ears is not too difficult once you are set up on final, but remember, your descent is greater than normal and your path is steeper. You will not have much speed retention so a roundout is out of the question. Your flare timing is more critical. A full flare should be produced by releasing the big ear lines and bringing your hands down all the way in one smooth motion when your feet are about 3 feet (1 m) above the ground. Don't wait too long, for you will be descending faster than normally. Of course, a little wind helps reduce forward motion.

When you hit the controls, the big ears will probably pop out and help slow your descent. From this point you can see that making this transition too early is not desirable. Once you touch down, all remaining procedures are normal.

A pilot flies with big ears (intentional tip fold) in his wing.

✓ Obstructed Landings

Sometimes a landing field presents an obstacle course due to a tree or two placed just where you don't want them to be. Also, cars, other gliders, spectators, cattle, ditches, surprise fences, etc. can suddenly complicate your landings. All pilots must learn to improvise a bit on landing to deal with the unexpected.

The first rule of defense in such a situation is to always allow enough height during landing setup to detect such potential problems. Also, allow a high enough final so that you have options such as overflying or maneuvering around obstructions. Finally, try communicating with people on the ground if they can remove the obstruction.

A very real and common reaction of inexperienced pilots when confronted with an obstruction such as a lone tree in a huge field is to watch it and unconsciously steer into it. We have a tendency to go where we look.

A lady pilot was soaring a ridge in Pennsylvania with a tall radio tower protruding through the trees. She described how the tower seemed to attract her glider and the closer she got, the more the pull until finally she crashed right into it! Luckily she escaped injury, but was bewildered as to what had happened.

Our lady friend was a victim of *object fixation.* Some people experience it more than others, but it is a common enough phenomenon that all pilots must be aware and beware of it. To avoid object fixation remember:

Safety Tip: *Always look* where *you want to go,* not *at what you want to avoid. We have a strong tendency to fly toward where we are looking.*

As an obstruction gets closer, the tendency is stronger. You can keep an object in your peripheral vision, but *never* look directly at it unless you are about to hit it.

✓ The Last Act

The final thing to do in the whole landing process is to move your glider off the field. This is both a safety procedure and a courtesy for other pilots. Accidents have been caused by pilots neglecting to move from the landing area. Don't be the cause of distress and stress in another pilot and glider.

A common mistake occurs when pilots move their gliders off the field, but each successive glider extends a little farther out than the one before. To protect your glider and avoid this landing field shrinkage, always move as far as you can off a field if others are apt to come in.

Pro Tip: *Many pilots have run out of landing field due to an unconscious desire to land near the staging area which may be at one end of the field. This problem is solved by moving the cars and packup area alongside the midfield if possible. In any case, be sure to land midfield and carry your glider to the packup point. It's better to do that than run out of runway and possibly have nothing left worth carrying away.*

If you happen to have a bad landing, get up and walk around as soon as possible, so pilots in the air, at launch, or across the landing field will know you are not injured. This is a matter of courtesy and common sense. Lying down near your glider in the middle of a landing field can cause unnecessary alarm.

SUMMARY

Landing is the last act of a three-act play. The takeoff sets the scene, then the action unfolds in flight. Finally, we have the conclusion where the principal actor attempts to awe the audience with his imitation of a bird landing. Perhaps your flights aren't as dramatic as all that, but your landings are as important to a good flight as the last act is to a play.

We have explored landings from the bottom up because that's the order in which we learn landing skills. But all aspects of the landing program are equally important to a pilot wishing to fly a variety of sites or conditions. Part of being skillful is being versatile. With the techniques presented here you can land safely in almost any designated field. Learn them well.

In the course of your experience you will be adding to your landing abilities. Still to come is landing back at takeoff in soaring conditions. But the main thing for you to focus on at the end of every flight is to perform flawless fearless landings using the judgement techniques we have discussed.

TEST YOURSELF (*Answers in Appendix II*)

1. We can break final approach into ____ phases.
2. The two main types of landing setups are: ____________ and ____________.
3. Which of the setup types is best for your first high flight? _______________
4. Which one is best for winds over 8 mph (13 km/h)? _______________
5. Which one is most versatile? _______________
6. In order to properly judge height and position, we must learn to judge ___________.
7. When landing in wind the potential dangers are _________________________.
8. In higher winds our flare control should be:
   a. ¼ control c. ¾ control e. Just enough to stop forward speed
   b. ½ control d. full control
9. Crosswind landings require last minute turns to land safely. *True or false*
10. Uphill landings are easy because the wind is moving up the slope.
    True or false
11. When landing on a side slope it is important to flare hardest with the (*uphill or downhill*) control line.
12. Big ears should be released on final in order to slow our descent.
    True or false

Chapter 9

LEARNING NOVICE SKILLS

"We do not see first, then define. We define first and then see."

Walter Lippman

Now it's time to take the next step. You are on your way to becoming a pilot able to make your own decisions and to fly in a variety of conditions at many sites. In order to graduate from the training hill you have to learn to fly with perfect speed control, perform efficient turns, handle more wind, operate at higher altitudes and understand the nature of stalls and folds. Also, you must develop good judgement.

Does this sound like a lot of adventures? Yes, there will be adventures because learning to paraglide is fun and exciting every step of the way. For example, once you master turns you can place yourself where you wish within the limits of your hill. Once you begin flying at altitude, the sky is your new world to explore.

As you progress through all these novel experiences and new ideas, constantly reflect on how the theory presented applies to what you encounter in the air. Go back and forth from experience to idea to experience so that your learning is both in the body and in the brain. Different types of individuals learn in different manners, but for best progress, a certain amount of both practice and understanding is necessary.

All through your flying career read, ask questions and think about your flights. That way mysteries are revealed and new insights develop, helping you to perform with ease. Use your senses. Be conscious of what you see, feel and hear as you glide along. Your awareness is your key to successful flight in all learning phases.

COMFORTABLE FLYING

In this chapter we are going to cover all manner of canopy control including efficient flying, turning and handling turbulence. However, let's first begin with getting comfortable in simple straight flight.

Your Goal: *To learn to relax in the harness, sit back properly and maintain a steady canopy.*

✓ *Easy Chair Flying*

Up to now on the training hill you may not have had the time to get really situated and comfortable in the harness. When your airtime is measured in seconds there's not much time to do anything but concentrate on controlling your path. Now, as you start flying higher, we want you to start relaxing. Sit back in the harness. If your rear doesn't go all the way back, you may have the leg straps too loose. If you find it necessary to use a hand to scrunch back in the seat while you are in the air, do so after you have stabilized in flight and are out of traffic and there is no turbulence. *Do not* let go of the controls, but put them both in one hand while you make the adjustment as shown in figure 9-1. In the drawing we show that you can apply speed control (up and down) and turn control (side to side) with the controls in one hand. Don't fly too long in this mode as control is reduced.

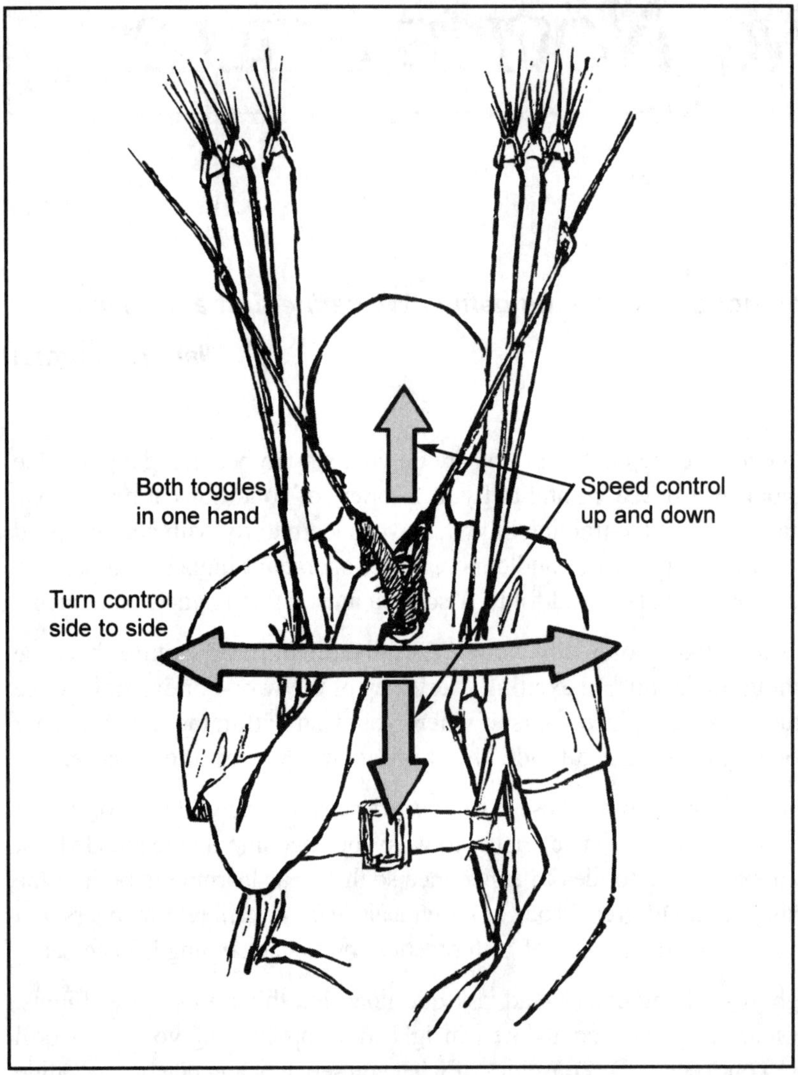

Figure 9-1: Controlling with one Hand

Spend some time in a simulator getting your harness adjusted. Have your instructor help you. If your harness is properly fitted it should be possible to sit back into it after launch without using a hand to assist you.

Be sure to get comfortable leaning back in the harness—that's what the back support is for. This position allows you to look at the canopy more easily. You should not be staring at the canopy, but you may glance at it occasionally, especially when you are learning to correlate the feel of the control pressures with the canopy position. Your main eye focus should be on what's happening around you, namely, where the other gliders are and how much clearance you have from the terrain. Be sure to take in the view as well. As you get higher you'll have more time to appreciate the serene beauty of flight rather than be solely overwhelmed by general rush. Cross your legs and have fun.

✓ Controlling Oscillations

The canopy is moving through a textured fluid—the air. It will react to variations in the air by slowing, surging, rocking from side to side, accordioning and possibly folding a tip. It is your job to dampen out extraneous canopy movement.

The first matter to attend to is to maintain adequate pressurization. Do this by flying more slowly. Even though flying faster actually increases dynamic pressure of the air, flying slower (at a higher angle of attack) presents the leading edge openings more directly into the airflow (we discuss this matter more in Chapter 12). Once you are away from the terrain and have more experience controlling and judging airspeed, you will begin flying at a slower speed. Keeping the controls in a position equal to your arm weight is a good initial pressurization position.

The next matter to control is fore-and-aft oscillations. When the canopy moves back above your head, the angle of attack increases and you should feel more pressure of the control lines (except when it stalls as we see below). When the canopy surges forward the opposite effect takes place (see figure 9-2). It is your job to learn to feel the control forces relating to canopy position as soon as possible. This practice should begin during ground handling so that you can automatically feel the canopy position by the time you are flying high. For best control and sensitivity watch the horizon—don't look up or down.

Figure 9-2: Feeling Canopy Position Through Control Pressure

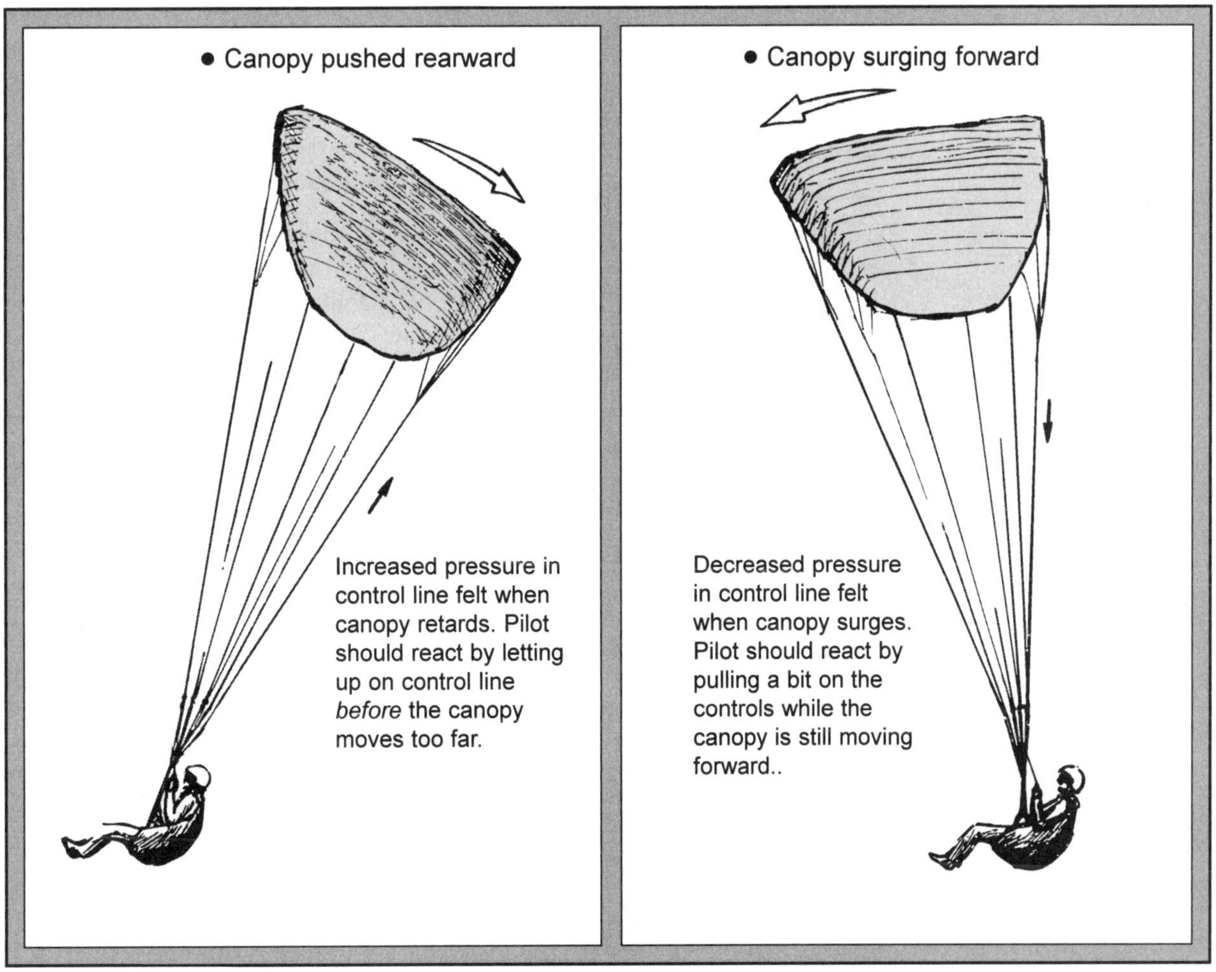

Practice canopy sensing until you can automatically feel the canopy position. While you are learning this, start picking up on more subtle clues such as the riser position and the pull in your seat. When the angle of attack increases, you should feel more G forces (a push in your butt) as the glider climbs momentarily. When it surges you should feel a short decrease in seat pressure. Using the seat and brake pressure feelings is the secret to detecting canopy behavior.

Right along with "feeling" the canopy you should learn to control it. Perhaps the best way to do this is to maintain an even pressure in the controls. If you feel them lighten (a surge in the canopy), increase pressure as indicated in the figure. If you feel them increase force (dropping back of the canopy) you should lighten up. With the proper timing (no delay), a smooth action and the proper amount of control movement, you keep a rock-steady canopy in all but the strongest turbulence. The proper amount of control depends on how vigorously the canopy is moving. Generally the amount of control input will be less than 3 inches (7.5 cm).

As you gain skill, you will experience fewer oscillations because you will catch canopy excursions more quickly.

This constant control procedure is called *active flying*. It is the process that all pilots must learn and use in order to fly safely in varying air. Think of it as balancing the canopy in a rolling, flowing fluid. We discuss active flying again in Flying in Turbulence later in this chapter.

Side-to-side oscillations dampen themselves out fairly readily. They are caused either by turbulence or too abrupt a turn (one control line) input. Avoid such oscillations by making smooth, gradual controls. You can dampen out side-to-side oscillations by gently pulling on the controls of the lower wing just as your body passes the bottom of the swing as shown in figure 9-3. Oftentimes a yaw (one wing moves forward) occurs during side-to-side oscillations. In this case you should steer with the proper control to establish the desired heading.

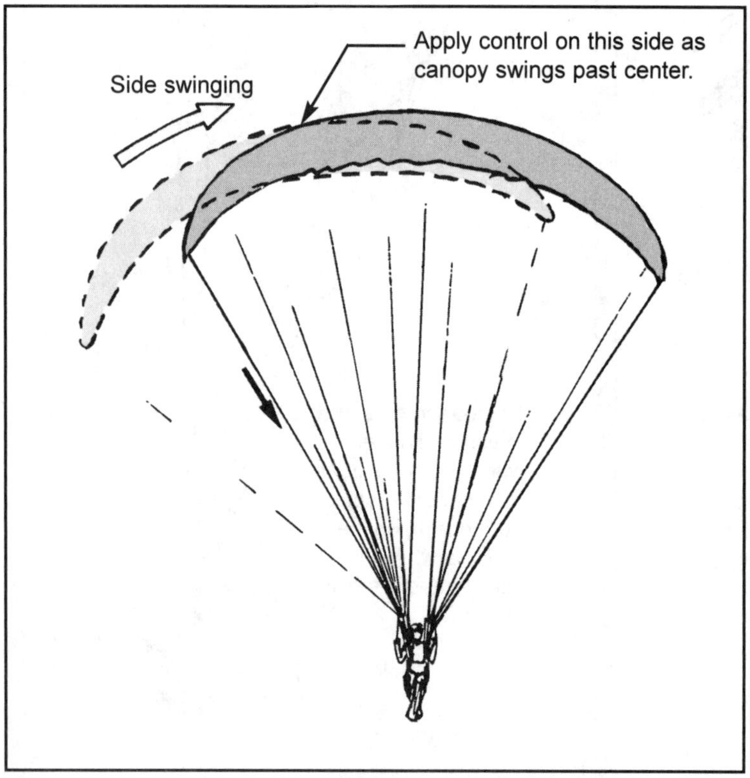

Figure 9-3: Controlling Side-to-Side Oscillations

It is important to keep your hands away from the risers so the hands can float. Don't stiffen your arms—relax. Your arms should act like shock absorbers in a continually active suspension system. Your hands should seek a constant pressure—about 4 lbs (2 kg) or the weight of your bent arms—that serves to recenter the glider and keep it stable.

Practice maintaining a well-pressurized steady canopy in all your flying from now on. If your timing is off, you may make the canopy movement worse, but that is why you are practicing. With a significant canopy displacement you may have to apply a couple of controls such as a quick let up, then a pull before returning to the normal flying position. We will deal with folds later in this chapter.

Skill Check: *You have learned to stabilize the canopy when you can fly through mildly varying air without the canopy moving a foot (30 cm) or more from its normal position.*

✓ Harness Adjustment

Your harness adjustment has an effect on the stability of your canopy in side-to-side action. Here's how it works: your chest strap holds the two sides of your harness together in the front. The tighter you adjust the strap, the closer the two carabiners are that attach the risers. Close carabiners mean a triangle is formed by the canopy and the lines that is less apt to be disturbed as figure 9-4 illustrates.

With close carabiners, if one side of the wing gets lifted your whole body has to swing to the side against the force of gravity and your inertia. With spread carabiners only a tilt of your body is necessary for a wing to lift. The glider remains more steady the closer the carabiners. On the other hand, you have less weight shift control (see below) with close carabiners. We suggest novice pilots begin with a tighter chest strap adjustment and gradually use more spread adjustment as they gain experience, skill and comfort.

Chest Strap Adjustment

- *Tight adjustment or close carabiners*—Glider is more stable in varying air but less responsive to weight shift.
- *Loose adjustment or spread carabiners*—Glider is less stable, but responds to weight shift.

Figure 9-4: Harness Adjustment Effects

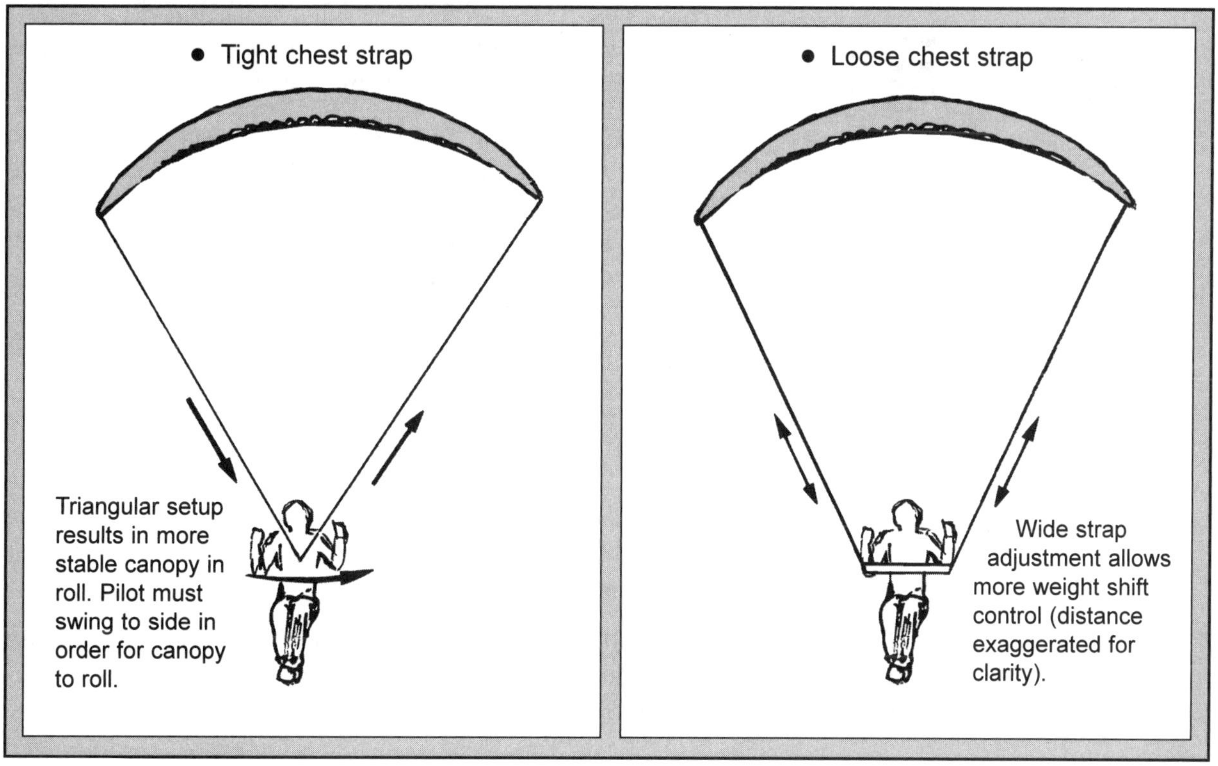

SPEED CONTROL

Up to now we have been carefully flying with our controls in a position initially for good response, then perhaps later for good pressurization and better efficiency. Now it's time to really get a handle on this speed thing and understand the limits as well as how we can maximize our performance.

Note that you cannot explore the full range of your glider's speed capabilities at the training hill. Your ground clearance isn't enough for very slow flying or faster flying. Naturally, we expect you to learn your glider's speed range gradually as you gain altitude and experience.

Your Goal: *To learn to fly comfortably in the entire speed range of your glider and to know where it performs best.*

✓ Slow Speed

You have been mainly remaining in the faster regime of your relatively slow wing. So let's slow down. First you need a general picture of the range of speed controls. The illustration in figure 9-5 shows the pilot control positions and how they relate to typical glider speeds. At the bottom of the figure we have drawn a graph of the speed range of a typical beginner/novice glider. Note the speed labeled minimum sink and best glide on this graph. These factors are explained in more detail in this chapter and Chapter 12. Also be aware that control line adjustment affects glider reaction to controls. We cover this matter in Chapter 11.

For now let's look at minimum sink speed. This is the airspeed at which you will descend the slowest. In other words, your flight will last the longest from a given height. This is the speed that soaring pilots often use to climb in ridge lift.

Safety Tip: *In thermals we recommend flying faster than minimum sink speed to avoid inadvertent stalls or spins caused by thermal turbulence.*

In flight you must constantly control your speed as if the Georgia highway patrol were watching.

To achieve minimum sink speed you must fly about 16 mph (26 km/h) which relates to a control position about at your shoulders (1/2 controls). Note that you are only a couple miles an hour faster than stall speed. For this reason, you should not fly minimum sink speed until you are well clear (at least 100 feet or 30 m when you first try it) of the terrain. Approach this speed gradually by first becoming familiar with the speeds and feel of the glider from trim to ½ control.

Note: We will deal with stalls under the next heading below.

✓ Best Glide Speed

The next speed to precisely define is best glide speed. Chances are you have been flying at this speed for it usually occurs with the hands (controls) near ear level. However, let's be more precise.

First you must understand what best glide means (see Chapter 12 for more). As you fly along in still air you are constantly descending. How far you move forward as you descend depends on your glider design and at what speed you fly. The

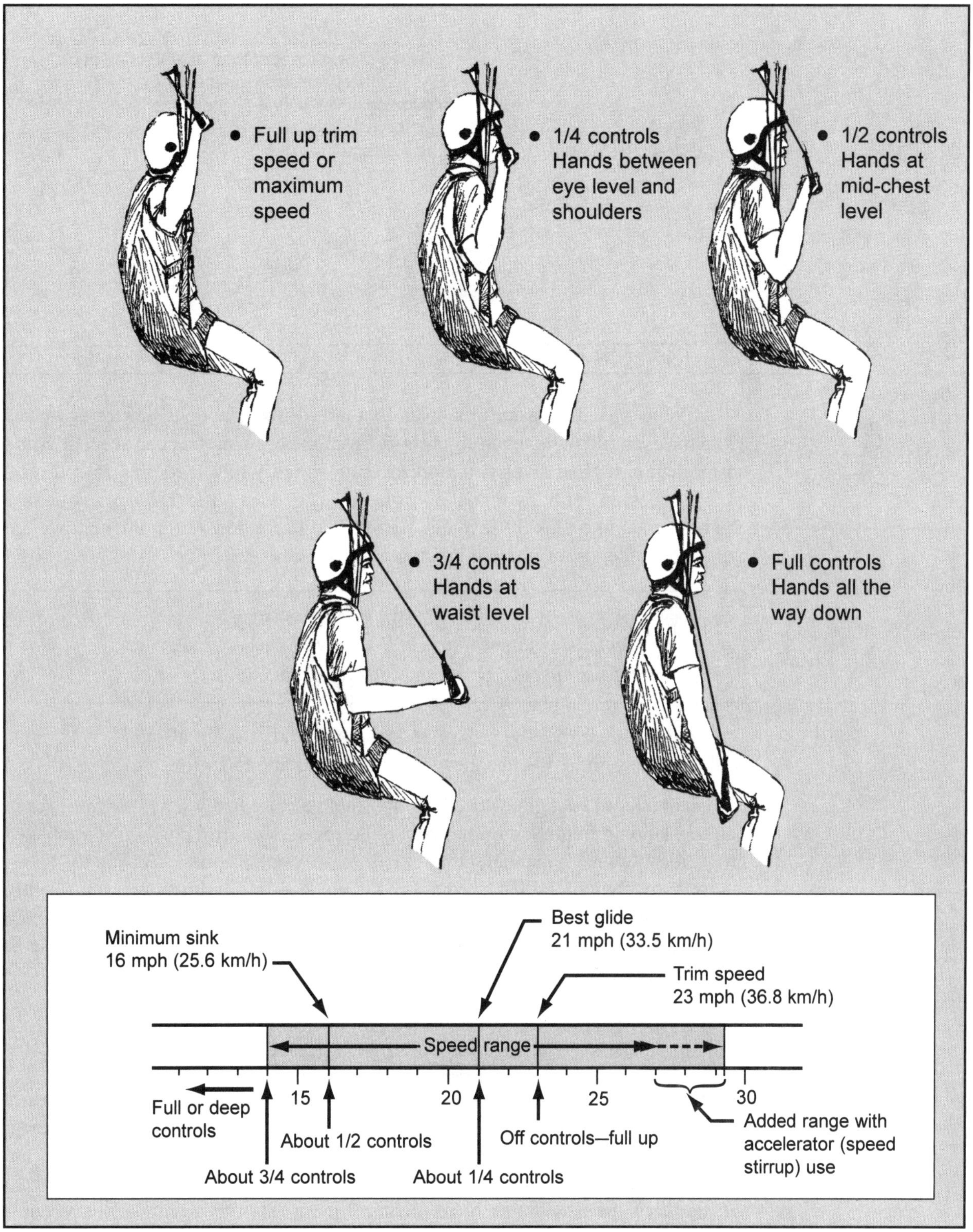

Figure 9-5: Glider Speed Control

fraction of forward distance divided by vertical descent is called the *glide ratio.* There is one angle of attack and related airspeed which allows your glider to go the furthest forward for a given vertical descent. This speed is your *best glide speed,* and your *best glide ratio* is the best possible forward distance covered divided by the vertical descent.

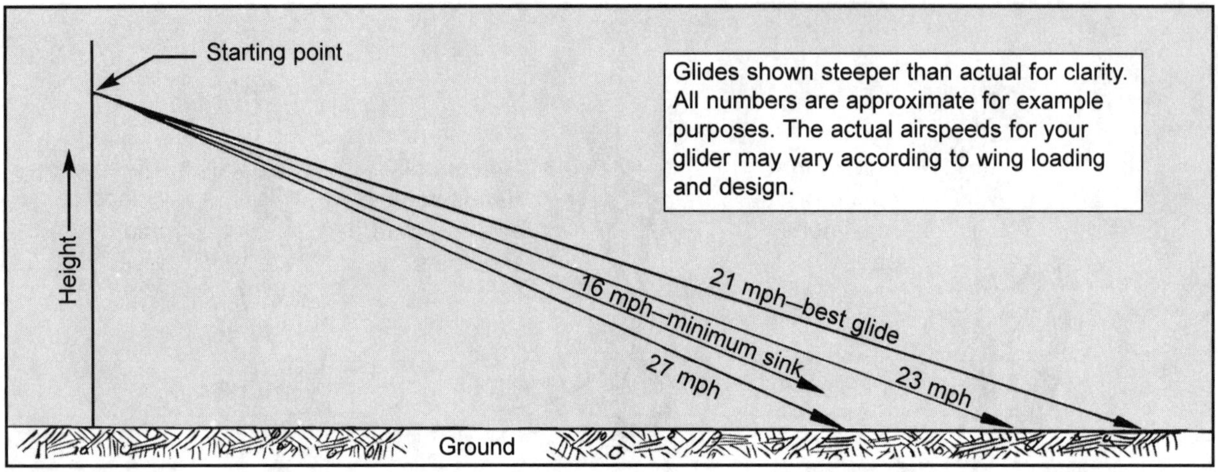

Figure 9-6: Effects of Speed on Glide

All other speeds—faster or slower—would result in landing shorter as we see from the example numbers in figure 9-6. From the drawing you can see that going faster than best glide dives you more steeply while flying slower makes your vertical descent slower but you make less headway. If we consider the descent along the best glide line took 30 seconds, we see that if we flew minimum sink we are still a few seconds away from landing. Our conclusion:

Still Wind Performance

- Fly *minimum sink* airspeed to get the slowest descent rate.
- Fly *best glide* airspeed to get the furthest distance.

We'll see later that the minimum sink rule holds true in all wind conditions, while the best glide rule changes with head wind and tail wind.

The matter of best glide speed is important because many times we must reach out to a landing field. Sometimes best glide speed or control position is published in a glider's owner's manual. Often it isn't however. So you should find the speed yourself by flying in still air from a moderate size hill multiple times at slightly different airspeeds (control positions) held steady as long as possible to see which one gets you the furthest. Note this best glide position well, for it will be one of your primary control and airspeed references.

✓ Fast Speed—Speed System

We have already discussed fast speed (hands nearly all the way up with only the slightest control pressure) in our earlier lessons at the training hill. However, there are a couple of wrinkles to add. They concern glider designs. If you are using a glider of a bit more performance than a trainer, chances are it will have a greater dive speed with nearly zero control input. This speed might not be comfortable close to the ground, so get used to it with ample clearance (no less than 100 ft or 30 m).

The next point is that many gliders from novice level on up come with a speed system. This system consists of two lines that route from a *stirrup* for your feet up through a couple of pulleys to attach to the A and B risers (usually) as shown in figure 9-7. By pushing on the stirrup you pull down on these two risers and lower the angle of attack of the entire wing. Other names for the foot stirrup are the *accelerator* or *speed bar*.

Once you are familiar with your speed system, fly with both feet resting in it so it is ready for use. When turning it is normal to lift one foot up for weight shift.

Gliders with this setup are frequently trimmed to achieve best glide speed with the controls nearly in the "hands off" position (all the way up). So to go faster you have to push on the speed bar. Such a setup is fine because you can automatically achieve best glide by lifting the controls up. The only problem is, without any control input you cannot feel the canopy, so we recommend holding at least 1 pound (1/2 kg) of control pressure.

✓ Speed System Use

You now know a speed system speeds you up, but what are the limits of its use? The fact is, the more you push on the speed system, the less stable the canopy will be. That means the more likely it is to suffer a frontal fold. The reason for this effect is that with the speed system engaged, the leading edge openings are not aligned to the airflow and the front of the leading edge may be indented.

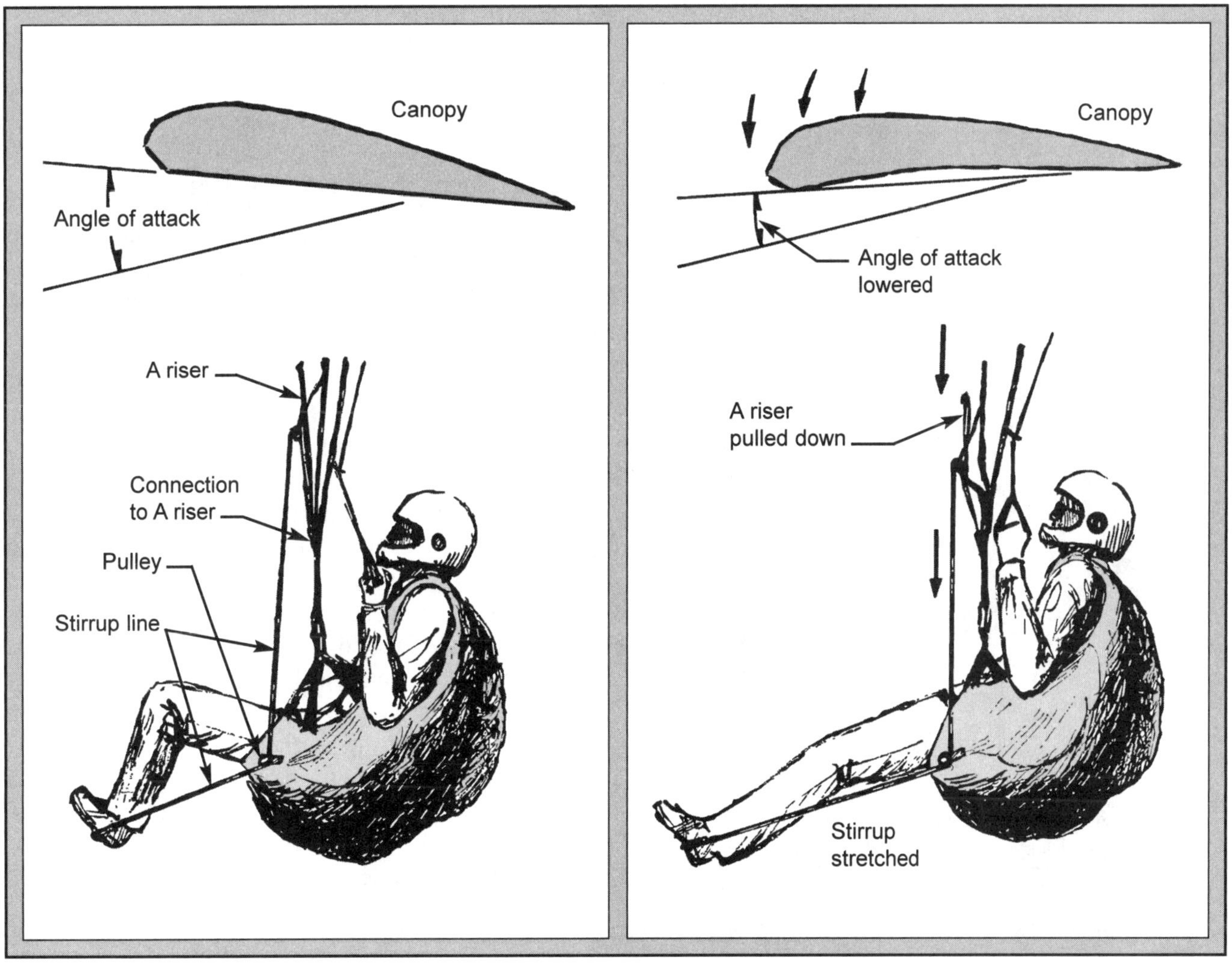

Figure 9-7: Stirrup Effects

Caution: *Do not use a speed system in turbulence, close to the ground (below 300 ft or 90 m) or around other gliders, due to the possibility of a fold.*

If you feel a lightening of the pressure on your feet, immediately pull your feet back to release the speed system and let the risers extend. Lighter pressure is a sign of an imminent fold. Once you are off the stirrup, apply a little control pull to repressurize the wing.

Caution: Do not *use control inputs when using a foot stirrup to full acceleration.*

The reason for this caution is that the application of the controls actually depressurizes the wing with the accelerator fully pushed since pulling the controls doesn't realign the leading edge openings while it slows the canopy.

It should be clear from these points that the speed system is used mainly for going in a straight line at speed. Despite our cautions, it is a safe enough system on beginner to intermediate gliders and you should become familiar with it to greatly increase your versatility. As with all the learning we promote, learn to use a foot stirrup gradually by starting with little inputs.

Pro Tip: *If you have short legs, add another crosspiece or cross line to the stirrup so you can get full extension in two steps.*

When both pulleys of the system touch you are full on.

A pilot's view of his speed system and spectacular scenery

✓ Launching With a Speed System

The lines and speed bar hanging around your legs can be a hindrance during launch. Some pilots ignore this matter, but it is far wiser to arrange your speed system so it is out of the way until you are stabilized in the air.

There are several ways to stow the speed system. The simplest is to stuff it between your harness and body. This practice isn't satisfactory unless you can find a place to put it where it absolutely won't fall out as you move around during takeoff. A better solution is to sew a small piece of Velcro on the harness and put one on the speed bar (use sticky back Velcro) so it can be stuck to the harness until needed. Another solution is to have a small bungee around your lower leg into which you stick one end of the speed bar. The advantage of this method is that you can use the other foot to stick in the stirrup and have the speed system ready to go without letting go of the controls. You can make a similar arrangement with a Velcro band around your leg and a Velcro patch on the speed bar.

✓ Speed Variation

The last thing to practice is a smooth transition from maximum speed to minimum speed and back. The idea is to learn to do this feat smoothly without any excessive diving, surging, ballooning up, retarding canopy or getting near stall. The secret is gradual controls with the correct speed of control. Practice until you can readily change speeds without displacing the canopy more than several inches. That's the secret of efficient and graceful flying.

Skill Check: *When you can automatically set the cardinal speeds (minimum sink, best glide) and change between them as well as accelerate to top speed you have learned to control your glider's speed with skill.*

STALL AND SPIN CONTROL

Now we arrive at two important subjects to all pilots (of all types of aircraft)—*stalls* and *spins.* We discussed the nature of stall in Chapter 3. We'll learn a little more about them here as well as their offspring, spins. We'll concentrate on how to prevent them.

Your Goal: *To understand the cause and prevention of stalls and spins and to recognize when your glider is near a stalled condition.*

✓ Full Stall

Unlike most aircraft, a paraglider can have two distinct reactions to a stall. The first we'll call a full stall which is the traditional reaction of a wing to a stall: The angle of attack increases, the smooth flow over the top of the wing breaks down into eddies, drag increases dramatically and lift is lost. When this happens on a paraglider there is often an abrupt loss of canopy shape and it will usually sud-

A canopy produces a rearward horseshoe when the pilot pulls too much and too abruptly on the controls.

denly drop behind as shown in figure 9-8. The pilot then falls until the canopy catches him, reinflates and surges forward. The experience is dramatic to say the least. It is also dangerous if it occurs near the ground or in turbulence which can further compromise the canopy shape or position.

To avoid such a stall, first learn the control position where a stall can occur. This position is usually with ¾ control or more. Your owner's manual or instructor can help you be more precise. Even modern gliders will stall abruptly if you make

Figure 9-8: Full Stall Effects

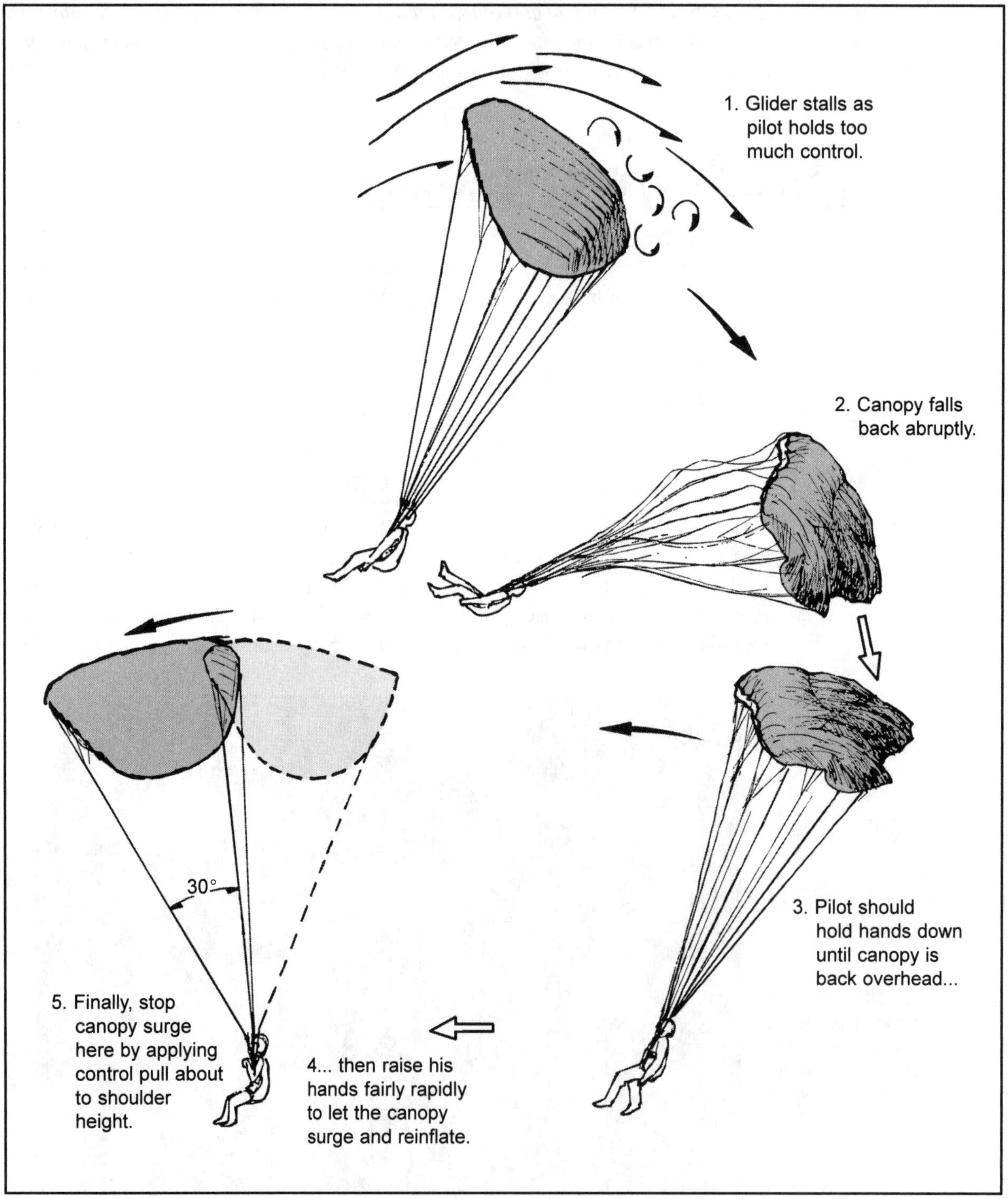

abrupt control motions and go way past this point. Again we emphasize the importance of performing smooth, gradual controls. CAUTION: *A stall can be extremely dangerous and should only be performed in a maneuvers clinic run by a qualified instructor. Much altitude can be lost very quickly in a full stall.*

The signs of a stall can be very subtle on a paraglider. It may be difficult for newer pilots to detect an impending stall. For that reason we recommend avoiding flying deep in the controls until much experience is acquired.

The most important signs of a stall are a feeling of diminishing airflow past your helmet (feel and listen), and a change in control line pressure. If your airspeed gets too slow (things seem to be getting quiet) after a deep control application, raise the controls *immediately*. If the feel of the controls suddenly lightens despite being deep in the controls, it's a sign that you have lost flow at the trailing edge and a stall is about to happen.

Signs of an Impending Stall

- Sudden loss of or very little airflow.
- Reduced sound or feel of wind passing by.
- Sluggish control reaction of the glider.
- Lightening control pressures.
- Buffeting of the canopy felt in the controls.

Note: Normally our control force feedback for canopy stability requires us to increase control pressure when we feel the line force lighten. In the case of a stall when you are deep in the controls, this is not the case. You will not feel a reduction of seat G forces (as you do with a surge) when the line force lightens in a stall, so that's one of your control clues.

If a full stall does happen to you, do not quickly move your hands up to relieve the control pressure. This action can result in the canopy reinflating with a sudden and rapid forward surge. It's best to hold the controls down until the canopy has stabilized over your head, then release the control equally in a fairly fast manner, being ready to stop a surge once it reaches 30° or more in front of you as shown in figure 9-8. Do not stop the surge too soon (the canopy must surge to recover) or a deep stall or spin (see below) may occur. The forces on the control lines may be strong and abrupt so it's best to hold your hands under your seat to lock them down while waiting for the canopy to settle above you. If you raise the controls asymmetrically one side can fold or a spin may result. Finally, if the canopy folds rearward (rearward horseshoe), letting up on the controls will bring the wings back.

Airspeed and control position awareness are your best defense against stalls.

Stall Control Summary

- Recognize stall signs and move the controls up without delay.
- If control line pressure is lost, the stall has occurred—be cautious of quickly releasing controls.
- Once the canopy has stabilized over your head in a semi-folded state, release the controls evenly, medium fast, and watch for the canopy to surge.
- Stop the surge once it reaches 30° forward using moderate control input.

If all else fails in your control process, hold the controls *evenly* at shoulder height and the canopy should sort itself out. It should be clear that it is highly desirable to avoid stalls throughout your flying.

✓ *Deep Stall*

Other names for *deep stall* are *parachutal stall* or *parachutage* (from the French who first discerned the problem). In this case the canopy doesn't experience an abrupt stall at the rear and drop, but enters a steady, nearly vertical descent with the whole wing stalled equally (see figure 9-9). The canopy is fairly stable in this mode. The signs of a deep stall are dramatically increased sink rate and the feeling of essentially no wind in your face. Control pressure may not go away. The descent rate is at least 1000 FPM (5 m/s) or 11.4 mph (18 km/h). If you hit the ground in a deep stall, this descent rate is equivalent to jumping from a height of at least 4.3 feet (1.3 m). Remember your PLF!

Figure 9-9: Deep Stall Effects

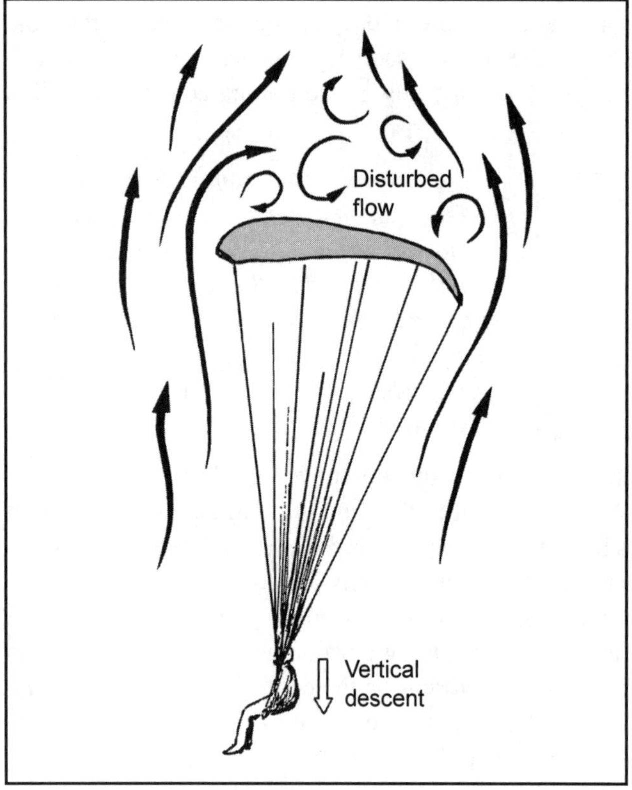

One problem with deep stalls is the descent rate noted, your lack of control and the possibility that you won't recognize the condition until it's too late. That's one reason why we advise maintaining ample flying speed near the ground. Another problem is a severe surge which may leave the canopy below you. The latter situation usually only occurs with an aggravated deep stall, i.e., one that you hold in with the controls.

In general, modern canopies will recover from a deep stall as long as the pilot puts his or her hands up all the way. This procedure should be your first reaction. If the glider does not recover immediately, either apply a quick push forward or short pull (down and up) on the A risers. Another popular method is to step on the speed stirrup. The idea is to lower the angle of attack and get the canopy to move from straight above you so airflow along the surface is reestablished. A little surge is what you want. These control actions pull the canopy forward a bit. An older technique of quickly pumping the controls to pull the canopy back then allowing it to surge forward is not recommended for if it is performed asymmetrically, a spin can result.

Deep stalls are caused by *slowly* pulling the controls too much, a slow release of pulled B lines (see Chapter 13) or the improper return to an open canopy from big ears (we cover this topic below under the heading Flying in More Wind). In general, modern canopies are resistant to deep stalls and you may never see one. However, you should be well-versed in how to recognize and get out of a deep stall. When canopies get older the wing gets more porous and lines can stretch or shrink. These changes render the glider more likely to enter a deep stall.

✓ *Spins*

A spin occurs when only one wing is stalled. That stalled wing retards because of the extra drag and the other wing rotates around it as shown in figure 9-10. The rotation can occur suddenly and the subsequent dive can be very steep. You have no control in a spin and will descend very rapidly to the ground with a fast swinging motion that usually results in a hard impact.

Obviously you must avoid spins at all times. To do this, avoid overly hard pulls on a control line to initiate a turn while you are flying slowly. Just as with a straight ahead stall, too low an airspeed is the culprit.

Spin recovery on a modern glider consists of putting your hands up and waiting for the canopy to slow its rotation and then surge to reestablish proper airflow (see figure 9-10). Let it surge about as much as with a deep stall. *Do not* apply too much control pull here, or you'll enter a stall.

If you have an older canopy, you are near the ground or the canopy is not slowing its rotation, you may try pulling the control toggle on the outside (unstalled) wing to stop the rotation. Immediately move this control up when the rotation stops to avoid spinning in the other direction and to allow the canopy to surge. Consult you owner's manual, instructor or manufacturer for the proper procedures if there is any doubt.

One of the serious problems of spins is that the rapidly turning wing may rotate faster than your body and twist up the risers and lines. In this case you must untwist them before attempting recovery by reaching above the twist (if possible)

Figure 9-10: Spin Recovery

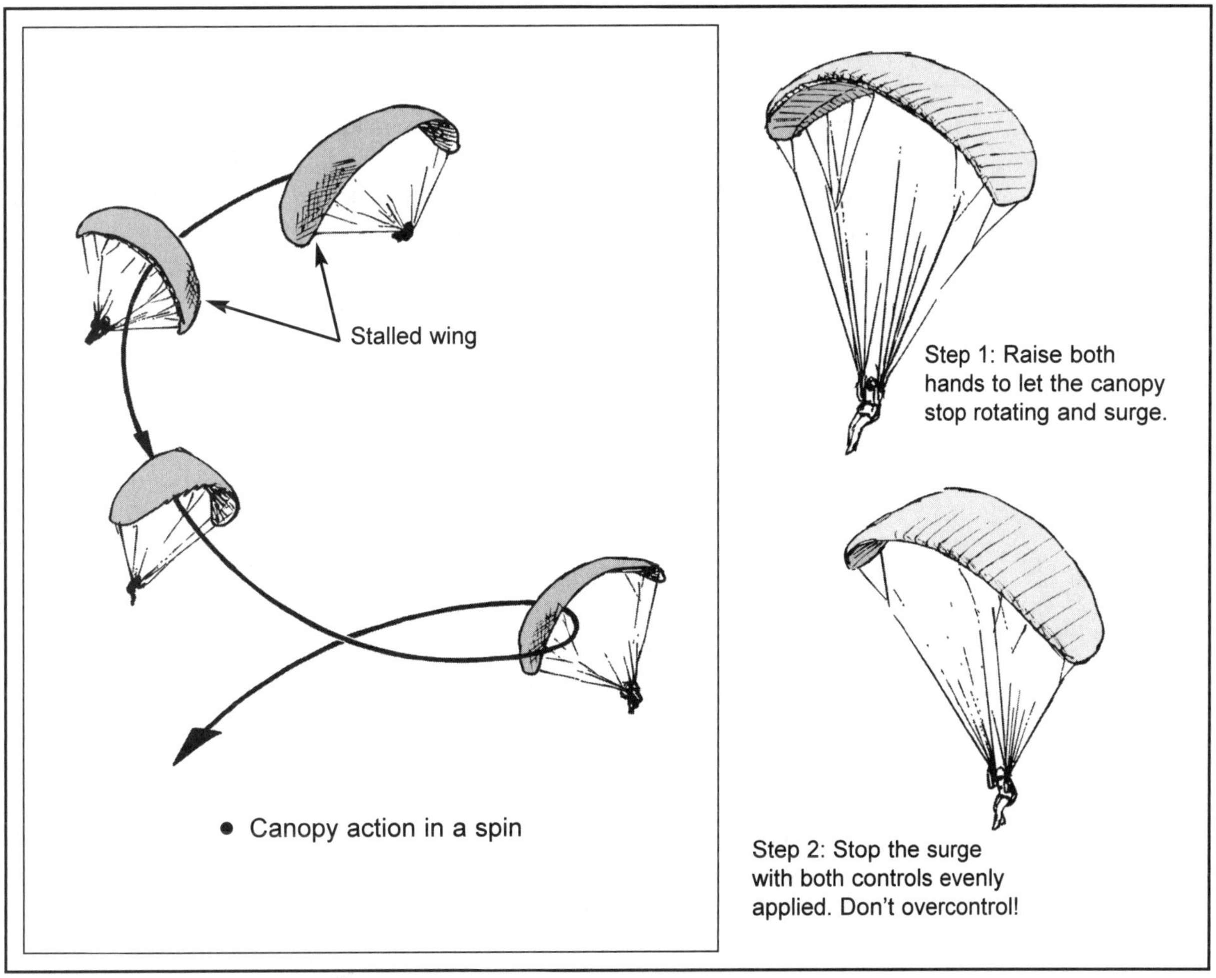

and spreading them apart with your arms while allowing your body to twist around faster. Needless to say, all this recovery input takes time and thus altitude. *Do not* hesitate to throw your reserve parachute if a spin is entered low or you can't seem to sort the problem out.

Another problem with spins is disorientation. The rapid rotation and G forces that occur in a spin can confuse your center of balance if you aren't used to the effect. You may not know what is happening or which way is up. Moving the head in any direction adds to the confusion problem—try to move only your eyes if you need to look around. If you begin to feel any disorientation in a spin, immediately look for your parachute handle and whip it out. Remember the rapid descent.

At the first World Air Games in Turkey I watched two pilots spin a few hundred feet above launch. The first pilot failed to recover and eventually hit the ground hard below launch. The second pilot threw his parachute immediately and landed without incident to pack up and refly.

Caution: *Below 500 feet (150 m) always consider throwing your parachute as soon as a spin occurs.*

From the foregoing you can see why we warn you to carefully avoid spins, even when you are experienced.

PERFORMING TURNS

In Chapter 5 we discussed turn corrections and beginning turns. Now we'll get into turns more deeply. There are various types of turns you can do and different turns have different applications. Naturally, you will have begun your turn practice at the training hill, but you were probably unable to get more than a 90° change of heading turn. So your true development of turning skills increases dramatically as you begin to fly higher.

Your Goal: *To understand the principles behind how your glider turns and to become adept at all sorts of useful paraglider turns.*

✓ Turn Types

The first thing we'll develop is a clear picture of what happens in a turn. Here's the terminology: A *bank angle* is the angle your wings make with the horizon as shown in figure 9-11. A large bank angle is known as a steeply banked turn. A *heading change* is the change in direction that your glider experiences from the start of the turn to the end as shown. A 180° turn results in you going the opposite direction from which you started. A 360° turn is a complete circle.

Perfect turns allow you to fully exploit a flight's potential.

Turns can be fast or slow, diving or efficient. Fast turns are usually steeply banked turns (and vice versa). A diving turn is also known as a spiral and we cover it below. An efficient turn means two things: first the turn occurs at a speed where the inside wing angle of attack is just below stall. Secondly, an efficient turn is non-slipping. Slipping is a falling towards the inside of a turn, which most aircraft can do. Slipping is not common on a paraglider unless we "cross control", which will be explained later.

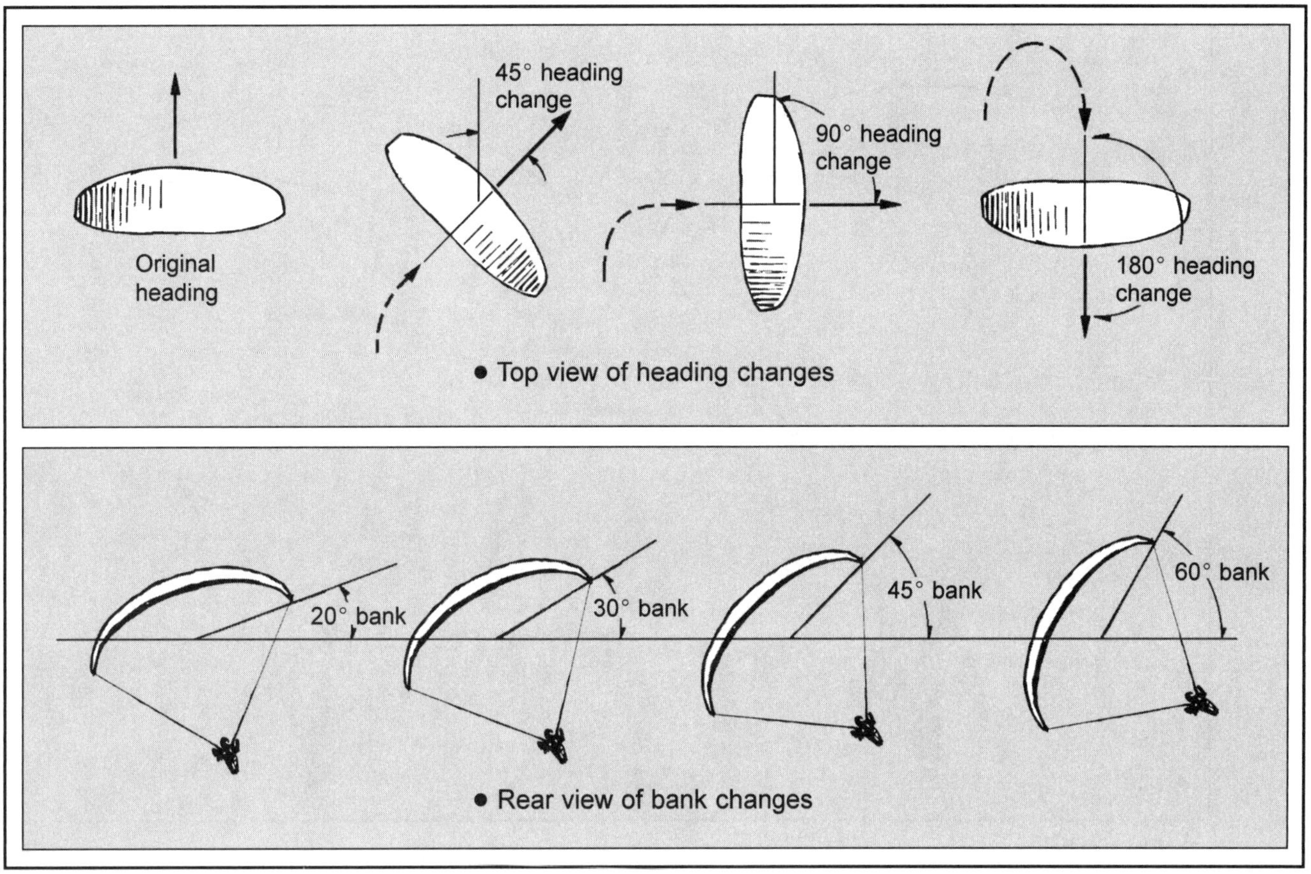

Figure 9-11: The Meaning of Heading and Bank

✓ Efficient Turns

When you started turning you were perhaps taught only to pull on one control line—on the side you want to turn towards. However, for more efficient turns you later were taught to pull down on one side while simultaneously moving up the opposite side hand. The result was a more coordinated turn and one which produced the same rate of turn with less control action (see figure 9-12). With this method you imagined your hands on a see-saw so that one goes up equally with the other going down. Such a turn keeps you further away from a spin, incidentally.

To learn to perform efficient turns, start with heading changes of between 90 and 180 degrees. Less heading change doesn't provide enough time to settle into the turn. Use a medium speed and later work on gradually slowing the canopy. Remember the spin! Your key to efficiency and safety is airspeed. Make your control inputs gradual and smooth and note the airspeed. The airspeed should be steadying once your control inputs have stopped. If airspeed is continuing to slow, you have too much control pulled—let up immediately. Learn to feel the combination of control pressure, bank angle and airspeed as you try different efficient turns at different angles of steepness.

Why do efficient turns? All turns increase your sink rate, so efficient turns are necessary for soaring (see Chapter 14). Also, efficient turns conserve your altitude and prevent over-speeding when you are maneuvering over the landing field

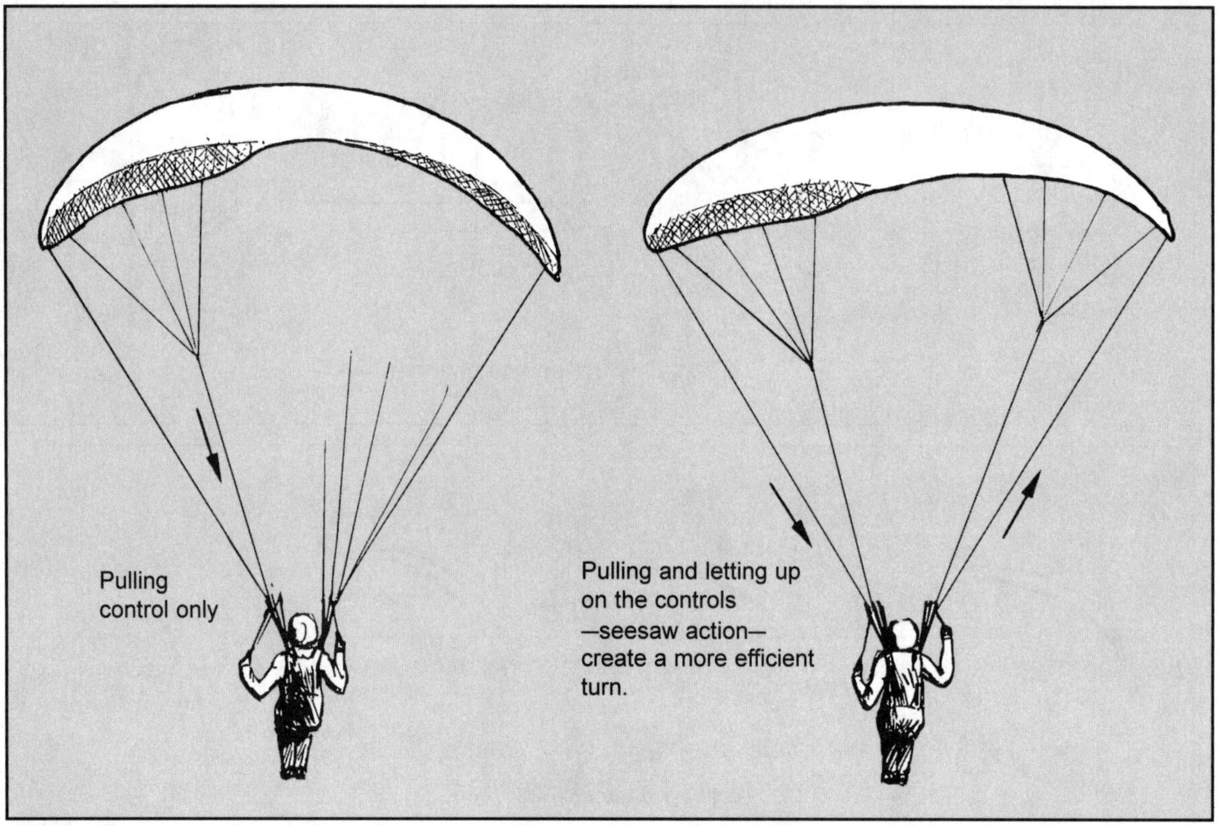

Figure 9-12: Producing Efficient Turns

(remember, we do not recommend faster turns on your downwind, base and final approach and figure 8s).

Also, note that G forces build up in turns due to centrifugal effects. You've felt this many times in a car or on a bike or a snow board. In a paraglider performing efficient turns, the G forces build up more with more bank angle. For example, at a 60° bank G loading is two—you feel twice as heavy!

The addition of G forces increases your sink rate and flying speed, just like adding more weight to your harness. As a result, we can chart the effects of a turn as shown in figure 9-13. It should be clear from this figure that steeper banked turns take less space and result in a faster change of heading and less altitude loss for a given heading change up, to a 45° bank. Beyond 45° more altitude is lost as bank angle increases. These figures do not take into account the time it takes to roll into or out of the bank angle.

Two other points should be well understood. With a steeper bank angle and higher G's on the glider you are stressing all the components more. Of course, your glider can handle more G's than your body can, but continued high G maneuvers require more frequent inspection of your canopy and lines.

The second point is that with a steeper bank you are less likely to spin because even though stall speeds increase, you must be much deeper in the controls with much more force than with shallow banked turns in order to stall. Remember this point when you are flying in turbulence or using shallow banked turns.

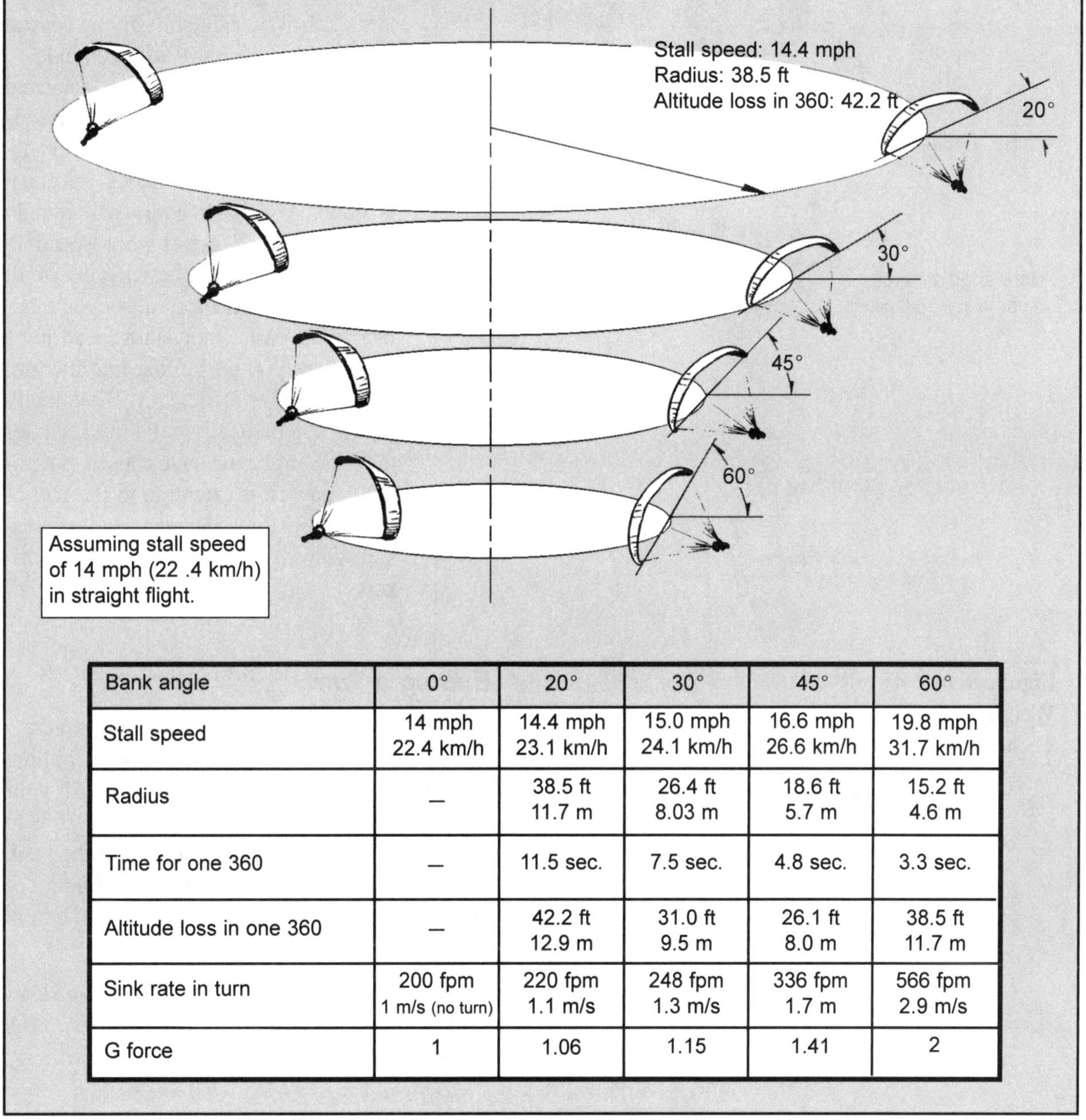

Bank angle	0°	20°	30°	45°	60°
Stall speed	14 mph 22.4 km/h	14.4 mph 23.1 km/h	15.0 mph 24.1 km/h	16.6 mph 26.6 km/h	19.8 mph 31.7 km/h
Radius	–	38.5 ft 11.7 m	26.4 ft 8.03 m	18.6 ft 5.7 m	15.2 ft 4.6 m
Time for one 360	–	11.5 sec.	7.5 sec.	4.8 sec.	3.3 sec.
Altitude loss in one 360	–	42.2 ft 12.9 m	31.0 ft 9.5 m	26.1 ft 8.0 m	38.5 ft 11.7 m
Sink rate in turn	200 fpm 1 m/s (no turn)	220 fpm 1.1 m/s	248 fpm 1.3 m/s	336 fpm 1.7 m	566 fpm 2.9 m/s
G force	1	1.06	1.15	1.41	2

Figure 9-13: Turn Factors at Various bank Angles

✓ *Weight Shift Turns*

Another way to improve the efficiency of turn initiation is to use weight shift. Weight shift means rolling on to one cheek of your hind end—left cheek for left turn, right cheek for… you get the picture (look at figure 9-14 if you don't). Here we show a left turn with differential control input, and a roll or tilt of the hip to put the weight on the left. Most pilots lift the outside leg to help accentuate the weight shift as shown. You can practice this process sitting in a chair or a simulator. Do all the motion including the arm controls and leg lift (we suggest a private place if you embarrass easily).

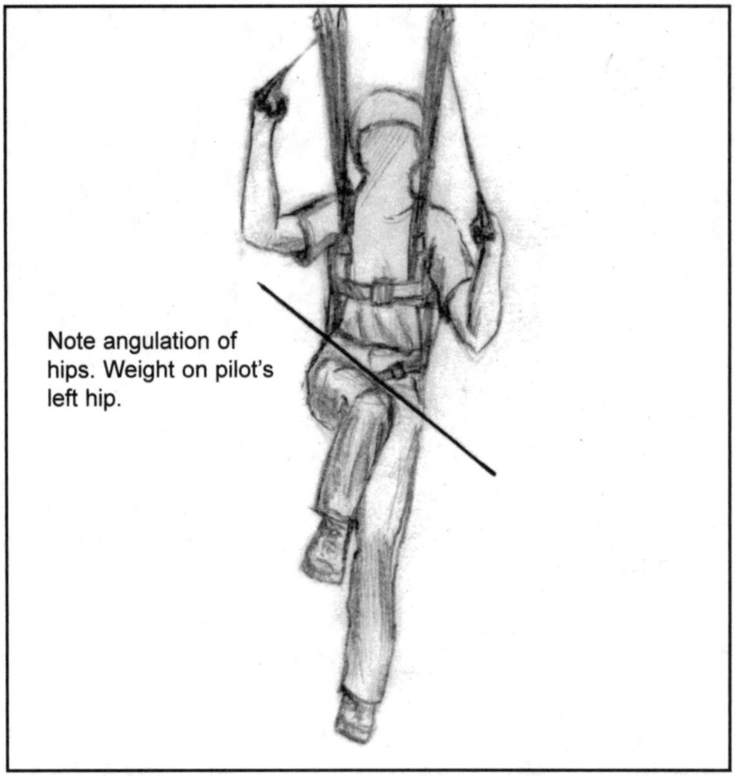

Figure 9-14: Weight Shift Technique

As long as you keep holding a turn control (weight shift or one control line down), the turn will continue to steepen. So once you have achieved your desired bank angle, all controls should be centered, your weight evenly distributed and your hands in equal position. The bank angle will remain and you should adjust your speed to produce an efficient turn if you aren't trying to do a diving turn. As you practice, try to learn to adjust for speed at the same time you are rolling into the turn for maximum efficiency. Eventually you'll automatically set the speed for the bank angle you have chosen, but initially, pay close attention to the feel of airspeed, G forces and control line pressure as well as your angle to the horizon.

✓ *Ending a Turn*

To stop a turn you must level the wings. There are two controls required—a roll and speed input—just like there is when beginning a turn. First begin pulling down with the outside hand while letting up with the inside hand. Also shift your weight to the outside. At the same time release the overall control pressure to allow more speed. Remember, more force is required on the controls as the bank steepens, so you have to back off a bit as you flatten out the bank. The amount of control and timing should be similar to that which you applied to get into the turn. Practice your turns until you can roll in and out smoothly.

Skill Check: *When you can perform maximum efficiency turns with control lines and weight shift in both directions up to 180° heading change without oscillating or surging, you have acquired good turn control skills.*

✓ *Rear Riser Turn*

In your ground handling practice you learned how to steer the canopy using the risers. The same effect takes place in the air. However, we caution you about using the front risers due to the possibility of creating a wing fold. Use the back risers only. Go ahead and try it the next time you fly and have clearance from the ground and other pilots. Pull the risers one at a time—left for left turns, right for right turns—and see how much control you have. Now try the same thing using weight shift. More control, right? Practice this control until you are very familiar with it and comfortable setting up landings with just riser and weight shift control. This is a safety procedure in case you lose a control line. Be sure to make your riser controls gentle to avoid a stall.

✓ Performing 360's

A complete circular turn is a useful maneuver for controlling altitude, staying in small patches of lift (such as a thermal), or just plain having fun. We call such a turn a 360 because one complete circle results in a heading change of 360 degrees.

Pilots can get into trouble when they first learn to perform 360's because of perception problems. These problems are vertigo, wind drift and loss of visual contact with the landing field. Vertigo, or loss of spatial judgement, is a factor of the body's balance mechanism as we discuss below. Wind drift is also an important consideration with 360's and we look at the problem in the section Turns in Wind. Finally, you can deal with loss of visual contact by starting your 360 practice high so you can always orient yourself to the landing field once you have completed your turns.

Practice 360's in both directions. Most pilots favor one turn direction. If you detect this, turn the opposite way three times for every one in the favored direction. A good pilot is versatile.

To avoid the above problems, do your first 360 in very light or calm winds near a large landing field area with at least 500 feet (150 m) of ground clearance. This altitude is necessary in case you cannot perform an efficient 360 and to re-orient after the turn.

Your first 360 should be a single 360 at about 20º bank. If you started very high, you can try another. Learn to do well-coordinated 360s one at a time in this manner before progressing to multiple 360's. Next proceed to a 360 and a half (540 degrees heading change). Try this in both directions, then you may perform two 360's in a row (a 720). After perfecting two, add full 360's one at a time until you can do as many as you wish linked together.

Front view of a pilot in a turn. Note weight shift and differential hand position.

Try to enter the 360 smoothly, which generally requires adjusting speed along with your roll control. Your airspeed in the 360 will be faster than in level flight, but your 360's will be smoother if you don't let the glider pick up too much speed during the initial part of the turn.

Gliders react differently to different bank angles. On some designs you may have to add a little control on the outside in order to prevent the glider from banking more steeply or "winding in." On beginner gliders you may have to hold the inside control to keep the glider from rolling out. In any case, concentrate on maintaining a constant bank angle and constant airspeed.

The next step is to perform 360's at steeper bank angles. To do this, again use the gradual approach. And constantly monitor your control position to produce a steady airspeed. Again proceed cautiously with plenty of altitude until you can perform multiple 360's up to about 45º bank. After this practice, try reversing 360's from one direction to the other.

A pilot turns left and weight shifts by lifting his right leg.

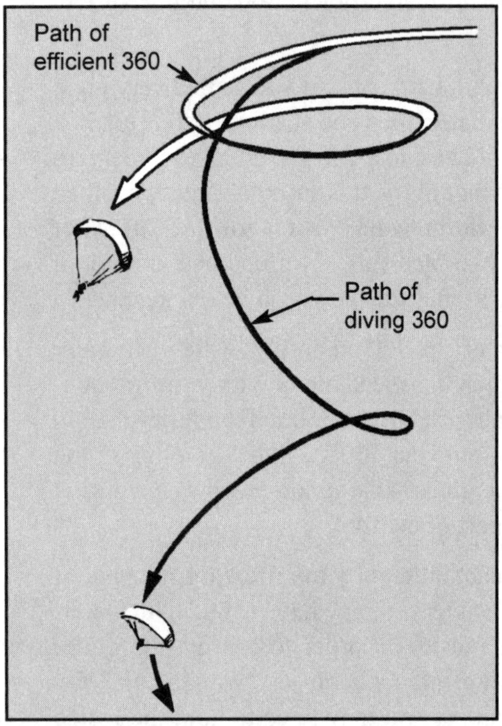

Figure 9-15: Performing Spirals

Once you are confident of your control and judgement performing multiple 360's in no wind, you should proceed to try them one at a time in light wind, then in gradually increasing wind (see Turns in Wind). Learn to perform continuous turns, in a step-by-step manner and you should have no problem scoring circles in the sky.

Skill Check: *Your 360 skills are optimum when you can perform multiple 360's in both directions at all bank angles with maximum efficiency airspeed maintained throughout the entire turns. Also, reversal of 360's should be smooth and well-controlled.*

✓ Diving Turns—Spirals

Diving Turns occur when we enter a bank with lots of speed. In fact, if you are flying normally and pull on one control line and make no speed adjustment, the turn will eventually dive more steeply downward as the bank increases. You'll be turning rapidly and G forces will build up considerably. We call continuous diving turns a *spiral* (see figure 9-15).

Spiral turns have an important use for losing altitude quickly (see the discussion concerning cloud suck in Chapter 13) and should eventually be learned. However, they must be learned gradually and carefully, starting with shallow spirals first. CAUTION: *Do not try your first spirals with less than 1000 feet (300 m) of ground clearance. Learn spiral turns under the guidance and supervision of an instructor, preferable in a maneuvers clinic over water.*

To perform a spiral properly, pull down one control while letting the other go up. Do not adjust for speed, but let the glider dive as it turns. Soon the G forces will build up, so don't start with steep spirals. To begin, use the roll control input you normally would for a bank angle of about 30º.

Many gliders tend to steepen their bank once they are past 30 or 40 degrees. The danger is that you can get very steep very fast. The G forces and fast rotation can be disorienting. To prevent such an occurrence, build your tolerance by gradually increasing the steepness and number of rotations in your spiral practice. Stop the spiral any time you feel the bank or speed increasing without your intended input.

To stop a diving turn or spiral, a definite weight shift must be applied to the outside of the turn along with opposite control. Make the control gradually in order to avoid oscillations or wingovers (aerobatics that are well beyond the capability of new pilots). Be aware that the control forces may be quite high.

A spiral turn may occur inadvertently while you are exiting a stall, spin or asymmetric wing fold. If you find the rotation rate and forces beyond your previous experience, stop the spiral immediately. If you're close to the ground or are not able to stop the spiral, immediately throw your parachute. Remember, a spiral loses altitude rapidly.

✓ Spatial Judgement

Some people have problems judging position in three-dimensional space. These problems mount when they start turning with more than gentle 90° heading changes. Others experience disorientation or vertigo. These problems are mainly physiological (due to the inner ear balance mechanism) and for the most part can be overcome with training. If you ever have a feeling of confusion or disorientation during flight—even at the training hill—let your instructor know immediately. He or she will suggest exercises to help you overcome this problem.

One thing you can do is practice turning. Spinning around like a top or going to the local playground and working the merry-go-round is the best training. Does the thought make you sick? Easy does it and build up your tolerance gradually. Start slowly and add turning speed later (do the same thing in the air). Your brain will learn to accommodate if you practice.

Flying tandem is a good way to learn spatial judgement.

When you turn, look towards the inside wing. Don't focus on the ground spinning around. Also, do not move your head in a turn. The extra input can be very confusing to your senses. Eventually you can build up tolerance to these items as well, but when you begin turns, follow this advice carefully.

The situation is complicated by the fact that the eyes automatically flick ahead and the fluid in your inner ear eventually settles in a continuous turn. The result is a change in response after about 20 seconds. Thus, you may be fine for a 180° turn or a 360° turn, but when you add a second or third 360, things might start to go haywire as your reality check bounces. For that reason we highly recommend you learn turns gradually as outlined in the 360 section. We provide more details of this matter in our advanced book *Performance Paragliding*.

✓ Clearing Turns

Signaling a turn while flying is just as important as it is when driving a car. Be predictable!

A very important part of every turn is clearing your intended path. Since you cannot readily see behind you, you may not know who's (or what's) lurking on your tail. Clear your turn first by looking over your shoulder in the direction you intend to turn. The very act of looking informs other pilots what you intend to do. If they are so close that they could have a conflict with you they will (or should) be watching you carefully. Multiple turning of your head is the best way to assure others see your gesture. Also try communicating by voice—yell it out—when in doubt. Sometimes a short arm pump on the side you want to turn towards is a useful addition (see figure 9-16).

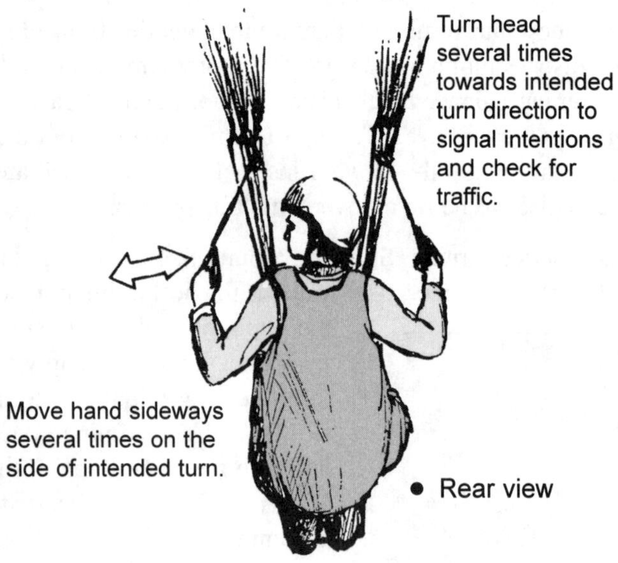

Figure 9-16: Clearing Turns

The first step in every turn is to clear it.

Initiate your turn only after you have determined you have plenty of room to do so without causing a close call. Midairs have happened which resulted in pilot injury or death. Clear *all* your turns, even if you think you are alone in the sky so it becomes a life saving habit.

✓ Turn Practice

Here we'll outline a program for learning turns gradually and safely.

1. *Start with shallow banked turns,* working your way up to 90° heading change.
2. *Progress to linked turns.* This means coming right out of one turn and entering another banked in the opposite direction. Begin with linked 45° heading change turns, then progress to 90° heading change as shown in figure 9-17. Turn reversals essentially involve coming out of one turn and smoothly entering another in a continuous flow.
3. *Produce 180° turns* one at a time. Remember, these turns are nothing more than linked 90s going the same way. After 10 or more single 180s, link them up as shown in the figure.
4. *Work on steeper banked turns* in a gradual manner. Make 90° turns first, then 180° turns with bank angles up to 30°.
5. *Next, learn 360° turns* like doing linked 180's going the same way. Do this learning first at a high site away from the hill (over the landing) in nearly zero wind. Do one 360 at a time at first (both directions) until you have performed at least 10 individual 360's, controlling entry, speed and exit. There should be no tendency to stall or surge throughout the maneuver. Concentrate on feeling the speed, G forces and control forces as well as hand position. Remember your eye focus and keep a steady head.
6. *Progress to multiple 360's* by adding one additional circle at a time. Work your way up gradually with shallow turns until you can do as many 360's in a continuous manner as you wish. Make them efficient. Remember, the steeper the 360, the more disorienting it is due to the faster rotation and G forces.

Practice 360's until you can come out exactly on a desired ending heading. Try opening the turn (shallower bank) and tightening it all in the same continuous series of 360's. All this practice will be useful for losing altitude, but also for thermal soaring which you'll learn later. Under the next heading we'll explain about 360's in wind.

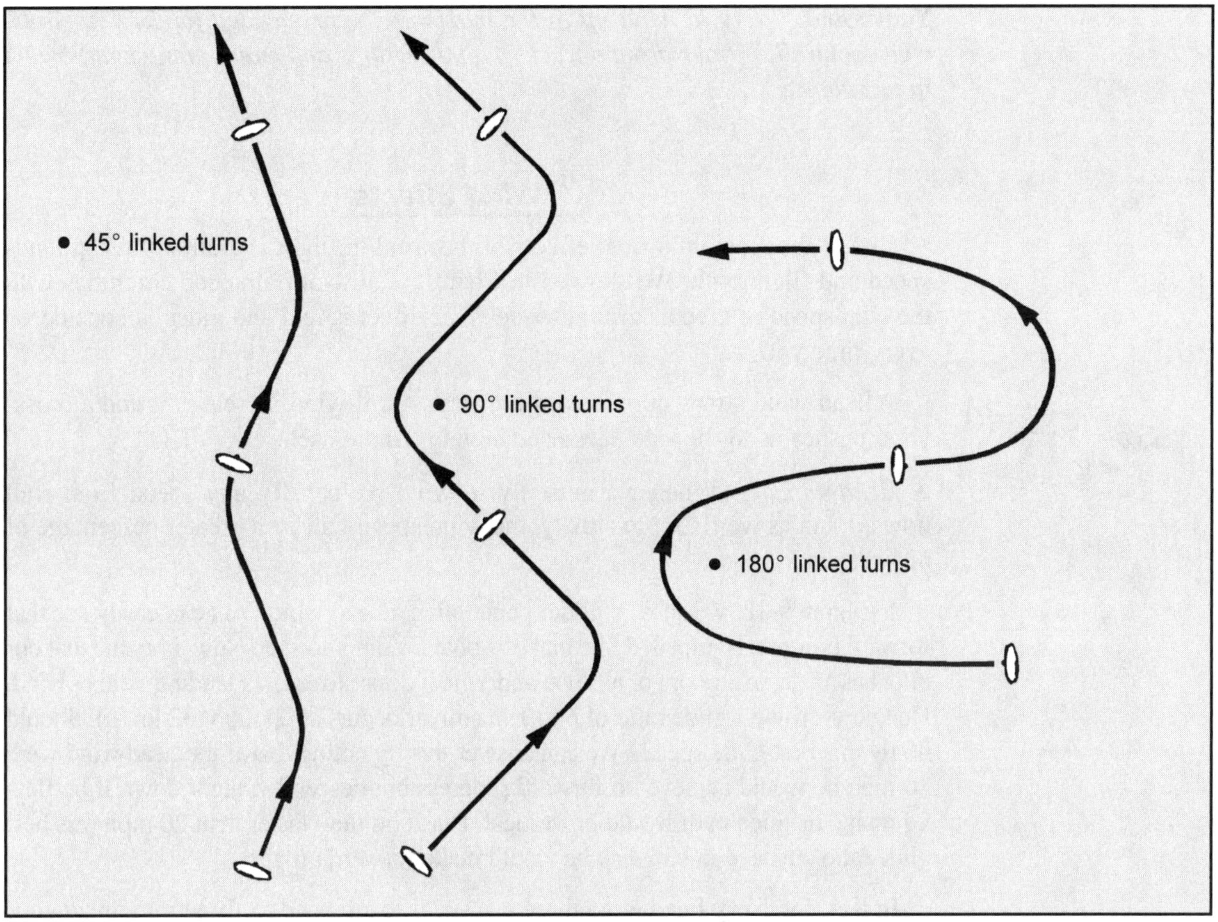

Figure 9-17: Linked Turns

7. *Finally learn spiral dives* in a gradual manner at a maneuvers clinic. Spirals are the most disorienting of all turns, so be well acclimated to multiple slow 360s before you try spirals.

Skill Check: *You have mastered 360s when you can produce them at a steady rate in smooth air without a change in airspeed. This mastery also includes entering and exiting them exactly where you intend without canopy oscillations, dives or zooms.*

Flying in More Wind

Wind is a fact of life on the earth's surface, and we aviators encounter plenty of it. It is to our great advantage to be able to handle more windy conditions. You can think of your training as taking place in a still lake of air. Later you learn to handle a flowing river, then eventually graduate to white water (turbulence)—within limits of course. Here we'll discuss about the complications that wind presents.

Safety Tip: *Before flying in more wind, review the material on ground control of a canopy in wind including pulling rear risers and the control lines (but not when the canopy is inflating in higher winds). This material can be found in Chapter 4, Building a Wall. Also, review the material on emergency canopy fold (Chapter 5, Dropping and Stopping the Canopy).*

Your Goal: *To fly in wind up to the maximum recommended for novice pilots, while controlling your ground track in all directions and maintaining equilibrium in turbulence.*

✓ *Wind Effects*

One of the most important effects of the wind in flight is to alter your groundspeed and flight path. We showed in Chapter 3 how our airspeed combines with the windspeed to give us groundspeed. This effect is real and must be considered every time you fly.

Wind complicates all aviation.

A head wind slows our forward progress, a tail wind increases it and a crosswind pushes us to the side. Let's see how to handle each.

▲ *Head winds* — When we must fly in wind we can fly any speed from stall upward, but as we fly more slowly, the wind speed will be a greater percentage of our airspeed.

In Figure 9-18, we show a glider penetrating a head wind. You can easily see that forward progress is impeded like that of a boat heading upriver. Now imagine that our pilot has a certain amount of altitude and a great desire to reach a landing field upwind. He knows his best glide ratio of 6 to 1 in still air occurs at 20 mph (32 km/h). Should he fly this best glide speed? We can answer this by noting that if the head wind were 20 mph he would achieve no forward progress but descend straight down if he flew 20 mph. His glide path would be vertical. But if he flew faster than 20 mph (his best glide ratio airspeed in still air), he would make forward progress.

In fact, for every head wind there is a specific airspeed to fly to maximize your glide ratio over the ground. Finding this exact airspeed is beyond the scope of this book (see *Performance Paragliding*), but we can generalize by noting that you should *speed up in a head wind or sink. The greater the head wind, the greater the increase in airspeed.* This concept and all the others relating to maximizing performance in varying air—lift, sink, head wind, tail wind and crosswind—is known as *speeds-to-fly.*

▲ *Tail Winds* — If the wind is from your rear it will carry you along. *The rule of thumb is to slow down in a tail wind or lift to travel the farthest.* You still must maintain sufficient airspeed, of course.

Caution: *The addition of a tail wind to our airspeed increases our groundspeed. If you are not paying attention, you may erroneously think you are flying too fast. Do not mistake groundspeed for airspeed or you may inadvertently slow down to a stall in a tail wind. Always pay attention to airspeed and your control positions. Remember, it is too high an angle of attack that causes a stall.*

The air is a fluid and the wind is its flow. Learn to judge it, work with it and master it like an old sailor on the sea.

In fact, you should never slow as much as your minimum sink airspeed in a tail wind to go as far as possible. The proper speed is between best glide and minimum sink rate.

When you choose to fly downwind, make sure you have ample clearance from all obstacles and plenty of altitude to turn back into the wind eventually. Your landing setups will occasion some downwind flying, so practice flying with an awareness of airspeed and control position in all winds.

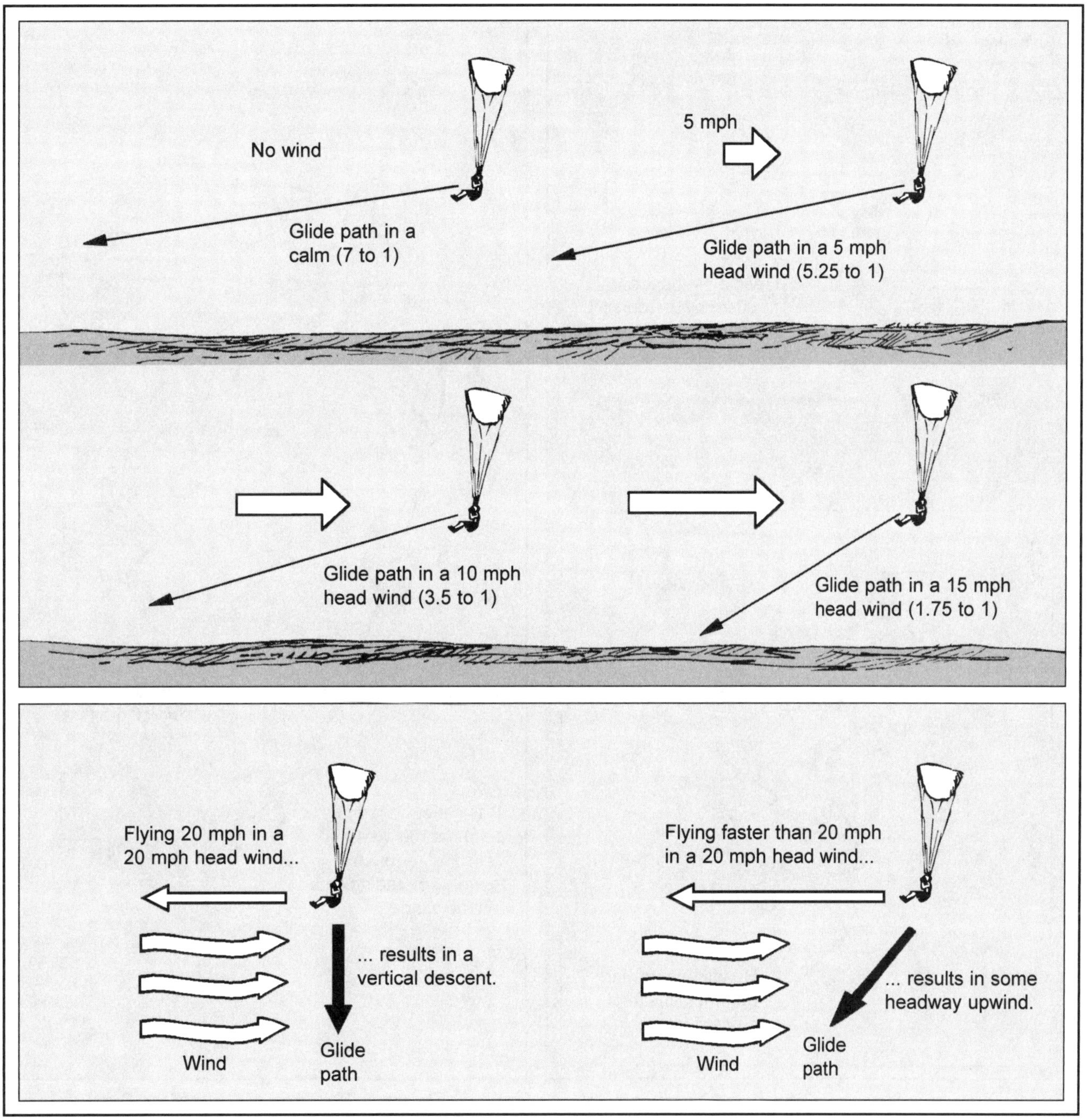

Figure 9-18: The Effects of a Head Wind

▲ *Crosswinds* — Crosswinds are interesting because they can not only slow your progress, but also require you to perform an action known as *crabbing*. If you've ever watched a crab scuttle sideways across the sand you'll know why the term applies. In order to go straight across the wind you must aim partially into the wind to keep from drifting sideways as figure 9-19 shows. From the ground it looks like you are flying sideways. The action is similar to a boat crossing a fast flowing river, as shown.

How do you perform a proper crosswind crab? It's really quite simple. Look at your desired goal, then draw an imaginary line along the ground to your present

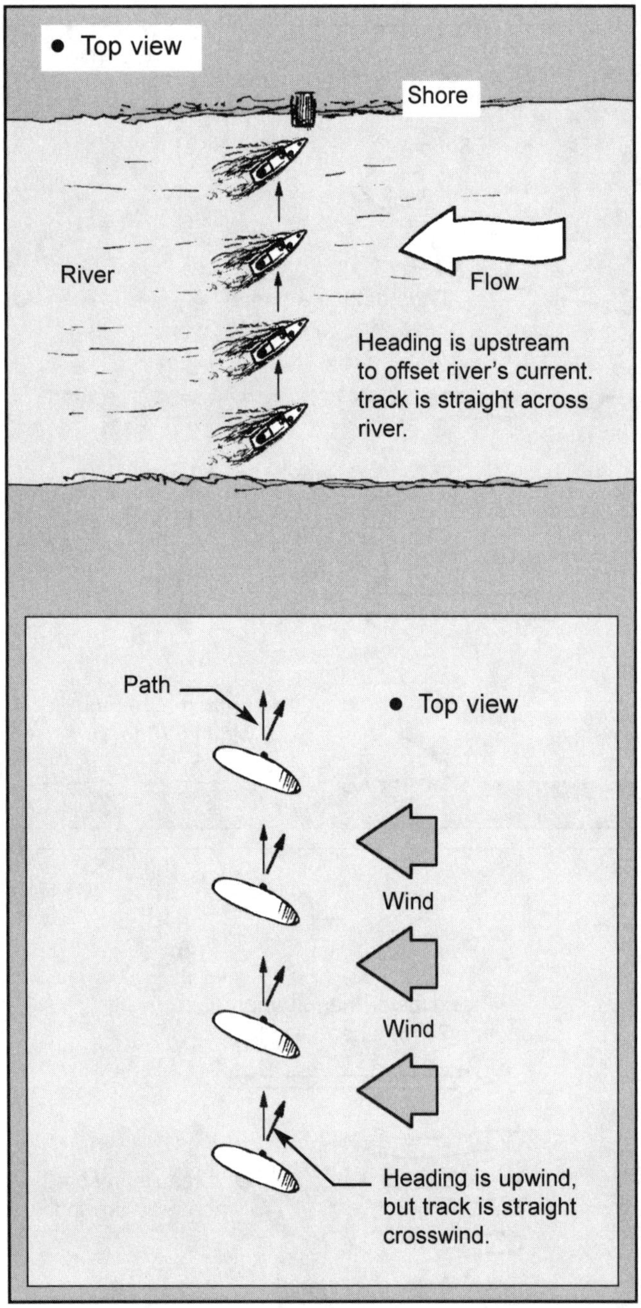

Figure 9-19: Crabbing in a Crosswind

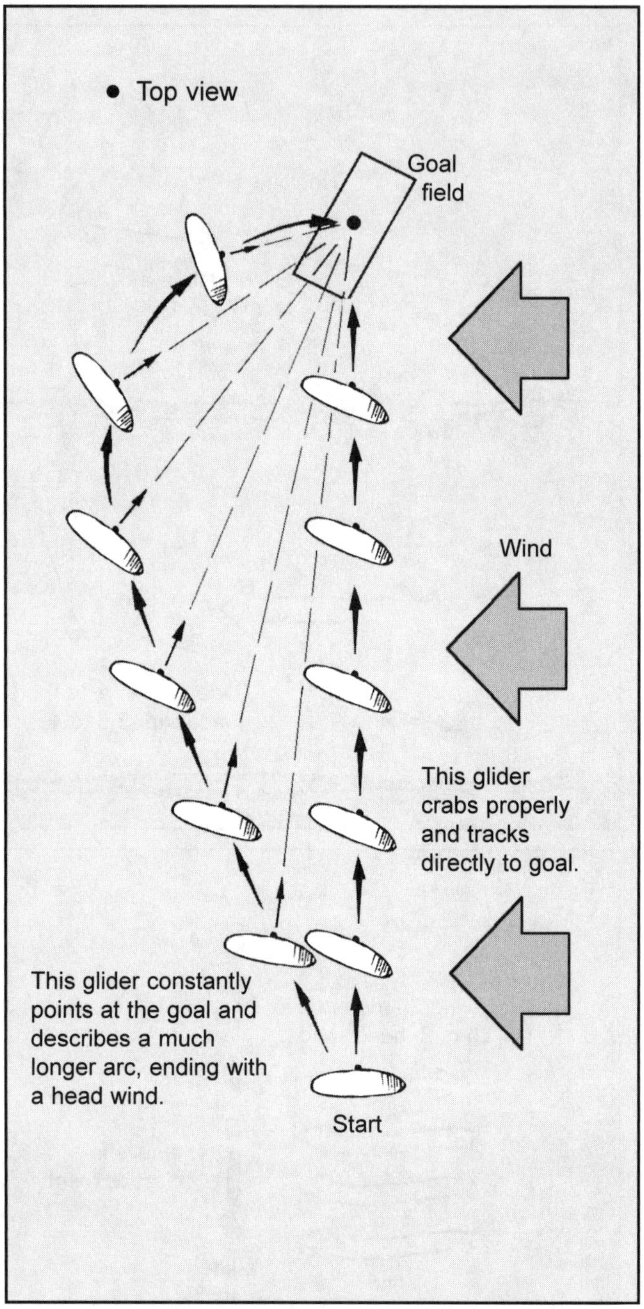

Figure 9-20: Proper Crabbing Track

location. That line is your ideal ground track. Turn partially into the wind and follow your line. If you are being blown a bit downwind turn a little more. If you are moving upwind of your path turn a bit less.

If you do not crab properly but simply point your glider at your goal you may eventually arrive there, but your track will be much longer and you will lose more altitude, as shown in figure 9-20. You should note that the straight line to any goal is always the most efficient (that is, loses the least amount of altitude) in any uniform wind. Also the faster the wind the more you'll have to crab while the faster you fly the less you'll have to crab.

As with all learning in paragliding, you should start gradually by experiencing only light crosswinds (5 mph or 8 km/h) then gradually progress to the stronger stuff.

Skill Check: *You are able to fly crosswinds efficiently when you can fly straight to a goal without wandering upwind or downwind of a direct line from your starting position to goal.*

✓ Wind Limits

Every pilot has a limit of wind strength in which he or she can safely fly. At the end of this chapter we give wind limits for novice pilots, but of course you haven't yet achieved all the novice skill and experience. You should always be conservative when deciding the maximum wind you will tolerate. As your experience grows you will find that more advanced pilots choose to fly in lesser winds, for in light winds they can work thermals better and get higher. With modern gliders we can soar in moderate conditions, so there is no reason to risk flying in higher winds. Remember, the force of the wind increases with the square of the velocity—if the windspeed doubles, the force quadruples.

High winds and paragliding are like oil and water—they don't mix!

There are two dangers associated with high winds:

1) Penetration problems
2) Control problems due to turbulence

Let's look at each in turn.

✓ Penetration

As you saw above, a head wind retards your forward progress. It is entirely possible that a landing field that is easily reachable in light winds will be unattainable in stronger wind, especially if you encounter sinking air. To prevent this from happening, always hedge your bets by remaining upwind of your landing field whenever possible.

When you can't remain upwind you should definitely remain within the range of your glider's penetration. Figure 9-21 shows the meaning of this range. In calm conditions, you can reach any point in a cone whose base is described by your maximum glide ratio. As the wind increases, you can reach farther downwind, but much less far upwind as shown. The figure illustrates that your cone of penetration is somewhat distorted and shifted downwind. It should be clear that at some wind strength you will not be able to reach an upwind landing field so you may have to head cross or downwind to an alternate field.

> *We watched a pilot land in a swamp in very high winds even though he started out his landing descent only about 150 feet (50 m) downwind of the landing field. You could see the look of dismay on his face as he descended nearly straight down. His glider, harness and ego were covered with mud.*

Always be aware of your glider's cone of penetration. You can learn it by practicing in various wind strengths. Before turning back to set up the landing approach fly upwind a bit and note the point you would reasonably reach. Etch these estimations into your mind and use them in wind with a good dose of extra slack.

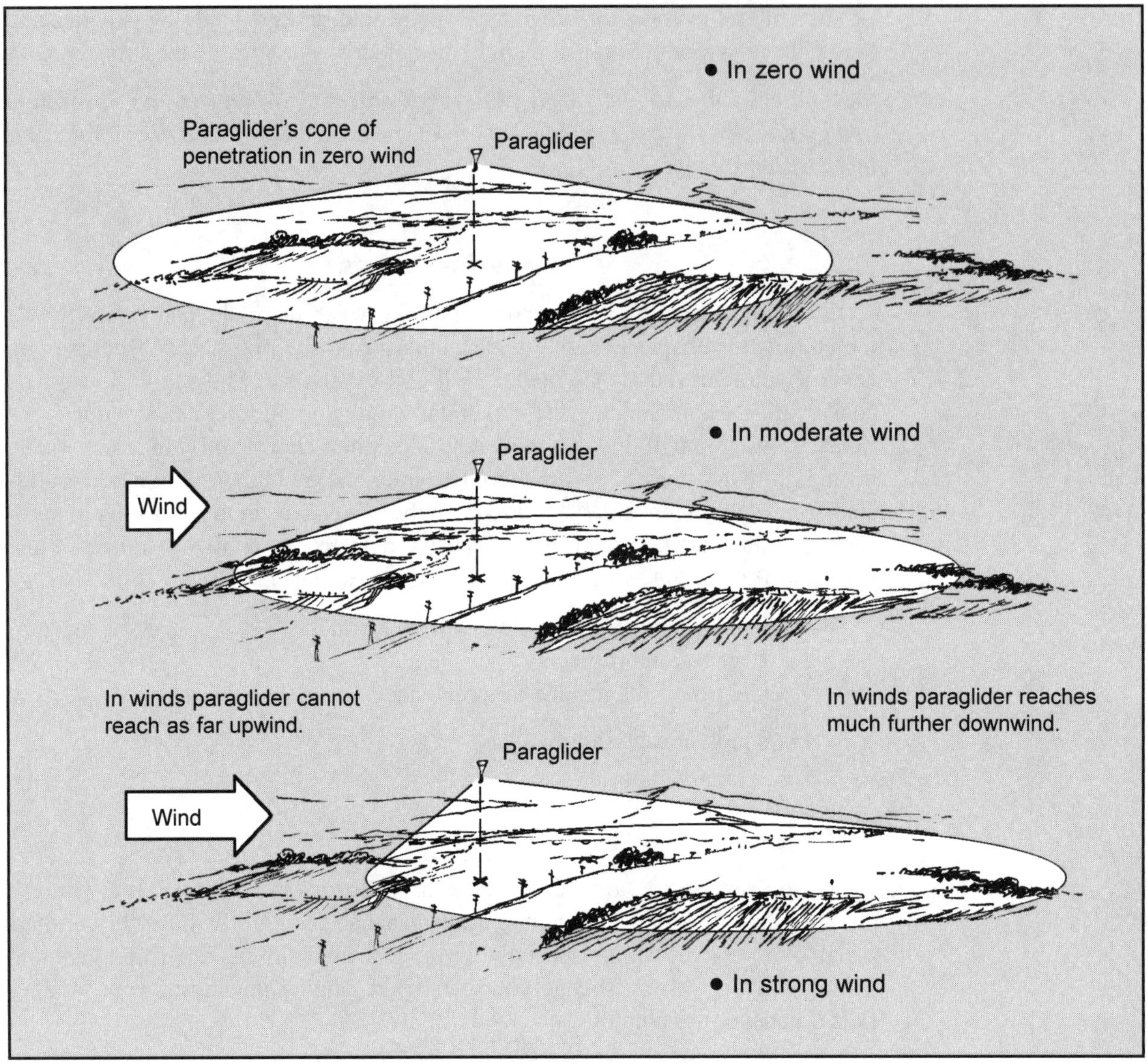

Figure 9-21: A Glider's Cone of Penetration

There are times when a reverse gradient is set up. That means the wind is stronger near the surface than it is at, say 1000 feet (300 m) up. These conditions often occur in hot areas with mountains nearby, for they usually combine a slope and heating effect. Know your area and be prepared by allowing extra altitude if such conditions are common.

✓ *Venturis*

You may have heard the term venturi in relation to a carburetor. It is a narrowing of the carburetor throat which causes the air to speed up. This effect happens in nature as well. Look at figure 9-22 and you will see that when the wind passes over a ridge line or through a gap in a ridge, it can accelerate where it is constricted. The higher the hill and the higher the wind, the more this effect takes place. A gap in a tree line or buildings can have a similar effect on a smaller scale. See Chapter 13 for more discussion of venturis.

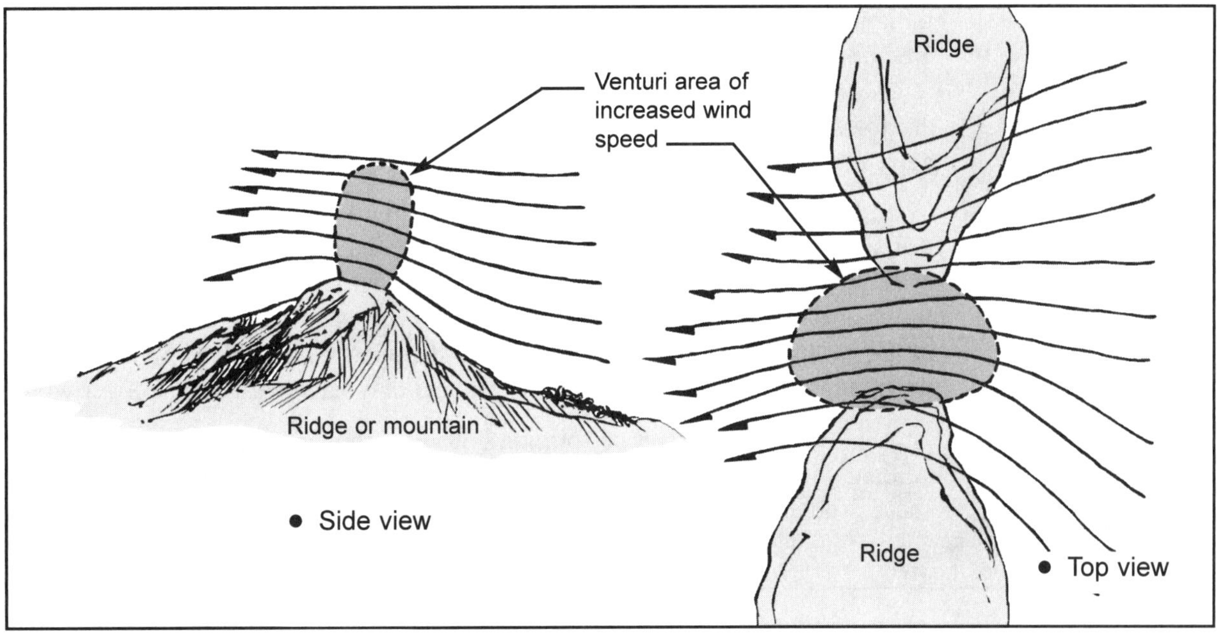

Figure 9-22: Venturis in Nature

Think about your experience on the training hill. Did you notice more wind at the top than at the landing field? That's physics in action. Study our examples and think about where venturis can occur where you fly. Then resolve to always avoid a venturi in a paraglider—your forte is slow flight!

✓ *Turns in Wind*

When you are above the effects of wind gradient you must still consider your upwind and downwind turns. There has long been a standing argument about whether or not an upwind and downwind turn are the same. The fact is, to the glider in the air out of the gradient caused by the influence of ground, there is no difference. Like for a boat turning in a current far from sight of shore, all forces and controls are the same. If you are blindfolded or very high so you can't perceive the ground movement, you cannot tell a difference when turning upwind or downwind.

Figure 9-23: Ground Tracks in Wind

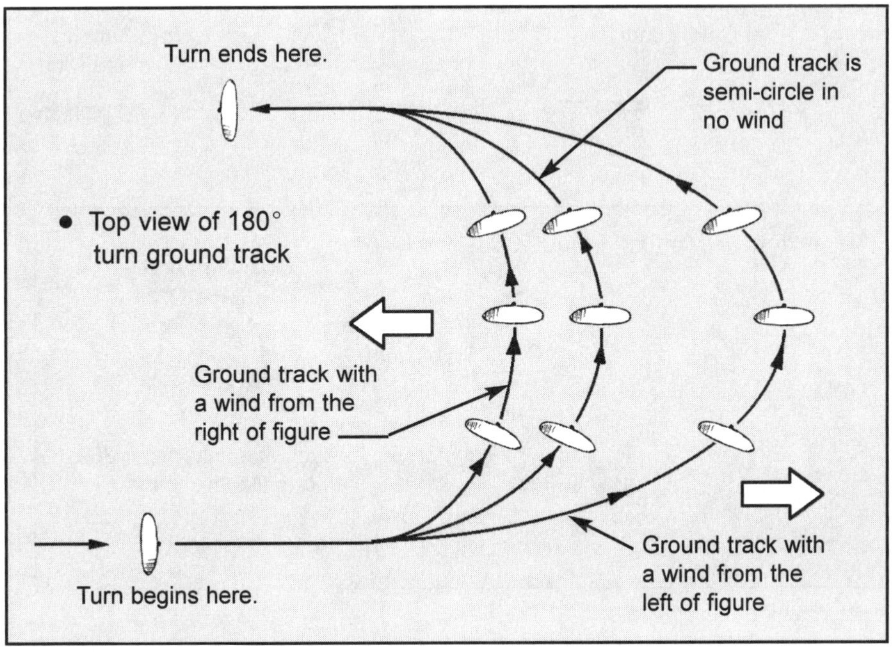

Of course, your ground track *is* different when you turn upwind or downwind, as shown in figure 9-23. It is somewhat distorted from the perfect arc and displaced downwind in relation to the point where you initiate the turn. The only place where this anomaly comes into play is when you are turning around a point such as a pylon in wind. Do not get fooled by the differences in your ground track when turning in wind. The forces and the controls are the same whether you are turning upwind, downwind or in no wind. Any apparent acceleration is an illusion.

Wind will always alter your ground track unless you offset the wind effects by changing heading.

Turns of 360 degrees or more are affected noticeably by wind because they drift along with it. In figure 9-24 we show a 360 in zero wind and one in a 10 mph (16 km/h) wind. In the calm situation (a), the 360 describes a neat circle in the air. In wind (b), the pilot performs the same turn controls but the downwind portion of the turn (labeled X) is elongated due to the wind. In the side view you can see the pilot got dangerously close to the hill and may have hit it in a stronger wind.

Another problem when performing 360s near the terrain is the perception that you are going too fast on the downwind portion and mistakenly slowing to stall speed. You can imagine the consequences if you stall at point X in the drawing.

Figure 9-24: The Effects of Wind on 360° Turns

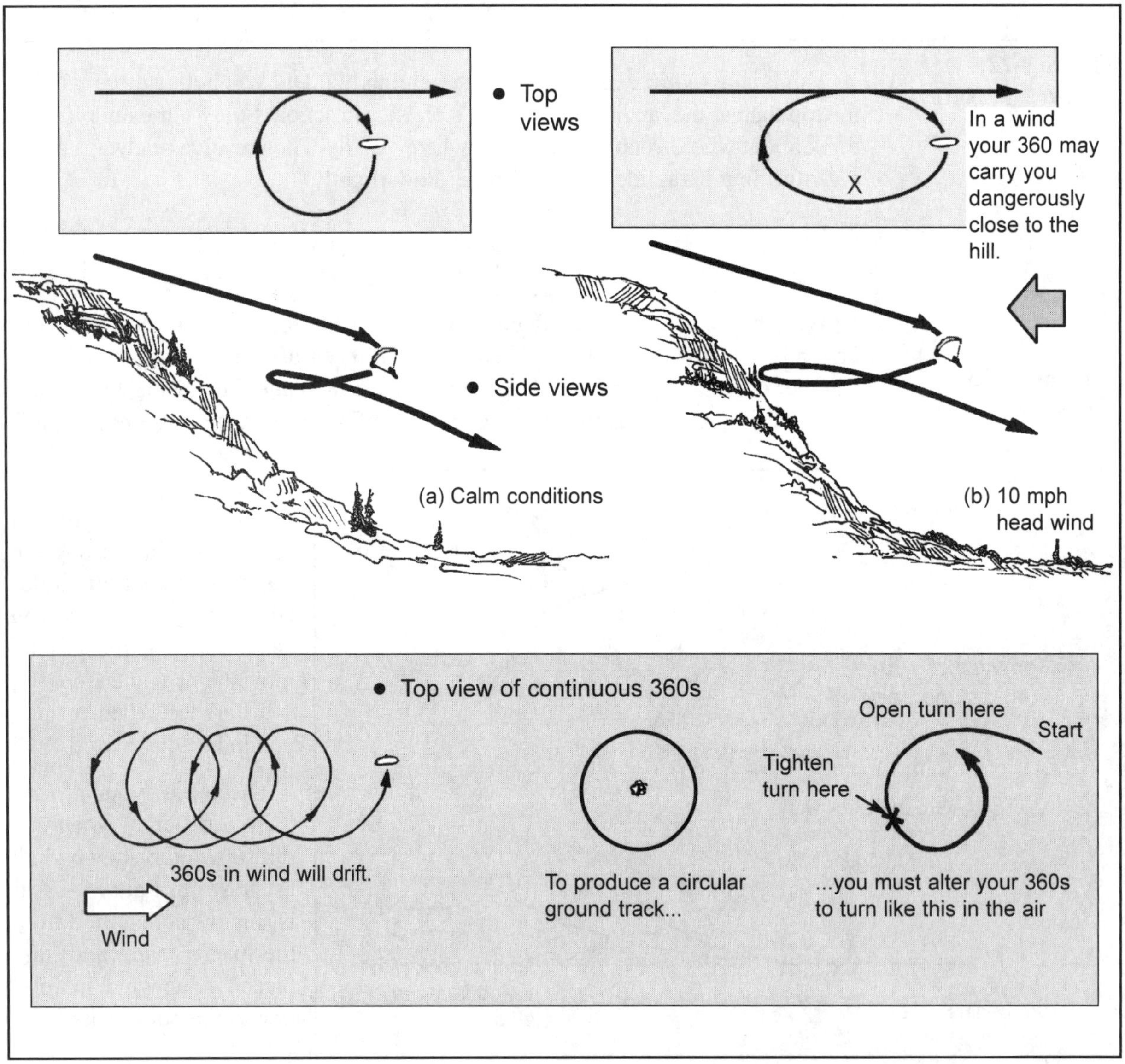

You can reduce the stall problem as well as the drift problem by turning a faster (tighter) 360. However, such turns near the terrain are advanced maneuvers and should be left to the experts. Learn your 360s in wind after you have perfected calm conditions 360s well away from the terrain in a gradual manner. It takes lots of practice to be able to predict where your circle will end up in different amounts of wind at different bank angles.

When near a ridge or mountain, do all reversing turns away from the hill. 360's near a hill are very easy to misjudge

When you begin flying higher in more wind, an important part of your training and gaining experience should be to practice turns without drift. Especially learn to do continuous 360s while staying over a point on the ground (a windsock or flag pole, for example). This feat requires elongating your upwind portion of the turn by banking less steeply and doing the downwind portion tighter by increasing the bank angle as shown in the figure.

Flying in Turbulence

Turbulence is a fact of life in the wild blue yonder. Turbulence, as we pointed out in Chapter 6, has several causes and is varying air, usually in the shape of swirls. Such variations are felt as gusts on the ground, and in the air they can greatly alter the airflow on our canopies. The result can be wing folds in strong turbulence. Here we'll discuss how to handle these matters.

Your Goal: *To gradually learn to fly in more turbulence while maintaining complete canopy control, and to remove wing folds and other canopy distortions.*

✓ Active Flying

Active Flying refers to the overall technique that adept pilots use in turbulence: they detect small changes and perturbations in the canopy and make a correction or adjustment before big changes or alterations can occur. The changes we are speaking of here are mainly tip folds, and occasionally front folds. Active flying requires that first you become sensitive to what the canopy is doing, and second you know the right control input for the situation. Look back at the opening of this chapter and you will see that we were developing canopy awareness and positioning. That is the first part of active flying. The second part is preventing folds.

Active Flying

The meaning of active flying is to join your senses with your wing so you actively control its behavior the instant the air alters it. Active flying can prevent most wing folds—both symmetric (frontal) and asymmetric. You must:

1. Maintain the canopy's position directly above you as shown in figure 9-25.
2. Fly at a good pressurization airspeed without over-controlling.
3. Anticipate the presence of turbulence or potential changes in the airflow.
4. React to beginning folds instantly with the correct surge and roll control.
5. Correct folds without delay in the safest manner.

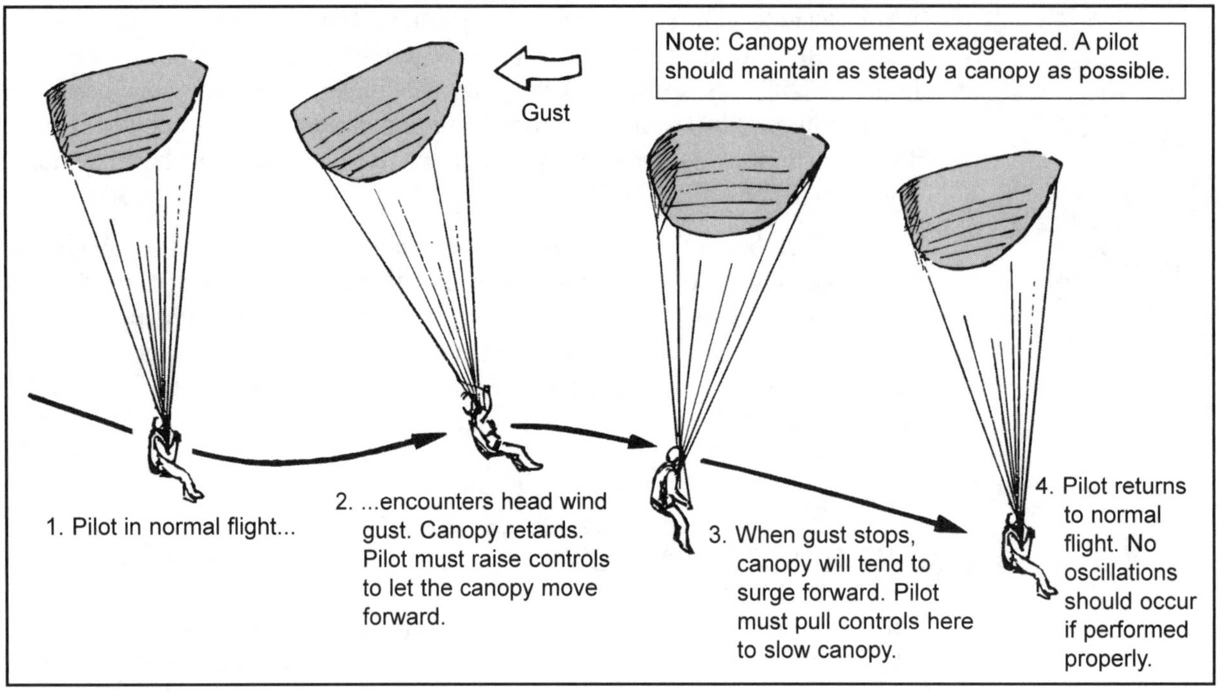

Figure 9-25: Maintaining Canopy Position

Active flying can thus be viewed as being on top of things. You look ahead, know what to expect but prepare for the unexpected. The idea is to become one with our canopy and eventually become one with the air like a bird.

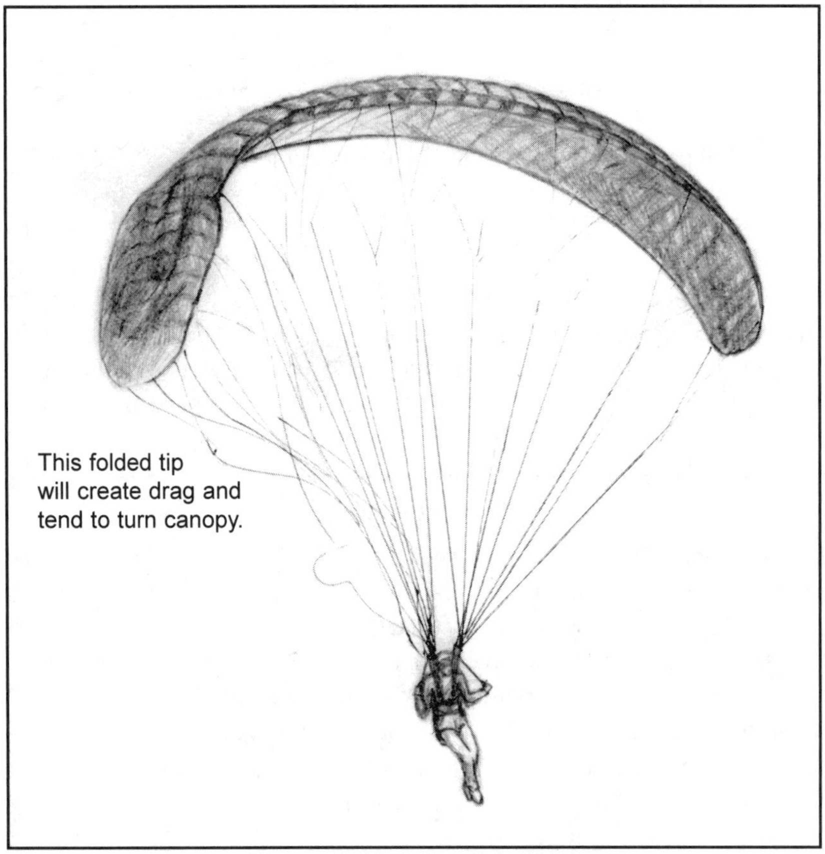

Figure 9-26: Wing Tip Fold

✓ *Wing Folds*

An eddy in the air usually doesn't hit your wing symmetrically. The most common form of canopy distortion is a wing tip fold as shown in figure 9-26. Here a portion of the wing is folded down and blows back against the lines due to disturbed airflow past the wing. The fold may be very small or include 50% or more of the wing. Normally such a large fold occurs only in severe turbulence which, as a lesser experienced pilot, you should avoid by flying in more stable, less windy conditions. Remember to check the air's gustiness before taking off and respect the limits of 5 mph (8 km/h) change in less than 5 seconds.

Wing tip folds can occur wherever there is turbulence. From the guidelines of Chapter 6, that means near the terrain (mechanical turbulence), around thermals, in shear layers, behind another glider and in rotors. Normally you will only encounter the first two. Just after you takeoff, as you approach for landing and when you encounter a thermal are the times to be most vigilant. If you fly towards another pilot thermaling, maintain extra pressure as you reach the thermal area. The stronger the thermal (indicated by how fast the other pilot is climbing), the stronger the expected turbulence.

✓ Controlling a Fold

The prevention of a tip fold consists of maintaining good canopy pressurization (slower flight) by increasing the control pressure slightly. Equally important, is actively pulling on a control line when the pressure lightens. Previously we discussed maintaining an even pressure in the lines to control canopy position. When the air varies the angle of attack along the wing, the same changes occur in control line pressure, but they are different on the left and right side. Thus you must *use each hand independently to keep the correct pressure on each side.*

A tip folds because the local angle of attack gets too low. You can detect this factor by sensing a lowering of your control pressure. You can counteract this effect by pulling the control to increase the angle of attack and rolling your hips away from the folded side. The amount of control you apply should be that required to regain the normal line force. Return the control back to normal as soon as the line force returns to normal.

If a fold does occur, the glider will tend to turn to the direction of the fold because of the extra drag on that side (see figure 9-27). The most important thing of all to attend to is making the glider go straight (failure to do this may result in you flying

Figure 9-27: Controlling a Fold

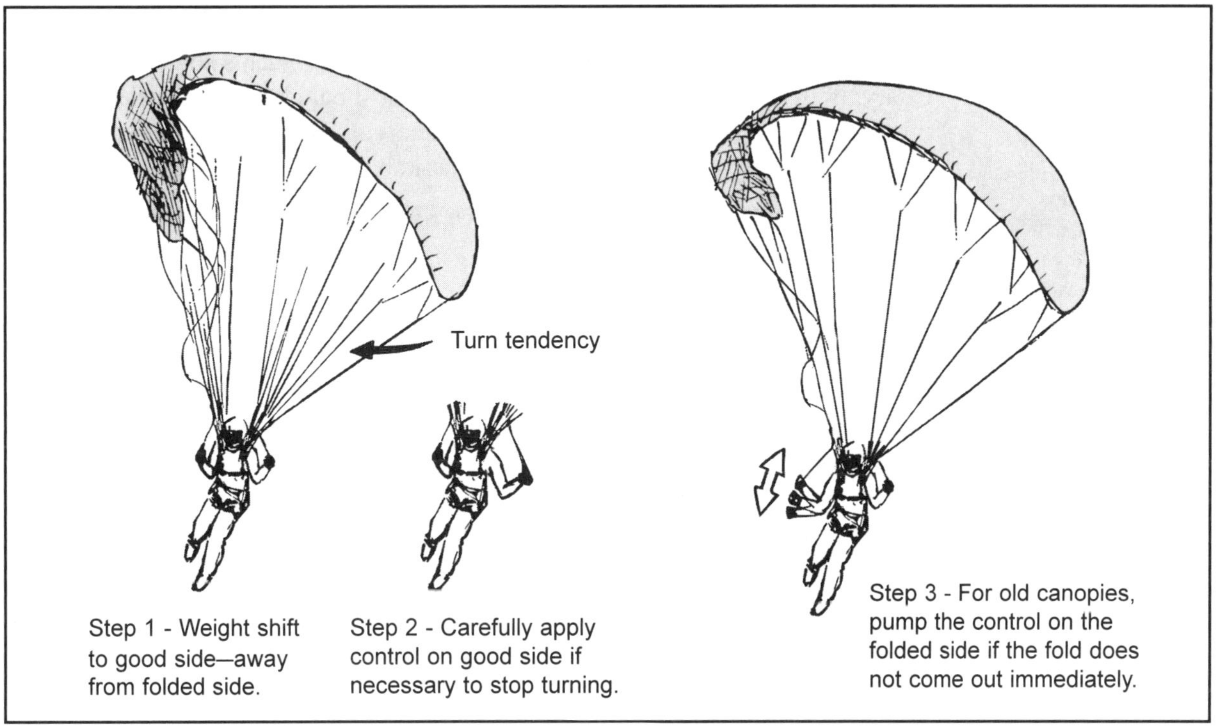

A pilot with an asymmetrically folded wing lifts his leg on the folded side to shift weight and fly straight.

back at the hill or entering a fast spiral). This correction control should first be a full weight-shift input (rolling the hips). On modern canopies there is no great tendency for a glider to turn with a mildly folded tip, so often simply moving your hands up and pausing while watching the canopy (as well as traffic and the terrain if applicable) is all that's necessary. Again we remind you: *don't overcontrol.* In most cases you will make matters worse with gross control inputs.

With a folded tip your wing is essentially made smaller, and the center of lift is shifted toward the open wing. The result is your stall speed is increased and pressurization better due to the higher wing loading. This factor serves to help you open the folded wing.

Occasionally the turbulence and sink that occur around a thermal oppose a quick reinflation or tend to increase the turn effect. In such a case, you may have to counteract the turn with a bit of control pull on the open wing (remember, the folded side is dragging more and the open side tends to accelerate forward). Such a control must be applied with great care, for too much control can result in a stall—an event much more severe than a tip fold. The control should be smooth and progressive with full weight-shift applied. Normally the pull should not be past the carabiner.

Here's a summary of the folded wing recovery steps for a modern glider:

Tip folds are common enough when flying in thermal turbulence, so practice controlling them and you'll be prepared to move up in your flying.

1. *Steer the canopy to fly straight.* First use weight shift, then a control pull if necessary. Both the weight shift and the control pull should be added gradually. With this action most folds will automatically come out. The idea is to slow the turn. With a large fold it may not be possible or desirable to completely stop the rotation.

2. *Evaluate.* Now, check to see how much of the canopy is folded and determine the appropriate action.

3. *Give the canopy time to recover on its own.* Very often, over-exuberant (adrenaline prompted) control actions by lesser experienced pilots results in unwanted glider reactions.

On older canopies (and competition models), a fourth step is recommended. This is a slow pumping action on the control of the folded side as shown in figure 9-27. This input should be like a landing flare, but not quite as far. The fold will pop out with this input. If part of the tip is caught in the lines, maintain slow flight and heading while pumping until it frees itself.

Safety Tip: *Throughout the reinflation and pumping process, focus on your flying path and fly straight. We used to preach "steer then clear". Now we preach steer with caution and clearing will probably be automatic.*

Remember our forward facing ground handling practice? Kiting your glider can help you learn to feel its reactions and the control pressures as well as to make the proper stability or recovery controls. Your instructor can induce tip folds while you are kiting so you can practice controlling and removing them.

✓ *Frontal Folds*

Figure 9-28: Dealing with Frontal Folds

Sometimes a turbulent swirl will hit mid wing and fold the front of your glider. The leading edge typically folds under partially or all the way. The fold may be symmetric or asymmetric as shown in figure 9-28.

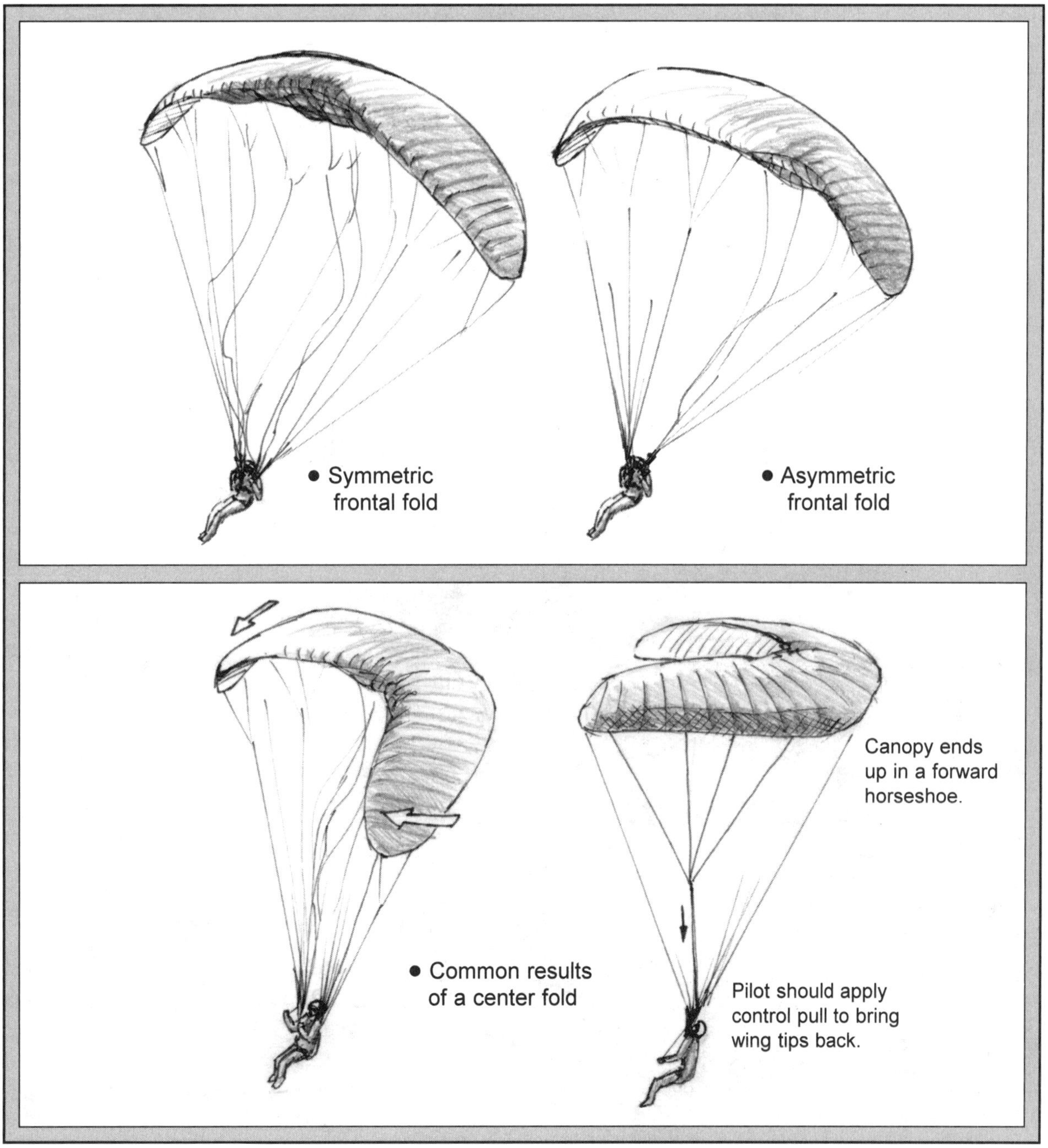

The prevention of frontal folds is essentially the same as with tip folds: Maintain good wing pressurization and a constant pressure in the control lines. When a frontal fold occurs the control line pressure usually reduces. On more advanced gliders, a major fold in the center retards that portion, so the wings fly forward and create a front horseshoe as shown in the figure. This is not a particularly dangerous or unstable configuration as long as it doesn't start rotating.

The recovery from a frontal fold is usually quite simple. First, stop any turn or rotation as with a tip fold (the glider will have much less tendency to turn with a frontal). Then the glider will recover itself as long as you keep good control pressurization. Your only job is to keep the canopy from oscillating fore and aft. It may tend to surge after recovery. You must avoid over-controlling by using light controls with your hands near trim position.

✓ Turbulence Security Position

In strong turbulence that seems beyond your capability to control your wing, there is an all-else-fails security position. This position is with the control handles in your hands while you are keeping them at about shoulder height with our standard 4 lbs (2 kg) of pressure. Let your arms float with this pressure for optimal security. If the glider is moving around above you the line pressure will vary, so your hands should move up and down to maintain an even pressure as much as possible. Most pilots new to strong canopy-warping turbulence tend to apply too much control input and often don't allow their glider to regain proper flight configuration.

A pilot has pulled in nearly 50% big ears.
Novice pilots should not induce such large folds.

If maintaining a steady pressure seems impossible, your last resort is to grasp your carabiners or shoulder straps with both hands while holding the control handles. This position pressurizes the wing well and avoids putting in any extraneous or ill-timed controls. You must hold on tightly as the control will want to pull your hand up in some violent situations. In general, this position will maintain a steady canopy. If a fold occurs, it will come back eventually. Note that this method doesn't work once a spin has occurred.

✓ Big Ears

We now discuss an important canopy folded position which you can induce yourself. *Big ears* is simply a symmetrical folding in of the wing tips. It is a relatively benign configuration as long as you heed certain warnings. Big ears are often used to descend rapidly and handle turbulence. We'll first show how to produce big ears, then how to use them safely. Try them the first time under supervision with plenty of altitude (at least 1000 feet—300 m).

To produce big ears, start with a stable, fast flying canopy moving straight ahead (if you are flying slowly your release of the controls will cause a surge). Keep ahold of the toggles but let go of any wraps on the controls if you have wound them around your hands. Reach your hands up and grasp the outside split on the A risers. Note: Some designs have tabs or pulleys on the lines for pulling big ears. Pull these risers (outside lines) down gradually and steadily until the wings fold in. The further you reach out on the riser or lines and more you pull, the greater the fold (see figure 9-29).

Caution: *Do not pull the lines to the point where you start pulling the rest of the A riser down. Always consult your owner's manual for the proper way to perform this or any maneuver.*

Pulling on the A riser will cause a front fold. If you do this by mistake, release the line immediately and go to the controls. To help prevent such an occurrence, watch the riser as much as the wing tip. Also, you should grasp the line with your fingers outside and use your hand in a twisting fashion as shown in the figure. Modern gliders have split A risers so you can pull the outside one down with no problem. It is recommended that you pull big ears one at a time to avoid the sudden increase of drag on the canopy rocking it back into a stall (check your glider's owner's manual on this matter).

✓ Flying With Big Ears

In flight, big ears produce a stable condition. The wing loading is higher so you have greater pressurization at a given angle of attack. However, the canopy also flies with a steeper path at the same attitude so the angle of attack is increased. Thus, it is advised to use the speed system in conjunction with a large amount of big ears in order to avoid stall problems.

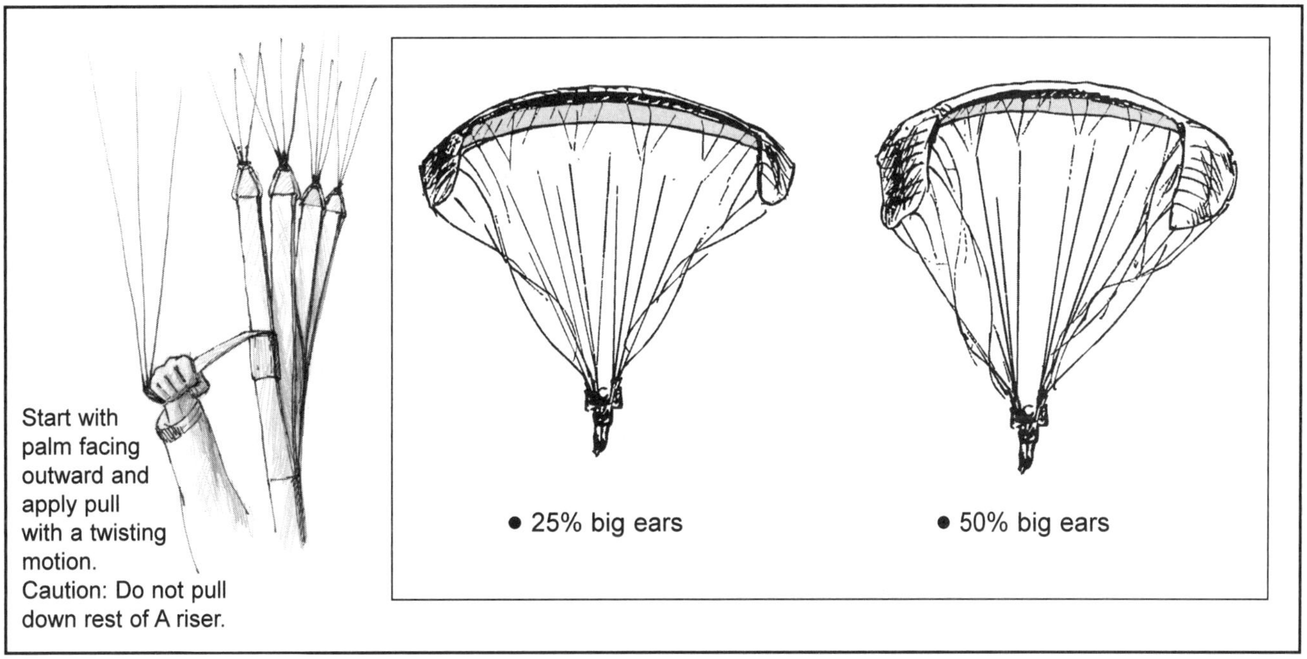

Figure 9-29: Big Ears Controls

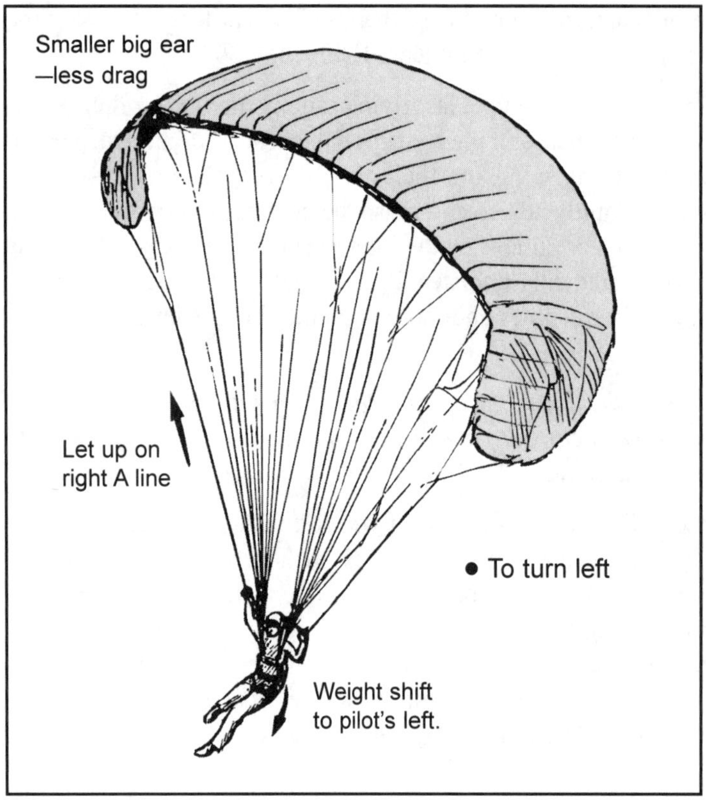

Figure 9-30: Turning with Big Ears

The best way to use the speed system (accelerator) is to have both feet on it before you pull on big ears. Keep your knees bent to avoid pressing the bar inadvertently. Now pull in your big ears. Finally, smoothly depress the speed bar. Only use about half the speed system control applied with about 25% big ears to avoid frontal fold. With greater amount of big ears the full speed push may be used, even in turbulence. Be sure to release the speed slowly before coming out of big ears.

With big ears your direction control is reduced. Remember, your hands are holding down the lines. You still have weight shift and you can pull the lines down differentially, but *don't* pull down an A riser! It's better to let up one side than pull on the other. If you want to turn left, shift weight left and let up on the right side lines as shown in figure 9-30. You will not have as much control as with the control lines and you will not be able to fly actively. Also, your roll dampening is less without the full canopy above you, so be careful with turns near the ground. Learn all the characteristics of your canopy with big ears up high long before you try it lower.

Even though you may be going faster with big ears, your steeper path means you are not moving over the ground as fast. Therefore, *never* use big ears in a situation where it's doubtful you will reach the landing field or in a venturi. Despite all these warnings, however, big ears are useful and even recommended for fast descents (perhaps to escape a cloud) or in turbulence with the use of the speed bar. You can normally double your descent rate. The full speed position can be used with big ears as noted above, but some manufacturers recommend inducing big ears with the speed system off, then adding the speed. Note that the higher wing loading on the non-folded portion of the wing means increased suspension line tension. Try to avoid excessive turning while in big ears to prevent line stretch.

Practicing big ears helps you familiarize yourself with tip folds, especially if you pull them in one at a time.

✓ Removing Big Ears

To get out of the big ears configuration, first release your speed system smoothly and when the canopy has settled, let up on the A lines one at a time and the tips should slowly unfold. If an "ear" doesn't come out automatically, simply add a slow smooth pull of the control. *Do not* hit the controls hard, for a stall may result.

The dangers of big ears pointed out above makes it clear that you *should not apply or remove them close to the ground.* Stalls can result. Primarily use big ears as a high maneuver in the situations stated above. Some pilots use big ears to descend through the lift and turbulence on a top landing, or in a normal field on a windy day. You are more susceptible to wind gradient and wind shadow with big ears, so keep your speed up.

If you do land with big ears, *do not* remove them while you are still in the air. Land with them and be prepared to do a PLF. If you try flaring with the rear risers, the big ears will probably pop out which can pull the canopy back in high winds. Be ready to turn around and run or produce an emergency fold to prevent a drag.

FLYING HIGHER

The reward for all your previous practice and study is the chance to fly higher and enjoy new perspectives and more airtime. You'll feel more and more like a bird the higher you fly. But just like so much in aviation, along with the rewards come responsibilities and complications. Begin your high flying carefully with planning and preparation.

Your Goal: *To achieve controlled flight from takeoff to landing at any altitude in a wide variety of conditions.*

A pilot's view of big ears. Note the outside A line is being pulled down.

✓ High Flight Preparation

The first thing to prepare for high flying is you, the pilot. Make sure you're ready by studying all the material and information your instructor gives you. This material includes the landing setup and launch factors given in the previous two chapters. Your instructor will guide you in these matters.

What we mean by high flight really depends on the area you learn in. For some lucky few (or those towing), hills are available that allow you to launch from any altitude, so there is a very gradual increase in height and thus no dramatic transition to high flight. However, for many others, leaving the training hill and flying with 150 feet (45 m) or more of clearance will be considered a step to high flight, because from these heights ground reference is reduced.

The best way to prepare for your first high flight from a particular site is to watch other pilots fly it. It is helpful to watch those just above your skill level so you see the same flight paths you will be taking. Then go home and dream about it. Live and breathe the flight. Image it by putting yourself in the picture and imagining the sensory inputs as you perform a complete flight. This practice is most realistic if you do it at the site, on both your observation day and on your flight day.

In addition to the pilot, we must consider the site and the conditions.

✓ The Site

The site of your early high flights will be carefully chosen by your instructor. Ideally, it will be perfect, but in the real world there may be complications which you will be prepared for with special practice or information. We will generalize for the ideal site so you know what to look for later when you are making your own decisions. The subject of flying new sites is covered in Chapter 13.

More altitude gives you more time to fly correctly.

▲ *The Launch* — A good launch area for your level of flying will be a slope as steep or a little steeper than the training hill you are used to. Ideally the grade will be uniform and smooth. The launch area should be wide open and free of obstructions in front of it. In areas where trees are prevalent, they will likely be present in front of launch, but they must be well clear below (about a 2/1 or 3/1 path as shown in figure 9-31). You should not launch in slots in the trees unless the wind is straight in or very light. This feat requires excellent control and is for pilots of intermediate or better skills.

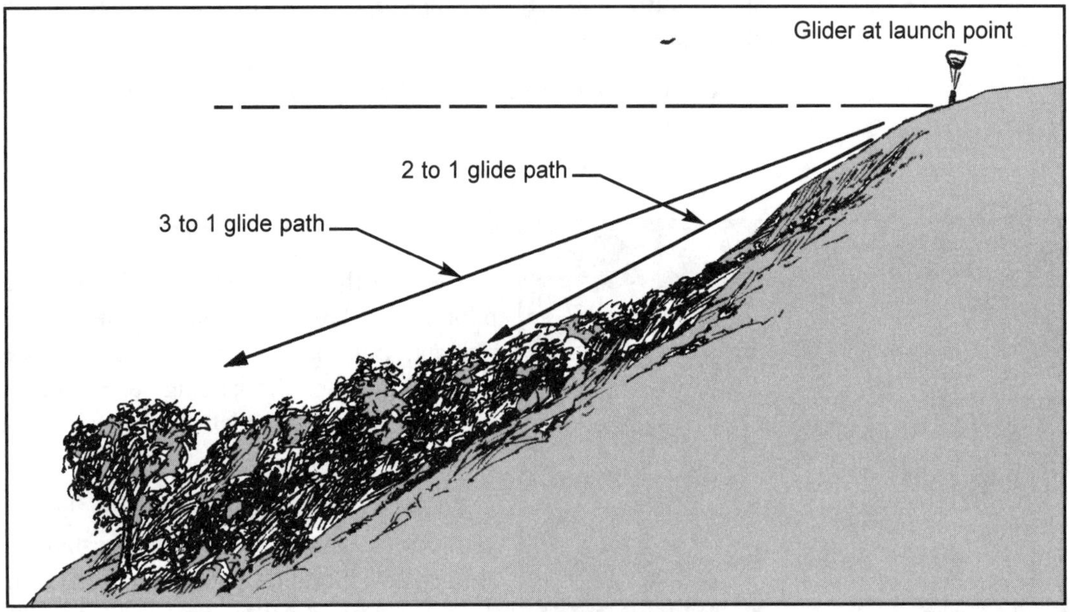

Figure 9-31: Safe Launch Site Limits

▲ *The landing* — The landing fields for your early flying should be large and unobstructed. No barriers such as trees, houses or power lines should exist from a point halfway from takeoff to landing. If your field is long and narrow it should be oriented with the long axis parallel to the wind. If this is not the case, you should wait until the wind changes or it is calm.

The best thing you can do every time you fly is to visit the landing area. Before each of your first high flights walk the landing zone to get a feel for its dimensions, and imagine your positions as you come in to land. Your instructor will show you setup patterns—where to be at what height and where to lose altitude—during your first high flights.

As you walk around the landing field, look for high weeds and hidden problems like ditches, fences or large rocks. Look for slopes and their relation to the wind direction. Check for power lines near the field. Their support poles may be hidden in trees with only the lines stretching across the field. These lines will be hard to see from the air unless you have located them beforehand on the ground.

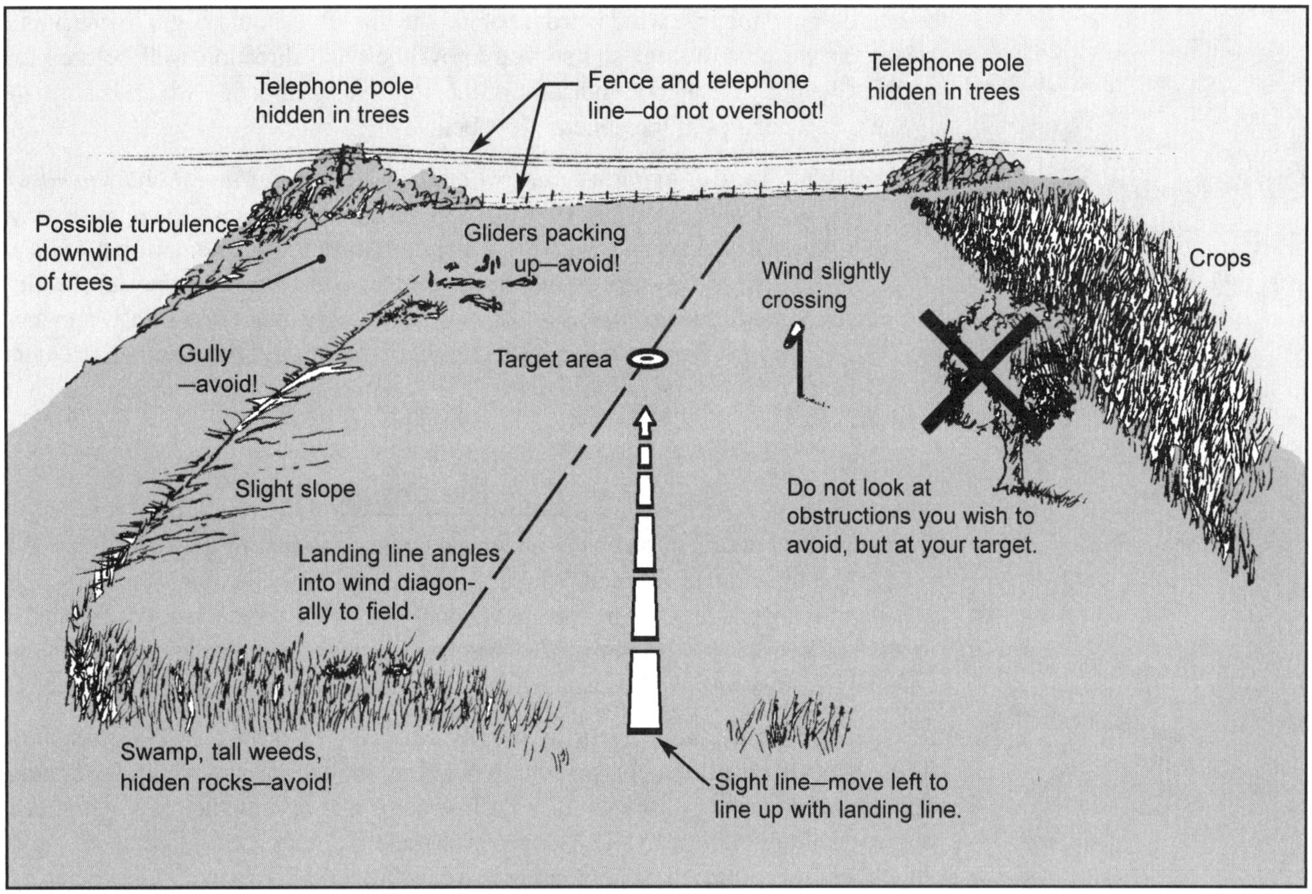

Figure 9-32: Judging a Landing Field

The final thing to do is look up at launch and imagine the direction you will be approaching, and repeat the landing setup in your mind. Figure 9-32 shows a landing approach example and some of the matters you will consider when you land at new fields.

Safety Tip: *First high flights should be performed with the maximum of a few experienced pilots in the air and no inexperienced pilots. All pilots should know this is an early high flight so they can give you plenty of clearance.*

✓ The Conditions

The ideal conditions for your first high flights are with light, smooth winds blowing in at takeoff and along the ground in the landing zone. The normal time for such conditions to exist is during the morning or evening. Middle of the day excursions should be attempted only after you've had a lot of light wind and high flight experience.

To indicate the conditions, there should be wind socks or streamers at launch and in the landing field. The indicator in the landing field should be clearly visible from launch. It is not uncommon for the wind in the landing field to be as much as 90° different from that at takeoff. This effect may be due to channeling from the terrain or to heating.

If the winds are quite different from takeoff to landing, you should question the safety of flight. Perhaps there is an inversion or a shear layer present. Perhaps you are in rotor and the true wind is from behind the hill. You should begin to check the winds on the way to your site so you know the wind direction well before you arrive. Cloud drift is a good indicator of the upper wind direction which is normally within 30 degrees of the general flow below.

In our first year of flying we had but one site. It was northwest and was ideal for our experience level. We flew it with enthusiasm every weekend, no matter what the wind direction. One day many pilots were flying and taking off into a gentle breeze. About halfway out, however, they were encountering severe turbulence. One pilot was tossed into a steep dive and landed in a small clearing. We decided to check the back of the mountain and found the wind was actually about 15 mph from behind takeoff! We learned about rotors on that day.

✓ Dive Syndrome

One of the important matters to understand when you first fly away from ground reference is attention to airspeed. You may be subconsciously judging airspeed by the ground moving past in your peripheral vision. If this is the case, when you get higher the apparent lack of motion will have you unconsciously speeding up. This is called the *dive syndrome*. It is a speed perception problem, as figure 9-33 indicates.

Always monitor your airspeed throughout your entire flight.

You can avoid this problem by simply attending to proper airspeed sensing. These sensations are the control toggle position, the feel of control line pressure, and the sound or feel of the wind. Your instructor will review the proper airspeed and control position to fly. This speed will most likely be best glide speed. You should set this airspeed as soon as you are stabilized after launch. On your way out to the landing, constantly check your airspeed and keep the canopy solid above your head.

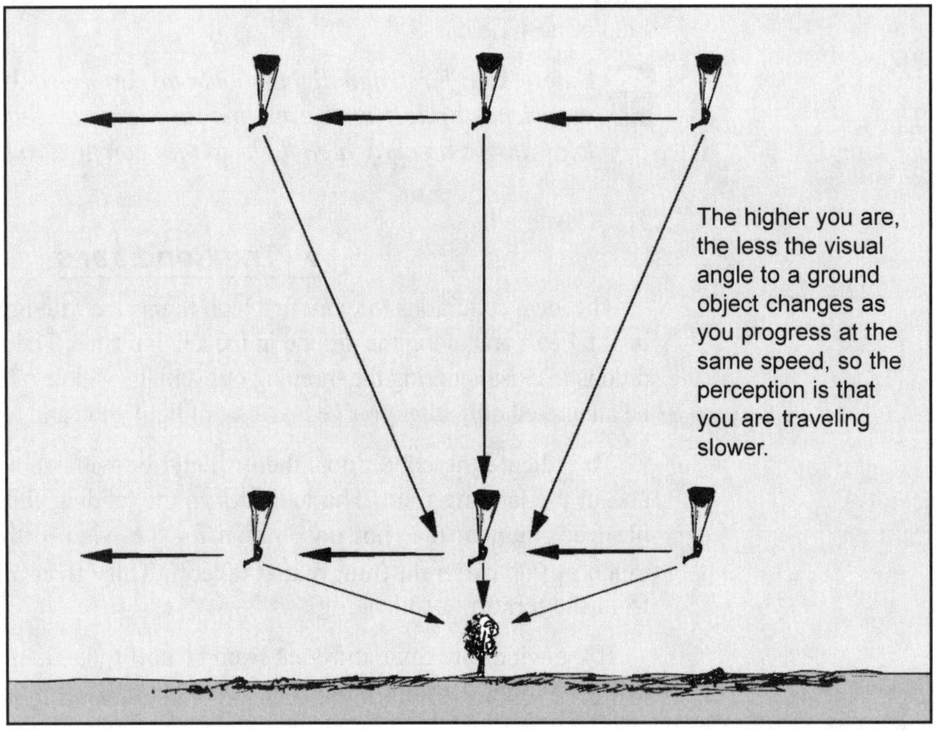

Figure 9-33: Dive Syndrome Perception Problem

✓ Flight Judgement

Early input allows early decision-making, "flying ahead of the aircraft" as airplane pilots say. Make decisions with time to spare and time slows down—the pilot can relax.

As you fly, it is necessary to constantly assess your progress and positioning. Some of the questions you should ask are: "How am I doing?", "Where should I go next?", "What point will I reach?"

The eye and brain team are not very well equipped to judge straight line distances without some form of reference. For instance, we judge the position of far objects by relative motion of objects in the foreground and background. If you are looking down at the ground, there is nothing to tell you how far it is. You can imagine looking down at a sea of trees. Nothing in your visual experience tells you their size or distance from you. Remember the secret to judging height position and trajectory is judging angles and their rate of change. Look forward and down rather than straight down.

✓ Judging Glide Point

Figure 9-34: Judging Glide Point by Viewing Angles

An important judgement tool is to see where your glide path is aiming. As figure 9-34 shows, the point you are going to reach (the point you are flying directly at) will remain stationary in your field of view. It doesn't move up or down, it

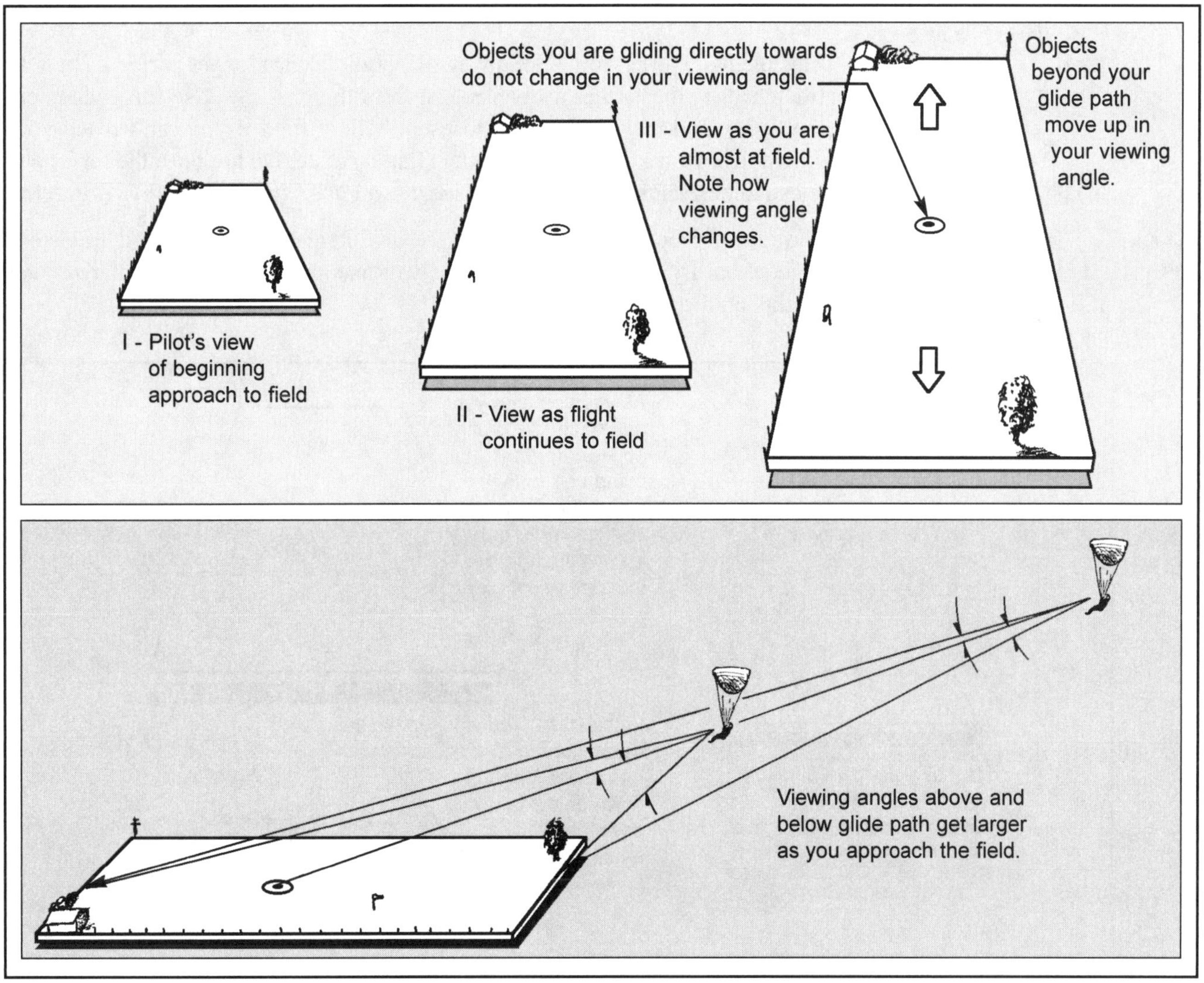

Our eyes judge angles and relative motion best. They are easily fooled about distance.

just gets bigger. On the other hand, any point beyond your glide path will move up in your view while any point below your glide path will move down.

This isn't a very precise technique, for objects move very slowly up or down, but you can use this trick for long glides by taking a sighting every so often and comparing it with your previous sighting.

✓ Judging Obstacle Clearance

A much more useful technique is judging whether or not you are going to clear an obstruction such as a power line, a row of houses or trees. The process is similar to the previous one, except that now you have a sighting device—the obstruction. As figure 9-35 shows, you must clear the power line in order to reach the landing field. Will you make it? Look at the bush that just peeks above the line. If you are going to clear the power line, the bush behind it will continue to move above the line. If you are not going to clear it (your glide path is below the angle to the power line) the bush behind it will move down below the line. If the object is solid like a building or hill the method still works. If objects behind the hill come into view, you will clear the hill. If not, start looking for alternate landing fields.

This method works from a long way off, but it takes a longer period of time to see the changes the farther away you are from the obstacle. Use this judgement method often and make a decision to use a bailout field if you can't reach your goal, or adjust your speed to get a better (flatter) glide. Do not push the limits and get close to obstacles. *Always keep at least two glider spans away from obstacles.*

The above processes are what we use to judge our position and trajectory throughout our flying, whether consciously or unconsciously. We summarize their highlights on the next page.

Figure 9-35: Judging Obstacle Clearance

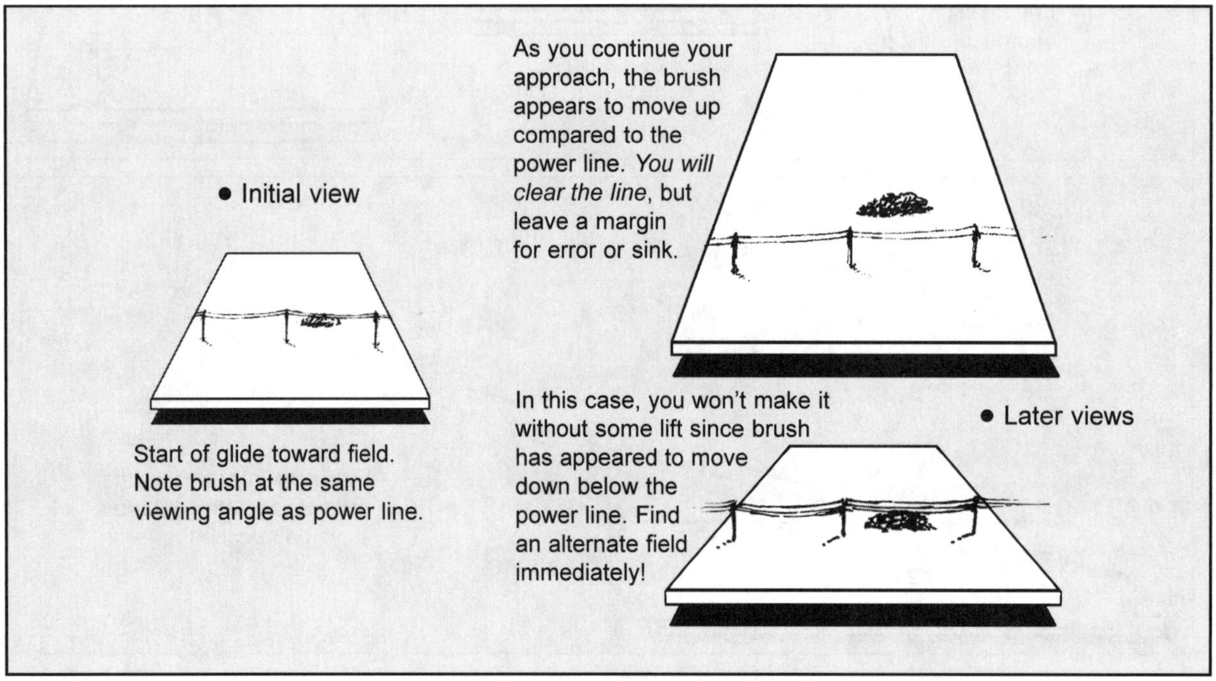

Judging Positions and Glides

- Always judge angles rather than distances and heights. It is very difficult to judge how high you are by looking straight down.
- Your position is given by your angle to a field or other feature and your vertical ground point.
- Your glide path will reach the point that is stationary in your viewpoint. A point beyond your reach will move up in your vision. A point you will pass moves down.
- To tell if you will clear an obstruction from far away, watch the area behind it. If objects appear to move up behind it you will clear it. If they move down you will not.

✓ Planning a Flight

Once you are equipped with information, good basic skills and the right environment you are ready to take a flight. First form a flight plan, then proceed.

▲ *Flight Plans* — Every flight you take for the rest of your flying experience should be preceded by a flight plan. As a new pilot your plan should be elaborate. Later on it may simply consist of an assessment of takeoff and landing conditions and a determination of the direction to head to find lift—but it's still a plan.

A high flight plan for inexperienced pilots should consist of determining the flight direction from launch (straight out), the turn point and direction to head for the landing field, the position and pattern to lose altitude, and where to enter the landing approach. The wind direction and wind awareness should be part of your plan. Your instructor will review your plan with you and offer amendments if necessary.

Your first flight from on high should be straight away from the hill to the landing field. You need about ten of these flights in order to learn judgement of your position and how to set up a landing.

The end of a perfectly planned flight is a perfect landing setup.

Here's how to proceed on your first high flights:

1) Prepare your flight plan and file it away in your mind.

2) Lay out and preflight your glider. Recheck the conditions and your flight plan. Then run through the flight in your mind—imaging.

3) When it's your turn to go, move your glider into position, hook in, perform your checks and get ready.

4) When conditions look good, take several deep breaths to relax and concentrate on takeoff. Do not think of the flight or landing while you are taking off. That's what the flight plan is for, and you'll pull it into your conscious mind once you are safely away from the hill.

5) Perform your usual good launch, stabilize in flight, then recall your plan. Fly to the landing area, take up your position in the *staging area* to lose height if necessary, and do so. On your early high flights there should be little or no wind, so the choice of approach will mainly depend on the size of the landing field. (A detailed description of losing altitude in the staging area and the different approaches was provided in the previous chapter.)

6) Set up the landing approach and have a perfect landing.

7) Move your glider from the landing area immediately so other gliders can come in. Despite your exhilaration, don't forget this last point, or other pilots may have more complicated landing. They won't be in any mood to share your joy in that case.

High Flight Guidelines

- Direct your vision to the landing field—that's your goal.
- Judge angles to your field and the landing point to assess your position.
- Do not look directly down at the ground—it will tell you nothing and may confuse you.
- Relax and feel your glider's responses to the air.
- Check your control position and airspeed frequently.
- Do not turn away from the landing field or downwind until you've gained much more experience.

✓ Busy Sites

Often the early stretching of your wings will occur at a busy site. As a new high-flying pilot your instructor or other pilots will direct and accommodate you so that you don't disrupt the flow at takeoff too much. Possibly you will fly in the evening or during the week when fewer pilots are present. At any rate, prepare yourself well for a busy launch situation by practicing your layout, kiting and launch skills to perfection long before you need them at a "social" site.

Your best policy is to try to launch when there's a break in the action, perhaps after all the anxious or hot pilots have left. The following procedures are what you should follow for most sites.

1. Lay out and check your canopy in a sheltered place away from the ground traffic and wind.
2. Check your lines, preflight and climb into your harness.

Paragliding flight can be dreamlike but watch for traffic.

3. Attach the harness in the reverse position to the canopy and double check your straps.
4. Once everything is deemed airworthy, gather the canopy and move to the launch area to await your turn.
5. When you have room, lay the canopy down, open it as much as necessary, then stretch out the lines.
6. Do your last minute checks, then launch when the wind is good and the traffic is clear.

If conditions are safe, do not hesitate or you'll gain a reputation as a launch potato. Perhaps you can live with the embarrassment, but can you live with other pilots pushing in front of you so they won't be delayed?

✓ Flying at Altitude

Ideally you should progress in 50 foot (15 m) increments in altitude until you reach 500 feet (150 m). Then you can add 100 feet (30 m) at a time up to 1,000 feet (300 m). From that point on you can handle larger increases in altitude. Unfortunately, we can rarely follow such a program due to the lack of progressive sites. This is where towing has the advantage. The best plan is to proceed at a gradual pace and become entirely comfortable at each level of flying before moving on.

"When you learn to trust your glider and equipment, your fear will diminish or disappear."

As you acquire more altitude, you may discover a fear of heights you never had lower down. This is common and will be overcome as you gain more experience and confidence in your equipment. Keep reminding yourself that there is safety in altitude. The hard stuff (other than other aircraft) is near the ground. Generally the winds are smoother the farther away from the ground you are. Very rarely does the wind increase much above 1,500 to 2,000 feet (450 to 600 m) over the terrain, so you can put that worry out of your mind. Remember, the higher you are, the more you can relax and enjoy the view. And what a view it is!

JUDGEMENT

The last thing we'll address in this chapter is judgement, but it should be the first thing on your mind at all times. Here we are speaking of safety judgement, deciding whether or not you, the pilot, your equipment and the conditions are suitable for flight.

Your Goal: *To learn and practice good judgement—Always!*

"Fear not, provided you fear; but if you fear not, then fear." **—Blaise Pascal**

✓ *Risk Management*

In many walks of life we need risk management. Whether it's driving a car, drinking alcohol, watching television or engaging in active sports, we need to assess the dangers and weigh the reward-to-risk ratio. In paragliding we do this by checking as many factors as possible that affect flight.

One risk management system that does this is the Charts of Reliability designed by Michael Robertson. His system separates the glider, the conditions and the pilot—he calls them the wing, the wind and the "windividual"—and provides a rating chart for the different factors involved with each. He is teaching judgement. Appendix I provides the address for obtaining this system.

We highly recommend Michael's system. Here we will mention a bit about the pilot. Paragliding is a spectacular sport. It has the tendency to attract attention, so it also attracts those who seek attention. The problem is that those who overly exert themselves to gain the admiration of others may push their limits and endanger themselves. The presence of daredevils and thrill seekers cannot be avoided perhaps, but we can minimize their effect by recognizing them for what they are. Remember the old adage: there are old pilots and bold pilots, *but there are no old, bold pilots.*

Fly within your limits, progress steadily and learn the basics well. Gather all the information you can about flying and put it to work for you. By all means make the training hill your friend. Don't be reluctant to go back and polish up a technique or retrain after a winter's layoff. Practice recurrent training where you test yourself on your skills every six months or so.

Friends help friends fly safely.

Plato said, "Above all, know thyself." When you are above it all in your paraglider, know where your limits are and make a mature, safe decision to stay within them.

Moving up in Performance

At some point in your flying you will probably buy your own equipment. This may occur during your novice practice. It is always important to buy a glider appropriate for your skill level. Be sure to heed the recommendation of the glider's manufacturer or your instructor's advice in this matter. In general, higher performance gliders have less stability and are more difficult to inflate. Their launch is more complicated and they are less responsive in turns. In addition, they are less forgiving in folds, stalls or other mishaps. They also require a bit more room to land. A detailed guide to buying new or used gliders appears in Chapter 11.

Flying in more wind requires good reverse launch technique.

NOTE ON TOWING

If you are learning to fly using tow training, most of what you have learned to date directly applies. The material on takeoffs concerns launching from a slope, but the controls and procedures are the same if you think about it. Eventually you will probably try your hand at slope launching as well. Review the launch procedures at that time.

The principles, practices and potential dangers of towing are too involved for our discussion here. We highly recommend you obtain the book *Towing Aloft* by Dennis Pagen and Bill Bryden if you plan to do any towing whatsoever.

SUMMARY

When you have accomplished all the new techniques and absorbed all the new ideas in this chapter you have reached novice level. At this point you should be beginning to make some of your own judgements. Practice decision-making constantly by making an assessment, then asking an experienced pilot what he or she thinks, then reevaluating your ideas.

"Practice is the mother of skill."

Perhaps the major portion of your practice at this level will be with turns which include landing setups. Practice, practice, practice. Soon you will be learning to soar—to stay up on wind currents—and now is the time to acquire perfect control so that you can turn your glider on a pin and make every air molecule account for itself. You have nearly arrived as a pilot. But don't stop here, for as you continue to improve your skills, the freedom and joy expand in like amounts.

TEST YOURSELF (*Answers in Appendix II*)

1. The primary mean of maintaining good canopy pressurization is:_______________
2. Where are the toggles when you are flying at best glide speed? _______________
3. What are the dangers of flying "deep" in the controls? __________________
4. How do we stop a surge? ____
5. Signs of an impending stall are: _________________________________
6. Signs of a deep stall are: ____
7. The most important thing you do before you turn is ___________________.
8. An efficient turn is made as slowly as possible without reaching a stall. *True or False*
9. A dangerous situation when one wing is stalled and the glider rotates rapidly is called a _________.
10. Spirals are simple spins in the forward direction. *True or False*
11. When flying higher, you should carefully monitor your _____________ and _______________.
12. When in traffic or near a ridge, a wing tip fold should be handled by first ___________.
13. Big ears should never be used in turbulence. *True or False*

USHGA Novice Rating Checklist

General Description - A Novice paraglider pilot has the knowledge and basic skills necessary to fly and practice without direct instructor supervision but within significant operating limitations. The pilot understands the USHGA paragliding rating systems and recommended operating limitations.

The pilot shall use good judgement and have a level of maturity commensurate with the rating. Pilots must demonstrate Beginner level skills and knowledge before obtaining the Novice rating. All witnessed flights must be pre-planned by the pilot and discussed with the Instructor or specially qualified Observer.

Novice Rating - Foot Launch - Required Witnessed Tasks

A. Logged Requirements
   1) Attends a minimum of 8 hours of ground school.
   2) 25 flights.
   3) 5 flying days.

B. Demonstrated Skills and Knowledge
   1) Demonstrates layout and preflight of the canopy, harness, and backup reserve parachute.
   2) Gives a reliable analysis of general conditions of the site and self, and a flight plan including flight path, areas to avoid in relation to the wind flow, and obstacles to stay clear of.
   3) Demonstrates 5 consecutive forward inflations with a visual check of the canopy each time.
   4) Demonstrates 5 consecutive controlled reverse inflations with proper surge dampening.
   5) Demonstrates controlled kiting of a glider overhead for 2 minutes in a steady wind.
   6) Demonstrates 2 clean, smooth reverse inflations/reversals prior to launch.
   7) With each flight, demonstrates a method of establishing that the pilot is properly connected to the glider, with cleared lines and risers just prior to inflation.

8) Demonstrates 2 successful, aggressive, confident inflations/launches, where the wind is at least 15° cross to straight up the hill in wind not exceeding 5 mph.
9) Demonstrates 2 no-wind (0-5 mph) inflations/launches.
10) Demonstrates how to brief and instruct a ground crew and explain when an assisted launch is necessary.
11) Demonstrates 2 high-wind (10-15 mph) inflations/launches.
12) Demonstrates flight with smooth variation in airspeed, from above minimum sink to fast flight, while maintaining a heading.
13) Demonstrates flight showing the ability to comfortably and precisely slow the glider to minimum sink and smoothly increase to normal airspeed while maintaining a heading. The pilot should not slow the glider to near the stall speed.
14) Demonstrates flight(s) along a planned path alternating 'S' turns of at least 90° change in heading. Flight heading need not exceed 45° from straight into the wind. Turns must be smooth with controlled airspeed, ending in safe, stand up landings on a heading.
15) Demonstrates hands-off flying, weight-shift turns, and rear-riser turns.
16) Demonstrates the ability to judge and allow for proper clearance from a ridge and other aircraft.
17) Demonstrates 5 landings within 25 ft of a target, safe, smooth, on the feet and into the wind. The target must be sufficiently close to launch such that turns are required to set up an approach and avoid overflying the target. The target should be at least 100 ft below the launch point.
18) Explains proper strong wind landing procedures and how to keep from being dragged back.
19) Explains correct canopy maintenance.
20) Explains how to lengthen and shorten the flight path.
21) Explains the right-of-way traffic rules.
22) Demonstrates reserve deployment while hanging in a harness in simulated turbulence or malfunction conditions.
23) Gives a thorough verbal demonstration of knowledge of how to:
    a) Maintain directional control during, and correct for, an asymmetric wing fold of 25% of the wing span.
    b) Fly at minimum sink while precluding any chance of inadvertent stall or spin.
    c) Increase descent rate and/or forward speed.
24) Demonstrates proper and effective PLF technique.
25) Must pass the USHGA Novice Paragliding written exam.
26) Must agree to all the provisions of the USHGA standard waiver and assumption of risk agreement for the Novice rating and deliver an original signed copy to the USHGA office.

Novice Rating - Tow Launch - Required Witnessed Tasks

A. Must demonstrate the above-mentioned Novice paraglider rating foot-launching tasks except B tasks 8 - 11.

B. Must demonstrate system setup and preflight, including a complete discussion of all those factors which are particular to the specific tow system used and those factors which are relevant to towing in general. Must demonstrate complete understanding of both normal and emergency procedures, including checklists for normal procedures and the indications of an impending emergency and convince the Instructor of his ability to recognize and execute emergency procedures.

C. Demonstrates successful, confident, controlled launches and flight under tow to release at altitude, with a smooth transition to flying, with proper directional and pitch control resulting in proper tracking of the towline and appropriate maintenance of proper towline tension and airspeed.

D. Demonstrates understanding of all Novice-level Paragliding Tow Discussion Topics.

Recommended Operating Limitations for Novice Paragliding Pilots

A. Should exceed these limitations only after thoroughly mastering all required tasks, and after acquiring a full understanding of the potential problems and dangers involved in exceeding these limitations.

B. Maximum base wind of 12 mph.

C. Maximum peak gusts to 15 mph.

D. Maximum gust rate of 5 mph in 5 seconds.

E. Should not fly in thermal lift where peak climb rates exceed 200 fpm.

F. If foot launching, should launch only on slopes steeper than 4:1, where the wind is within 25° of being straight up the slope.

G. Visual contact with the landing zone.

H. Avoid application of either brake beyond 2/3 of the way from slack to stall position.

I. Limit turns to 30° of bank, limit speed in turns to 1.5 times the straight line, brakes off, cruise speed, and smoothly exit any spiral turn which shows a tendency to steepen or accelerate.

J. Should fly a canopy recommended by the manufacturer as suitable for Beginner to Intermediate pilots.

What a beautiful flying day, but watch the overdevelopment cloud on the horizon.

Chapter 10

RULES OF FLIGHT

"Few are those who err on the side of self-restraint."
— Confucius

Part of the attraction of paragliding is the freedom it offers. It is uncomplicated and the least regulated way to fly. The main rules we must follow are the laws of nature.

Experience has taught us a good deal about the safety limits of our sport. In this chapter we'll review these limits so you can benefit from the mistakes or experience of others. These are the rules we follow to stay healthy.

Other rules we impose on ourselves make sense in order to preserve sites or our freedom. Such rules as how to treat landowners or the public are just common courtesy. We cover them here. Finally, we'll look at right-of-way rules and government rules that all pilots must follow. These are the rules we adhere to in order to cooperate or meld in with other air traffic.

Most of these rules are simple and require minimal effort, but like driving on the correct side of the road, ignoring them can have a great impact.

RULES TO LIVE BY

Sometimes we can get away with breaking man-made rules. If no one sees us we don't get caught. But the laws of nature and physics aren't so easy to cheat; they are always enforced by unseen forces. If we wish to avoid accidents we must obey these laws.

Aviation, as is often said, is as safe as the pilot in command. If you wish to fly safely, you must make a conscious decision to do so. If you push the limits you are not flying safely and are more likely to have an incident or accident. The choice is yours.

Here we offer a short list of do's and don'ts that will help guide you safely onward and upward. The number one rule to live by is to *always fly one step below your limits*. If you are very aware of your skill limitations and leave a reasonable margin, you will fly safely. All the remaining rules follow this primary one.

✓ Launching Rules

Perform a careful pre-laucnh check.

▲ 1. *Perform a pre-launch check just before inflation.* Remember: shoelaces, leg straps, chest strap, carabiners, risers, helmet, traffic...CLEAR! This rule helps assure you will never fall victim of the dreaded "hanging by the armpits" syndrome.

▲ 2. *Assess conditions carefully before launching.* The nature of the air plays a major role in flying safety. The best policy is to check conditions before laying out, after preflighting, then again after moving to launch. (If you worry about conditions during canopy preparation, your attention is not where it needs to be.) If you move to launch before checking conditions, you may be pressured to go in less than proper conditions by your ego or other pilots. It is easier to avoid moving to launch than it is to back down once you are there. Never let others talk you into flying when your instinct says "don't."

Tail wind —don't launch

▲ 3. *Never launch in a tail wind.* Although some pilots get away with tail wind launches by virtue of their fleet feet, the margin for error is low and this feat is too close to the limits of safety.

▲ 4. *Do not fly in strong conditions.* At the end of the first two parts of this book we give weather guidelines for pilots of beginner and novice levels. In each case the pilot should have a reasonable cut-off point for wind strength. Remember, the force of the wind varies with the square of the wind speed. In addition, gustiness diminishes your control. For safety's sake, avoid conditions with a gust factor greater than half the mean velocity of the wind or with changes in which the velocity varies more than 5 mph (8 km/h) in less than 5 seconds. For example, if the mean value of the wind is 10 mph the allowable gust factor or variation is 5 mph

On a crowded launch, courtesy and common sense must prevail.

(half of 10 mph). Thus the wind can vary from 7.5 to 12.5 mph, but the variation from the lowest to the highest velocity and back down again should take at least 10 seconds (5 seconds to go from 7.5 to 12.5 and 5 seconds to go back down).

Check canopy

▲ 5. *Limit the crosswind component acceptable for launch.* About 30° crosswind component in ridge soaring conditions and 45° in light winds is one suggested guideline for better than novice pilots. As gustiness increases, these limits must be reduced.

▲ 6. *Check the canopy before committing to the air, every time.* A severe fold or tangle can greatly reduce your control in the air which can be dangerous near launch.

✓ In-Flight Rules

Paraglide only if you're 100%.

▲ 7. *Never fly with inadequate rest, when experiencing depression or emotional problems or under the influence of drugs (including alcohol).* Mental and emotional state is a major determinant of safe performance. Do not compromise this matter.

▲ 8. *Maintain proper airspeed at all times.* When operating near the terrain, use additional speed to avoid stalls in wind gradient. Learn the signs of a stall and the prevention and recovery techniques. Practice fast flying without oscillations so you can use speed when necessary.

Clear your turns!

▲ 9. *Clear all your turns and obey the right-of-way rules* (see below). Look carefully for traffic, fly defensively and assume other pilots will not necessarily offer you the right-of-way.

▲ 10. *Avoid high traffic areas.* As a novice pilot you should stay at least two spans away from another glider in smooth conditions and five spans away in turbulence. This rule is intended to avoid a conditions-induced midair. Do not fly directly above or beneath another glider, for a surge of lift or sink can raise or drop one glider more than the other. Flying closely in thermals or other types of lift demands great skill and coordination on the part of all pilots.

▲ 11. *Do not add more than one new thing at a time.* For example, don't try a new glider at a new site or in conditions never before experienced.

▲ 12. *Always remain within reach of a safe landing field.* This "rule-to-live-by" should be obvious.

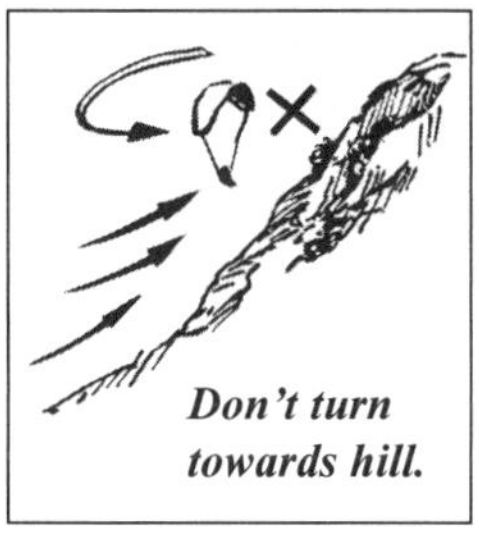

Don't turn towards hill.

▲ 13. *As a beginner or novice don't turn towards the hill.* Such an act can be very disorienting as ground speed increases and the slope of the terrain alters your perception.

▲ 14. *Make all turns efficient or intentionally diving.* Do not produce slow, stalling turns or a spin may occur. This rule is especially important after taking off, during the landing, or anytime you are operating near the terrain. Minimize your turning near the ground in turbulence.

▲ 15. *Use proper safety equipment and procedures.* Always fly with a parachute, an adequate helmet and proper footwear. Always preflight your glider and auxiliary equipment carefully.

▲ 16. *Try new equipment at the training hill first.* Learn new experiences gradually. Be sure the equipment you try is appropriate for your size and skill level. Never add more than one new thing at a time.

Avoid thunderstorms

▲ 17. *Exercise extreme caution in widespread lift greater than 500 FPM (2.5 m/s).* Stronger spread-out lift can signify thunderstorm buildup or general cloud suck, which may be difficult to escape. We will learn about this in Chapter 13.

▲ 18. *Never fly near thunderstorms or into clouds.* When visual contact with the ground is lost, you may become disoriented and lose all ability to control your glider (see Chapter 13).

✓ Landing Rules

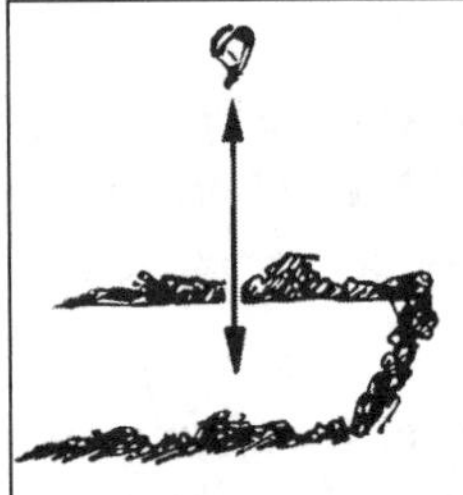

300 ft for arrival

▲ 19. *Plan to arrive at your landing filed with at least 300 ft (100 m) of altitude above the ground.* This policy will give you time to assess conditions, plan your landing pattern and position yourself properly. In addition, planning for such an altitude allows you a margin of safety if unexpected sink or a head wind is encountered.

▲ 20. *Below 300 feet (100 m) remain within a 45° angle (1 to 1 glide) of your landing field.* This is an important rule which assures you don't wander away from your safe field.

▲ 21. *Use a downwind, base and final approach in winds below 8 mph (13 km/h); use a figure eight approach in higher winds.* We discussed these matters in Chapter 8.

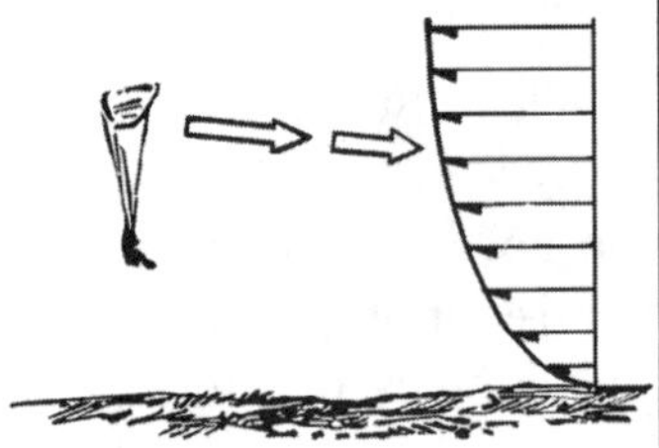

Maintain ample speed in gradient.

▲ 22. *When on final approach, increase your speed to offset wind gradient.* Lighten the control pressure slightly. *Do not* use the speed system below 300 feet (100 m).

▲ 23. *Flare to a no-step landing in most conditions.* The reason we advocate no-step landings is so a landing performed in less than ideal conditions (tail wind, downhill or in low density air) will involve the minimum amount of running. To perfect no-step landings you must learn to vary your flare according to the conditions (see Chapter 8).

▲ 24. *Never fly alone.* You don't have to be with another pilot—a ground observer or driver will do. The reason for this rule is to prevent a mishap from becoming more serious if you need assistance. Remember, with a safe approach you should never have an emergency, but a mature pilot is always prepared.

Before we turn our attention to other rules, let us present ways to enhance our safety rules.

EMERGENCY PROCEDURES

We must take all due precautions to avoid getting into situations that require extraordinary measures to extricate ourselves. However, the totally capable pilot should be aware of the possibility of extreme conditions or a lapse of judgement that can create an emergency. For this reason, we discuss the emergency situations

Paragliding brings you unbelievable new experiences.

that have most frequently occurred in paragliding. Chances are you'll never experience these foibles, but knowing what to do if they occur can be a lifesaver.

✓ Airsickness

Although motion sickness has never really injured anyone, you may feel like you are going to die... as it worsens you begin to wish you would! While we may joke about getting sick in the air, it is no laughing matter to the suffering pilot. Nothing can take the fun out of flight like a queasy stomach.

Motion sickness is the result of physiological and psychological effects that combine to confuse the brain. Commonly, our vision tells us one thing while our sense of feeling or inner ear balance mechanism tells us something else. Why this confusion should manifest itself as nausea is anybody's guess, but the side effects can be unpleasant and debilitating. Besides nausea, some people suffer from prolonged disorientation or *vertigo*.

The random motion of turbulence, G-forces and the visual problems related to continuous turning may cause airsickness. Other related factors are being too hot, too cold, dehydrated or fearful. The bad news is that practically everyone is susceptible to airsickness given the right situation. The good news is that we can become more and more immune to it with practice.

> *In 1979 I had my first experience with vertigo and severe airsickness. I was thermaling up beneath a large cloud, unaware that I was getting too close. I entered the cloud but soon flew out the bottom. As I regained visual contact with my surroundings, the wisps of clouds moving against the ground were disorienting and I experienced sudden severe nausea that lingered for hours.*
>
> *Another time I was in a competition near Crestline, California. We were dressed for high altitude, but I sank to the foothills. Down low the temperature was nearly 100 °F (38 °C) and the bubbling air was very turbulent. I was determined to get back up, but the price was sweating like a blacksmith and a definite green complexion.*

Here's how to prevent or reduce the effects of airsickness: First, dress appropriately. Clothes that can be adjusted in flight (a neck zipper and sleeves that pull up) are the wisest choice. Second, drink ample water, starting the night before, especially in desert conditions. Third, pay attention to your breathing. Quick, shallow breaths often accompany anxiety and lead to nausea. Take deep, regular breaths at the first sign of nausea and try to always maintain such a pattern. Fourth, avoid rapid head movements, especially when turning or in turbulence. Try to hold your vision on a distant horizon point and fly as straight as you can. If you must turn, look to the inside of your turn.

Experience in the air usually tunes your system so the brain is not so confused by conflicting inputs. Thus you become less susceptible to airsickness. Some pilots, however, never reach this point, so additional measures must be taken. Folk remedies such as swallowing ginger capsules or bands that apply pressure to points on the inner wrists may have some value and should be tried first. As a last resort, you can get motion sickness prevention drugs at the pharmacy. Some of these drugs have side effects such as drowsiness which are not compatible with aviation. Always try these drugs before you use them in flight to make sure you won't have a debilitating reaction in the air.

✓ *Helmet and Harness Problems*

It has so happened that pilots have had their helmet slip over their eyes during takeoff. This problem may occur due to a line snagging the helmet at the rear or simply a loose helmet. In such a dire occurrence, keep your head up but your eyes down so you can see the ground and direct your flight away from the hill. Once you are safely clear and stabilized, hold the controls with one hand and fix the problem with the other hand. Use a properly sized helmet and tighten it firmly to prevent such a blind takeoff. The helmet is part of the pre-launch check.

Forgetting to hook your chest strap is not a serious problem. You may have to let go with both hands in order to pull the two halves together, so fly straight away from the terrain and other gliders before you do this. Next time, include the chest strap in your pre-launch check.

Forgetting your leg straps is another matter—a serious matter. If you launch without having fastened them, chances are you will end up hanging by your armpits as shown in figure 10-1. Pilots have fallen from this position, so if it happened to you keep your arms pinned to your sides until the next step. Now reach *one arm at a time* up to grab the risers as shown. You will now be hanging by your arms, so grip tightly and don't delay... swing your legs up over your head and hook them around the risers. Keep your knees bent when you raise your legs. If you don't get it the first time, swing your legs back then up for momentum. If your arms get too fatigued, bring them back down and relock the harness in your armpits by holding your hands in front of you.

Once your legs are wrapped around the risers, reach down with one hand and pull the seat forward under your rear as shown (if you have the controls in hand, drop the toggle on this hand). Slowly lower yourself to normal flying position and correct your flight. Now fasten the leg straps and thank your guardian angel just before kicking yourself in the rear for a poor pre-launch check.

Forgetting your leg loops is dangerous and deadly. Don't assume it can't happen to you. Prepare for such an event by hanging a harness up high, and sliding out of the seat to hang by your armpits. See how quickly you can make the correction maneuver. Remember, your glider will be flying on its own through all this. Also, a parachute or instruments in your lap may complicate matters.

If you experiment with the described recovery technique and find you don't have the strength or flexibility to do it readily, it is better to form an alternate plan. In this case, immediately turn to parallel the hill and attempt a side slope landing.

Even a tree landing is preferable to falling out of the harness, but remember, if you hit a tree, you will have to hold yourself in the harness during impact and throughout the rescue if you are suspended. Grab your belt or clothes at your side to help hold your arms down in this case.

Caution: Never *throw your parachute in this situation for the opening shock may catapult you out of the harness.*

Finally, note that this problem is common enough that some harness designers have developed integrated leg loop/chest strap systems which can eliminate the forgotten leg

Figure 10-1: Recovering from a Leg Strap Emergency

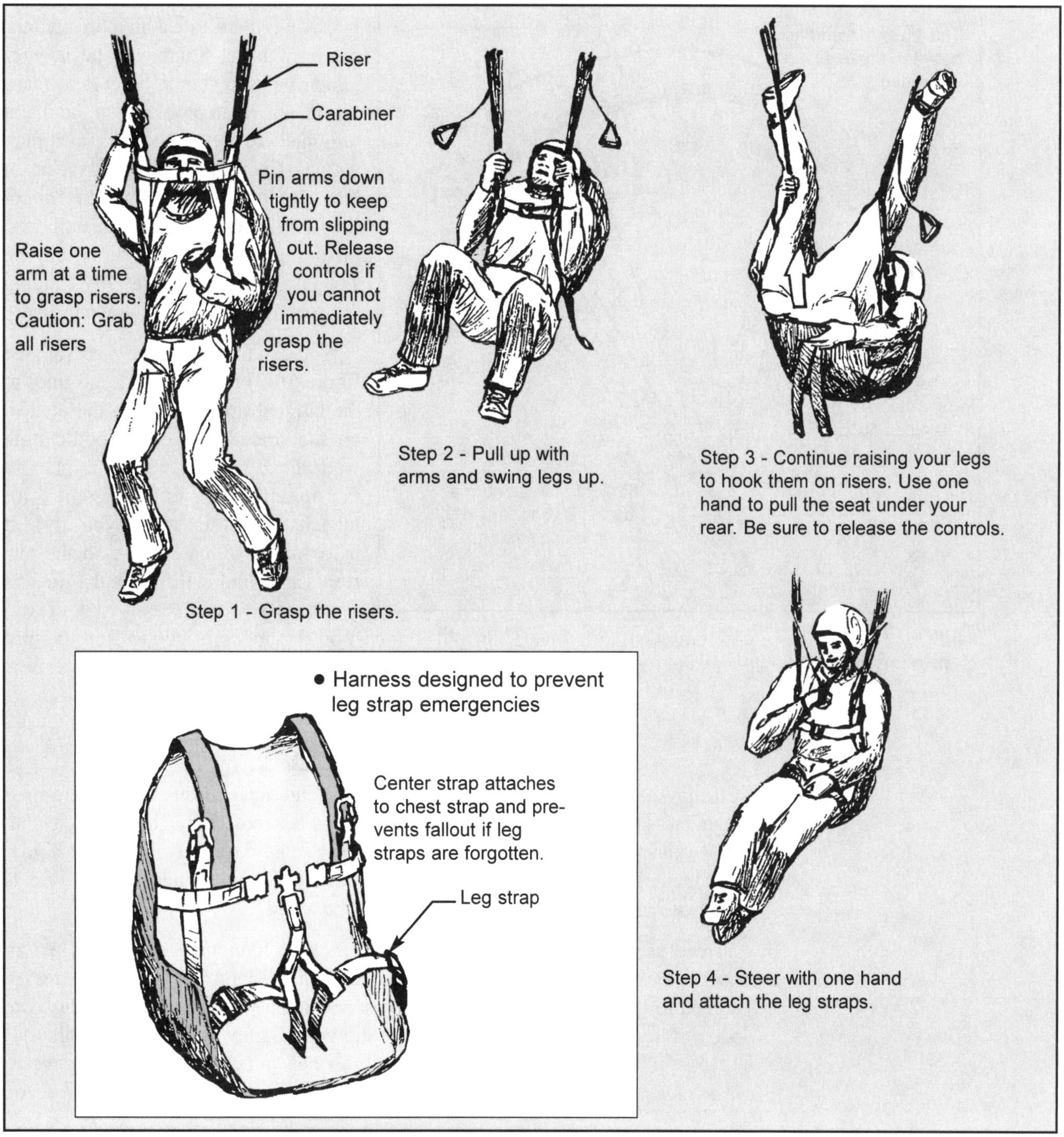

loops problem. When you attach the chest strap the leg loops are automatically attached as well (see figure 10-1). Consider these factors when you purchase a harness.

✓ *Control and Canopy Problems*

If you lose a control due to the line knot coming undone or the line breaking, all is not lost. You have already practiced weight-shift and rear riser turns haven't you? Do them and land safely. It's best to flare with both rear risers if only one line is lost.

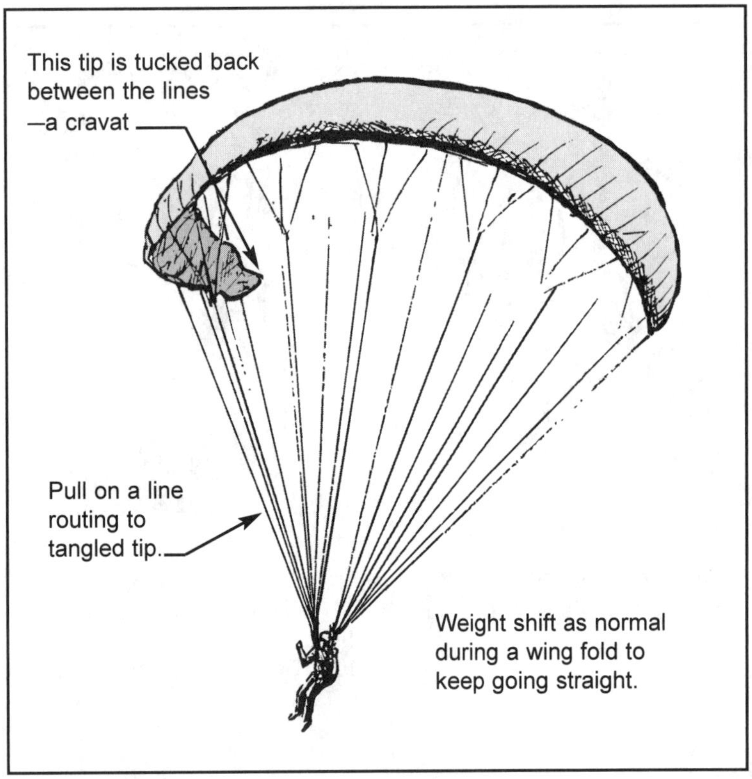

Figure 10-2: Removing a Cravat

Line tangles or a tip folded in the lines—a *cravat*—are a bit more serious, but only a bit. A turn will be in your glider, but you should be able to manage it. The main problem will be that a turn made to the side of the cravat may wrap up (increase fast) toward the problem side. Use turns to the other direction but carefully avoid a stall (too much control pull). The turns may be sluggish, so plan a wide DBF landing setup in as big a field as you can find.

You can try to remove a cravat (see figure 10-2) by pulling on the lines at the tangled tip or pumping the control on the folded side. You should note well ahead of time what line goes to the tip. Often it is of a different color (check it the next time you fly). It helps to be flying slowly, so the airflow isn't pinning the tip in the lines as much. If you can't remove a cravat, you may wish to induce a tip fold (big ear) on the opposite side to stop the turn tendency. Fly out to land using weight shift control as you do under big ears.

✓ *Tree Landing*

It may happen that you can't reach a safe landing field due to a severe misjudgment. The options will then be such things as trees, roads, buildings, water, etc. By far, the safest non-field option is brush or low trees, followed by higher trees, then calm water. Surf, narrow roads, buildings, power lines areas and fields with close fences are dangerous and should be avoided.

Trees are not without their own dangers which include getting poked or knocked by branches or falling out of the tree. If a tree landing is imminent, resign yourself to the fact and choose the stoutest specimen you can find. The idea is to do a spot landing on top of the crown, as shown in figure 10-3. Flare to limit your impact and immediately grab the biggest branches you can find. Again *the greatest danger is falling out of the tree.* If you are coming into the side of a tree you

Figure 10-3: Tree Landing

must flare to stop your forward speed, but not too hard or the canopy may be blown backwards in wind. The idea is to snag the canopy in the tree to prevent falling. Grab branches!

Pick a low tree if you have a choice—you will always be higher than you think from the air. Also choose a deciduous tree rather than a pine. Although pines may afford a softer landing, they break much more easily. Remember, it is better to land in a tree than risk catching a tip and being dumped into the field. Don't compound the problem of straying from a safe field by trying to land in too small an area and clipping a tree. Despite their inherent danger, trees have saved lives.

The emergency isn't over once you have perched in the tree, for you can still fall or the wind can blow the glider out of the tree. It may be useful to throw your parachute through the branches to act as a safety line (this may damage your parachute but will likely prevent injury).

Once you have freed yourself from the glider, work your way to the trunk, hold on securely and await rescue. If you can't reach the trunk, try to remain as still as possible. Do not attempt to climb down unless you are very skillful, for it's possible to land successfully in a tree but be injured in a fall while climbing down. A safety rope over a branch and connected to your harness is the best protection on your climb down. Some very high trees may require a hook and ladder truck from the local fire department. Carrying a small line to lower for a larger rope is a wise idea in tree-covered areas.

Some pilots fly with tree-extraction kits consisting of rope, webbing and a carabiner.

While you are up in the tree, you may consider the rescue of the glider. Firemen usually aren't interested in this matter. The only way to retrieve a treed glider without inflicting further damage is to cut the necessary branches one at a time to free the glider and lower it down gently. Also, try opening the quick links to pull the lines out more readily. Mark them with tape if you can to make their replacement less confusing. Remember, sawing the tree down usually creates a hopeless tangle.

If you've landed safely in a tree the problem has only just begun.

✓ Power Lines

It doesn't take too much imagination to realize that getting zapped or falling from power line height is hazardous to your health. Power lines should be avoided at all costs to the point of landing in trees or even water rather than hitting a power line.

The biggest problem with power lines is that they are hard to see, even if you know they are there. If power lines exist around your regular landing fields, you may be able to get the power company to put red attention balls on the wires. When you reach the skill level whereby you are flying cross-country, you must learn more about finding power lines from the air (see *Performance Paragliding*). Until then, learn the following power line procedures.

Never fly over a power line with less than 100 feet clearance. Sudden sink can drop you unexpectedly and many line systems have thinner (hard to see) lines on top. Do not fly under power lines, for it is hard to judge your clearance and a gust may raise you to contact the wire. Finally, if you do land in power lines, try not to touch more than one wire. Lie as still as possible in your harness while waiting for help to arrive. When help does come, the rescuer should not get near you (one pilot was electrocuted trying to rescue his buddy), but should call the power company immediately to cut power. There is no way to safely be rescued from a power line with the power still on.

✓ Still Water Landing

Landing in a lake or a pond with still water is generally a matter of escaping from your glider in good time. Some pilots have the foresight to unhook their straps and remove their gloves before reaching the water. This is a good idea if you know well ahead of time that a water landing is a sure thing, but most of the time a pilot will be maneuvering to avoid the drink until the last moment.

When you hit the water your most important concern is to avoid getting tangled in the lines. Make sure your hands are free of the toggles and do not take wraps. You should collapse your glider well behind you by using a full control pull, or better yet, let go of the controls and use the rear risers. Flaring a bit high is good (if the water is deep) for it helps assure that the canopy drops back. On the other hand, in deep water it's better to splash down with a tail wind so the canopy falls in front of you. In that way the leading edge submerges and holds air in the canopy. It can act as a floatation device for a long time in this manner, especially if you gather the leading edge under water to prevent air from escaping. However, only attempt this in still water, for entanglement is a real danger. Normally you should get out of the harness and away from the entire rig as quickly as possible. A hook knife can very quickly cut lines or risers if you get entangled.

Once you are dunked, take your gloves off, calmly unhook the straps as efficiently as possible. Seat belt type clips are much better than nested rectangles in this situation (see figure 4-10) and should be a consideration when you buy a harness. You may have to take a breath, look down and do one at a time. Once your harness and parachute get soaked you will be less buoyant. If you operate near water it is wise to use some sort of floatation device and carry a hook knife to aid your survival. Many air bag or foam harnesses that remain inflated float very well, but make it difficult to float on your back. They may be dangerous if they flip you face down and hold your head under water. In many cases it is better to unbuckle before hitting the water if possible.

✓ Surf and Moving Water Landing

When the water is moving you have a more serious problem. The massive weight of water on the canopy can drag it along and you with it. Several pilots have landed in knee-deep ocean water only to have a wave drag their glider and them out to sea.

Moving water is a dangerous landing choice.

Landing in surf is very dangerous, even near shore. Usually this event occurs when a pilot ventures too far out to sea and tries to make it back. This usually results in the glider dropping in front of or on the pilot with an onshore breeze. Waves then entangle the pilot and send him to a watery doom. It is not reasonable to fly into the waves, for then you would be going back out to sea, but detaching in the air and landing crosswind are the best way to prevent the waves' deadly action (see figure 10-4).

Even in very shallow water, don't plan on running into shore. The drag of the glider will probably prevent this act. It's much better to unhook first, then pull in the glider. The only way to haul the glider out is with the trailing edge (otherwise the interior will fill with water). However, this process can be dangerous itself in waves for the chance of entanglement is high. Your life isn't worth a paraglider.

Figure 10-4: Landing in Surf

A pilot lands near shore in still water. If waves were present this would be dangerous.

In the final analysis it is always better to land downwind on the beach with a PLF rather than in even knee-deep water if there is wave action and you have a choice.

In a river or stream, it is desirable to land facing upstream regardless of the wind direction so the canopy doesn't float over you. If you have a choice, land in as shallow water as possible, usually close to the shore or on the inside of a bend. If you land in deep water the procedures are about the same as still water landings.

While pilots experiencing water landings generally fare better than those in power lines or severe obstructions, such landings can be quite serious or deadly when things go wrong.

✓ Getting Blown Back

When you first begin soaring, later in your experience, you will be flying in more wind. You'll learn more about the dangers of high winds above the mountain. Unfortunately, pilots still get caught in a situation where they can't penetrate out from the mountain. What should you do? If you can make any headway, it's usually better to fly forward, for even if you lose all your altitude and end up in front of launch, it's no worse than if you just launched.

On the other hand, if you are pinned above the mountain and absolutely can't move forward, you must either descend to land on top or turn downwind to land behind the mountain (see figure 10-5). What choice should you make? If you are a minimum of 500 feet (150 m) over the top of the mountain and there are safe landing fields behind it, and no safe landing field on top of the mountain, your

Figure 10-5: Penetration Problem Escape

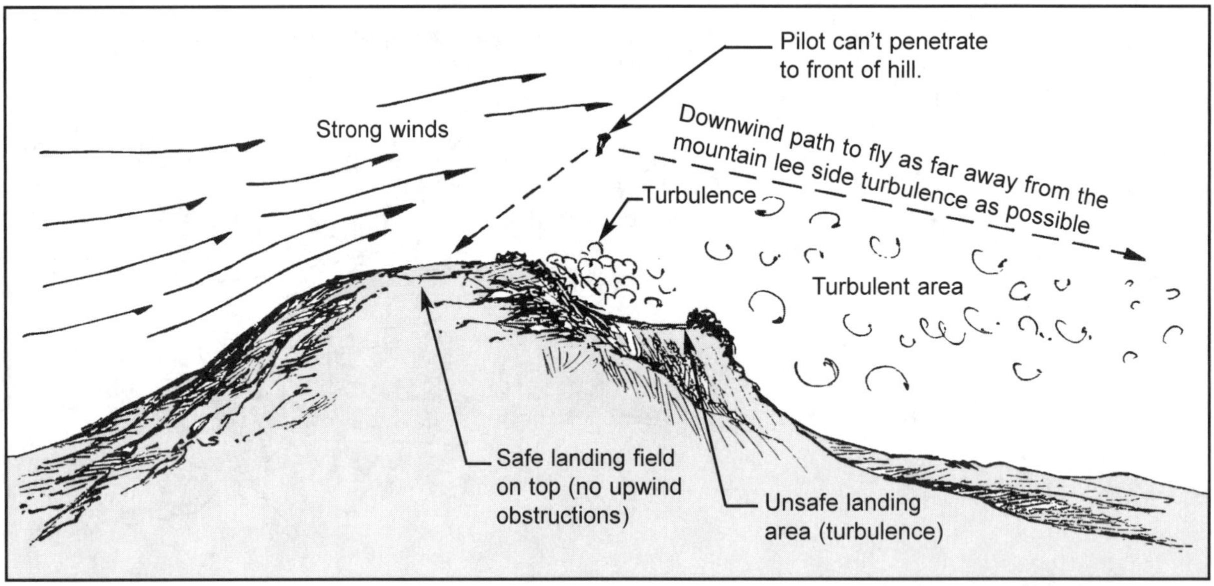

best chances of avoiding problems is to go over the back. To do this, slow your glider if you are in lift to gain as much altitude as possible before you go back. Once you've peaked out, turn around and run with the wind. Your quest is to get as far away from the mountain as possible. Once you get near mountain top level you should expect turbulence trailing downwind from the mountain itself. Maintain the best pressurization control position (4 lbs or 2 kg of actual pressure) all the way from that point to your final on landing. Flying slowly will get you as far downwind as possible unless you enter sinking air in which case you should fly a bit faster unless the air is turbulent. Choose as big a field as you can find as far away from the mountain as possible.

In soaring conditions with ridge lift, move more in front of the ridge if you're going up to avoid the venturi.

If there are large open places on top of the mountain it is usually better to land on top as long as the upwind side of the field is not sloping uphill which would induce a rotor or turbulence. Maintain good pressurization (and possibly big ears) all the way to landing in this case.

Normally we recommend a minimum clearance over the top of a mountain for going over it equal to the height of the mountain (*Performance Paragliding* provides more detail), but the situation here is an emergency. Needless to say, any choice you make results in a less than ideal situation. You should use your judgement and caution to avoid having to make such a decision.

✓ Getting Dragged

We have mentioned before the problems of landing in windy conditions (see Chapter 8). In fact, any time you are in your harness and it's hooked to the risers, you are vulnerable to a sudden gust. Such natural phenomena like thermals or dust devils can appear very suddenly on hot days, even in no-wind conditions. It is thus important for all pilots to know the emergency procedures for dealing with the situation of strong winds thrashing the canopy and taking the pilot along for a drag.

As we stated before, the best thing to do is pull on one control line hand over hand until you reach the canopy as shown in figure 10-6. Then pull in the material if you can. At any rate the forces will be manageable once you reach the canopy.

Figure 10-6: Getting Dragged

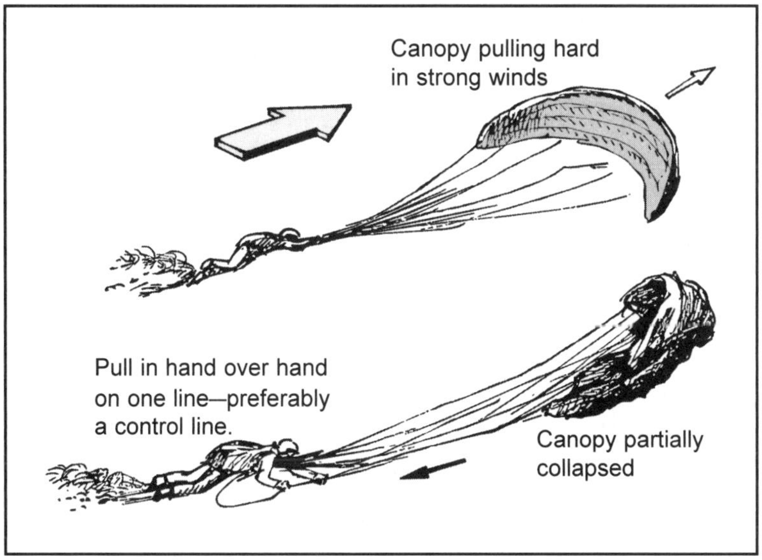

You may not have the luxury to locate a control line in some sudden problem situations. In that case, pulling any line all the way in will suffice to disable the canopy and render it reasonably docile. If you see a severe gust or gust front coming, grab a corner of the canopy ahead of time, or better yet, quickly get out of your harness. Pilots flying where strong thermals can lift off are wise to carry a hook knife. A hook knife is the only reasonable way to escape from a parachute if you've had to deploy one and touch down in high winds.

✓ *Parachute Deployment*

Very few pilots have ever "tossed silk", but a wise pilot is prepared for all emergencies. If you do have a spin, a midair, a glider that balls up, or some other unrecoverable situation, here are the important matters to consider:

When a parachute is thrown it will extend the full length of the bridle, then pull out the shroud lines from the deployment bag (see figure 10-7). Finally the canopy comes out to be filled with air. Figure 10-8 shows the parts of a parachute.

The reliability and time of opening depends on how hard your throw is, how fast you are moving in the airstream, and how recently your parachute has been packed. The last point is important, so have your parachute serviced regularly (at least every six months).

Here are the deployment steps:

Throwing a Chute

1. Don't hesitate. It's better to use it than lose it.
2. *Look* at the handle.
3. *Reach* for the handle.
4. *Extract* the chute by pulling it strongly out of the harness.
5. *Look* for the clear air away from the canopy.
6. *Throw* it mightily in the direction of the turn.
7. *Pull* back bridle and rethrow if it hasn't opened. Jerk it hard if the bridle has extended but the bag remains closed.
8. Once deployed, pull the paraglider canopy in by pulling a line and eventually pulling in the material to gather it in a bundle. Otherwise it may interfere with the parachute, but more importantly, may begin flying away from the parachute and increase the descent rate (downplaning) as shown in figure 10-9. Pull the paraglider in by first simultaneously pulling on the middle or rear risers to collapse the canopy if necessary, then pull on a control line until you can grasp the canopy and gather it in.
9. Remain motionless so you don't start oscillations. The emergency isn't over until you're safely on the ground.
10. About 10 feet (3 m) above the ground, prepare for a PLF.
11. If high winds are carrying you, be ready to cut the parachute bridle with a hook knife.
12. Once you're safely on the ground send your parachute manufacturer a bouquet of flowers and the details of your deployment so the manufacturer can continue development.

Figure 10-7: Parachute Deployment

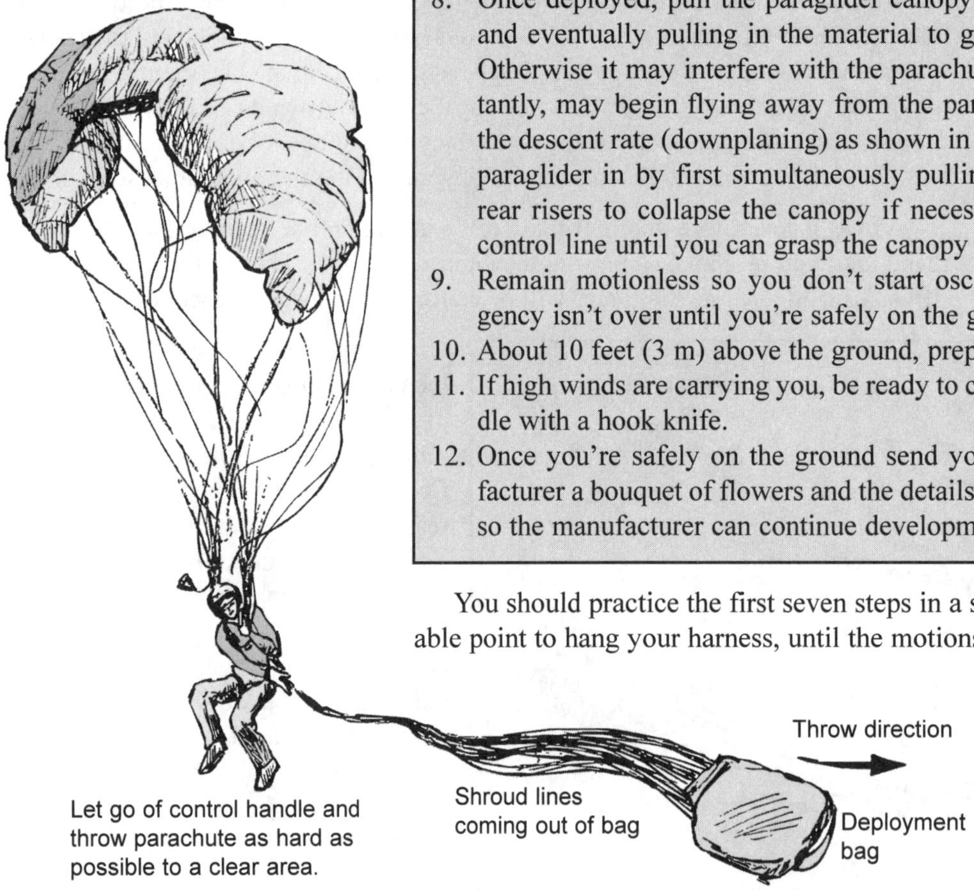

You should practice the first seven steps in a simulator, or any suitable point to hang your harness, until the motions become automatic.

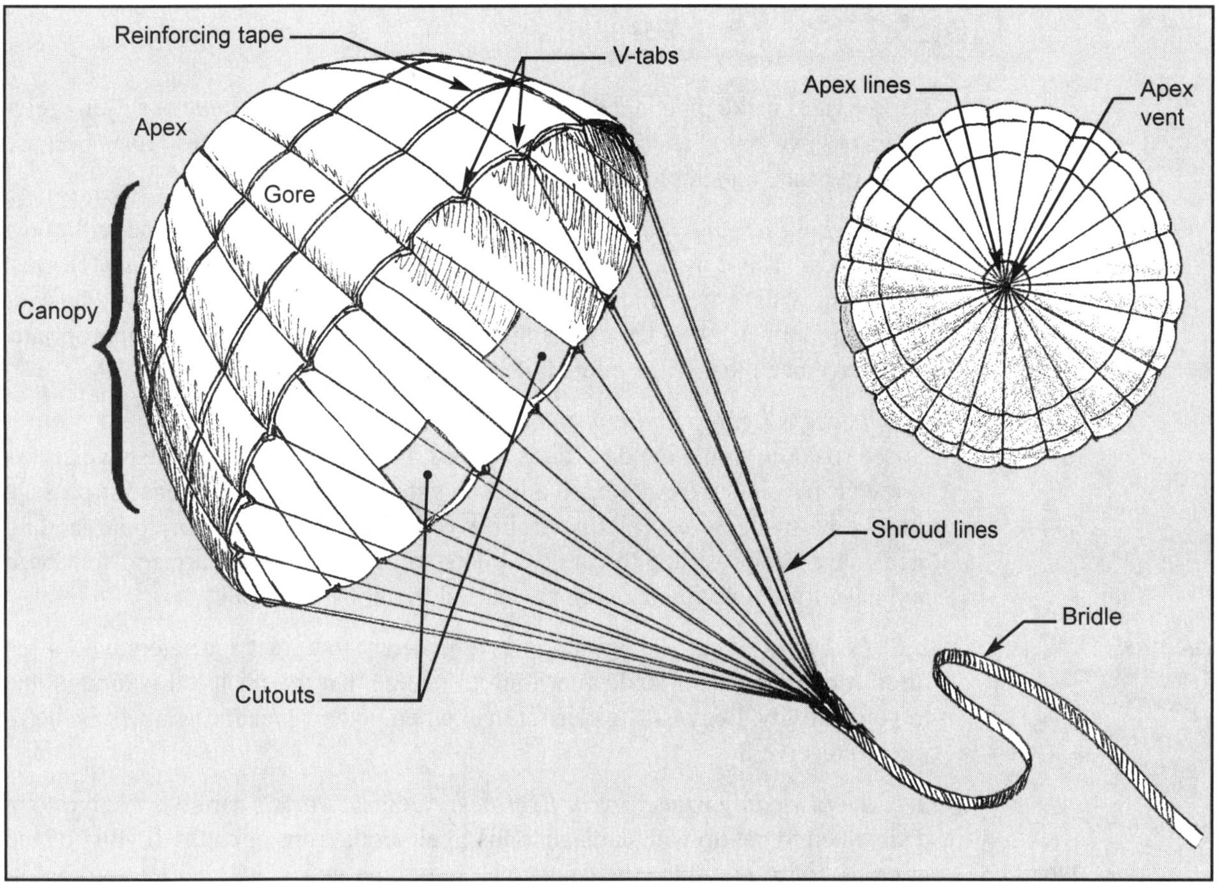

Figure 10-8: Parachute Nomenclature

Parachute deployment steps:
- *Look*
- *Reach*
- *Extract*
- *Look*
- *Throw*
- *Rethrow*

Figure 10-9: Downplaning

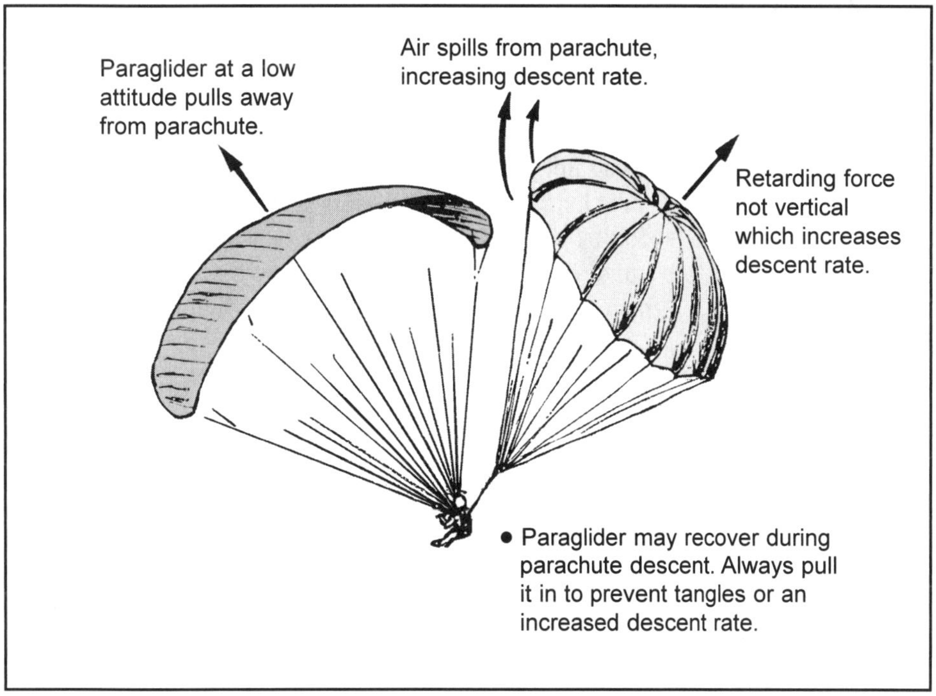

Courtesy Rules

Certain rules make plain old common sense if we wish to continue our flying privileges or have any friends around to fly with. We call these courtesy rules because they are based on simple human courtesy embodied in the golden rule.

▲ 1. *Always respect a landowner's wishes.* Most sites exist on the good will of the landowners. These owners derive very little benefit from our activity, so we must treat them with respect and honor their rules. Landowners should always be contacted for permission to fly and liability forms should be filled where appropriate. Sites have been lost when pilots failed to obey a few simple courtesy rules.

▲ 2. *Leave all gates in the position you found them.* If a gate is closed, keep it closed so cattle won't wander. If it is opened, leave it opened so cattle have access to the proper fields. Do not open a locked gate or future negotiations for passage may be harmed. Be especially careful when climbing fences, for some landing areas have been lost due to careless pilots damaging fences. Farmers work hard and have little time to make repairs caused by thoughtless pilots.

How to have fun paragliding:
- ***Don't be in such a hurry that delays cause stress and frustration.***
- ***Treat paragliding as a relaxation not a necessity.***
- ***Don't rue the weather, for you have no control over it and a certain number of miscalls will occur.***
- ***If you end up sitting on the hill, enjoy the view and ponder how much nicer it is than the view from a stuffy office window.***

▲ 3. *Pick up all litter.* Paragliding has a tendency to attract spectators who often litter. It doesn't take a landowner long to realize that the trash left behind is due to your activity. Leave a site cleaner than when you arrived and you will be helping to preserve it.

▲ 4. *Avoid landing with crops or livestock.* Again a farmer's hard work leaves him disinclined to put up with damage to his livelihood: crops or cattle. If you do land in crops, try to minimize the damage by removing your glider before you fold it. Be sure to inform the farmer of your mishap and offer to pay for the damage. Chances are he'll appreciate your frankness and refuse your payment, but your gesture is very important. Put yourself in the farmer's place and imagine someone damaging your property, then slinking off.

Cattle (cows) are very rarely overly excited about landing paragliders. They usually just stop and stare with big question marks over their heads until they remember their main business—eating. On the other hand, horses panic. Some horses are worth more money than you will see in a lifetime, so avoid landing in horse fields. Sites have been lost when a horse careened through a fence. If you must land in a field with horses, do so as far away from the beasts as possible and don't fly over them. Once you've landed, call to them to let them know that yes, you are indeed a human despite the bizarre canopy above your head. Note: Cattle have been known to chew through nylon. Don't leave your glider lying about unattended near cows.

▲ 5. *Obey site rules.* Most flying sites have some rules concerning ratings and landing areas. Don't press to fly sites that are above your rating level, for only by self-regulating these matters will we be able to preserve sites for your future flying. Some sites have fees and passes that must be attended to. Ask the local pilots about the rules, hazards and special conditions at the site. You will fly more safely and make more friends if you do. Remember, local clubs have probably worked hard and long to acquire the sites they have, and you can destroy all this work by a moment of carelessness or selfishness.

When more than one glider enter the airspace, right-of-way rules must be heeded.

▲ 6. *Obey right-of-way rules*. These rules are to keep pilots, including yourself, alive and healthy. They are discussed in the next section.

▲ 7. *Help less experienced pilots.* As your experience grows, you should pass along your knowledge and wisdom to those coming up, just as other pilots did for you. In this way we all benefit, for knowledge spreads and accidents are prevented. A large part of our learning comes through other pilots and we should all be part of this process.

▲ 8. *Sell your used equipment only to those capable of handling it.* It almost goes without saying that we are responsible for assuring that a glider or harness we sell will not lead to another person's injury due to his or her lack of skill. Such an occurrence hurts the pilot, the sport and us (if we have a conscience).

▲ 9. *Maintain a positive attitude*. Nothing is worse than a pilot who always carps about the conditions or is constantly dissatisfied with his flights. The idea of this sport is to have fun, so make every outing a pleasure whether the flying is great or not.

▲ 10. *Maintain pleasant public relations.* You are a representative of the sport. With a canopy over shoulders there's no mistaking what you are. Be courteous to spectators no matter how many times they ask the same questions. You never know where your next flying buddy or driver will come from!

▲ 11. *Take a first aid course*. Such a simple task can help save a fellow pilot's life. Indeed, this is not just a courtesy but a common sense way to improve the safety of our sport for everyone's benefit.

▲ 12. *Support your local school/instructors.* These people help preserve the sport through long-term dedication and are invaluable when it comes to site management and maintenance.

Right-of-Way Rules

When you begin to fly with other gliders, you must learn the basic traffic patterns called *right-of-way rules*. We can separate these rules into general rules and special soaring rules for convenience. You should learn all of these rules so you can blend harmoniously with other air traffic and gliders. These are the same rules that sailplanes and hang gliders follow. You should note that these rules are actually conventions that have been adopted over time through experience. They are generally not laws of the land and they do not apply in and cover all possible situations. We'll discuss some of the exceptions.

✓ General Traffic Rules

▲ 1. *When approaching head on with another glider, both gliders should turn to the right*. This rule is similar to that with automobiles in the U.S.—keep to the right (see figure 10-10).

▲ 2. *When two gliders are converging (coming together), the pilot on the right has the right of way*. This rule is similar to automobile traffic rules (see figure 10-10).

▲ 3. *When overtaking another glider, pass to the right*. This rule is the opposite to automobile rules (in the U.S.) and comes from general aviation where normal turns around airports are made to the left (see figure 10-10).

▲ 4. *The lower pilot has the right-of-way*. This rule is logical since the lower pilot of a paraglider cannot see the upper pilot if he is directly over the canopy. Also, the lower pilot has fewer options. This rule especially applies during landing. If you are the lower pilot in the landing situation, courtesy and safety demand that you burn off altitude quickly to give following pilots a clear shot at landing. We cannot emphasize this point enough, especially when many gliders are in the air at the same time. You must not use the fact that you are lower to force a conflict.

▲ 5. *Do not fly directly over or under another glider with less than 30 feet (10 m) clearance*. This rule is intended to prevent a midair in case the lower glider encounters lift.

▲ 6. *Paragliders (and hang gliders) in the U.S. must yield right-of-way to all other aircraft except powered ultralights.*

▲ 7. *Clear all turns by looking carefully in the direction toward which you intend to turn.*

✓ Soaring Rules

When ridge soaring we are often close enough to the terrain that freedom of movement is hampered. Therefore special rules apply. When thermal soaring we often must work close to other gliders, so again we need additional rules to coordinate matters.

▲ 1. *When ridge soaring, the glider closest to the ridge has the right-of-way.* Since the glider near the ridge cannot turn into it, the other glider must turn away from the

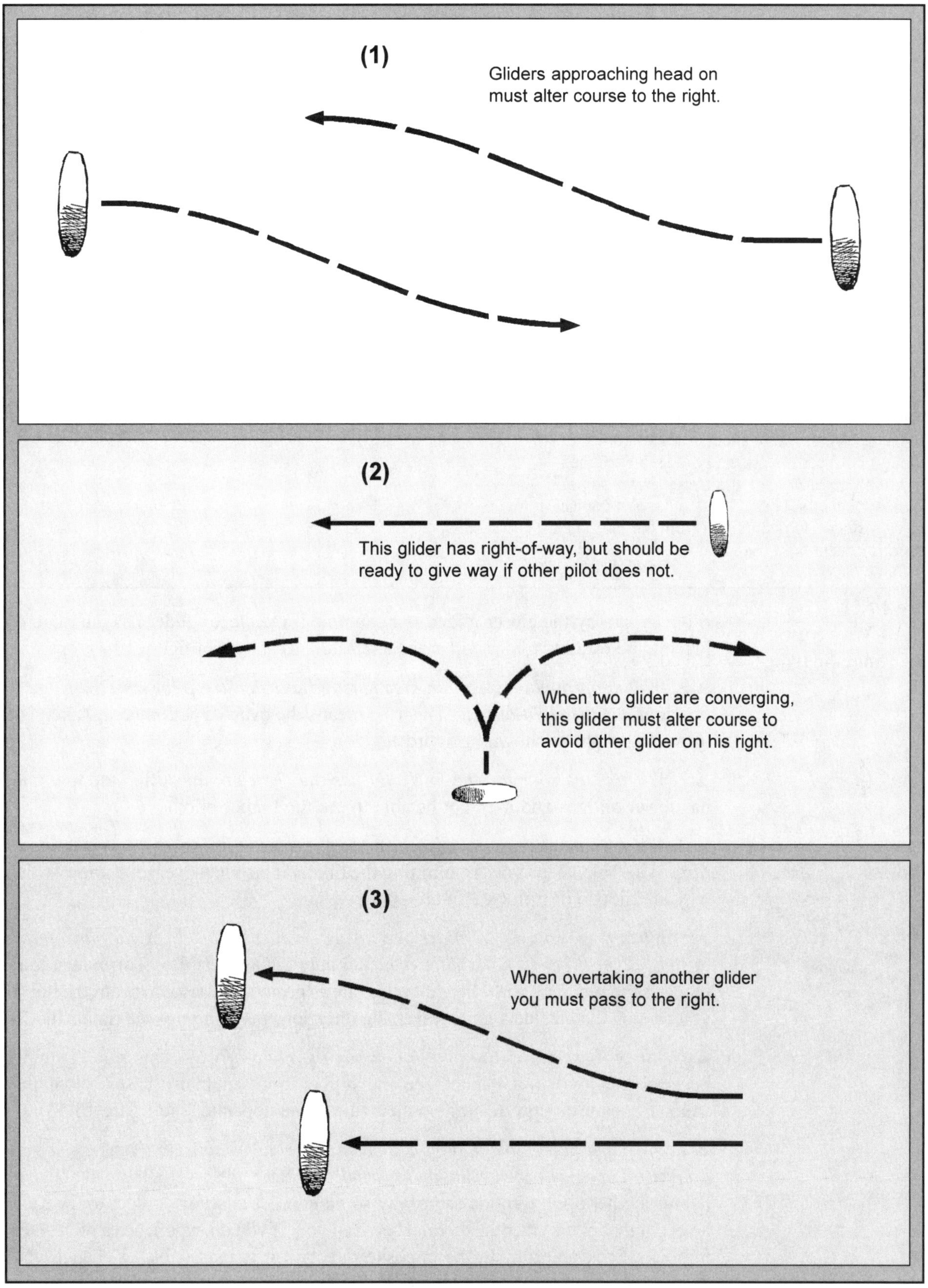

Figure 10-10: Right-of-Way Rules

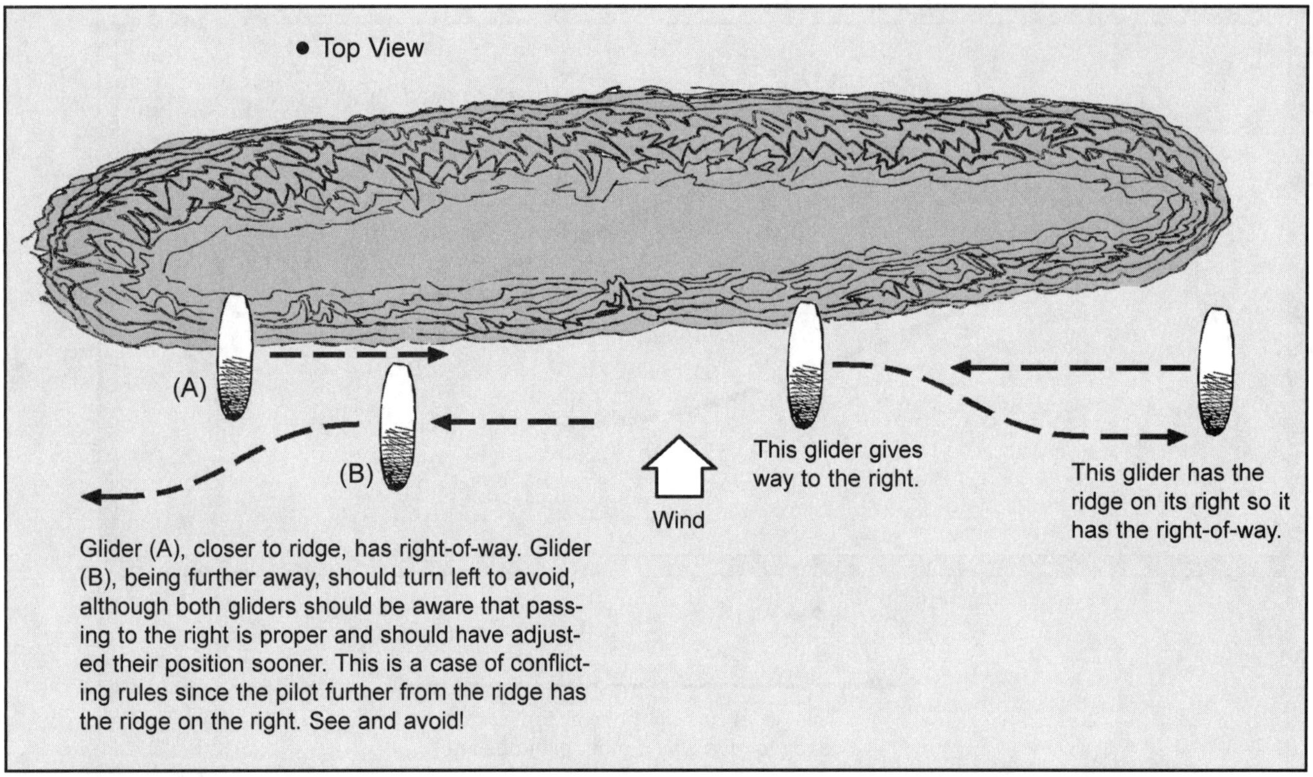

Figure 10-11: Right-of-Way Rules When Ridge Soaring

ridge to pass by the closer glider. For example if the closer glider has the ridge on his left, the evading glider will make a *left* turn, as shown in figure 10-11.

▲ 2. *When approaching head on during ridge soaring, the pilot with the ridge to his right has the right-of-way.* This rule means the evasive action being taken is a turn to the right as shown in figure 10-11.

▲ 3. *When ridge soaring, the low pilot has the right-of-way.* Often the low pilot has fewer options and may not be able to see the higher pilot.

Beware of wake turbulence! It is usually better to clear your turn and reverse direction for a while than to pass through or follow in another pilot's wake.

▲ 4. *When ridge soaring, always turn away from the hill when reversing directions.* This rule keeps you from hitting the hill, and also lets everyone know which way all pilots will turn (see figure 10-12).

▲ 5. *When ridge soaring, a glider overtaking another glider flying the same direction should pass between the ridge and the glider being overtaken.* This rule accommodates the previous one where all pilots are expected to turn away from the hill. If you pass on the outside a glider reversing directions may hit you (see figure 10-12).

▲ 6. *When thermaling, the low glider has the right-of-way.* This rule is simply because the low pilot cannot see the higher pilot in many cases. Note that sailplanes can see upward best so their rule is the opposite (see figure 10-13).

▲ 7. *When thermaling, the first pilot to enter a thermal establishes the turning direction for all other pilots entering that thermal.* Whether you enter a thermal above or below another pilot, turn the same way so no conflict arises if you get to the same level or other pilots enter the thermal between you. If you do experience a pilot entering below you and turning the opposite direction it is reasonable to change your direction to avoid conflicts as long as there are not other gliders near your level.

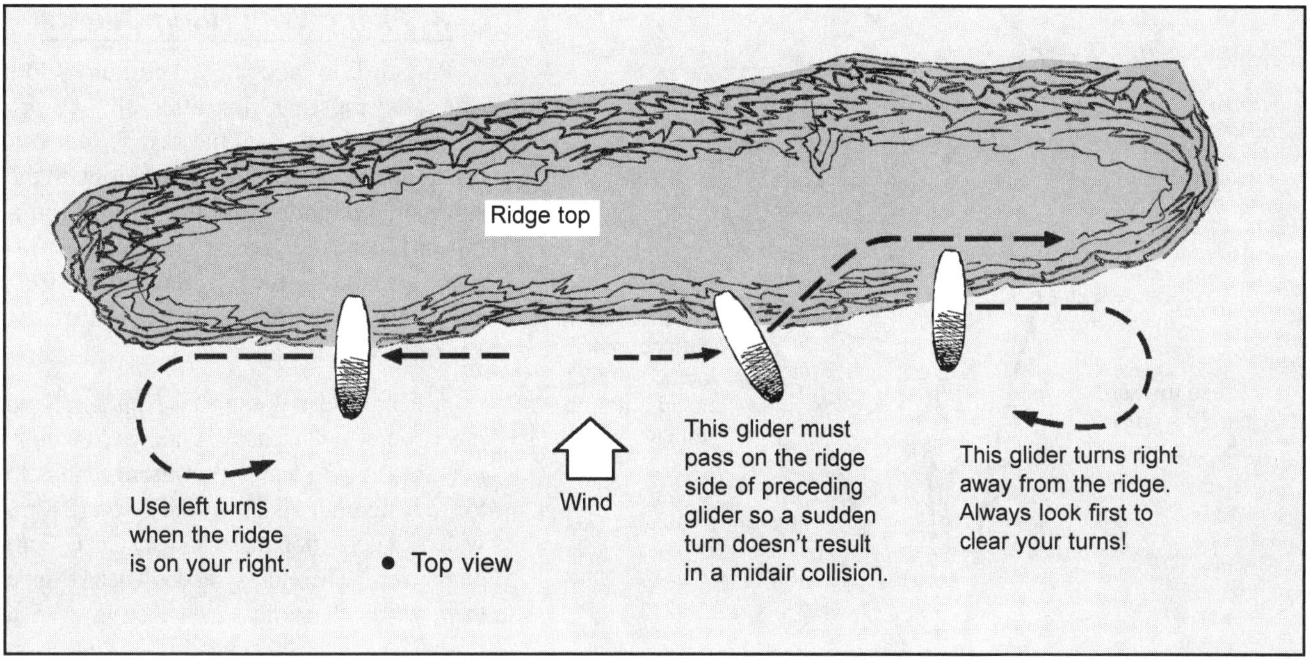

Figure 10-12: Reversing Turns When Ridge Soaring

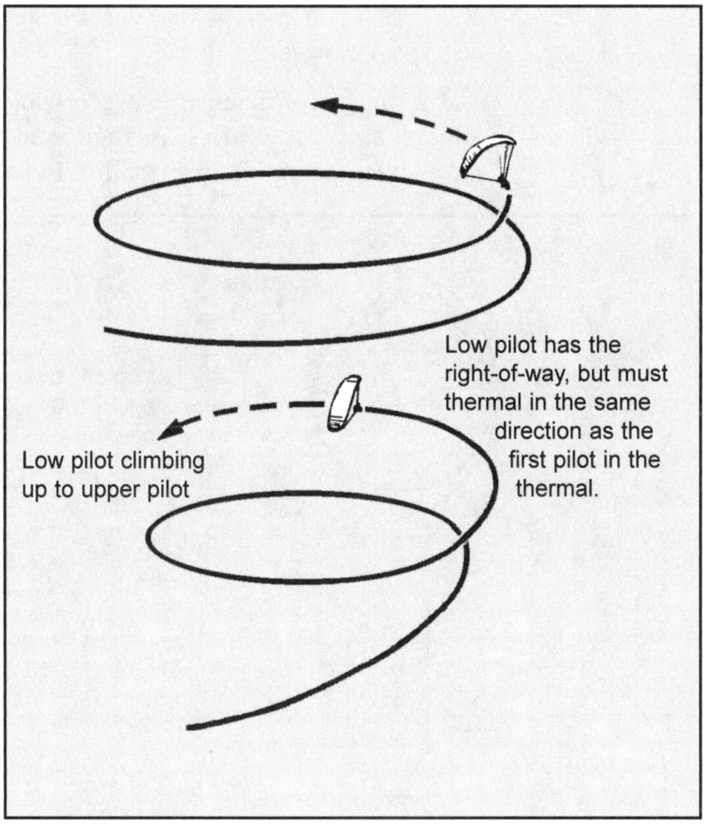

Figure 10-13: Right-of-Way in Thermal

▲ 8. *When thermaling, the person in the thermal has the right of way*. Once a pilot clears his turn and establishes a thermal, other pilots must give him clearance or join the thermal in a non-conflicting manner. The reason for this is that a thermaling pilot is constantly turning and cannot follow all traffic. In addition, most pilots in the vicinity will be coming to join the thermal and are in a better position to judge a smooth entry than the thermaling pilot. A pilot joining a thermaling circle must do so at a tangent to the circle in order to end up circling in the same direction (see figure 10-14 and rule 7).

Caution: *The right-of-way rules are conventions to help avoid close encounters and midairs. They cannot possibly cover all situations. The ultimate rule is to see and avoid.*

There can be situations where the right-of-way rules conflict or do not indicate a clear right-of-way. For example, if a pilot with the ridge on his right is approaching head on to a pilot lower or closer in, who has the right of way? The rules do not determine this matter and it is the equal responsibility of both pilots to take evasive action or more importantly, predict such a situation and avoid it in the first place.

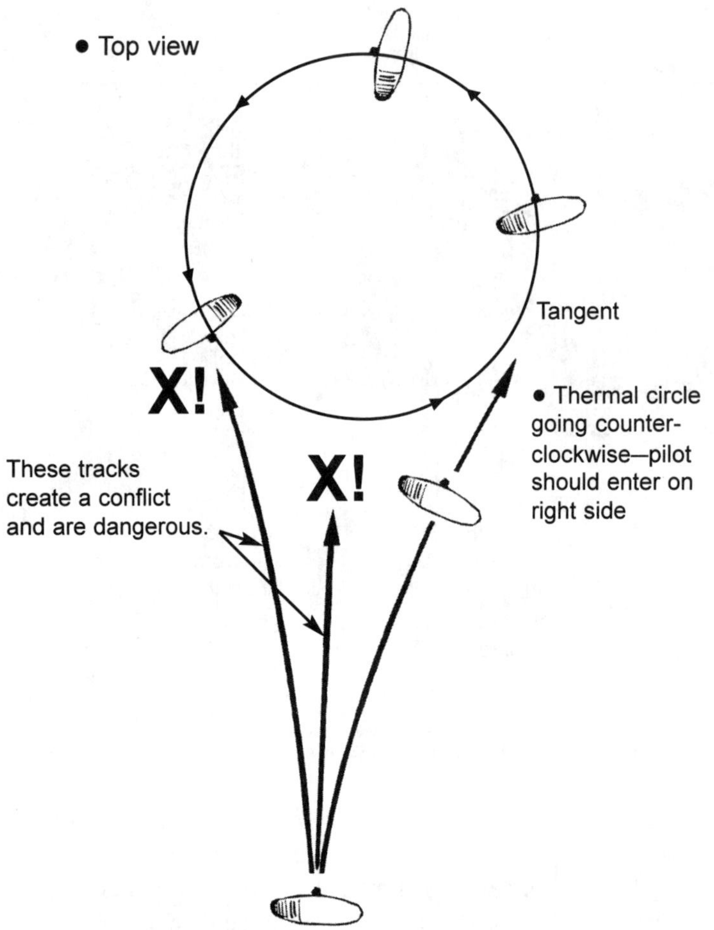

Figure 10-14: Entering a Thermal With Traffic

✓ *Flying With Hang Gliders*

Paragliding and hang gliding pilots often fly together, for after all, we are brothers and sisters of the sky. We are trying to do the same things, we belong to the same organizations and we're both fighting to repeal the law of gravity. The two types of gliders have a few differences, however, that affect how they must be considered in the air.

A hang glider doesn't take up nearly so much vertical airspace. While a paraglider may measure 26 feet (7.9 m) from canopy to toe, a hang glider is more like 8 feet (2.4 m) with a kingpost (top mast) and 5.5 feet (1.7 m) without one. Therefore, you don't need to give them as much vertical clearance as with a paraglider. On the other hand, don't depend on the hang glider pilot to know how much vertical clearance to give you if he or she hasn't flown with paragliders before.

The biggest difference between hang gliders and paragliders is their stall speed. Your paraglider may stall at 14 to

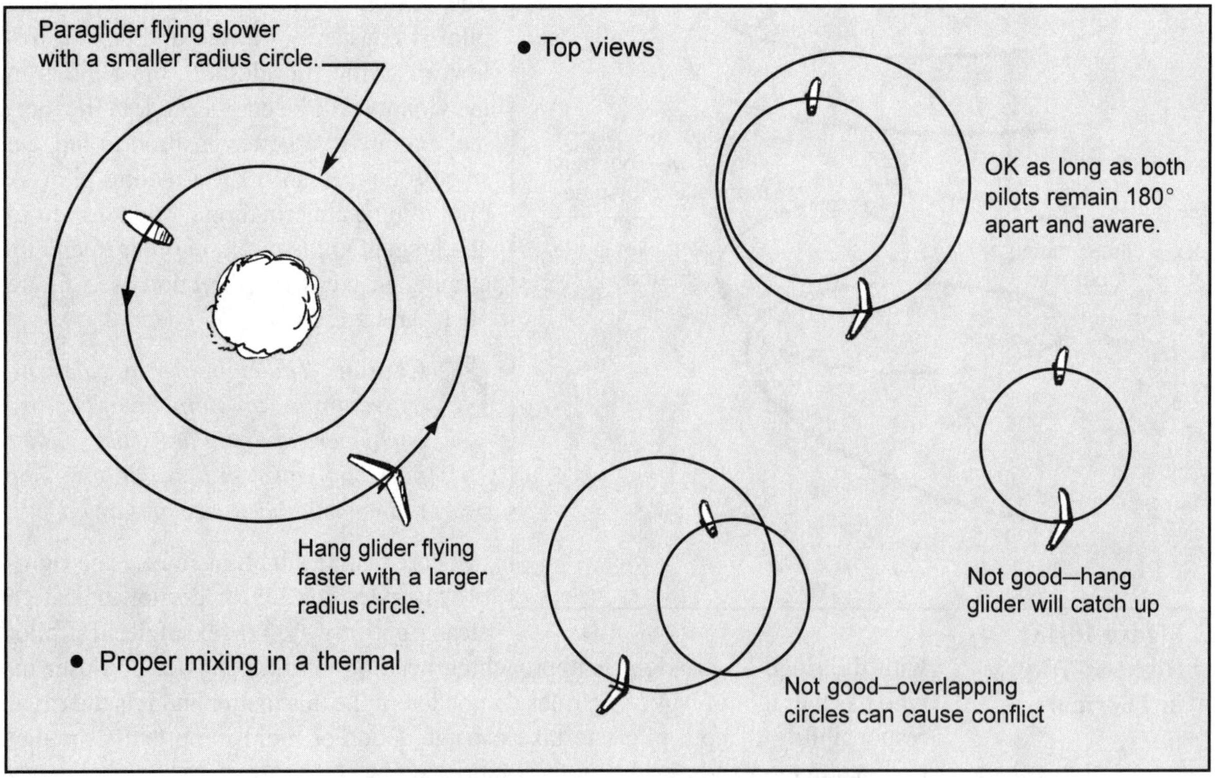

Figure 10-15: Thermaling With Hang Gliders

15 mph (22.4 to 24 km/h) while a hang glider typically stalls at 20 to 21 mph (32 to 33.6 km/h). The result is that the hang glider's minimum sink circle is going to be a larger diameter than yours. (For example, the hang glider circle will have a radius of about 79 feet (24 m) at a 20° bank while a paraglider's radius will be about 39 feet (12 m) at this bank angle.) The obvious problem thus comes during thermaling. But a little thought and experience indicates sharing a thermal can (and does) work just fine as shown in figure 10-15. The hang glider is going faster in the larger circle and the paraglider travels the smaller circle more slowly. As a result they can remain opposite each other and climb happily ever after. If the paraglider tries to follow the same circle as the hang glider, the hang glider eventually catches up, which causes a conflict. Try to learn such cooperation with an experienced hang glider pilot and you will see how it works.

On a ridge the hang gliders will be catching up and passing you. They should give way to you because you are slower. Remember to clear your turns and follow a predictable flight path. Watch out for their wake; it is similar to the wake of a paraglider since they are heavier but aerodynamically cleaner. Following close behind any type of glider is dangerous due to wake effects (see page 120).

Federal Aviation Rules

All countries have some government regulations controlling aircraft including paragliders. Recent changes have made the international airspace designations uniform in many countries. For that reason we present the U.S. laws here. They will be similar in many countries, but all pilots should learn the rules drafted by their specific government. If you don't heed these rules you will be subject to a steep fine at best, and end up straddling the tail of a Cessna or clogging the intake of a 747 engine at worst.

In 1982 the Federal Aviation Administration (FAA) in the U.S. instituted special rules governing powered ultralights, hang gliders and paragliders. These rules were designated as Federal Aviation Regulation (FAR) Part 103. Below are the main rules that relate to paragliding. All pilots should understand them well, for violation is punishable with a $1,000.00 fine for each infraction.

✓ *Applicability*

Section 103.1 of FAR Part 103 defines an ultralight vehicle. It turns out that paragliders are unpowered ultralights according to the federal government. There are several criteria we must meet as listed below:

a) Single occupancy. Paragliders may carry one person only except under an exemption the USHGA and ASC (Air Sports Connection) later applied for and received. To fly tandem legally you must meet certain qualifications and achieve a special tandem rating. Write the USHGA or ASC office for the forms and procedures.

b) Recreation purposes only. You cannot receive compensation for your flying in any way. However, exceptions that do allow you to make money are instruction, rental of equipment, sporting events, authoring books and receiving discounts on

Freedom is the essence of flight, but if rules and limits are not considered the freedom will be lost.

equipment. Aerial advertising is specifically prohibited. You *can* fly with a company logo and may even get a free glider in the deal, but you cannot enter into an agreement that specifies the location, number or pattern of flights. You cannot fly for hire in air shows or public exhibitions.

c) Airworthiness Certificate. If you want to fly for hire or otherwise avoid some of the rules in Part 103, you may obtain an airworthiness certificate from the FAA. This will take some doing, but it has been accomplished on several occasions. Write the FAA for Advisory Circular 20-27C.

d) Weight limits. A paraglider must weigh less than 155 pounds.

✓ Inspection Requirements

Section 103.3 says that you must allow an FAA official to inspect your glider for compliance with Part 103. This rule is mostly concerned with powered ultralights although some towing operations have been inspected. In the field, if an FAA official approaches you with specific requests of this nature, we suggest you exercise courtesy, for they have the power of arrest and the usual human reactions to questions of their authority.

✓ Waivers and Certificates

Section 103.5 says you can only conduct operations that deviate from Part 103 by obtaining a written waiver. To obtain such a waiver, contact your GADO (General Area District Office). Remember you are dealing with masters of red tape, so you need patience and luck. However, many of these officials are curious about paragliding and may go out of their way to help you.

Section 103.7 says that our gliders are not required to meet any federal government airworthiness standards nor are they required to be registered or bear markings. Furthermore, pilots are not required to be licensed or have any particular experience, knowledge, age or medical qualifications.

Since the above section was written, however, the FAA has pushed for the voluntary compliance with a three-part program: vehicle airworthiness, registration and pilot registration (presented in Advisory Circular 103-1). The USHGA has been administering this program for paragliding. The USHGA uses the DHV standards for airworthiness, your membership and rating card for pilot registration, and offers a system of registration numbers to be displayed on your glider.

✓ Hazardous Operations

Section 103.9 prohibits such operations such as flying or dropping of objects so as to create a hazard to persons or property. The interpretation of hazard is vague—intentionally so—so that you can be fined if your actions are perceived as a threat to the public. Certainly, diving at spectators on launch, crashing into someone's car, or landing in a mature corn field could lead to a fine under this ruling.

✓ Daylight Operations

Section 103-11 states that you can only fly from sunrise to sunset. That's legally defined sunrise and sunset which may be found in your local newspaper, an almanac or a GPS unit. All those moonlight flights or our favorite pastime of watching the sunset from the air are thereby prohibited. The only exception to this rule is a 30-minute extension before sunrise and after sunset if you display a blinking anti-collision light which is visible for three miles and you remain in uncontrolled (Class G) airspace (below 1,200 feet above the terrain in most places).

✓ Right-of-Way Rules

Section 103.13 determines who has the right-of-way in ultralight operations. Simply put, paragliders must yield right-of-way to *all* aircraft except powered ultralights and hang gliders (paragliders and hang gliders are on an equal basis). It doesn't make sense that paragliders have to yield to airplanes, but that's the only way the FAA could get such liberal laws passed. Know your local airspace and avoid airplane traffic.

✓ Congested Areas

Section 103.15 prohibits flying over a congested area of a city, town or settlement or an open air assembly of persons. This is another catch-all ruling. In the past, a congested area has been defined as one house. Thus, if you make it a habit to fly over nude sunbathers in the privacy of their backyard, they can complain and have you fined. Skirt habitations on your way to a landing field and avoid landing in city parks that require you to fly over buildings.

✓ Operation in Controlled Airspace

> *If we are going to mix with the big iron in the sky, we must know where they are and stay out of their way.*

Sections 103.17 and 103.19 prohibit the operation of paragliders in certain areas of controlled airspace. The definition of controlled airspace is any airspace where some rules apply. In the U.S. and most other countries we must remain out of Class A, B, C and D airspace. Class A airspace is anywhere above 18,000 feet (5,500 m) above sea level (we use the designation MSL as an abbreviation for Mean Sea Level and AGL as an abbreviation for Above Ground Level).

Class B airspace occurs around busy airports and looks like an upside down wedding cake, as shown in figure 10-16. Class C is similar but lower, as shown. Class D airspace is a cylinder of 5 miles (8 km) radius and 2,500 feet AGL around medium traffic airports, as we see in the figure. Class E airspace is controlled airspace usually 1,200 feet above the surface. It is designated in areas of likely air traffic. We are allowed to fly in Class E airspace. There currently is no Class F and Class G is uncontrolled airspace where we can fly with the least requirements (see Cloud Clearance below).

Two additional areas you must remain clear of are Prohibited and Restricted areas. These areas are found around national security areas and military activity areas respectively. You can enter all of these areas with prior permission, but such permission may be hard to get except in the case of air shows and the like.

Now how do you determine where these areas of airspace lie? You must use an aviation map known as a *sectional chart*, or sectional. These charts are available at most airports and on the internet.

The illustration of the sectional shows the different designations of airspace. Class A is everywhere above 18,000' and not depicted. Class B airspace is a series of concentric blue lines with numbers giving the layer heights in thousands of feet. Class C airspace is shown with magenta (purple) concentric circles. Class D

Figure 10-16: Controlled Airspace shapes

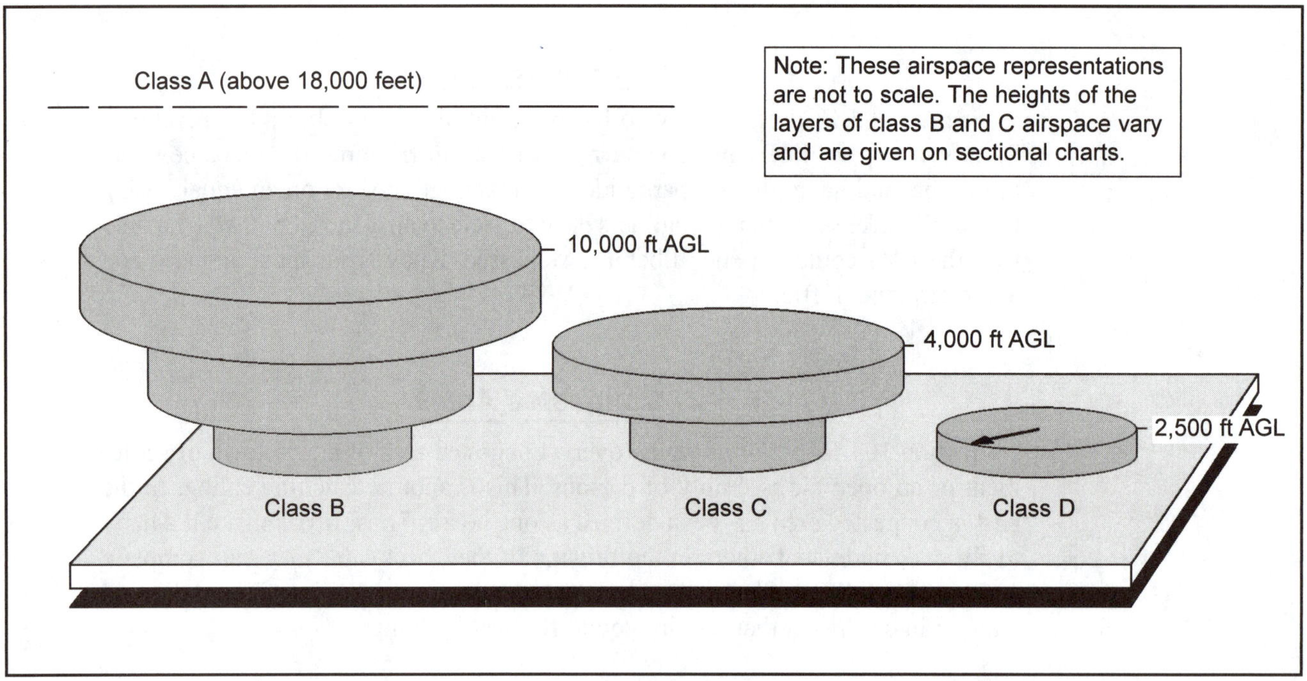

airspace is shown with dashed blue lines around airports. Prohibited and Restricted areas are shown with blue cross hatching on the sectional chart.

Each sectional has a legend that explains symbols and various designations. On the map you will see some long straight blue lines. These are "victor airways" which are frequently followed by airplanes and thus have more than average traffic. These airways are eight nautical miles wide. Another matter to notice is areas with magenta hatched lines. These are Military Operations Areas (MOAs) which paraglider pilots *are* allowed to fly in. However, in many of these areas flight may be hazardous due to the prevalence of fast-moving jets and other "heavy iron." All pilots should have access to the sectional of their area and note where the airspace restrictions are located. A good club activity is to review the local sectional to learn where you can and cannot fly.

✓ *Visual Reference With the Surface and Cloud Clearance*

Section 103.21 states that you must maintain visual reference with the surface at all times. Simply put, that means you *cannot* fly in clouds or even above a stratus layer or extensive fog bank.

Section 103.23 defines the cloud clearance requirements. There is some logic to this: pilots flying airplanes on instruments have their eyes glued to the panel

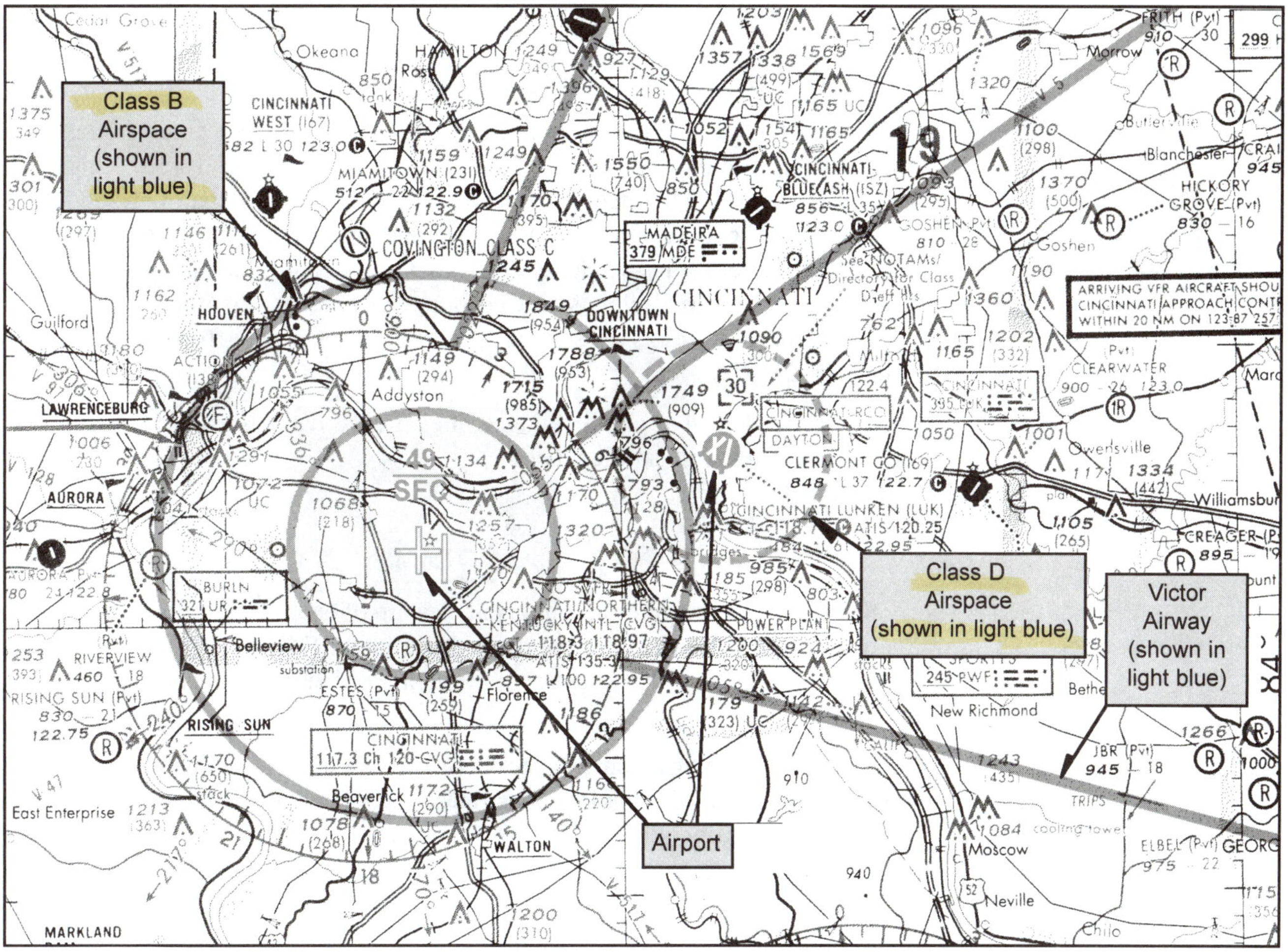

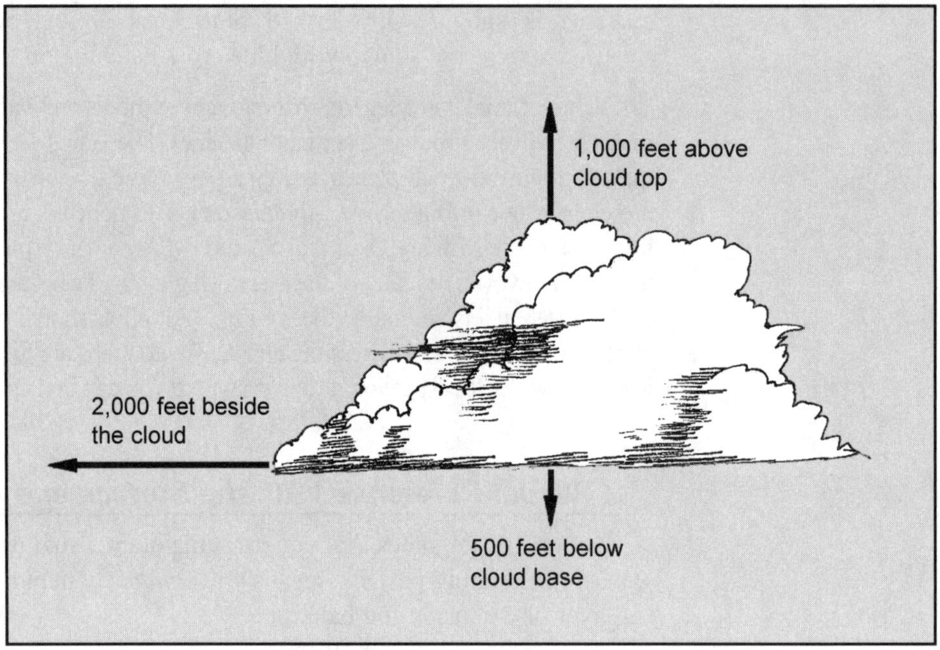

Figure 10-17: Cloud Clearance Below 10,000 feet

when they zip through the clouds. They exit horizontally, so it may take a good distance before they focus on their environment and start looking for other air traffic. You wouldn't have a chance if an airplane popped out of a cloud and aimed for you from a hundred yards away. The other considerations are that cloud bottoms tend to be fairly level while the tops tend to billow.

With these three criteria you can remember the cloud clearance values as: 2,000 feet horizontal, 500 feet below and 1,000 feet above (see figure 10-17). This holds true except above 10,000 feet and outside controlled airspace (below 1,200 feet in most areas). Above 10,000 feet, cloud clearance must be: 1 statute mile horizontal, 1,000 feet below and 1,000 feet above. Outside controlled airspace, you must simply remain clear of clouds and visibility must be at least one mile. In controlled airspace, visibility must be at least three miles below 10,000 feet and five miles above 10,000 feet.

Again we note that cloud flying is illegal (and dangerous). Also, getting to cloud base is likewise an illegal act.

✓ *NOTAMS*

Section 103.20 requires all pilots to be aware of NOTAMS (notices to airmen) by calling Flight Service (see the phone book under Federal Government, FAA). These notams may restrict airspace in the vicinity of important personages or public events.

From the preceding you can see there are several ways to end up having to spend your lunch money on insidious fines. The most likely rules you'll run afoul of are hazardous operations, flying over congested areas, and cloud flying.

SUMMARY

Our dream is to fly free from the fetters that control our everyday lives. In a large part this dream is realized with paragliding, but a dose of reality is needed to preserve our health and this very freedom we cherish. We cannot flaunt our skills and taunt the forces of nature without some unpleasant consequences. We cannot ignore certain courtesies when dealing with land owners, the public or other pilots. Finally, we cannot ignore the minimal government regulations or these minimal regulations will maximize. All of us have a responsibility to know the rules of flight and obey them to our best ability in order to preserve paragliding for ourselves and for others who follow our lead.

TEST YOURSELF (*Answers in Appendix II*)

1. One of the most serious equipment problems is launching without leg straps. What procedure was neglected if this happens? ____________________________
2. When trying new things such as gliders, sites or conditions, how many items are reasonable safe at one time? ____________
3. Put the following emergency landings places in order from least to most dangerous:

a. Still water	d. On a building
b. Power line	e. Ocean surf
c. Deciduous tree	f. Pine trees

4. When throwing a parachute in an emergency, the emergency is over once the 'chute is out and inflated.
   True or False
5. The pilot entering a thermal has the right-of-way as long as he's lower.
   True or False
6. When catching up to another glider on a ridge you should pass on _________ side of the glider.
7. Paraglider pilots are not allowed to fly in what controlled airspace?
   ____

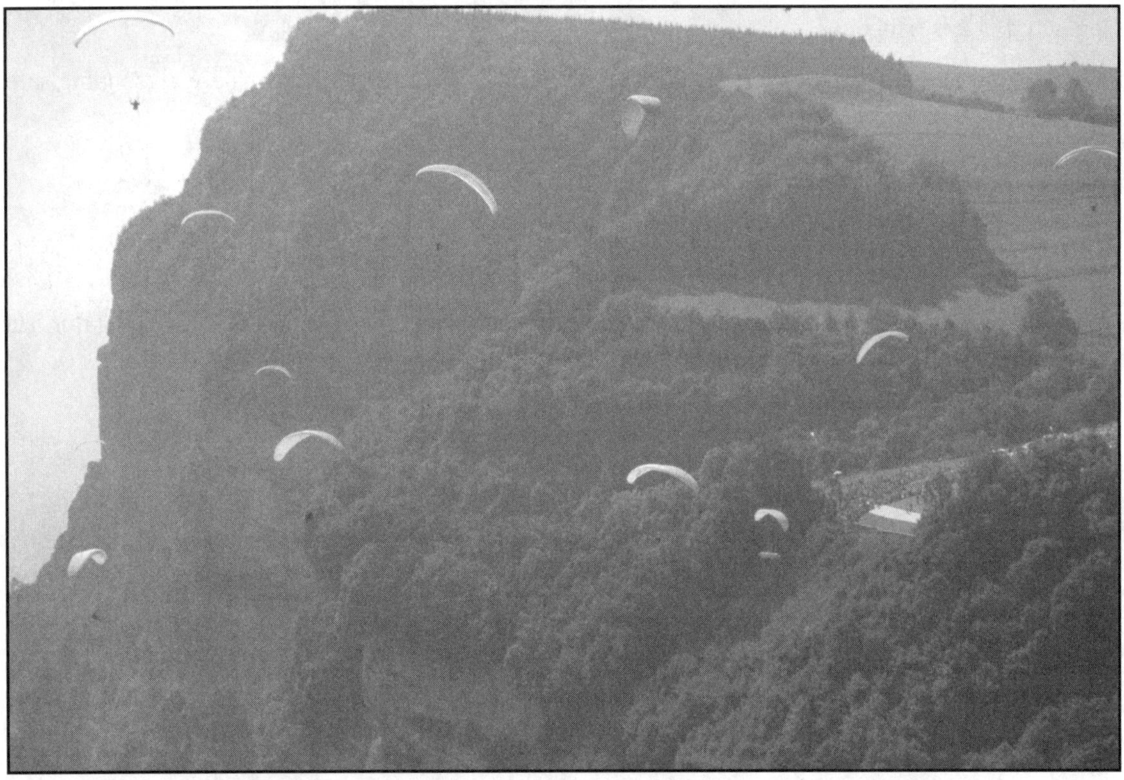

Pilots decorate the air at the famous St Hilaire (France) flying festival.

Chapter 11

MORE ABOUT EQUIPMENT

"One machine can do the work of fifty ordinary men. No machine can do the work of one extraordinary man."

— Elbert Green Hubbard

Around about the time you have your first fully-controlled thrilling flight, you start thinking about buying your own equipment. Flying your own personal glider is rewarding, for you soon form an attachment to the wing that carries you aloft. It becomes an extension of your body, and the wedding of man and machine is enlightening.

But when you get your own equipment you become the person solely responsible for its care, repair, upkeep and reliability. Your flying gear will only carry you safely through the tides and textures of the sky if it is in tune and you are in tune with it.

As our experience grows, your understanding of your equipment should grow equally. In this chapter you'll learn about glider construction, care and repair. We'll also present guidelines for buying novice and intermediate gliders. Finally, we'll discuss auxiliary equipment that completes your flying gear.

GLIDER CONSTRUCTION, CARE AND REPAIR

The first thing to do is learn more about our gliders' manufacture so you can better know how to use it, not abuse it. You want to get the most mileage out of your wings so your investment (and life) is preserved. We'll assume you've read the material in Chapter 2 on glider construction. If not, you should do so. Here we'll look a bit deeper into the nature of our wings. Your goal is to learn enough to make educated guesses when judging equipment to buy, whether new or used.

✓ Canopy Construction

Most canopies are made of ripstop nylon. The cloth comes from several manufacturers. A typical weight for material is 45 g/m^2 (1.3 oz/yd^2)* for the upper and lower surfaces and 48 g/m^2 (1.4 oz/yd^2) for the ribs. The thicker material in the ribs helps prevent stretch and hold the airfoil definition.

The highest stress on the canopy occurs where the lines attach and at the leading edge. Typically the canopy will (and should) be reinforced at these points. If not, it will not last as long. Ribs are normally stiffened at the leading edge with extra material, often made of low stretch Dacron (polyester). The line attach points may have the same treatment with flares or triangles sewn to the ribs in order to spread the load.

All paragliders are made from fabrics. These materials are light, conveninet and strong, but will suffer from physical abuse.

The performance of a paraglider greatly depends on the design of little things such as those mentioned above along with the choice of airfoil, number of cells, shape of canopy, and so forth. We'll deal with these latter matters in the next chapter. The longevity of a glider is even more dependent on the details of construction.

Gliders offering a bit more performance will have more cells. A beginner glider may have 30 or fewer cells and an advanced one more than 70. The reason is that more cells help define the airfoil better. The drawbacks of more cells are more construction cost and more canopy weight which is a factor in zero wind inflations. It is desirable to have fewer lines, because all the lines add together to be a major source of drag. One trick that helps reduce the number of lines needed is using diagonal ribs. As shown in figure 11-1, such rib arrangement spreads the pull of the lines further sideways to help eliminate some lines.

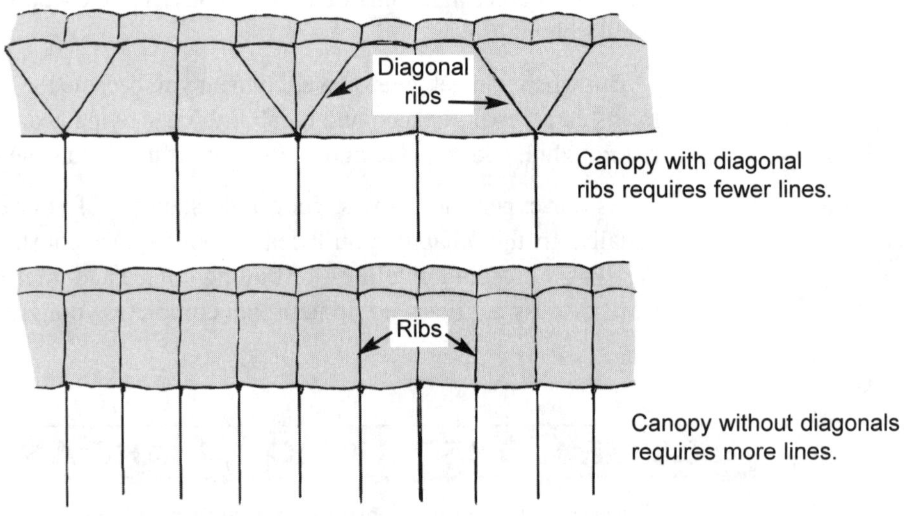

Figure 11-1: Diagonal Rib Arrangement

* Note: All material weights and line dimensions are normally given in metric units in paragliding. The English units are provided for comparison with sailboat and hang glider sails which typically use English units.

The load on a canopy is spread over the entire wing, but not evenly. The center section and the front area are more greatly loaded and thus are likely to stretch more. Some manufacturers are adding bands sewn to the lower surface from side to side between the line groups to help prevent stretch. A final item to look for in canopy longevity is the sewing. Ideally the threads on the upper surface should be internal. If not they are subject to wear on every layout and deflation and will warrant more frequent inspection and repair.

✓ Line Construction and Strength

We have already covered line makeup in Chapter 2. Here we'll add a few points for consideration. Fewer and thinner lines reduce drag and thus improve performance. However, the result is more loading on each line and thus a greater tendency to stretch. Thinner lines simply must be replaced more often.

The lines on a paraglider are certainly strong enough, but they must be of a certain diameter to account for the possibility of some loss of strength due to rough handling. Here is a chart of typical line diameters, strengths and locations on a cascaded suspension line:

Location on line	Material	Diameter	Tensile Strength
Lower	Superaramid (Kevlar)	2.15 mm	392 lbs/178 kg
Middle	Superaramid (Kevlar)	1.70 mm	268 lbs/122 kg
Upper	Polyethylene (Dyneema or Spectra)	1.10 mm	165 lbs/75 kg
Note: The outer sheath is for protection only. It is typically made from polyester (Dacron).			

If we add up the combined strength of the lines we get between 8000 and 13,000 lbs (3600 to 5900 kg) which is incredibly high. However, remember that we need some redundancy because of abuse and the fact that the connecting knots weaken a line—by as much as 50%. Also, overly-strong lines help reduce stretch. When all factors are considered (knots, angling lines, unequal loading) we can consider that the suspension line system on an undamaged glider is good for at least 10G's. That's more than most airplanes.

A rat's nest of lines will come out neatly as long as the risers do not slip through the lines

The rest of the system—quick links, risers, carabiners and harness are similarly strong. For example, each riser is typically rated for more than 2000 lbs (907 kg) and carabiners are typically good for 4000 lbs (1800 kg) each. The harness webbing can handle more than 5,000 lbs (2268 kg). The sewn loops depend on the integrity of the thread, but as long as they are not pulled out or cut, the system of overlapping renders the loops nearly as strong as the webbing.

✓ Canopy Care and Repair

We have already outlined the principles of glider care in Chapter 2. Figure 2-4 summarized these matters. Remember, the heat of a closed car in hot sunny areas can exceed 180°F (100°C). Such temperature can damage fabric and shrink Spectra lines.

Take care of your glider on the ground, and it will take care of you in the air.

Once you are more skilled at canopy handling (and have your own wing) you will want to take more care to avoid abuse. There are five main things to attend to: watch your layouts, keep it clean, keep it dry, keep it shaded and don't tenderize it. Tenderizing means pounding it to the ground in strong winds. Perfect your inflation and kiting skills to avoid this behavior.

Getting a canopy wet is sometimes unavoidable—you just had to get that flight in even though rain was predicted! If this happens, spread your canopy to dry as soon as you can—out of direct sunlight! In the shade with wind is the best place. *Never* store a canopy wet. Do not pull on a canopy when it is wet for it stretches more readily.

If your canopy, harness or parachute gets dunked in salt water, you must rinse it thoroughly in fresh water and let it dry as described above. Do this even if the equipment is already dry from the salt water dunking. The sharp salt crystals left when salt water dries can abrade and weaken lines, webbing and material.

If a canopy gets stained, the best thing to do is wash the area with warm water and a sponge, then let it dry. Do not use a brush on the delicate material. Stubborn stains or animal droppings should be immediately cleaned with methylated spirits which are then rinsed with water. Animal droppings are acidic which is damaging to nylon. Consult your owner's manual before using any cleaning substance.

If small tears (less than 4 inches or 10 cm) occur in the canopy you can repair them with tape as long as they aren't in a critical area such as a seam or line attachment. Use tape specifically designed for nylon which is readily available at sailboat, camping or paragliding shops.

Figure 11-2: Tear Repair With Tape

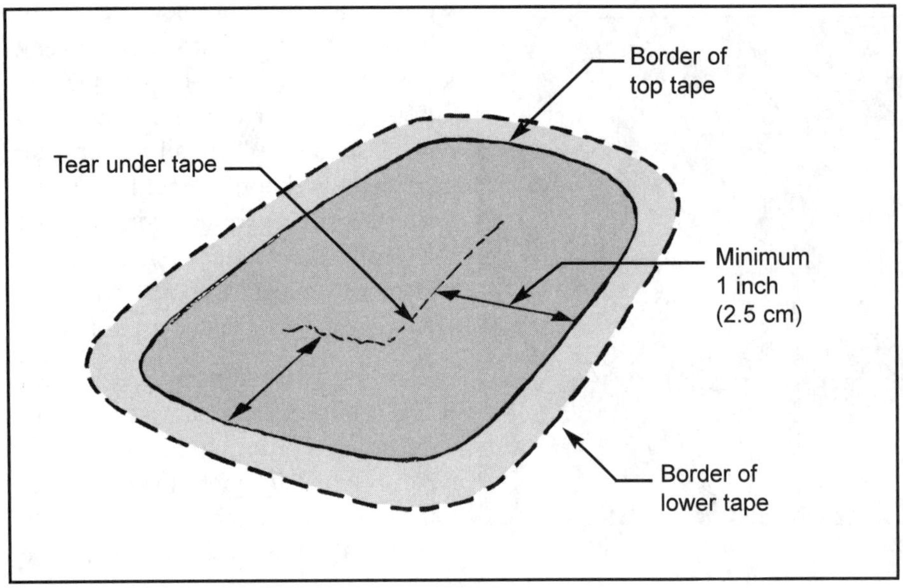

To do an expert tape job, start with a clean, dry area, measure the tape size and round the corners as shown in figure 11-2. Realign the ripped area carefully to prevent wrinkles. Make sure the tape overlaps the ripped area at least one inch (2.5 cm). Apply the tape to both sides of the fabric and make the patches of different sizes to help spread the load as shown. Now remember to give this area special attention on all your future pre-flights.

Larger rips, strategically placed rips or failing seams are beyond the repair capabilities of the normal pilot. Do not attempt these repairs, but take the canopy to a recognized repair facility or shop for a professional job.

Canopy porosity is the most serious and hard to detect problem in paragliding. Usually a canopy gets porous at the front half in the center area which is where most of the load occurs. When this deterioration happens, lift is lost in front and the glider has to be flown at a higher angle of attack to achieve the proper lift. Also, the balance changes because more lift is being produced rearward. Full stalls and deep stalls are more likely to occur on a porous canopy.

Sunshine is the greatest opponent of canopy longevity.

How do we check for porosity? There are devices that test it accurately by blowing air through a section. However, as a pilot, you can perform a crude test by locating an area on the canopy not likely to suffer loss of porosity such as the rear surface tip. It is loaded lightly and is exposed less to sunlight. Put your mouth to this area and blow. Note the resistance to air passing through the fabric. Now go to the suspect area (front, center, top surface) and repeat the process. Did you notice a difference? Repeat the test several times for better accuracy. Note that sometimes the same type of fabric is not used on the top and bottom surface of a wing. If this is the case, use the rear of the upper tip area as your "good" fabric.

If an increase in porosity is noticed, get a second opinion. If the conclusions are porosity degradation, replace the canopy. In any case, canopies should be replaced every 300 hours of airtime according to most manufacturers' recommendations. Of course, the longevity depends on how severe the sunlight exposure has been (high altitude and sunny, dry area flying is worse) and how much the canopy has been beaten.

The final canopy issue we'll address is the presence of debris in a wing. We have seen leaves, grass and gravel in our wings. Flying in a snowstorm can sometimes result in a bit of snow entering the canopy. Enough accumulation of foreign matter can pull down the trailing edge and affect glider behavior. Remove it periodically. Do this removal by lifting the trailing edge and shaking, then grabbing closer to the leading edge and repeating the process. The biggest problem is the wing tips because they do not have leading edge openings. We have found the only way to remove all the debris is to invert them through the holes in the cell ribs as shown in figure 11-3. Pull the fabric gently through the front rib holes, one at a time, starting with the first rib at the last leading edge opening and progressing to the tip.

Figure 11-3: Cleaning Canopy Interior

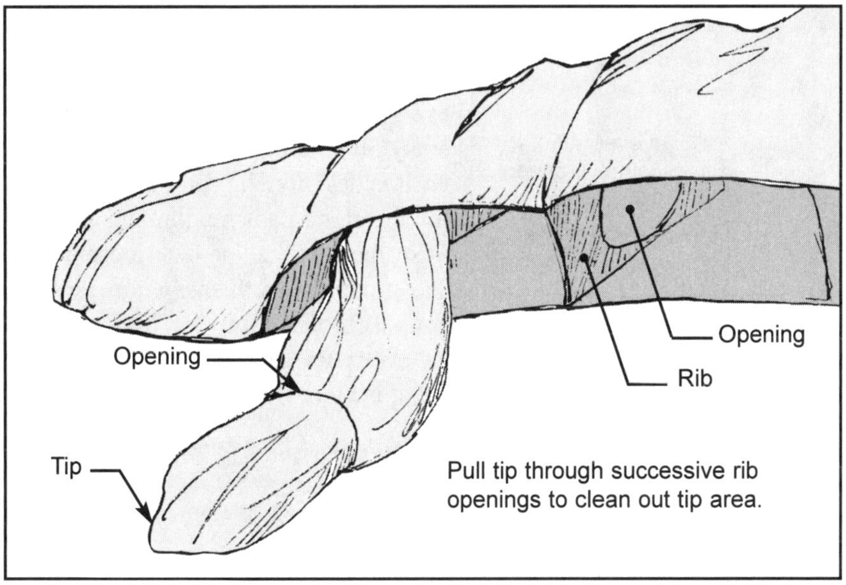

✓ Line Care and Repair

Now let's review the handling of lines. If they get wet, dry them immediately. Experiments have shown that nylon and other synthetics stretch a bit when wet and shrink as they dry to end up shorter than they were originally. Be careful the next time you fly to make sure the glider's aerodynamics hasn't changed. Eventually the lines will resume their original length, but you can't tolerate more then ¼ inch (.6 cm) difference from the original. Measure your lines after a thorough soaking and compare them with the originals.

Pro Tip: *Always measure the length of your lines when you get a new glider so you can compare their measurements in the future.*

Of course, some owner's manuals provide these measurements. The center and forward lines are the ones most likely to undergo stretch with use, so be sure to inspect them and know their proper length. On the other hand, if your canopy gets wet and all the lines shrink, the A and B lines which are more heavily loaded may go back to normal size while the ones in the rear stay shrunken. The result will be the wing flying at a higher angle of attack. Your top speed will be less and stalls or spins will occur more readily.

If you drop your glider into salt water, special attention must be given. As soon as you can, rinse it thoroughly in fresh water. Dry it out of the sun. The same treatment should be given to the parachute and harness. Salt crystals left in the lines weaken the lines, even when they are rinsed. Most manufacturers recommend replacing lines that have entered salt water.

Figure 11-4: Checking Line Integrity

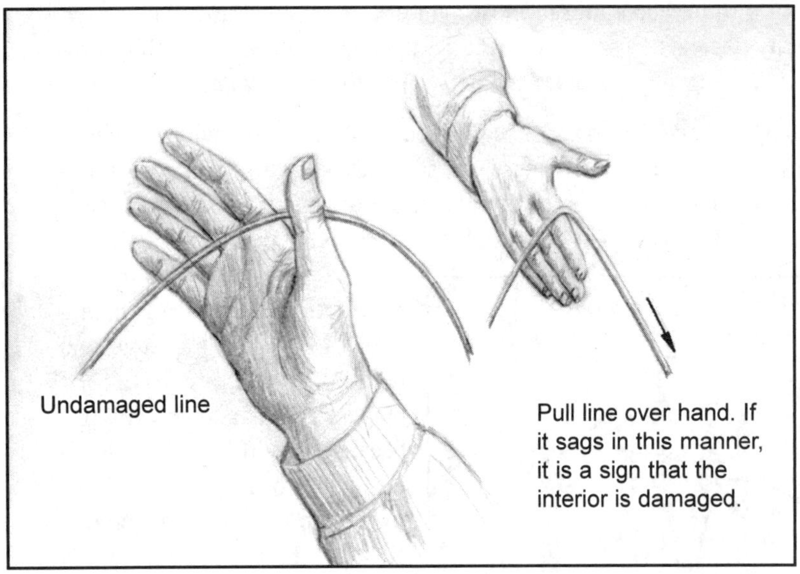

Recommended Line Replacement

Replace your lines:

- Every 100 hours
- When permanent stretch or shrinkage is evident
- When soaked in salt water
- When obvious damage occurs

Naturally lines should be inspected every pre-flight as well as annually or every 50 hours, whichever comes sooner. Besides stretch, you should inspect for sheathing damage, kinking and loss of stiffness. If the line feels loose or you can stretch a length easily, it means the core is broken and the line must be replaced. Figure 11-4 shows how to run a line over your hand to quickly check for core integrity. You are looking for any place that is more flexible than normal.

Caution: Never *try to repair a line by knotting it. Knots considerably reduce both the strength and the length of a line.*

✓ *Replacing Lines*

If you aren't entirely adept at knots and organizing details (how good are you at knitting, weaving or chair caning?), you should leave line replacement to an expert. The replacement of one line isn't too hard, but note that it must resume its proper place on the quick link. To avoid confusion we recommend placing a bit of masking tape to label each line's position order before removing it from the quick link.

Line removal consists of sliding down the quick link protector to expose the gate, opening the gate and taking off the lines. Now the line can be untied at each cascade one at a time and then at the canopy end. The knots at the cascades and canopy are a simple lark's head. The removal and replacement of a line at this knot is shown in figure 11-5. Be careful not to bend the new line unnecessarily. Once you have the knot nearly tightened, arrange it properly (look at the others) and pull it snugly (don't jerk it).

Figure 11-5: Removing and Attaching Lines

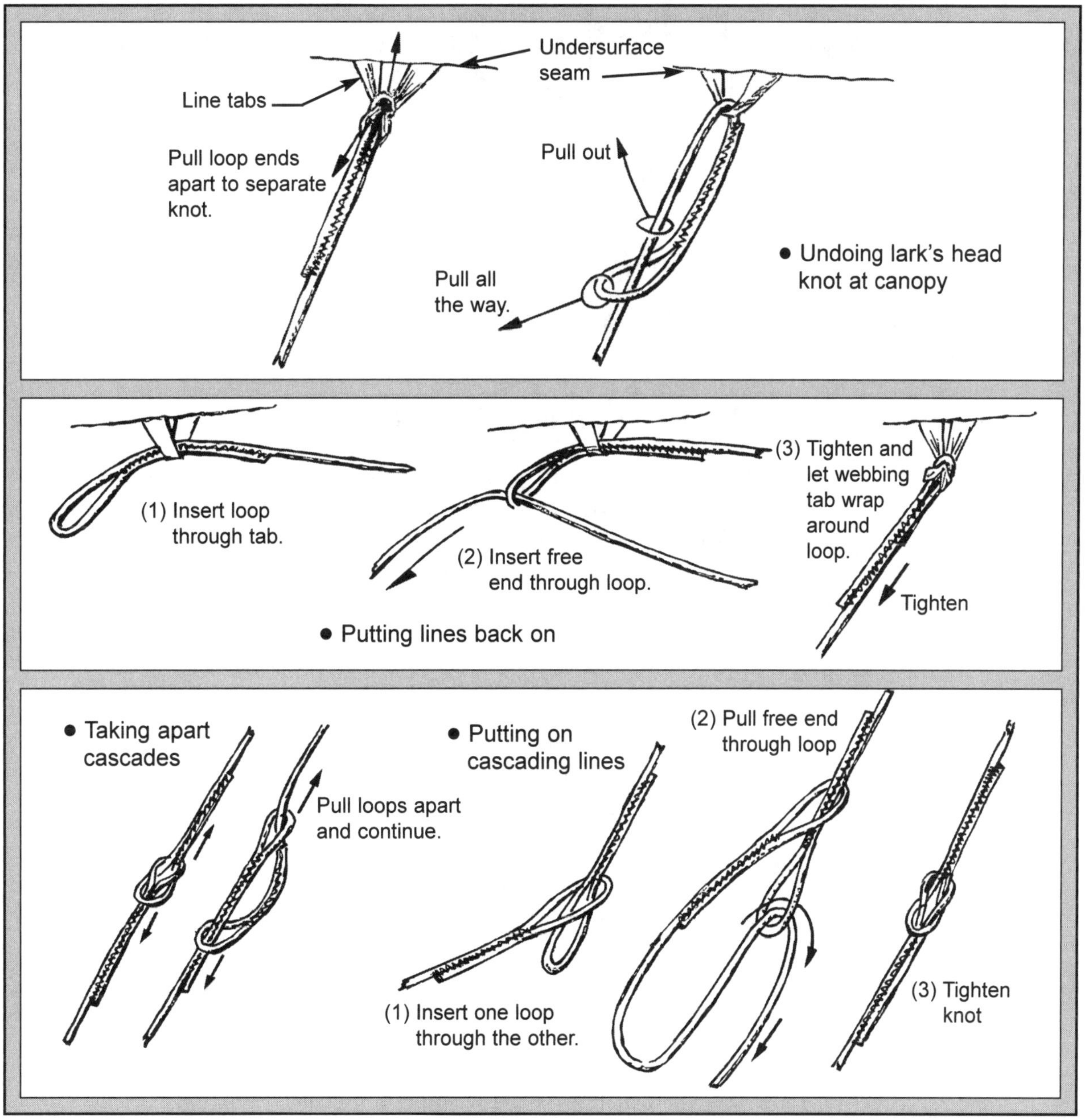

A pilot checks his lines before flight.

When replacing the lines on the quick link, work from left to right (or vice versa) so you don't twist lines around one another. Close the quick links as tightly as you can by hand if they are protected by a plastic oversleeve. If not, close them tightly by hand and add ¼ turn with a wrench. *Don't* forget to close the quick links and *don't* overtighten them.

If you are replacing all the lines, we recommend you hang the canopy on a wall (this is how the factory does it). Use spring clothes pins or padded paper clamps in several places. Complete the process systematically: start with one side A lines, then B lines, then continue. Work from the outside in and check your work frequently. Remember, you don't have to remove the last cascade of one line that routes all the way to the canopy. However, you still have to remove and replace a lot of knots—at least 350 for a typical canopy! Plan to spend a couple of days.

Buying Your Own Equipment

At some point early in your flying experience you will want to buy your own equipment. We'll provide you with some guidelines here for this all-important endeavor. We make the assumption that you will be buying a beginner to intermediate glider. For more details on test flying gliders and buying more advanced designs, see our companion book *Performance Paragliding*.

✓ New or Used

You will face an important question. Should I buy new or used equipment? There are benefits and drawbacks to each. Let us enumerate them to aid you in your decision.

▲ *New equipment* — You will buy a new glider or harness from a dealer. As such you will be receiving expert advice as to their appropriateness to your skill level and size. The follow-up service as well as the original guidance you receive from a dealer is no small reward, for safety and enjoyment depend on the nature of the equipment you buy.

On the other hand, new equipment is naturally the most expensive. Your economic situation may require that you shop around for used gear. Economics is the main reason not to buy new equipment.

▲ *Used equipment* — Besides price, used equipment's only attraction may be that it's available immediately from a local pilot or dealer. In the high season gliders can take a month or two from order to delivery, and this delay may put you back a season if winter sets in. The drawbacks of used equipment are that it may have been abused and it is possibly outdated. Remember, paraglider canopies and lines normally only last a few seasons.

If you decide to buy a used glider, get expert advice from a shop or very experienced pilot as to its quality. In truth, there are some very good gliders and good deals to be had on the used market. But you must know what you are doing to get the glider that's right for your skill level. Any glider manufactured within 5 years from the present should be a good, safe design as long as it's in proper shape. Of course, the only way a glider more than a couple of years old will be in good shape is if it hasn't been flown much.

✓ What to Look For

Whether you are buying new or used, you should be attending to the following items, listed in the order of decreasing importance.

A glider above your skill level will not make you fly better—it will do the opposite.

▲ 1. *Appropriateness for your body size and skill level.* As mentioned above, a dealer or the owner's manual can guide you in selecting the right design and size for your needs. In general, it's better to be in the mid-to-upper areas of a glider's acceptable weight range for safety reasons. The lighter you are on a given design, the less airspeed you will have at any chosen angle of attack and therefore less pressurization to combat folds. On the other hand, being lightly loaded improves your minimum sink rate (see the next chapter) a little bit, while it hurts your penetration. Best glide is not affected by wing loading. Typical wing loadings range from 2.3 kg/m^2 (.46 lbs/ft^2) to 4 kg/m^2 (.8 lbs/ft^2). A range of at least 2.5 to 3.0 kg/m^2 is normally desirable for beginner to intermediate wings.

There are currently many models and styles of gliders from which to choose. Don't forget to pick a pleasing color.

Wing Loading Effects

- Stability improves with more weight on the glider.
- Penetration into wind improves with more weight.
- Sink rate increases with more weight.
- Handling response usually improves.
- Best glide does not change.

Skill level appropriateness is extremely important. If you fly a glider rated higher than your skill level you will be risking your life needlessly. More advanced gliders suffer folds and spin more readily and are harder to land. They also require more finesse to turn. How do you know when a glider is appropriate for you? All gliders are rated by either the French AFNOR or the German DHV system. (These ratings will be combined in the future under the European Union CEN standards.)

The two systems equivalences are outlined in the box below.

Skill Level	Afnor	DHV
Beginner	Standard	1
Novice	Standard	1-2
Intermediate	Performance	2
Advanced	Competition	2-3

An outline of the AFNOR and DHV tests is given in our sequel book *Performance Paragliding*. Suffice it to say here that besides loading and strength tests, each canopy undergoes exhaustive testing of response to maneuvers and various forms of folds. The recovery rates and behavior are carefully noted and a grade is given on each item. The tally of all the scores determines a glider's final rating. For example, under the DHV system, if a glider gets even one score of 3 (which means pilot input is required for recovery), the glider gets a 3 rating.

Where do your skills lie? When answering this question, don't just look at your rating on paper, but consider your natural abilities, understanding, athleticism and the amount of flying you do. Make a choice of glider level that is conservative. You'll have much more fun, have fewer heart-stopping episodes and will be safer. Today the 1-2 gliders perform as well as performance or even competition models of yesteryear and are much safer. We strongly recommend you select these gliders.

▲ 2. *Ease of control.* All gliders within a given rating must control to a certain standard, but there still can be a big difference in the different designs. Some people like more solid handling, some people like right-now twitchy response. Vive la différence! Fly any glider you intend to buy to see how it suits your style.

▲ 3. *Ease of inflation.* Again, a glider's rating includes inflation, but there are differences in a class. Note that the more modern a glider is, the easier it is to inflate as well because of improvements in design. Try it and see how it works for you.

▲ 4. *Performance.* Certainly we all want more performance. However, experience has shown that one of the greatest detriments to performance is how comfortable a pilot is on his glider and how well he can make it obey his commands. If you

Taking time to inspect your glider in the landing field is a good practice.

get a glider that's beyond your skill level, you will not be able to realize its potential. Disappointment will follow. Go for performance only after the previous three items have been satisfied.

▲ 5. *Features.* Newer gliders usually have enhanced features including easy harness buckles and no-fault leg straps, specially designed quick carabiners, speed stirrup, convenient parachute position, color-coded lines, special big ear risers, canopy reinforcing and air bags (see below).

▲ 6. *Price and resale value.* Of course, the real cost of equipment is the price you pay minus what you will sell it for. A glider that's a couple of years old will be rather worthless after you put a couple more seasons on it. Only if you get an exceptional deal will buying used equipment be ultimately worth it. Try to look at used glider prices in magazines or at shops to get a general idea of value before you settle on a price for a used glider.

✓ Inspecting a Glider

If you are buying a new glider you shouldn't worry about its condition, but inspect it to make sure all quicklinks, carabiners, buckles and pulleys have been placed properly. Also, check for shipping damage. This is your dealer's job, but a backup check doesn't hurt.

With a used glider you must be more careful. Perform a complete check as you would a very thorough preflight. Don't forget to check inside the canopy for rips. It is especially important to check for porosity and line stretch or weakness (see above). It has happened that pilots didn't even know they had damage until such an inspection was made. Faded areas on the canopy are a good sign that it has seen too much sun (compare the top surface with the lower if they are both colored). Suspect a loss of porosity.

Any damaged or worn part on the glider requires replacement. You shouldn't necessarily reject a glider because of these deficiencies, but the need to pay for repairs or line replacement should be reflected in the selling price. Don't neglect to make these repairs before you fly your new wings. Be sure to check on the availability of service *before* you fork over your hard-earned cash. Some paraglider companies have gone out of business. Where will you get new lines?

Always try a new piece of equipment (glider, harness, helmet) at a familiar site. Never add more than one new thing at a time (new site, new equipment, new type of conditions).

✓ Test Flying and Transitioning

Flying a new glider at the beginner to intermediate level requires that you go to the training hill to become familiar with it. An instructor or experienced pilot should accompany you to help you work out any problems and advise you on controls.

Here are differences to expect as you go from a training glider to a novice, or eventually, an intermediate glider.

▲ *Better glide performance* — A higher performance glider will achieve a better glide path at all speeds and will travel much further during the landing final. Expect this and leave extra landing room.

▲ *Different flare timing* — Better performing gliders often have a shorter "flare window" of opportunity. This doesn't get critical until you graduate to very high performing gliders, but be alert to the glider's feedback so that you flare properly.

▲ *Faster flight* — Higher performing gliders generally fly faster because they are flying at a higher wing loading (see Chapter 12), and suffer less drag. Be prepared for this slight difference and maintain proper airspeed.

▲ *Difference in control* – Higher performance gliders often respond to a trim input more slowly. They have a wider span to rotate around. In addition you may feel more control forces due to the higher wing loading. Finally, steep turns in a higher performance glider usually lead to a "winding in" tendency which means the glider tends to steepen the bank leading to a spiral dive unless you counteract this effect.

A pilot with a permanently inflated air bag harness is in front of a pilot with an air bag harness which inflates in flight.

▲ *Less canopy stability* – The higher the performance of a glider, the more likely it is to experience a fold. In addition, the longer it will take to restore and the more pilot input will be necessary. Higher performance canopies have more span and a narrower chord. They are also not as thick between surfaces. All these points lead to easier folds. Spins are also more likely on more advanced canopies.

A final point we should make is some more advanced gliders have *trimmers*. Trimmers or trim tabs are devices on the rear risers that allow you to adjust the length of the risers in flight. The effect is to change the glider's angle of attack, similar to the working of a speed system. The fact that pulling on the trimmers lowers the rear of the wing means that it is more susceptible to stalls the more the trimmers are pulled. Another potential problem with trimmers is if one slips off unknown to the pilot, a turn will be induced in the wing. Finally, both hands are required to adjust the trimmers so your controls are a bit out of commission while this adjustment is made. Note that most manufacturers do not certify their gliders with trimmers, so their use creates an uncertified configuration. Trimmers are intended for more advanced flying and will be discussed in *Performance Paragliding*.

Keep your test flying of a new glider safe by using the gradual approach: start at the training hill and fly only in gentle conditions. It is possible that you won't be qualified to high fly a glider you wish to buy. In this case, you should enlist the assistance of an experienced pilot to test fly a used glider for straightness and trim, then proceed to get acquainted with it at the training hill.

Safety Tip: *Be cautious of a form of the Peter Principle in operation, whereby pilots rise to their level of incompetence by moving up to higher performance gliders too soon. Solicit honest advice from several pilots as to whether or not they think you are ready for the next step in gliders. (The seller is not the person to ask, for obvious reasons, unless this seller is a reputable dealer.) If you can fly a new glider design in a totally relaxed manner, launch and land it well (be honest!) then you are probably ready for its performance.*

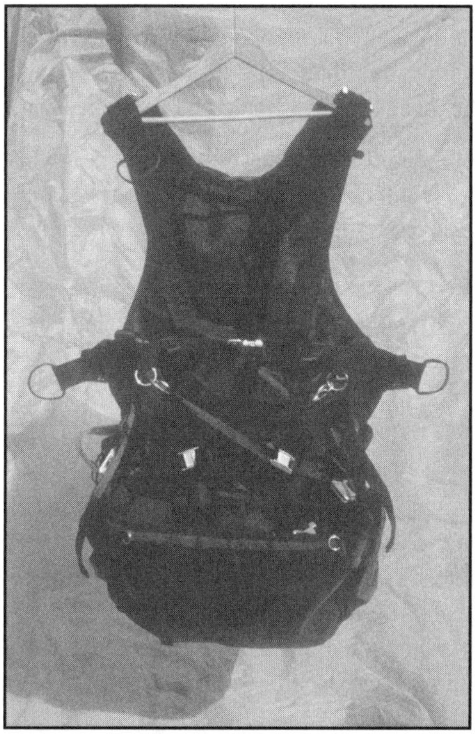

This harness employs cross straps which can be removed. Note the lack of air bag.

✓ *Control Line Adjustment*

The control lines on a glider can be adjusted by untying the knots at the toggle (handle) and either lengthening or shortening the line. The line should be adjusted to have no less than 3 inches (8 cm) slack since the trailing edge rises a bit when you use the accelerator. However, you may wish to make changes to position your hands up or down for comfort. Note that the longer the lines, the less pull you have for landing flare. Also you will have to retrain yourself as to the position for important speeds (minimum sink, best glide, stall, etc.) just as you do when you take wraps. Be aware that with shorter lines it will be easier to stall or spin the glider when using large control inputs. Also, higher hand position is fatiguing on long flights.

Most pilots use wraps (one or two loops of control line around the hand—see Chapter 4) to allow more range to control. Without wraps the controls can be off with the hands at ear level or above and with wraps the full control can be applied for landing or higher wind control. On page 75 we discuss using wraps when flaring for landing.

YOUR NEW HARNESS

As you graduate from the training hill you'll want more comfort in a harness. Light weight and convenience won't be as important, for you will not be climbing up hills as much. However, safety and protection are. Here are a few things to look for.

✓ *Selecting a Harness*

There are many harnesses on the market. Some companies specialize in only harnesses and they try to sell style, innovation, comfort and safety. Virtually all harnesses fit all modern gliders, but check just to be sure. New design features may make a difference with a new canopy and old harnesses or vice versa (some years ago "steering seats" were used which required different rear riser sizes).

Older harnesses often had cross-braces. These were straps that went diagonally from near the seat to the chest to prevent the seat from rocking. They actually limit the ability to weight-shift so are no longer in vogue. Some later harnesses had short diagonal braces below the chest strap and these work just fine. If you are transitioning from a cross-braced harness to one without, expect to feel the seat rock more than you are used to. After some familiarization you'll find you have more control and should become comfortable.

Hang around in the harness you intend to buy to see if it provides long-term comfort for your body shape and style.

There are a few things to check for when selecting a harness. First try it for size and comfort. The best way to do this is to hang it on a simulator and sit in it for a while. If you can read a chapter of this book while lounging in the harness without pressure points or discomfort, we say "buy it." Of course, the other features should be right as well.

Check out the buckle system. We highly recommend quick release buckles (seat belt type) for emergency situations. We also recommend a leg strap system which helps prevent launching without hooking up the leg straps as shown previously in figure 10-1.

One of the main things to consider is the type of protection on the harness. These items tend to be of two different types: either foam "mousse" bags (layers of thick foam) or air bags. They are designed to prevent spinal injuries by dissipating the forces of a hard impact.

Practice going back and forth from flying to landing in a new harness to make sure you can get in and out of it easily without using your hands.

Air systems usually are crushable volumes of air which act like an air bag on a car. Some air systems work exactly like your canopy. When you start flying they fill up with air through an opening or scoop in the front. During an impact the air cannot escape quickly through the one-way valve, so great cushioning is in effect. Figure 11-6 shows an air bag harness. Note that scoop inflating air bag harnesses are more bulky and often don't fill to offer maximum protection immediately after launch, particularly as they get older. Both mousse and air bags provide superior crash protection compared to older dense foam back pads.

Figure 11-6: Protection Harness

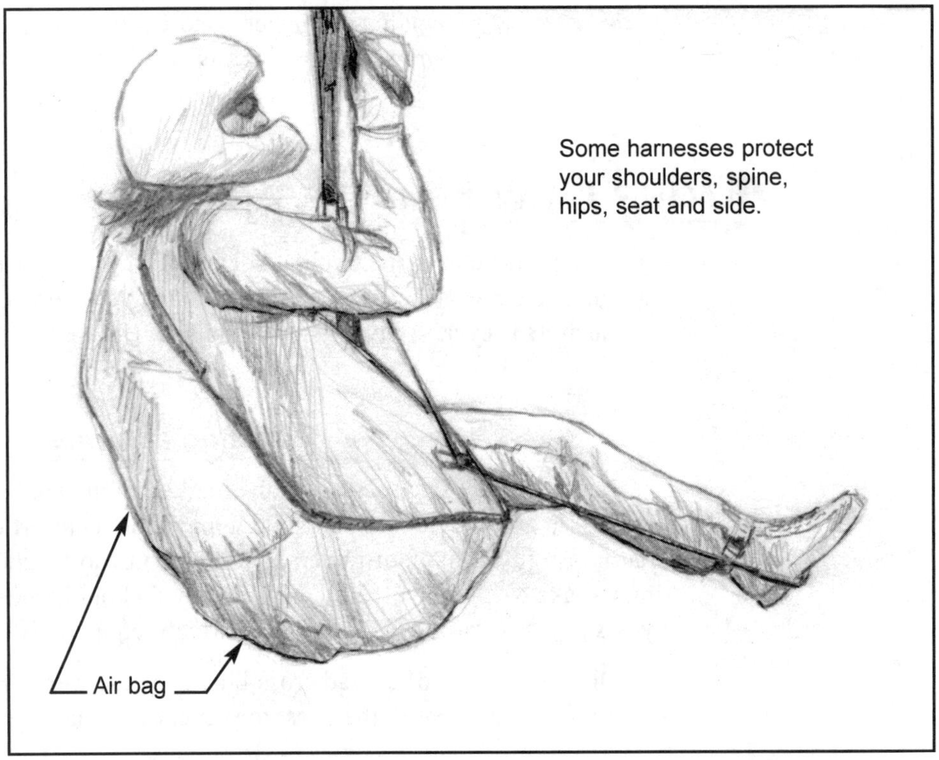

✓ Parachute Position

The next choice to make is where the parachute is mounted. Naturally, with an air bag-type harness the parachute cannot go in the bag. The choice is then on the side or in front. On the side you can only throw the parachute with one hand freely. A front mounted 'chute affords the best access, but poses a complication if a water or tree landing occurs. You have to unbuckle the parachute on one side to get out of the harness.

Usually it is better to stretch your budget and buy a new harness to get the latest safety features. Newer harnesses have side or front mounted parachutes which are much easier to deploy than older rear mounts. When buying a harness consider safety, comfort, convenience and cost in that order.

Used harnesses arc usually a deal only if they are of current design with ample protection. Just be sure to inspect a used harness carefully for integrity before you buy. Make sure the harness has an integral strap system by checking under the seat board. There should be a continuous webbing strap that would hold you if all the cloth tore and the seat board broke (buckles in the system are allowed). If a used harness has aluminum carabiners, we recommend replacing them with steel, for aluminum can lose strength if it receives a really hard knock. This event would generally only occur if a harness was tossed on a rock.

PARACHUTES

Once you leave the training hill, a parachute is your constant in-flight companion. In Chapter 10 you learned how to use your parachute. Here we'll present more information about its construction and care.

Figure 11-7: Parachute Types

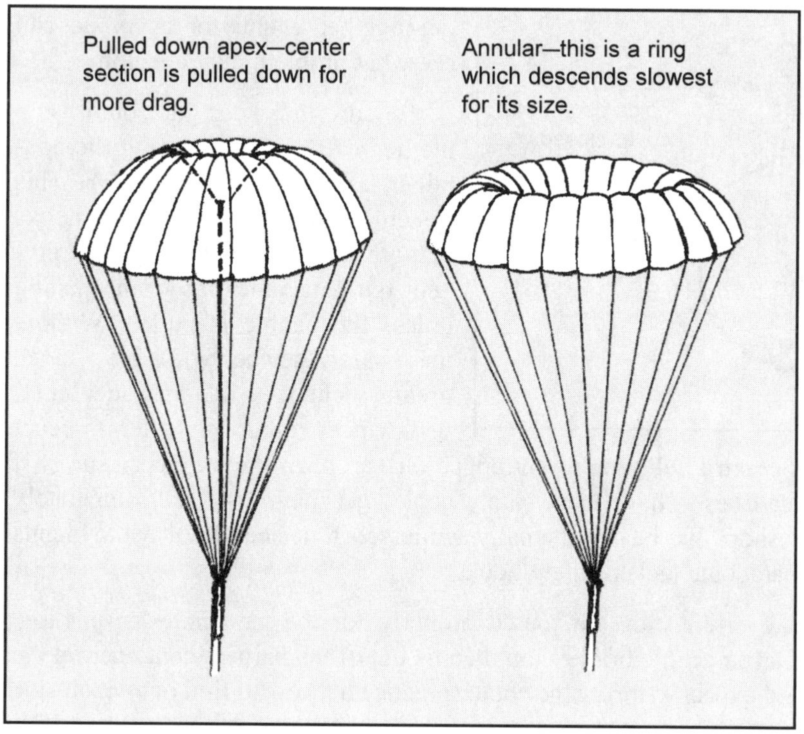

✓ Parachute Types

New developments occur in parachute design just as they do with gliders. Reserve or emergency parachutes aren't like the rectangular models that skydivers use, but they aren't simple round umbrellas either. Modern reserve 'chutes are either *pulled down apex* (PDA) or *annular* as shown in figure 11-7.

A PDA parachute has the center pulled down to afford a wider drag area for a given amount of material used to make the canopy. Such 'chutes also open faster than more conventional types. The apex pull-down is built into the design.

An annular parachute is a ring design which descends even slower than the PDA. The slower the descent rate, the less impact felt when you return to hard Mother Earth.

Your dealer will help you find the size and type of parachute proper for your body and budget. If you get too small a parachute, you will descend too rapidly. Used parachutes are fine as long as they are in good shape and well-packed, but older parachutes are often too small.

Your parachute is the last defense against gravity. Practice deploying yours, but fly safely so you never have to use it.

✓ Parachute Care and Function

In order for a parachute to work properly it must be repacked regularly—at least every six months. Eventually you should get proper training to pack your own so it gets done more regularly. The book *Performance Paragliding* outlines the packing procedures for paragliding reserve 'chutes.

When a parachute hasn't been repacked in a while, the cloth begins to stick together and the deployment rubber bands deteriorate. Both of these factors reduce reliability and increase opening time. In some cases you may not have reliability or time to spare. Never sit on your parachute for this will pack it tighter and make it stick together and be less likely to deploy properly.

Whenever it's time to repack your parachute you should practice throwing it. Do this practice by hanging up a harness in an open area and have someone time you as you *look, grab, extract, look, throw and pull bridle*. It lends good realism to this exercise if an assistant is jerking the harness around as if you were being jostled by a flailing canopy. Grade yourself on how quickly you got the parachute out and how far away the canopy landed. This practice is a great get-together event, for everyone can see what problem may develop.

Figure 11-8: A Modern Parachute System

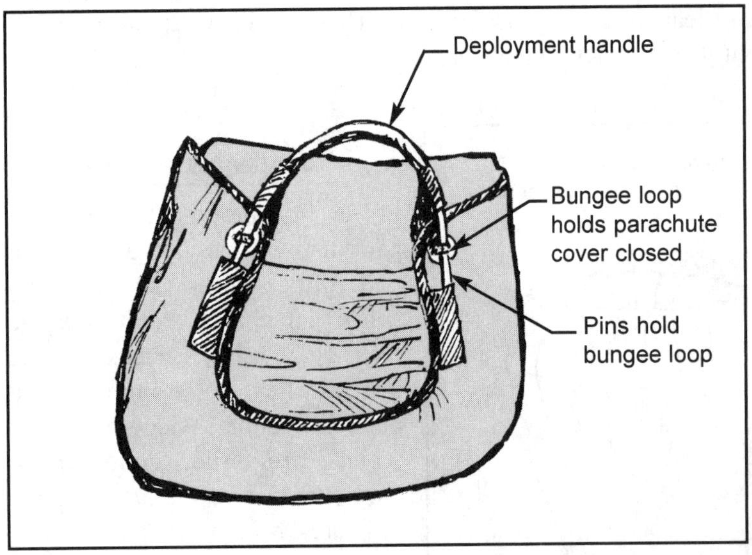

Be sure that your parachute has a method of positively closing the container, as shown in figure 11-8. This system consists of one or two hooks that are routed through loops that prevent the parachute flap from opening unless the handle is pulled. Without these safety devices you have a much greater chance of an accidental deployment, especially after a recent repack, for the extra bulk due to entrapped air can open the Velcro closures. If your parachute doesn't have these safety hooks, get them installed immediately by a capable shop. We have personally witnessed four accidental deployments because the parachute lacked safety hooks.

A parachute bridle must be routed properly for the parachute to function. Preferably your parachute bridle should come out of the harness container on the side you would expect to throw the chute (practice a throw to find out which side feels most natural). Route it up one of your main support straps and attach it to the loop provided on the harness. Most harnesses have a Velcro sleeve to route the bridle and protect it from ultraviolet damage.

If your parachute gets wet, open it as soon as possible and hang it to dry, preferably inside, but definitely not in direct sunlight. If it is dunked in salt water, rinse it in fresh water and dry it thoroughly. If you must clean it, use the same methods as for a harness (your parachute is nylon so acid materials will damage it). Make sure it is absolutely dry before repacking, or mildew may form which can deteriorate nylon. Do not store your parachute near heat or it will weaken in time. Take good care of your 'chute and it will let you down gently when you need it.

Instruments

The instruments you may wish to add as your skills and experience improve are an altimeter, a vario and an airspeed indicator. Let us briefly review each of these.

An instrument deck strategically placed is an essential part of paragliding equipment.

✓ The Variometer

A *Variometer*, or "vario" for short, tells you how fast you are going up or down. It works by measuring the minute changes in air pressure at different levels. Most modern varios are extremely sensitive. Usually a variometer will have an audio output (a beeping) to let you know when you are in lift, as well as a visual display to show you how fast you are going up or down. The audio output will change its frequency as your climb rate varies. This sound is a great help in traffic, for you must always keep an eye on other gliders, not your instruments.

You can pay a lot for a full-blown vario with all the bells and whistles, or you can find a basic used one for a very reasonable price. Try any instrument in actual flight before you buy it. A vario is a valuable aid in soaring flight and we recommend that you get one as soon as you can afford it.

✓ Altimeters

An altimeter tells you how high you are above a predetermined set point. This point may be your landing field, takeoff or above sea level (MSL). Pilots flying locally often use the landing field setting until they learn to soar, then they set zero for takeoff so they know how much altitude they gained. Pilots flying cross-country or near air traffic set for MSL so they can follow the airspace rules. We recommend that all pilots use MSL settings for safety and legal reasons.

An altimeter detects the change of air pressure at different levels and works just like a barometer only with a height scale on the dial. Many varios have one or more altimeters built in. If you can afford it, these are the ideal instruments, for everything you basically need is in one package.

An altimeter is a safety device as well as a reporter of achievement, for knowing your altitude is very important for judging your position in relation to landing fields and the front of a mountain.

✓ The Airspeed Indicator

While we mostly judge airspeed by feel and relate it to control toggle position, an airspeed indicator can help us fly more precisely. We mentioned speeds-to-fly in Chapter 6. For the best application of these techniques we use an airspeed indicator. These are matters for more advanced pilots and are covered in *Performance Paragliding*.

SAFETY EQUIPMENT

There are a few items that are important for continued safe flying. All pilots should know a bit about their properties and use.

✓ Helmets

One of the most important items in your equipment bag is a helmet. There has always seemed to be a philosophical side to arguments pro and con to wearing helmets, whether in the sports of rock climbing, motocycling, paragliding or many others. Perhaps this is because a helmet seems so confining of our basic personal space. However, there is no doubt that helmets save lives. It seems to us that a pilot who enjoys flight would realize that maximum protection on the thinking end of his or her body is sane and sensible.

Any pilot with a good head on his or her shoulder should protect it with a good helmet.

It certainly makes good sense to us to recommend a full-face helmet. The modern ones offer more protection than any other kind, they are lighter and they afford better peripheral vision. Don't leave the ground without one.

A good, safe helmet should not have ridges or holes, for they can catch on irregularities and prevent the helmet from glancing off whatever it hits. Bicycle, hockey and rock climbing helmets are not adequate for paragliding. A helmet should be fairly rigid. You can check for this factor by pushing it in at the sides and from front to rear. A proper helmet will have crushable foam to dampen the shock of a blow. A 1-inch (2.5 cm) layer of foam is usually the minimum considered adequate.

Finally, a helmet should fit snugly on your head. You should not be able to turn your head significantly inside the helmet. The chin strap should be snug, with room for only one finger between the strap and your chin. If a helmet is a bit too large, you can add some high-density foam inside the inner liner to form-fit your head. Likewise you can shave a bit of the helmet's foam away—not too much!—to get a better fit. A proper helmet uses the foam to absorb shock, the hard shell to protect from sharp objects and deflect blows, and a snug fit to maintain the protection in the event of repeated blows.

Anything worth doing... requires a helmet.
— Stan Koszelak

Take care of your helmet as you would any other part of your equipment. Never sit on your helmet or throw it to the ground. Inspect it as you would your harness or glider. If it is dented or cracked, replace it for it has lost much of its protective purpose. Also make sure the chin strap is sound and secure.

Protective footwear can help prevent a twisted ankle.

✓ Protective Footwear

Since you land on your feet every time, you should give some attention to protecting your feet and ankles. Most highly recommended are the boots specifically designed for paragliding and available through paragliding shops and magazines. These boots have thick soles to prevent bruises when landing on rocks, and stiff ankle support to help prevent twists and sprains.

These boots aren't cheap, but they can pay for themselves through medical bills avoided or peace of mind when landing in rough ground cover. This consideration is especially important when you begin flying varied sites or going cross-country.

✓ Hook Knife

A hook knife consists of a handle with a razor blade armed hook designed to catch lines or webbing to cut them efficiently and readily. The idea is to use the hook knife in an emergency to escape from your canopy or reserve parachute. In case of a water landing a hook knife may save your life. In some situations, a diver's knife is recommended in order to poke a hole in the canopy to breathe if it lands on top of you. Most likely a hook knife is preferred in moving water and a normal blade in still water.

If you ever throw your parachute with a ground wind present, you will be a believer in the wisdom of having a hook knife on board. Your parachute can take you for an unpleasant drag across the (rough) countryside. One pilot was dragged across the Australian plain and was only stopped by a lone windmill or else he may still be heading across the Outback! Note that you can't reach a line to collapse the reserve canopy, so a hook knife may be your only salvation in this situation. Use it to cut the bridle and save your body. Hook knives are sold along with other paragliding equipment.

Risk Management

Your equipment is one of the three items to assess in your risk management process. This is an important part of a safe attitude and thus safe paragliding practice. The way to apply risk management to your equipment is to have a regular schedule of care, maintenance and checking.

The most frequent check occurs every time you fly, in the form of a preflight. Also, you should have a seasonal maintenance program and a longer-term replacement program. Finally, you should have a policy of carefully checking for damage after every mishap and replacing any suspect part.

All this attention to equipment is only rendered effective if it is done on a regular schedule. The best way to assure regular attention to your equipment is to use a logbook to record your procedures. Write down all work you do on your equipment, and record the date so you can see how closely you stick to a reasonable schedule.

Club-organized maintenance or parachute repacking clinics are ideal ways to get everyone thinking about the safety of their equipment. That's risk management.

Summary

The subject of equipment used in paragliding could fill a whole book in itself, but still all the information would not be covered adequately since new developments occur all the time. However, we have hit on the main points for a pilot improving in skills and looking to upgrade his or her equipment.

There are many gliders to choose from. The ideal practice is to fly any glider (or harness) that you intend to buy. However, if your skills are not developed to the point that you can judge the nature of a particular glider, you must trust the judgement of those more experienced than you. This is where reputable shops come into their full importance, for they not only offer advice but also can find you the glider to fit your budget and skill level. The one good thing to know is that all recent designs of the same category fly well with very little differences in performance and handling.

Paragliding is truly a cooperative effort between pilot and glider to capture the air's energy. Once you are in harmony with your equipment, you can go on to find more possibilities in the nuances of the air's ebb and flow.

Assessing risk includes air traffic, conditions and terrain considerations as well as equipment and pilot proficiency.

TEST YOURSELF (*Answers in Appendix II*)

1. Name three (or more) things that can go wrong on a paraglider with use, abuse and time.
   ____
2. What repairs can you do on a canopy?
3. When is it necessary to replace a line?
4. Snow or debris in a wing can make it fly faster because of the added weight.
   True or False
5. The most important matter when buying a glider is:
   a. Cost
   b. Color coordinating
   c. Country of manufacture
   d. Size
   e. Skill rating
   f. Availability
   g. Intuition
   h. d and e above
6. If a parachute is too small for your size, your descent will be ____________.
7. A parachute should be repacked at least every ____ months.

A modern canopy and harness provide a pilot with comfort, safety and great flights.

Chapter 12

HOW OUR WINGS WORK

"Explain all that," said the Mock Turtle. "No, no! The adventures first," said the Gryphon in an impatient tone: "explanations take such a dreadful time."

— Lewis Carroll

In your earlier experience you learned some of the basics of how a paraglider flies. Now that you have felt the forces of flight and gained more familiarity with gliders in general, it is time to learn more about how they work. The proper name for this study is *aerodynamics*. Aerodynamics is a vast science and we cannot hope to do more than provide a simple explanation of the principles of flight in this book.

A paraglider is unique in the aviation world because it is flexible and collapsible. Thus, we will concentrate on how a paraglider operates. In the course of our presentation we'll discuss what the limits of performance and safety are and how to maximize controls. Our interest is to relate what we feel in the air to the causes. You'll find that the more you learn about our fascinating aircraft, the more you'll be integrated with your personal wings.

PERFORMANCE FACTORS

There are a number of things that contribute to the performance of our gliders, the most important of which is to produce lift at the minimum drag expense. As we explore the wing's behavior we'll show how the interplay of lift, drag, control and stability is what largely determines how our gliders are designed.

✓ The Airfoil

If we take a cross-section of a paraglider wing along a line parallel to the cell seams we get a curved figure we call an airfoil (see figure 12-1). We show the airflow around it and identify a few terms. You don't need to know these terms to fly safely, but you will occasionally hear discussions where they are used and can refer to this material so you're not left in the dark. We will relate some of these ideas to practical matters below.

In Chapter 3 you already learned that the *chord* is the line drawn from the furthest forward point (leading edge) to the furthest back point (trailing edge). Next we add the concept of *mean camber*. This is the line that connects all the points along the airfoil that are midway (the mean) between the upper and lower surface. The significance of this line will become clear when we discuss stability below. Also, the *high point*—the highest point on the upper surface— and position of *maximum thickness*—the fattest part of the airfoil—will come into our stability discussion.

Before we move on, note the smooth flow of air above and below the airfoil. There is one point—the *stagnation point*—where the flow seems to stop. It does for the most part, and this is the point of greatest dynamic pressure of the air.

Now let's learn more about lift, drag and performance.

Figure 12-1: Airfoil Characteristics

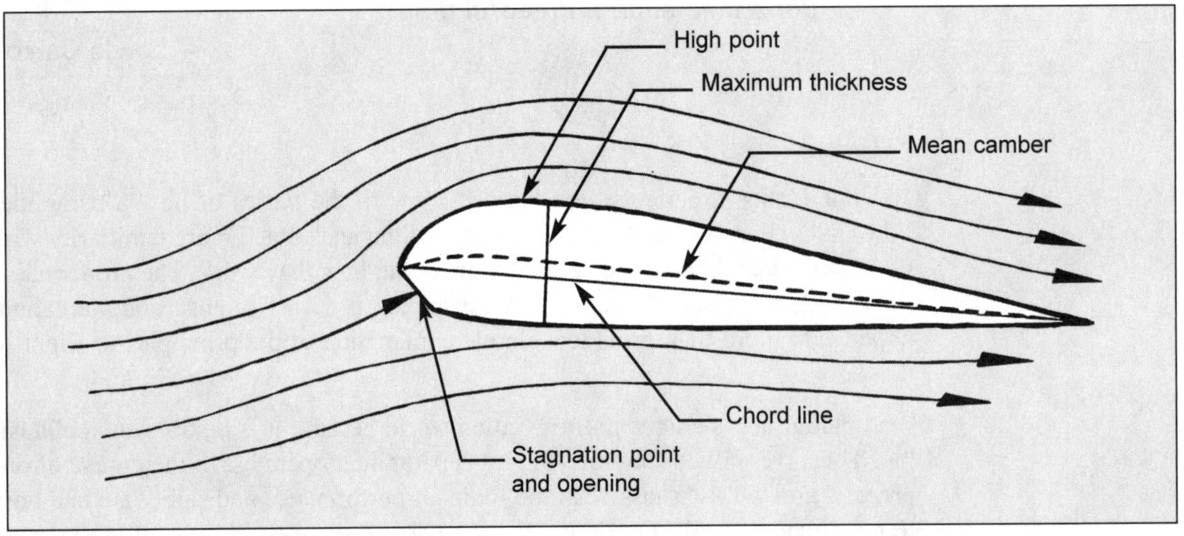

✓ Flight Angles

Now we'll use an airfoil section to define more exactly the different aspects of flight. In figure 12-2 we show an airfoil section gliding through the air. When the air flows around it, certain forces are felt on the airfoil as it deflects the air. These forces retard its descent and allow it to plane forward along a specific path. We call this path the *glide path*, and this path will always be the same for a given airfoil at a given angle of attack as long as the air is completely still. If the air moves upward (lifts) or comes from behind the airfoil, the path will be flatter. If the air moves downward (sinks) or comes from in front of the airfoil, the path is steeper.

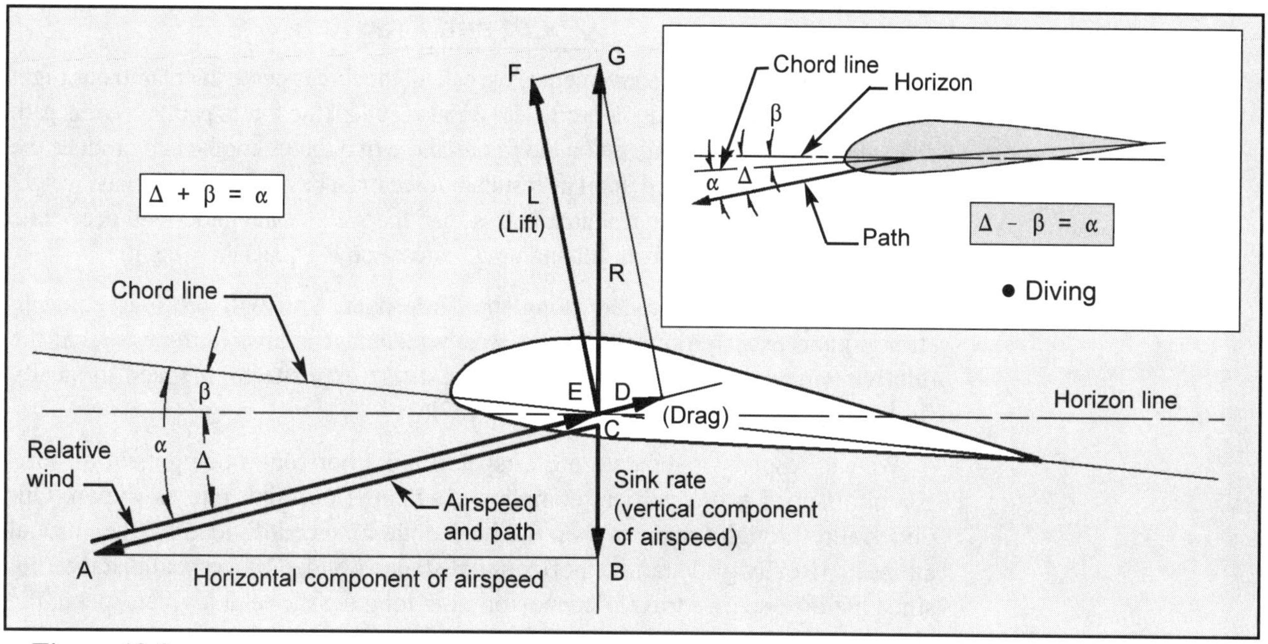

Figure 12-2: Lift, Drag and Glide Angle

The attitude of the airfoil (β) is the angle it makes with a horizontal line. We always use the *chord* of the airfoil to measure this angle. The angle of attack (α) of the airfoil is the angle it makes with the relative wind or glide path. The glide angle (Δ) is the angle between our flight path and the horizontal line. Thus we see that attitude angle plus glide angle always equals our angle of attack ($\Delta + \beta = \alpha$). If we alter our angle of attack we also alter our glide path and thus glide angle.

If our airfoil is tilted downward (as when diving or in a surge), the relation still holds. In this case, the chord line is below the horizon so the attitude angle is negative as shown in the inset of figure 12-2. So the attitude angle, β subtracts from the glide angle, Δ to give us angle of attack, α. That is $\Delta - \beta = \alpha$.

Some people think that a paraglider wing needs to be tilted downward in order to fly forward. However, that is not the case. You can prove this to yourself by watching hang gliders in slow flight. You'll see that they are very definitely angled upwards. The fact is, the airflow around the front of an airfoil creates some forward thrust as shown in figure 12-3. This forward thrust is balanced by the drag as the wing flies steadily along.

Figure 12-3: Forces on an Airfoil

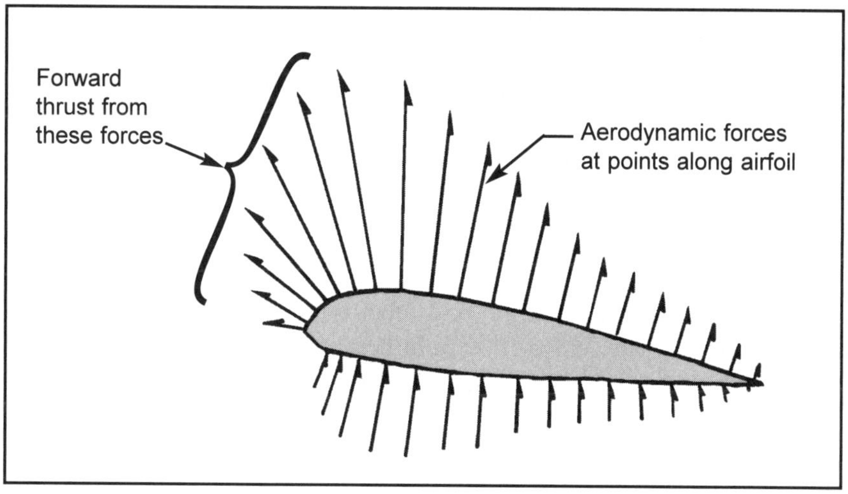

It may appear that a paraglider is tilting nose down (and it does at lower angles of attack) in order to go forward. But this appearance is due to the fact that on a paraglider the glide path angle (Δ) is very close to the angle of attack (α). For example, a 7 to 1 glide is about 8° which is a typical airfoil angle of attack at best glide. Thus, β is zero and $\Delta = \alpha$.

✓ Lift and Drag

By convention and for convenience, we call all the forces perpendicular to our flight path *lift forces* (Labeled L in figure 12-2). We also call all the forces parallel to this path *drag forces* (D). As you can see on the figure, these two forces combine to produce the total force or *resultant force* (R). The resultant force must be exactly equal to the weight of the entire system. If the resultant is less than the weight, the glider will accelerate along the glide path until both lift and drag increase so R equals the weight.

Now look at our airspeed along the glide path. You will see it is equal in strength and exactly opposite to the *relative wind.* It is always the case that the relative wind, or air the wing "feels", is exactly opposite to airspeed in steady flight.

We can resolve or separate our airspeed into a horizontal component or horizontal airspeed and a vertical component which is our sink rate as shown. Our glide ratio through the air is then the horizontal airspeed divided by the vertical airspeed. (Really glide ratio is horizontal distance divided by vertical distance but since the distance we travel depends on how long we move at a given speed, the airspeed ratio is the same as the distance ratio.)

Finally, if we look closely at the figure we can see that triangle ABC is similar to triangle EFG. This means that they have the same angles and thus have proportional sides. From this relation we can see that:

$$\frac{\text{horizontal airspeed}}{\text{vertical airspeed}} = \frac{\text{Lift}}{\text{Drag}} = \frac{L}{D}$$

This is why our glide ratio is often referred to as L over D or L/D.

To have good performance we want to increase the glide ratio or L/D ratio. To do that we must reduce D, the drag for a given angle of attack or airspeed. On a given glider, the only way we can change the L/D factor is to change our angle of attack (unless we reduce our drag by pulling our arms in and pointing our feet into the airflow). In figure 12-4 we show the forces, airspeeds and L/D ratios at different angles of attack.

At a higher angle of attack, the airspeed is slower and the glide ratio is 5 to 1 for example. Let's assume this is our minimum sink angle of attack. At a lower angle of attack (let's say best glide), all speeds have increased, including the vertical speed (sink rate). But notice the proportion of horizontal speed to vertical speed has increased, as has the ratio of lift (L) to drag (D) in an equal manner. Our L/D ratio has improved and our glide path is better. That's why at minimum sink flying speed we descend slower but don't go as far as we do at best glide speed (see figure 9-6 for a comparison of flight progress at different speeds).

Now, if we dive the glider, we don't improve our glide, because *form* drag goes up significantly. (We explain about this and other types of drag in the next section.) The L/D ratio is lower. If we slow from best glide speed the *induced* drag goes up, so best glide always occurs where the total drag is least.

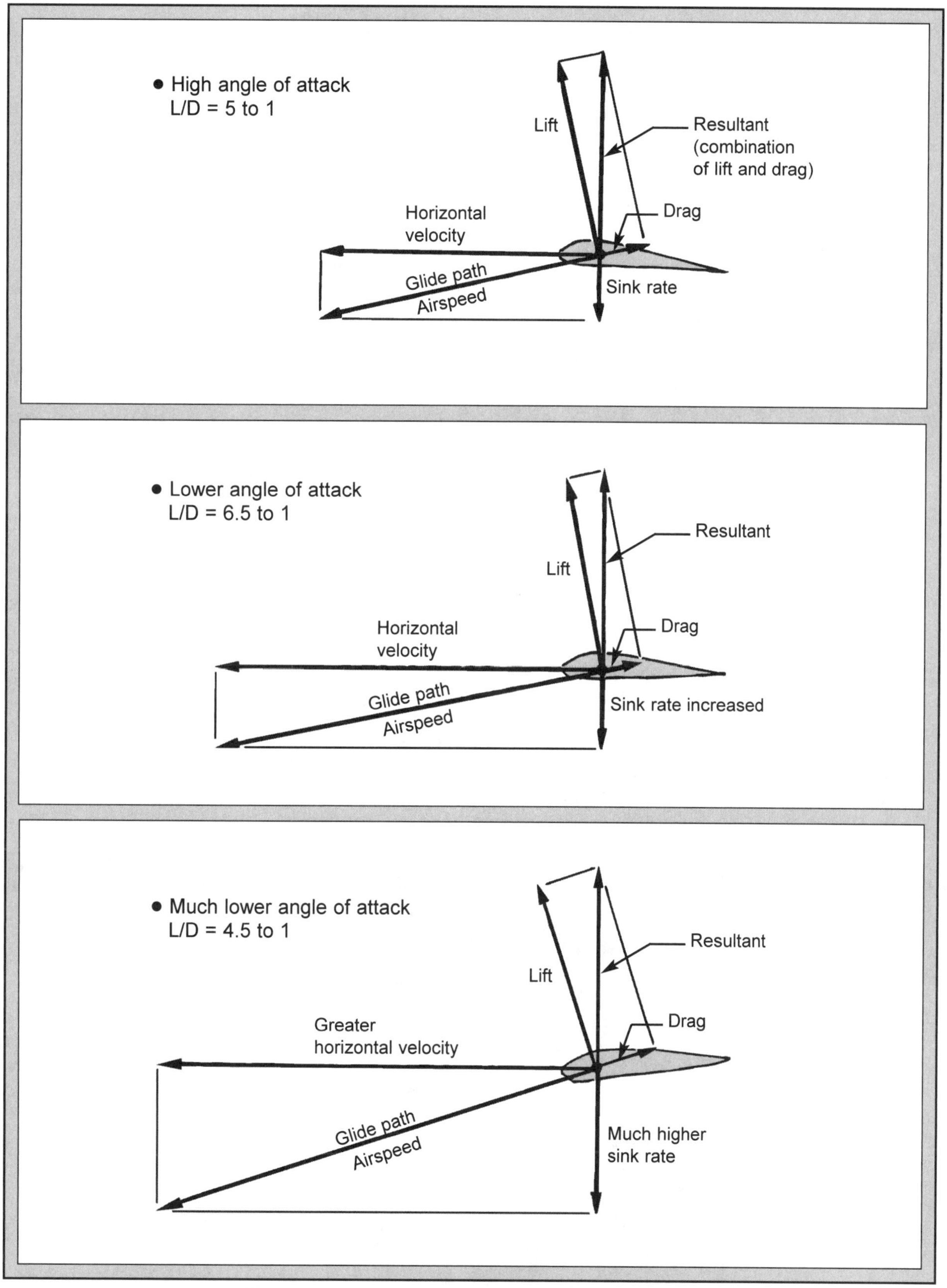

Figure 12-4: Effects of Changing Angle of Attack

✓ Drag Types

Now let's look at where this drag comes from on your flying system so that you can understand the limiting factors to performance. There are two main classifications of drag: parasitic and induced.

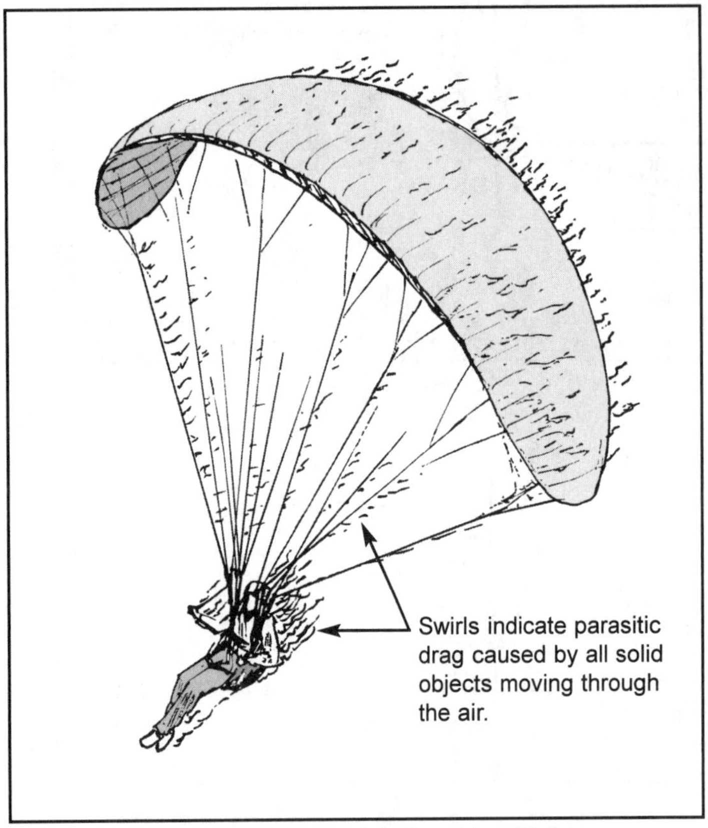

Figure 12-5: Parasitic Drag in Flight

▲ *Parasitic drag* is that produced by solid forms moving through the air, and is essentially due to all those air molecules hitting an object head on (see figure 12-5). We can further break down parasitic drag into *profile drag* and *form drag*. Profile drag is the friction and blockage of the airflow which occurs due to the wing itself. *Form drag* is the drag forces due to everything else, including the pilot, harness, risers and lines. Designers try to reduce profile drag by using low drag airfoils, clean sail surfaces, small openings and curved tips. Form drag is reduced with a streamlined harness and helmet, a minimum of lines, and a proper body position.

▲ *Induced drag* is a loss of efficiency due to the escape of airflow out the sides of the wing. This sideways flow produces *wing tip vortices*, as shown in figure 12-6. The best way to reduce wing tip vortices is to use a wider and thinner wing (more span and less chord). This is known as having a higher aspect ratio (see box on next page). Sailplanes reduce their induced drag to a bare minimum by having such a high aspect ratio.

Your body and harness are the biggest drag contributions to the flying ensemble. Keep your profile to the wind small if you wish to improve performance.

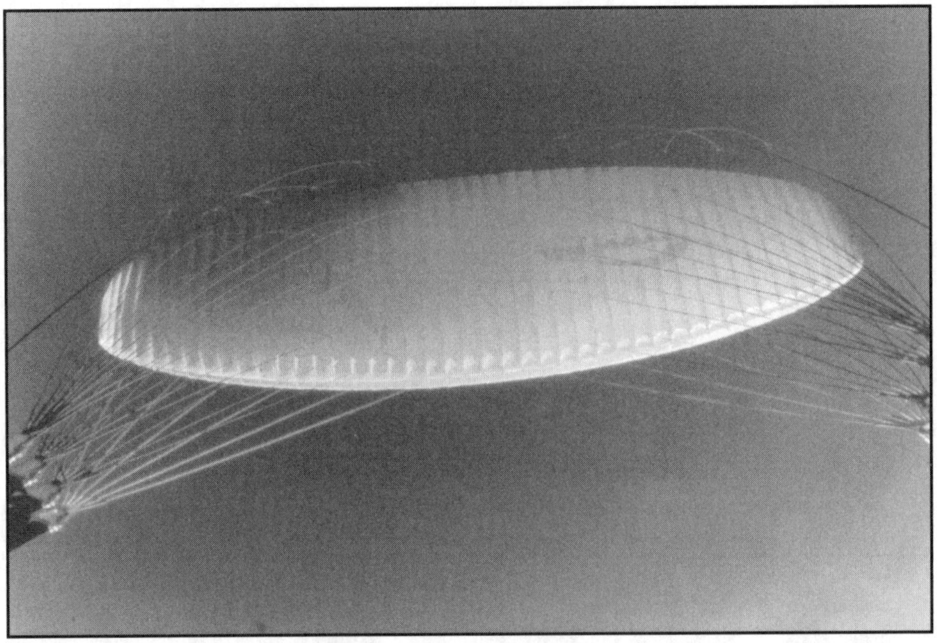

Drag comes from all parts of the flying system. There is considerable drag on the lines as can be seen from their curved shape in flight.

Aspect Ratio

The ratio of a wing's width (span) to its fore and aft dimension (chord) is the aspect ratio. However, with an elliptical-shaped wing such as a paraglider's it is difficult to decide what the chord is. So we use a trick. Since area (S) of any rectangle equals the width (b) times the chord (c) we have S = bc. But aspect ratio (AR) equals b/c. So we can solve for c in the first equation and substitute it in the second equation. Thus AR = b^2/S (aspect ratio equals the span squared over the area). Paragliders typically have aspect ratios between 4 and 6.

The aspect ratio of a wing lying flat on the ground is higher than it actually is in flight, since the ends get pulled in considerably when the wing assumes its familiar arc shape. The area of a wing viewed from above in flight is known as its *projected area* (it is the area of the shadow it would project if it were suspended just above the ground). Using the projected area in the aspect ratio formula gives the *projected aspect ratio*. The projected aspect ratio is the important thing to note, for that is the factor that helps determine flying performance.

Increasing aspect ratio on a paraglider has its limits, for wider spans or smaller chords create longer, thinner wings which are more prone to fold. There is a practical limit to a paraglider aspect ratio so designers seek to reduce drag elsewhere.

✓ Wing Loading Effects

Wing loading is the amount of weight our wings carry per unit of area. This factor has some important bearing on performance. We calculate wing loading by dividing the total weight of the flying ensemble—pilot, glider and gear—by the surface area of our wing. Typical wing loadings on a paraglider are .5 to .7 pounds per square foot (2.4 to 3.5 kg/m^2) for lower performance glider and .6 to 1 lb/ft^2 (2.9 to 4.8 kg/m^2) for higher performance models. The reason we load up higher performance gliders more is that they are more efficient at producing lift at a minimum of drag.

Here are the effects of changing wing loading:

Figure 12-6: Wing Tip Vortex —The Source of Induced Drag

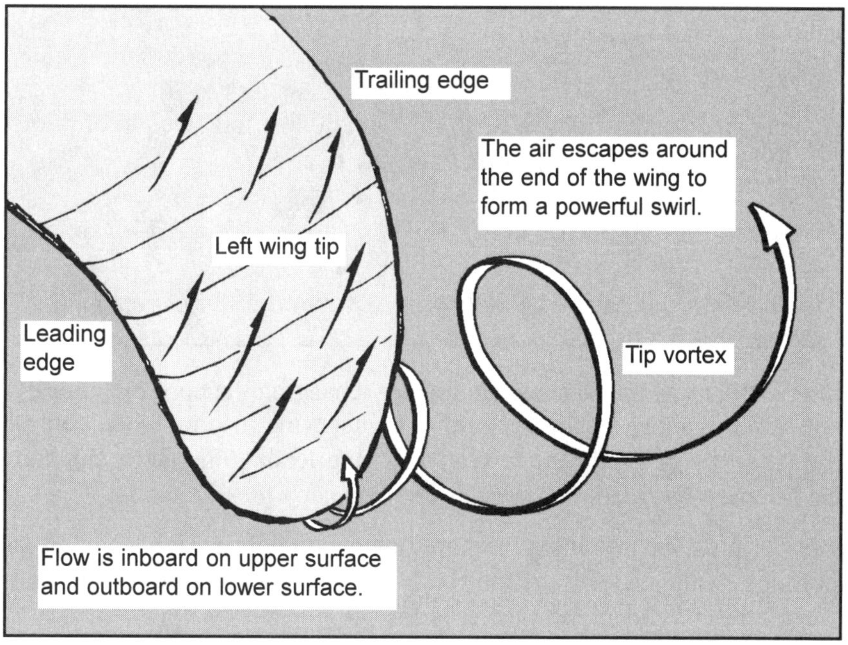

▲ *Airspeeds* — As you increase your wing loading, you increase your speed at any given angle of attack. As a rule of thumb, at the slower flying speeds you will add about .4 mph for every 10 pounds you add in weight (.7 km/h for every 5 kg). All speeds are proportionally increased including that for stall, minimum sink and best glide ratio. Even though *the angle of attack and control position remain the same* for all these cardinal speeds, the resulting airspeed increases as we load up our wings.

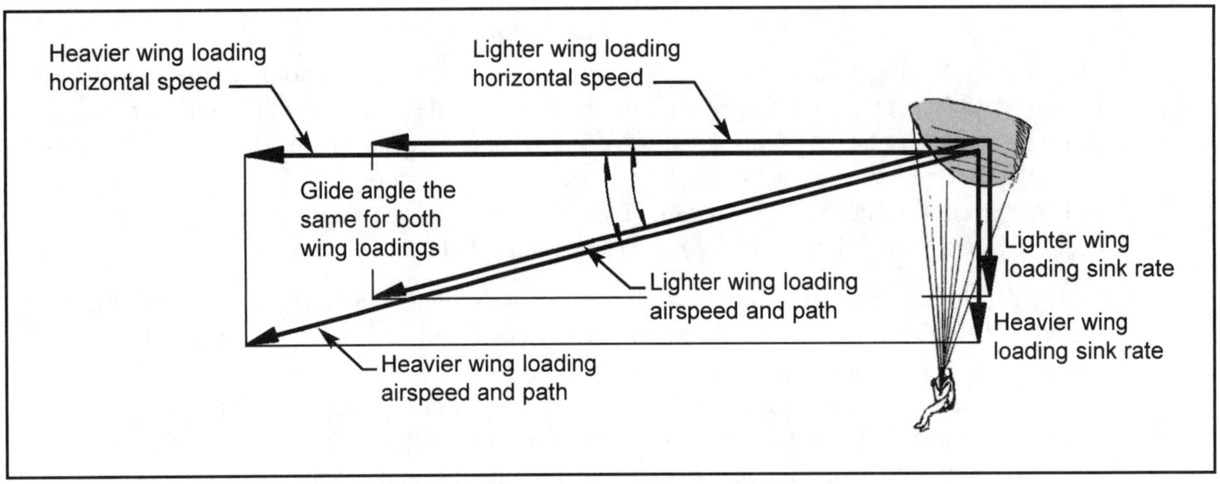

Figure 12-7: Wing Loading Effects

Because our sink rate is simply our vertical airspeed, it increases if our wing loading increases, as shown in figure 12-7. However, our glide ratio is the horizontal airspeed divided by vertical airspeed and this ratio does not change since all airspeeds increase proportionally. We summarize wing loading effects in the box below.

Wing Loading Effects on Glide and Sink

- Increased wing loading increases our sink rate about 4.5 feet per minute for every 10 lbs (0.023 m/s per 5 kg) added.
- Decreased wing loading does the opposite.
- Changes in wing loading do not alter our glide ratio in still air at any given angle of attack, including best glide.

For those interested in working with wing loading effects, the formula is:

$$\frac{V_N}{V_O} = \sqrt{\frac{W_N}{W_O}}$$

Where: V_N = new airspeed
V_O = original airspeed
W_N = new weight
W_O = original weight

For example, if our original airspeed and weight were 16 mph at 200 lbs (total weight of pilot and glider) and you added 20 lbs for a new weight of 220 lbs, the relation is:

$$\frac{V_N}{16} = \sqrt{\frac{220}{200}} \text{ so } V_N = 16.8 \text{ mph} \qquad \text{In metric units: } \frac{V_N}{25.6} = \sqrt{\frac{W_N}{W_O}} \text{ so } V_N = 27.4 \text{ km/h}$$

You can use this formula to find sink rate changes as well by using your original sink rate for V_O.

▲ *Handling* — Wing loading affects handling by changing the control authority. Usually the glider is more responsive with a higher wing loading while control forces are higher. So generally the heavier the wing loading the better the handling. Also heavier wing loading promotes better canopy pressurization.

Lesser performing and training gliders are flown with lighter wing loadings, so their takeoff and landing speeds are slower. Handling problems are not an issue because these types of gliders have lower spans.

Competition pilots sometimes add ballast (up to 20 lbs or 9 kg). They take a very small penalty in sink rate for big gains in penetration and speed.

▲ *Altitude Effects* – You know that as we go higher the air gets thinner. Therefore our wings have to fly faster to create the necessary amount of lift. They do this automatically for every angle of attack we set. The airspeed effects are exactly the same as when changing wing loading. All our speeds increase as we get higher and they all decrease as we get lower. The amount of change depends on the air's density change which in turn depends on the temperature. However, we can generalize and say that speeds increase by about 1.5% per thousand feet (300 m) of altitude gained.

Just like with the wing loading situation, horizontal speed increases, vertical speed (sink rate) increases but the proportion remains the same. Thus, glide ratio through the air remains the same at all altitudes. Unlike wing loading effects, however, handling doesn't change with altitude.

✓ Design Effects

A designer has to make choices which include glider size, aspect ratio, number of cells and lines, airfoil shape, wing shape, curvature, opening positions, and so forth. Clearly, paraglider development has been a matter of experience gained through trial and error. But many aerodynamic principles apply. So here we'll present a few design factors so you gain a better understanding of your wings.

▲ *Wing shape* – The wing shape when viewed from above is known as *planform.* It has been proven long ago in aerodynamics that a wing with a planform the shape of an ellipse is the most efficient. So why aren't airplane wings elliptical? Because they would be too hard to manufacture with all the compound curves required. But not so with paragliders. You just crank up the computer and the laser cloth cutter makes the ribs and surface pieces to be sewn. So most paraglider planforms are elliptical or an approximation thereof.

Recent NASA studies have shown that a modification to a regular ellipse is even better. If the leading edges are curved back from the center to the tip, slow speed performance is improved a bit. Figure 12-8 shows such a planform. Recent designs of paragliders, hang gliders and sailplanes have incorporated these ideas. Fast fish and birds have used the design for eons.

Figure 12-8: Curved Tip Planform

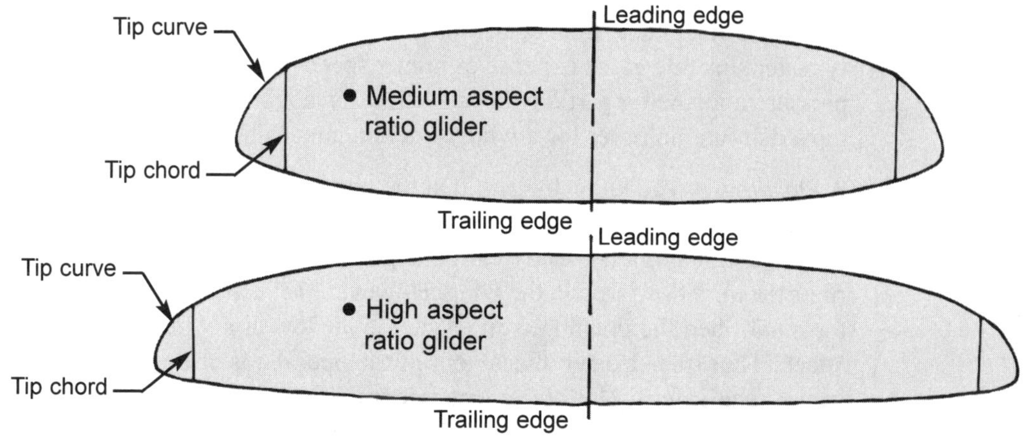

We have previously mentioned the performance benefit of having long, thin (high aspect ratio) wings. Wings of two different aspect ratios are shown in figure 12-8.

A designer is an artist working within the confines imposed by the limits of materials and natural laws.

The final shape factor we'll note is the curve arc of the wing when viewed from the front or rear. This bow is necessary for stability as we'll show below, but too much bowing hurts performance. The reason can be seen in figure 12-9. The areas out at the tips produce lift which is directed outward more than upwards. Thus, they pay a drag penalty without much lift benefit. Of course, this outward pull is necessary to keep the wings spread and inflated. In addition, there is an actual reduction in tip losses—and thus induced drag—with down-turned tips (look at a seagull in flight to see a natural example). Our conclusion is that the curve of the wing helps performance at low speeds but hurts at higher speeds.

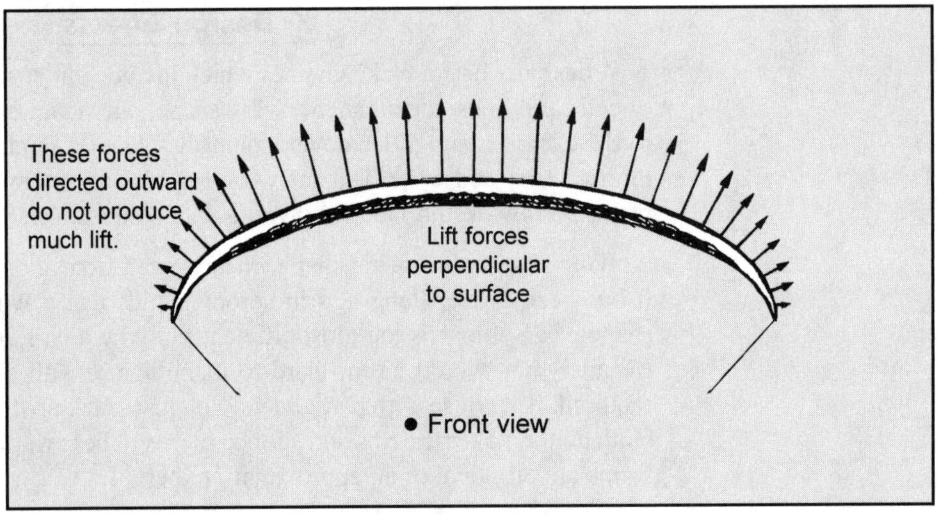

Figure 12-9: Effect of Canopy Arc

▲ *Airfoil* – The airfoils used on paragliders are relatively stable. That means they want to maintain a reasonable angle of attack and not tuck under. This feature is desirable with a non-rigid wing. What makes it so is the far forward high point and relatively thin profile. The high point on a typical paraglider airfoil is around 20% of the chord back from the leading edge whereas on an airplane or sailplane it is 25% to 40% back.

On the other hand, thicker airfoils tend to fly more slowly and produce a slightly better sink rate at the expense of higher speed performance. They also maintain pressurization better partially because they have more volume of air that has to be moved around in order for a wing to deflate and fold.

▲ *Openings* – You know by now that the openings in the leading edge are important to inflate the wing. They also cause drag, especially if they are large because the stagnation point is exactly that: one point. Any other point on the opening disrupts the air's flow around the wing. However, if the designer makes the openings too small then the opening won't face the airflow at a wide variety of angles of attack. Thus, the size and placement of the opening is critical to canopy pressurization and safety. The placement is optimized for slow speed angle of attack because at higher speed the flow is faster and creates more dynamic pressure. Beginner gliders have large openings for safety reasons.

▲ *Lines* – We have already mentioned the desirability of having fewer and smaller lines from a performance viewpoint. The problem with fewer lines is greater loading on the lines, of course, which limits how thin they can be. You can readily see a tradeoff in effect. Furthermore, fewer lines makes it harder to control the smooth airfoil and curved shape of the canopy. So the designer makes the best compromise for safety and performance. A design breakthrough occurred when designers began using cascaded lines (e.g. one line splitting into two, then three). This trick allowed fewer lines running full length from riser to canopy.

The use of more cells does essentially the same thing that more lines do: controls the airfoil shape better. High performance gliders may have twice the number of cells that beginner gliders have. The fabrication labor is twice as well. One trick designers use is to slant the cell walls (ribs) so that some sideways pull is put on the upper surface. Such a design method allows for fewer lines while maintaining the same canopy shape and pressurization. You will see this setup mainly on higher performance canopies.

Here's a summary of design performance factors:

Flying Regime	**Performance from Design Factors**
Slow speed flight (Minimum sink performance)	Thick airfoils with lots of upper surface curve and high mean camber. Large surface area.
Medium speed (Best glide performance)	Medium thick airfoils. Low drag lines and harness. Low drag tips pulled down as little as possible. High aspect ratios.
High speed (Penetration and high speed glide)	Thin low-drag airfoils. Low drag lines and harness. High aspect ratio does not help speed performance significantly because of the excess drag on the tips at low angles of attack.
Canopy integrity (Preventing folds)	Stable airfoil, low aspect ratio, large openings. Very curved canopy (as viewed from the front).

This glider laid out in a horseshoe with a wall shows how the cell openings meet the airflow.

PARAGLIDER STABILITY

Stability in flight is necessary for safety and comfort in all aviation. Paragliders have their own special stability problems and solutions. It is necessary for you to understand the basics of stability matters so you can feel more confident, but also so you know why alterations can be dangerous and equipment maintenance is so important.

✓ The Control Axes

We separate the different motions a wing can undergo into three separate axes so we have a clearer picture of what we are talking about. Note, however that a wing can turn around any two or all three axes at once. The three axes are shown in figure 12-10. Rotation through the lateral axis is known as *pitch* changes (the wing tilts leading edge up or down). By now you may recognize that when we change pitch we are changing angle of attack.

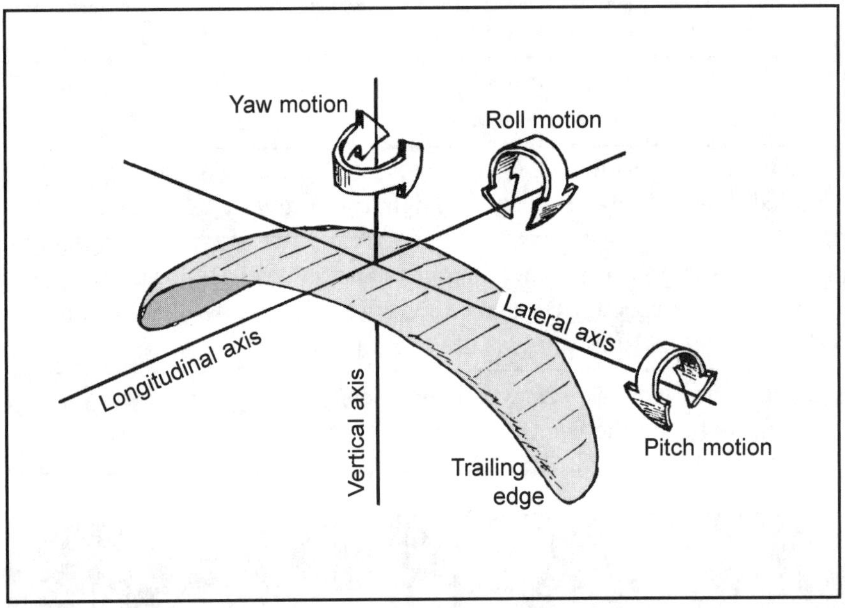

Figure 12-10: Wing Control Axes

Rotation about the longitudinal axis is known as *roll.* One wing tip goes up while the other goes down. We roll a paraglider when we bank it to turn. Finally we have rotation about the vertical axis known as *yaw*. We also yaw a wing back when we pull one side control to turn.

You know that a pilot can control the wing in all three axes, but to be comfortable to fly, a glider (or any aircraft) must also control itself somewhat in all three axes. That is what we call stability.

✓ Pitch Stability

Birds are unstable in pitch, but their quick tail reactions keep them flying just fine.

A paraglider achieves pitch stability through two means: the airfoil and the pendulum effect. We discussed the airfoil's role previously in Design Effects. The pendulum effect is due to our body hanging so far below the canopy. As long as the wing doesn't fold, our body seeks the low spot which sets the wing's angle of attack according to the suspension lines and our control input. However, our body can swing and move separately to the canopy. Thus, if the canopy is suddenly slowed (we hit the controls for example), our body keeps moving due to its momentum until the canopy stops it. But by then we have swung ahead, pulling the canopy, which has a tendency to overshoot our position (surge) as figure 12-11 shows. The result is a somewhat unstable situation that is only resolved by dampening the canopy. In other words, pendulum stability is only effective with slow, small changes in canopy

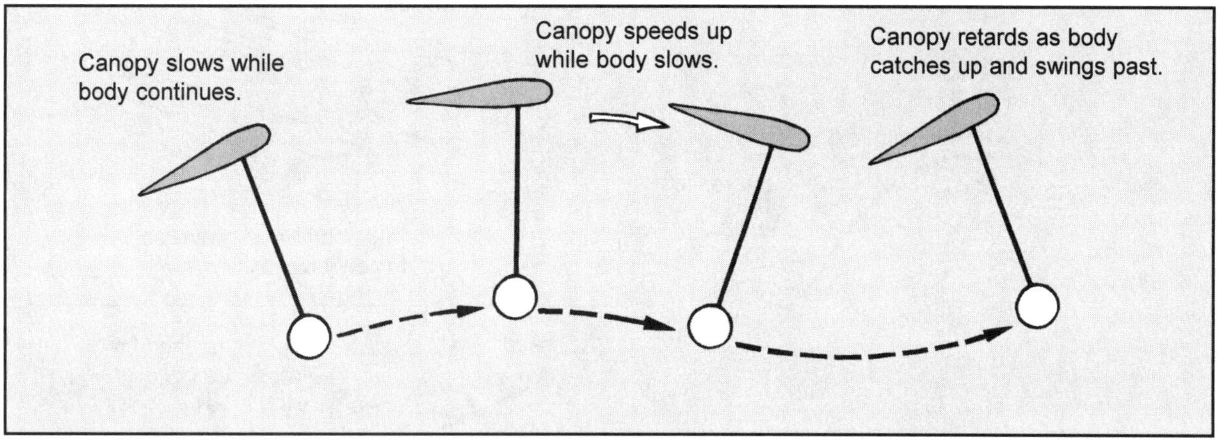

Figure 12-11: Pendular Stability and Instability

position or angle of attack. With larger or faster changes, the pendulum effect actually adds to the instability.

Fortunately, the canopy is dampened in its movement by its drag. However, that isn't always enough and pilot input is sometimes needed to stop a serious surge. Pendulum pitch stability is just barely adequate in a paraglider.

✓ *Roll Stability*

Roll stability also comes from the pendulum effect as you can see in figure 12-12. In this case, there is much more dampening in the canopy movement so it doesn't tend to overshoot its vertical position if it gets out of place and swings back. Only occasionally does a canopy go so far to the side that a tip folds, but this usually occurs in the process of doing aerobatic maneuvers. Roll stability is very adequate in a paraglider.

Figure 12-12: Roll Stability

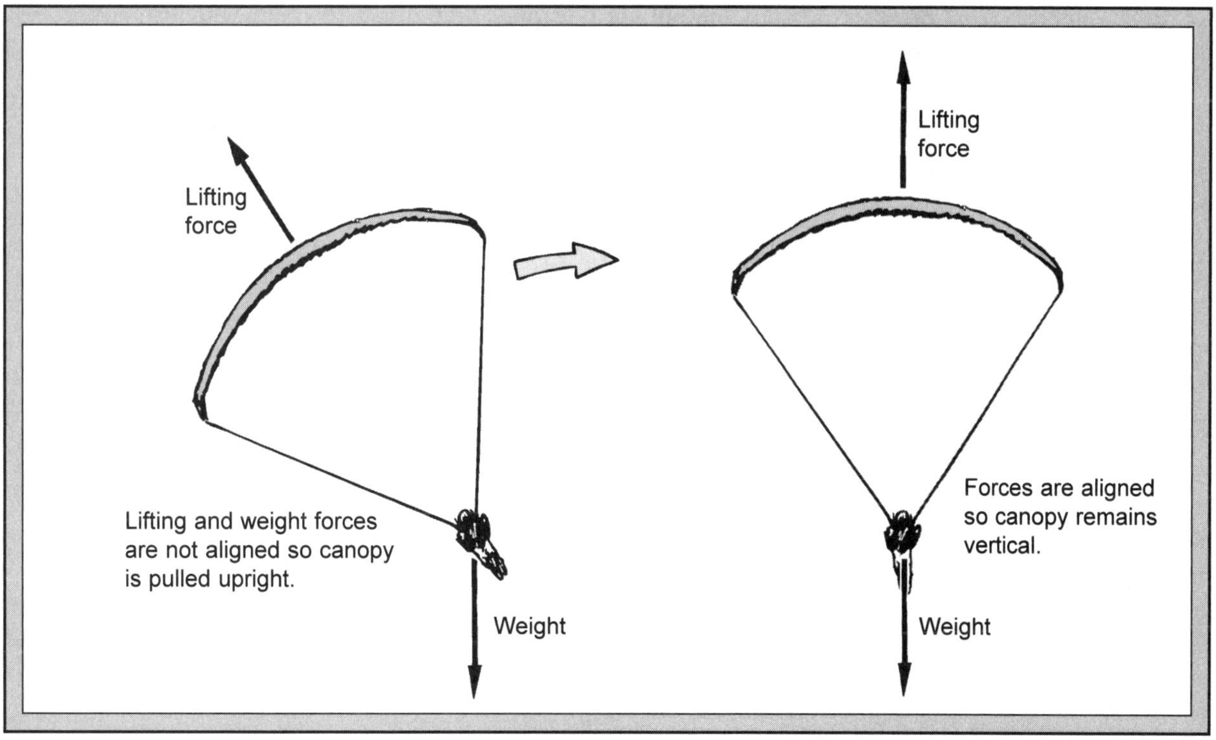

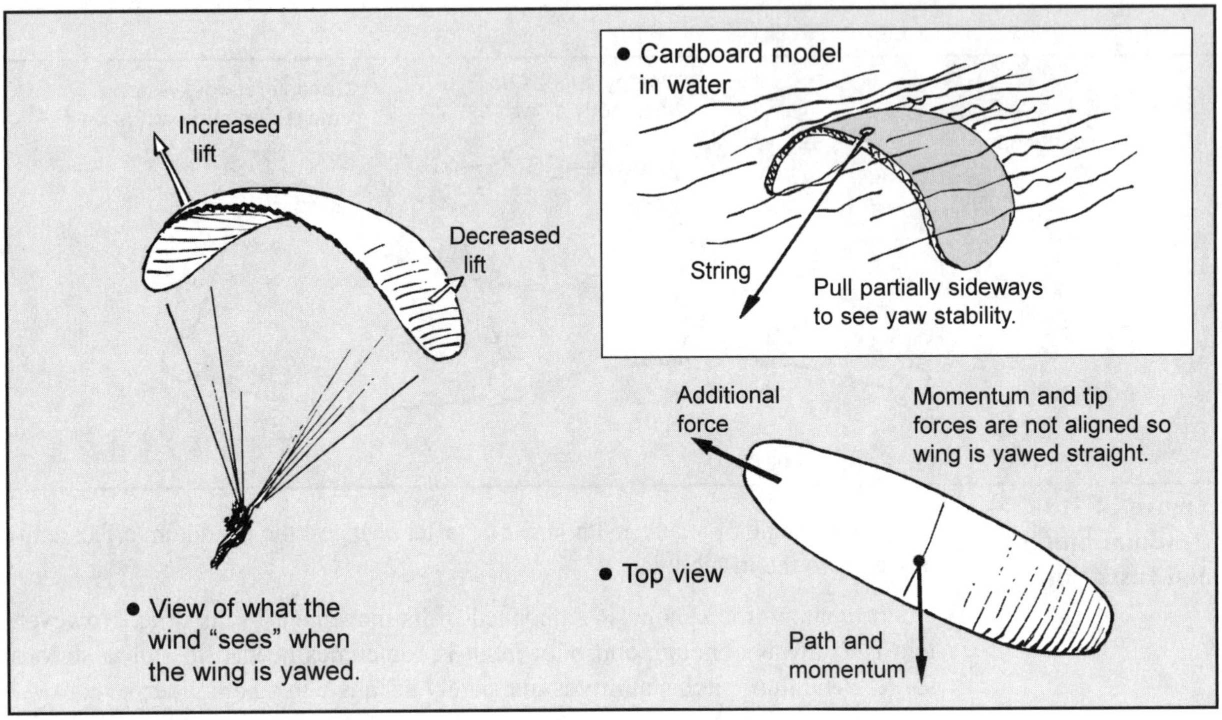

Figure 12-13: Yaw Stability

✓ *Yaw Stability*

Our gliders have good yaw stability as well, due to the curved shape as shown in figure 12-13. If a wing drops back, the airflow "sees" that wing tip at a much higher angle of attack so more lift is created on that tip while it's reduced on the other tip. Since that tip is now behind the center of pressure or gravity (where your body is located), its outward pull creates an imbalance that rotates the wing forward. In a greatly yawed condition, roll also takes place which adds to the restoring motion. You can experiment with this principle by making a little curved wing out of cardboard and pulling it through water on a string attached to the forward center of the wing (see figure 12-13).

All these stability factors combine to provide a wing that is pleasant to fly. As with most systems, too much stability results in poor control. The designer must again work his magic to give us the best compromise.

TURN EFFECTS

The last topic we'll investigate in our look at how our wings work is the way our gliders turn and what the limits are to turning. It is important to understand these matters so you can maximize your control efficiency but also remain within the limits of safety.

✓ *How a Turn Works*

Let's step through a turn and see what happens. Figure 12-14 illustrates the process. Before a turn is initiated the glider appears in (1) with the typical forces on its wing (lift and drag add up to the net resultant force, R). When the pilot pulls a control on the left side (2), more drag is created and the net resultant force

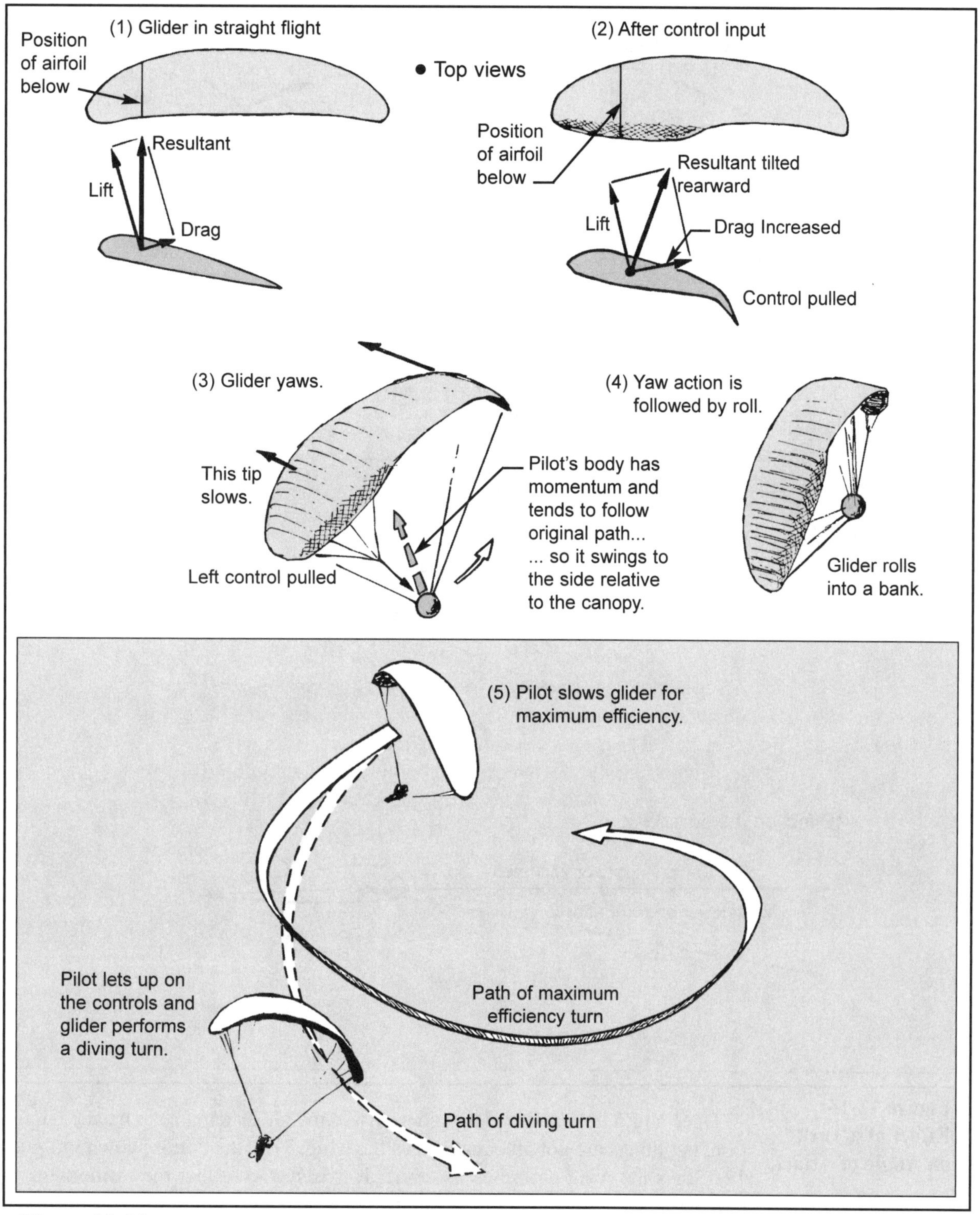

Figure 12-14: The Elements of a Turn

moves rearward and temporarily tilts backward. This backward force slows the wing and the other wing begins rotating forward (3). With the outside wing moving faster than the other, it develops more lift while at the same time the pilot's body swings away from the canopy. The combination of these effects rolls or banks the wing (4). This process continues until the pilot lets up on the control. If he adjusts for minimum speed, he will achieve a maximum efficiency turn (5). If not he will be in a diving turn which will become a spiral if he lets it continue.

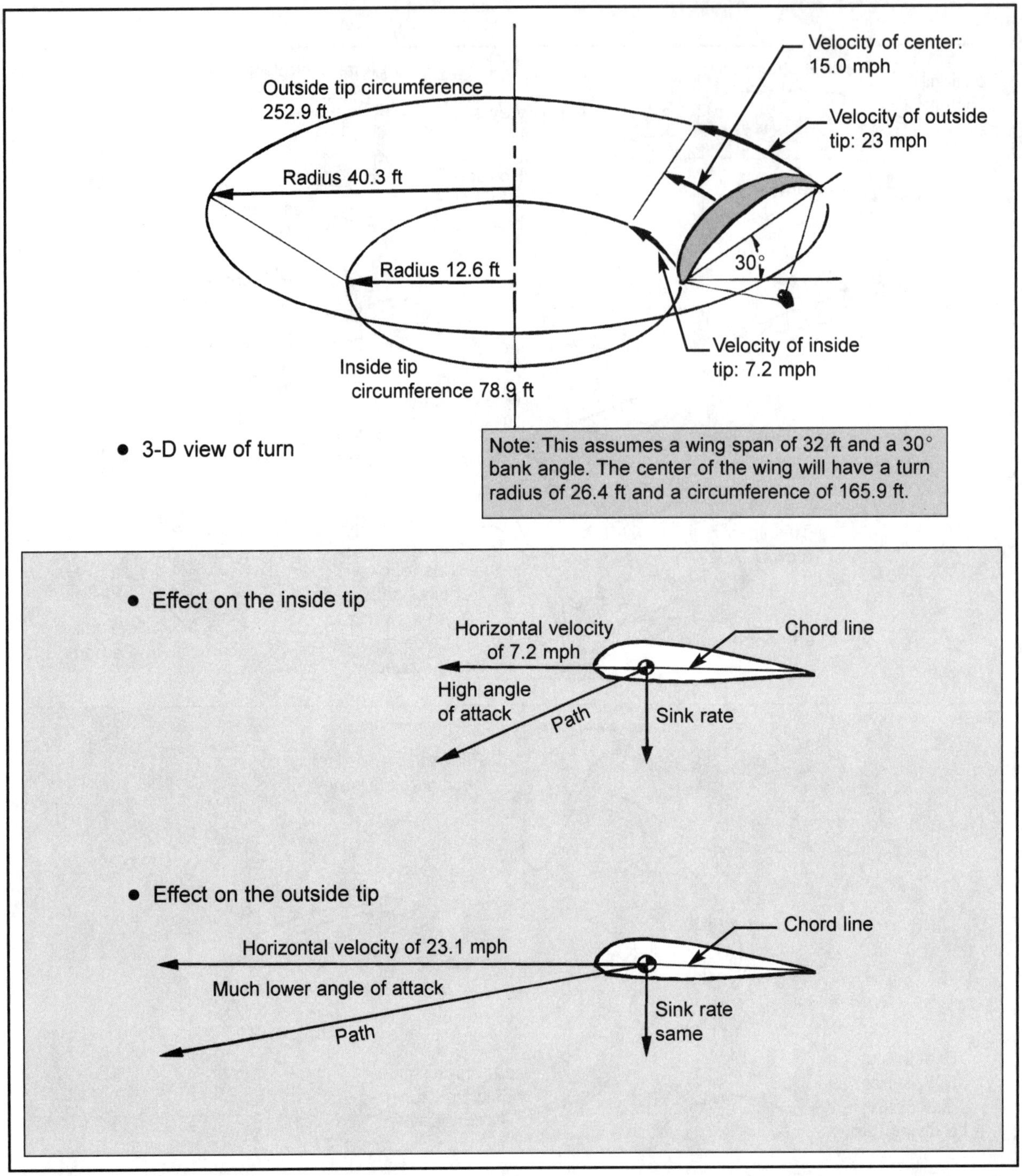

Figure 12-15: Effect of a Turn on Angle of Attack

Once a turn is established and coordinated the glider remains in balance in the turn, but things are not all equal across the wing. The situation is shown in figure 12-15. As the wing describes its circle, it is easy to see that the inside wing is moving much slower than the outside wing. Since they both have the same vertical velocity (sink rate) we can look at their relative wind and realize that the inside tip area is at a *much higher* angle of attack than the outside area. Thus, if we slow the wing to a stall, the inside wing will stall, never the outside wing.

From this discussion you can see why wings spin when they stall in a turn. Not only is only one side stalled and dropping back (a stall creates a large increase in drag), but the stalled area is way outboard so it has a long lever arm to act and rotate the glider. Spins are generally not friendly because the pilot has little control.

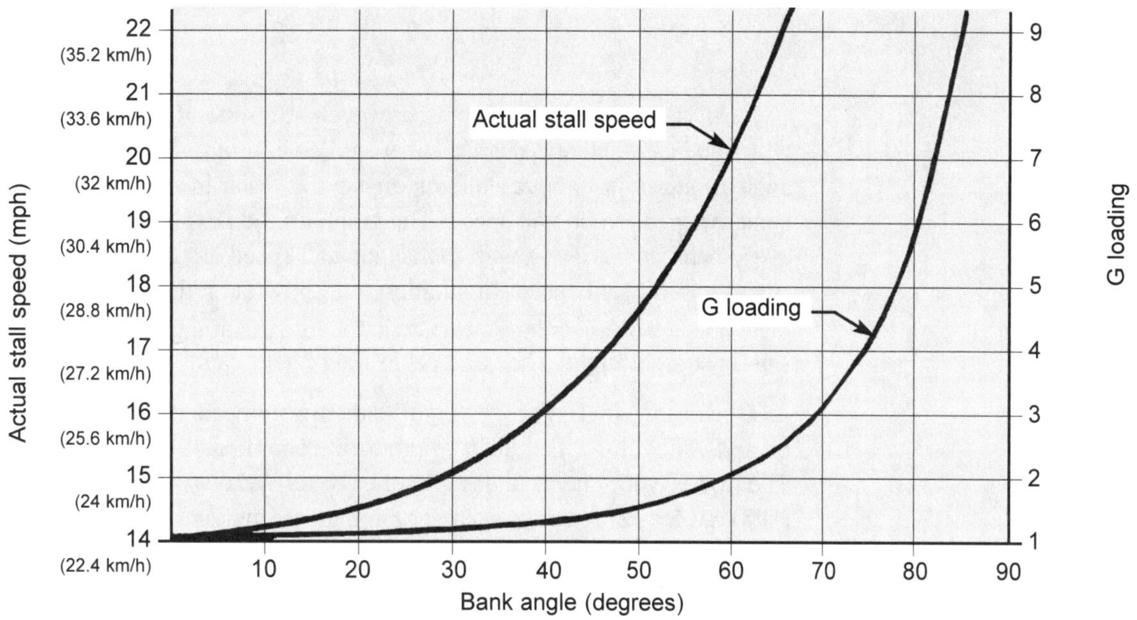

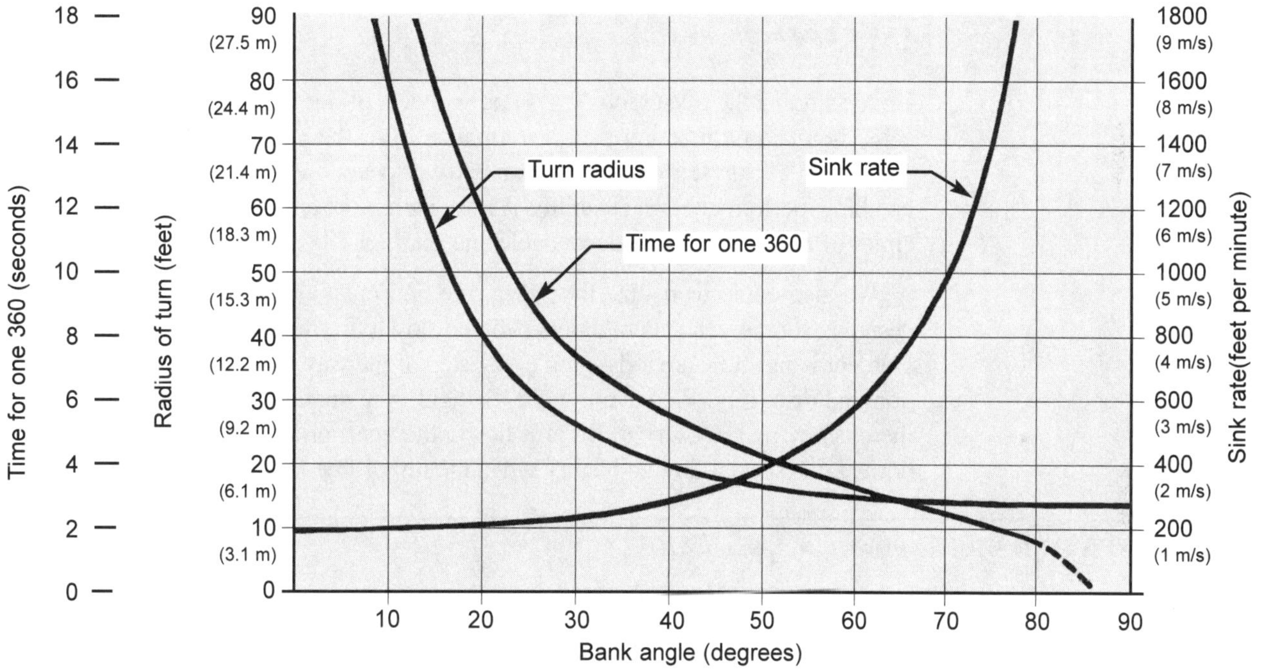

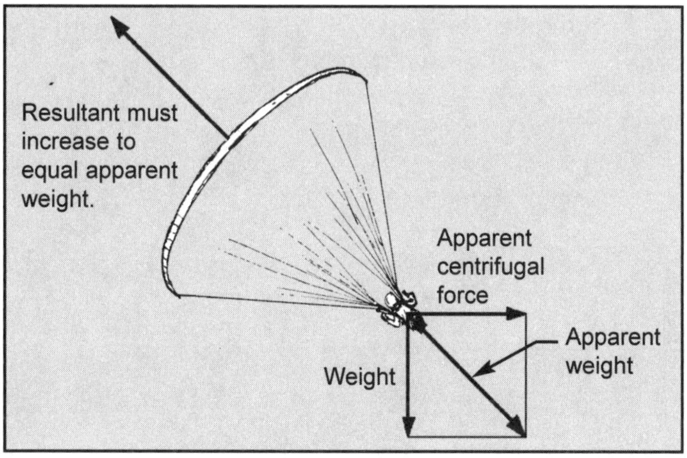

Figure 12-16: Forces in a Turn

✓ *Turn Limits*

When we discussed turns in Chapter 9 we showed how your turns get tighter as your bank angle increases, even though your airspeed is faster (see figure 9-13). The reason your airspeed increases is shown in figure 12-16. Here you see the rear view of a glider in a bank. As the glider follows a curving path, the (virtual) centrifugal force combines with the weight to make the pilot appear heavier. This apparent increase in weight is known as *G forces*. If we pull two Gs, we appear to be twice as heavy. This same effect can be felt on a roller coaster or a bicycle on a curving path.

Now you know that an increase in weight on the glider increases wing loading, which increases all speeds. This also applies in a turn. Even though we slow to the angle of attack just above stall, our airspeed is faster in a turn. In fact, the steeper the bank the greater our stall speed. The graph on the next page shows the G loading at every bank angle, along with the actual stall speed assuming a 14 mph (22.4 km/h) straight ahead stall speed. In addition, we show the radius of turn, the time to complete one 360° turn and the sink rate in the turn assuming 200 FPM (1 m/s) minimum sink in straight flight. The numbers in figure 9-13 came from these graphs.

Obviously, you can't carry all this data in your head. However, study these charts a bit and you will see the effects of producing coordinated turns at various bank angles. The more you understand turns the better you will be at turning efficiently. *Performance Paragliding* discusses more turn techniques for thermals and maneuvers.

SUMMARY

Understanding the mysterious ways and wonders of our wings helps us know how to coax the maximum action and performance from them. Come to think of it, these wings aren't so mysterious after all, once we discover the subtle little tricks that effect handling, performance and stability. As we learn more and more we see that the principles of flight are logical, dependable and our best ally as we pursue safe flight.

We also come to realize the great care and creativity that went into our glider designs over the years. Paragliders evolved slowly from basic parachutes to the sleek crescent wings they are today, but every step of the way was the offspring of inspiration and discovery. We have the gift of flight in a wonderful convenient package. If there is any real mystery to flight it lies in the question "why we are so lucky to be born at this special time in history when the joy of free flight became possible?"

TEST YOURSELF (*Answers in Appendix II*)

1. A paraglider airfoil is generally *stable* or *unstable.*
2. Lift on an airfoil is always ____________ to the airflow.
3. Drag on an airfoil is always ____________ to the airflow.
4. The ratio L/D is always equal to ____________ through the air.
   a. Progress
   b. Relative wind
   c. Head wind
   d. The resultant force
   e. Glide ratio
   f. 12 divided by 4.
5. List at least two types of drag: ____________________________
6. Aspect ratio is a measure of ______________________________
7. The axis that results in one wing moving forward or back is ________.
8. Pitch stability on a paraglider comes from ________________________.
9. As bank angle increases, G loading _____________.
10. Slowing too much in a turn can result in a stall of the __________ wing and then a _________.

Chapter 13

Soaring! Introduction to Intermediate Skills

"Nothing can cure the soul but the senses, just as nothing can cure the senses but the soul."

— Oscar Wilde

All your fledgling training, your study and careful practice pays dividends. Your investment of time and effort is rewarded with longer flights and more airtime. You can't say that the learning hasn't been fun, but now you are ready to stretch your wings.

One of the finest things a human can do is to be borne aloft on the heft of the air and there remain for hours. The vistas are unlimited, the rewards are unbounded, and the excitement is unparalleled in any other pursuit. You are now nearly ready to perform such a spectacular feat, but before you reach this pinnacle of flight there are a few more skills, there is more knowledge, there is more experience that must be gained.

Your novice skills have led you to higher flights in more challenging conditions and more varied terrain. When you put your heart and soul into flying at this level you learn to perfect your judgement, timing and control. Now, take all these elements of flight and direct them to becoming a true master on high. You will do this by learning finesse on the controls and sensitivity to the air. Your ultimate goal is to soar.

Within the scope of this chapter you will learn to soar, and that's exciting. You'll also learn other tricks and practices, such as spatial orientation and landing on top, to round off your skills. As with all of your learning, this step to the next level must be taken with care, for safety is your guiding principle. Learn gradually by adding new experiences in increments. Your success as a pilot of long-term skill depends upon your success in the intermediate phase. Learn this material well, have fun, and earn your place in the sky.

You, the Pilot

As you reach the level where you are learning intermediate skills, you should pause to take stock of your personal attitudes and practices. In this section we'll discuss a few ideas that will better prepare you for your role as a pilot making your own decisions based on your own judgements.

Your Goal: *To adjust your psychological approach to the level of your present and future flying skills or practices.*

✓ Recurrent Training

Once you have "arrived" at the level that allows you to fly a variety of sites in a variety of conditions, you may feel you have left the training hill far behind you. Yet occasionally bad habits or sloppy techniques may creep into your performance on takeoff and landing. These bugs are best worked out at the training hill. Also, a spring tune-up of your skills and a review after any long layoff from flying are good times to revisit the training hill and your favorite instructor.

Kiting practice should continue throughout your flying.

Make the training hill your friend. It has its demands which will hone your skills. In addition, your experience and examples are a good source of information for pilots just learning. It's fun to help their progress and reflect on how you progressed from their level.

In *Performance Paragliding* we present the Annual Pilot Review. This material is exactly what recurrent training is all about. Take time once in a while to examine and perfect your basic skills and judgement. Your flying will progress safely if you do.

✓ Decision Making

All through your learning you practiced making more and more decisions on your own. Now you are at the point where most decisions are yours alone to make and act upon. Are conditions safe? Should I fly? Where should I go? Am I getting too low? Is this a safe landing field? All these are questions you have (or will have to have) asked yourself and answered. Naturally some of these questions can and should be discussed with others, but you will be fully responsible for the outcome.

To make the proper, safe decision, recall the risk management material we have presented in this book. Also refer to the Robertson Charts of Reliability. These systems help you make wise choices based on simple observable facts, which should be the basis for all sound decision making. Also be sure to read our book *Understanding the Sky* to acquire the background that lets you evaluate weather conditions with insight and understanding.

Flying over a snowy landscape will change your perspective.

As you continue to paraglide, you will occasionally find yourself last on the hill. When you are alone intending to fly is when the peak of your judgement and practical skills come into play. Conditions can change in short order. Make sure your decisions are sound. The best way to assure this is the case is to practice making decisions and reviewing their results, possibly with the help of more experienced pilots. One of the best decision learning tools is discussions with others, especially if they are on site to observe along with you.

✓ Fear of Flight

It is not unusual for a pilot to develop an unreasonable fear at some point in his or her flying career. If you find yourself as nervous as Pinocchio at a bonfire, it's time to take stock of your attitudes and practices. Fear in itself is not unreasonable until it prevents enjoyment or makes you overtimid.

The best way to deal with fear is to talk to other pilots about the aspects of flying that make you fearful. You'll find that in many cases their assessment of the situation will indicate that the conditions or terrain is perfectly safe. The positive support of other pilots you trust can do a great deal to reduce fear.

In the early days of hang gliding we used to fly gliders that were barely capable of ridge soaring. If we got 500 feet over the hill we were ecstatic. On the rare days when general buoyancy or large thermals carried us higher I recall getting more fearful the higher I got. At 2,000 feet over I would be white-knuckling the bar. Eventually I had the chance to talk to pilots from other areas with higher sites and stronger lift. Their stories of thousands of feet of altitude and the fun they had made me envious. But the moment that turned my thinking around was when someone declared, "There's safety in altitude." That made sense because there are only a few stray airplanes aloft. It's the ground effects—rotors, mechanical turbulence—and the ground itself that present the real danger.

Soon after that my fear of getting really high dissolved and I eagerly lusted for altitude. Then we learned thermal soaring and I couldn't get enough.

Fear is a reaction intended to keep us safe. But just as in my case, fear isn't always rational. If you have fear, control it and eventually eliminate it by using the following steps:

Learn from the pros and you'll fly like a pro.

▲ 1. Identify your fear specifically. It isn't paragliding that scares you, but a particular aspect of it. You aren't afraid of laying out or carrying the glider. Are you afraid of takeoffs? Turbulence? Landings? Put your finger on the exact cause(s) of the fear you feel.

▲ 2. Work on that cause specifically. If it is a phase such as taking off or landing, go back to the training hill and get your basic skills perfected. Work with an instructor, for outside observation and expertise is the only sure way to efficiently perfect your technique. If it is turbulence or altitude, fly in smooth conditions for a while and gradually add the challenge (more turbulence or height). Learn more about the particular item that scares you. For example, talk to pilots with experience and you'll find that they handle much more turbulence than that which begins to disturb you. You'll learn that modern gliders handle turbulence well (up to the point of raging conditions that shouldn't be flown in) and many pilots enjoy the thermal kind, for it's like surfing in waves rather than paddling on a smooth lake; like skiing moguls rather than a groomed slope.

▲ 3. Look at the situation logically. If the challenges are well within your skill level, then be confident that you made the right decision to fly. On the other hand, all learning should be in gradual steps. When you bite off more than you can chew, fear can jump in and take command. Often this doesn't occur the first time you overstep the safe bounds. But if you make a habit of aggressively overachieving, you may suddenly find yourself anxious on every flight because you had periods of terror due to bad decisions on earlier flights.

Don't get peer pressured into a stressful situation.

Give yourself time to let your judgement catch up to your control skills. Judgement often develops slowly; so don't keep pushing to fly in more challenging conditions. Ask the advice of other pilots as to the appropriateness of the situation for your skills. Repeat the same flight as often as you can, and you'll find that fear will fade away as everything becomes familiar. The truth is, once you have much experience and set reasonable limits for yourself all fear disappears in normal situations. This result is because you gain self-confidence which comes from knowing that your skills are adequate for the situation. This gain of self-confidence in the air is one of the most rewarding aspects of paragliding for it carries over to other aspects of our daily life.

We discuss more about fear in *Performance Paragliding*. We summarize the steps to allay fear below.

Dealing with Fear

1. Listen when it appears.
2. Quantify exactly why it is there.
3. Focus on the matter that causes it
4. Determine whether or not it is legitimate.
5. Assuage it by logical assessment.
6. Repeatedly practice within your limits.
7. Add new challenges (higher wind, higher altitude, new landing sites) very gradually.
8. Give your judgement time to catch up with your skills.

✓ The Intermediate Syndrome

Often when we get to a level where we have met and mastered many situations we begin to feel we are invincible or at least up to anything that nature can throw at us. This little attitude problem is known as the *intermediate syndrome*. It is basic overconfidence (the opposite of the fear problem discussed above).

The very fact that it has the name "syndrome" indicates that it happens to enough of us that it is a troublesome source of accidents. If you have a strong desire to push the limits or to "get real good, real soon" then you may be a candidate for intermediate syndrome. Learn to prevent or cure this dreaded disease by exercising patience. The mountains and sky will most likely stick around long after you're fertilizing the earth. Paragliding will probably do the same. There is no rush, for if you nip any intermediate syndrome in the bud you will remain healthy

Dune soaring has its own charm. Caution: Watch the waves.

to develop your skills many years into the future. The great pilots of today started building their skills many years ago. Learn from their experience and develop solid basics and a policy of slowly expanding limits. Your fellow pilots will respect you and your mastery of the sky will come all the easier.

Flying New Sites

There will come an occasion when you graduate from your local friendly hill to a strange new site. This can be an anxious experience, but flying new sites is also one of the joys of our sport once you get the hang of it.

Your Goal: *To become adept at judging the appropriateness of a new site for your skills, and to become comfortable flying new areas.*

✓ Judging Safety

Safety is a book not a word, an attitude more than an action, it is a practiced pattern not a single step, awareness of a broad perspective and attention to fine detail. Safety is managing risk.

There are three aspects of a new site that you must judge: the safety of the takeoff, the conditions, and the landing field. The takeoff should be within the range of what you have experienced previously (not much steeper, flatter or shorter). Obstructions should not appear threatening. If your launch skills are fine-tuned and you see other pilots take off with a good margin of safety, then you can consider going. Otherwise, do not attempt a questionable launch. The place to learn new launch techniques is at a carefully controlled site under the guidance of an instructor (see Chapter 7), not at a new site.

Concerning the conditions, you should always talk to local pilots to learn the reasonable limits of the site for a new or inexperienced pilot. Find out what tricky conditions lie lurking and what is the standard landing setup in the given conditions.

Remember: one new thing at a time. Don't try a new glider at a new site without expert skill.

The landing field must be close enough for you to reach with plenty of altitude. Walk the field to look for ruts, obstructions, fences, bushes, power lines and sloping terrain. Take time to walk your intended landing flight path and imagine the controls you will make. Be sure to choose a conservative approach, not necessarily that of the local hotshot pilots.

At takeoff, carefully assess the conditions again and continue to request guidance from the local pilots. Then create a flight plan. In short, apply all the principles and procedures you learned in Chapter 9 about flying high for the first time.

Coastal soaring on the Isle of Wight.

✓ Flight Procedures

Your flight plan should consist of a careful outline of what you will do during your entire flight. It should include performing a fast, clean takeoff, slowing to the proper speed (faster than best glide until you are clear of the hill), and turning to head in the desired flight direction. The flight plan should include contingencies for unexpected wind, sink or lift. It should also include arriving over the landing field with at least 300 feet (90 m), as well as the landing setup with an alternate plan for landing in case the wind changes direction.

In the air, be aware. Judge your position constantly, and situate yourself properly for landing as soon as it is expedient. Probably your early experiences at new sites will occur before you learn to soar, so flying directly to the landing area is in order. Set up the landing carefully and land gently. Review what you accomplished, what you did right and what you can improve on. Then, assuming that you are flying on a gentle day, go back up and repeat your flight.

Each time you fly a site, learn more about its particulars and how far you can venture before positioning for landing. In this way you develop your experience and judgement.

You should be aware of the different flying conditions generally found in different areas. In desert or dry conditions, most inland winds are created by the sun's heating of the surface, so penetration (reaching landing fields) is usually not a problem since winds are confined to the proximity of the mountain or the valley floors.

In moister areas where fronts pass frequently, horizontal wind is more common and higher winds complicate the entire flight. There are, of course, exceptions to these rules, with strong horizontal winds generated in deserts due to pressure systems or heating effects near the mountains or sea. Coastal sites in general tend to exhibit much smoother winds than inland sites due to the uniform temperatures over water. These conditions render coastal sites ideal for learning to fly. However, strong gradients can exist in stable coastal air, so the old admonition to "maintain thy airspeed" upon landing is in force.

✓ Model Behavior

When flying a new site, consider yourself a guest of the landowner and the local pilots who work hard to maintain the site. You may have to obey some flying restrictions based on your rating. This is reasonable, since accidents at a site can jeopardize its use. Be prepared to sign a waiver and perhaps pay a fee. Don't grumble too much, for many sites are leased and consider the user fees for skiing or golfing.

Be sure to pick up all litter (including that of spectators), close all gates and stick to the designated roads. Learn and obey the protocol of the site. Generally, senior pilots take precedence at a site if only because they best know the conditions. By all means, don't be a launch potato. Don't move out to launch and wait for an overlong length of time before launching. If you don't have your launch skills down pat, wait till those who do take off, and then you can sit at launch all you want. Note that the training hill is the place to perfect launch skills, not the high sites.

FLYING IN TRAFFIC

To be sure, part of the fun in flying is sharing the air with other pilots. When you fly in traffic, you must do it with caution and respect for the airspace of your fellow pilots. We have already covered right-of-way rules and soaring traffic rules in Chapter 10. Also see Landing in Traffic in Chapter 8. You should review them so they become as familiar as traffic rules on the highway. Here we'll point out some other traffic considerations.

Your Goal: *To fly near other gliders with comfort and full awareness while avoiding being a hazard to yourself and other pilots.*

When the skies get crowded, you must carefully observe traffic rules. Communicate!

✓ Turns in Traffic

One of the most important things you can do in traffic is to clear your turns as we discussed earlier. When you begin soaring the matter becomes more critical. Look carefully in the direction you intend to turn to make sure the road is totally clear *before* you turn. In fact, one of the ways you broadcast that you intend to turn is to turn your head in an obvious manner several times in the intended direction. Another unmistakable sign is a few short sideways motions on the arm on the side you want to turn (do not do this if you are flying near stall speed).

When you are trailing another glider you should watch for turn signals and position yourself so you don't block the way. More importantly, if you are following another pilot along a ridge in light conditions, be sure to turn before the end of the ridge so the pilot in front of you can turn back before running off the end of the ridge.

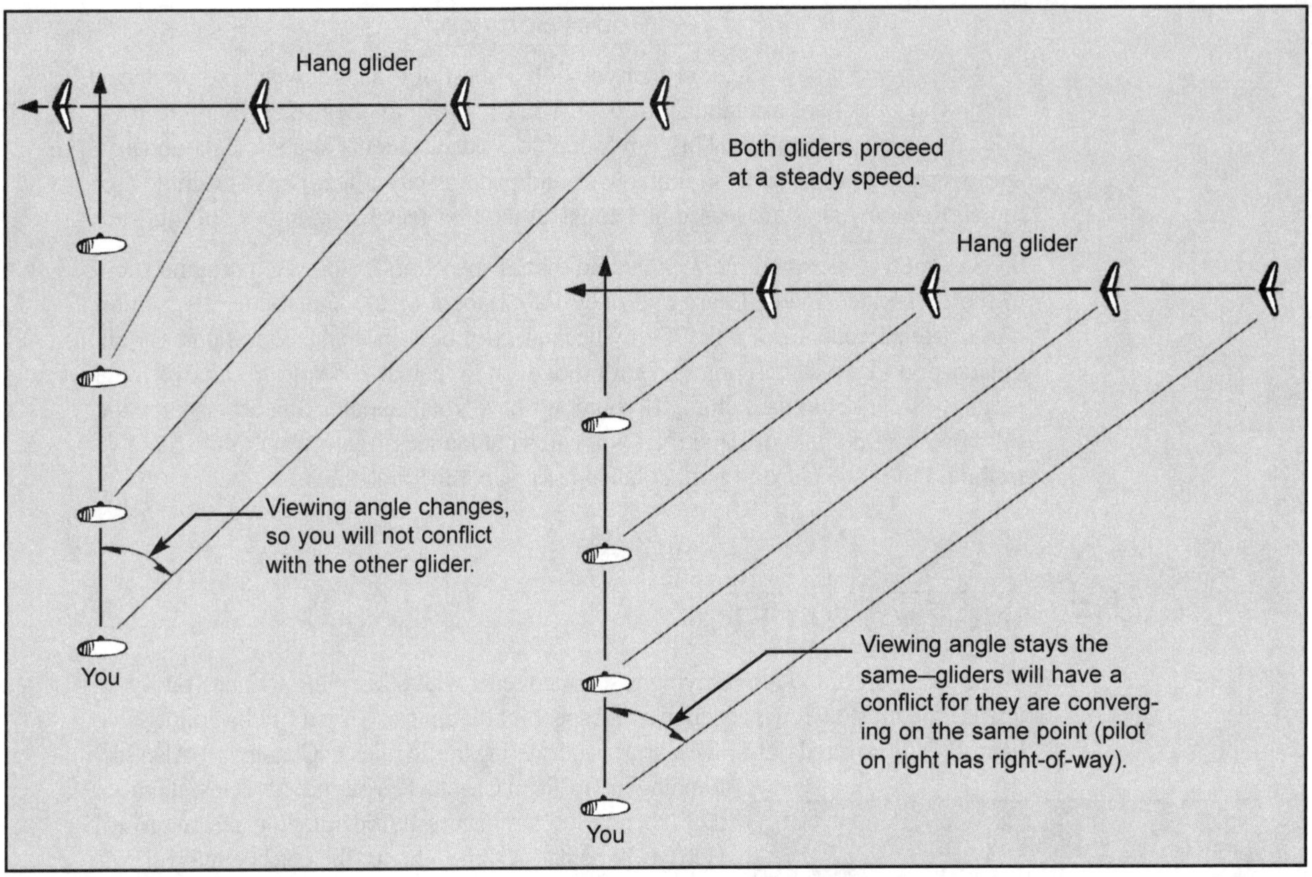

Figure 13-1:
Judging Traffic Conflicts

✓ *Judging Traffic*

Develop an awareness of the position and movement of all gliders in the air with you—even those well above and below. Watch for others.

When you are flying across the path of another glider (or airplane), you can easily judge whether or not a collision is likely by watching the angle of the other craft. As figure 13-1 shows, you are in a paraglider with a faster hang glider approaching on your right. Will you have a conflict? If you are on a converging course the angle remains the same. If you are likely to miss, the angle to the other craft changes. Of course taking evasive action early is always the best policy.

Finally, be aware of wake turbulence (vortices) left by other aircraft. Note where the path is and avoid it as much as possible. This awareness is especially important during the landing process (see Turbulent Landings and Landing in Traffic in Chapter 8).

LEARNING TO SOAR

Now we have come to one of the highlights of paragliding: soaring. This is where all your skills pay off, for on one good soaring flight you can often equal your total airtime up to that point. Soaring is exciting stuff, for as you take off into an azure sky your thoughts are no longer on the landing field but on the drifting clouds above. Your focus is upward, not downward. Soaring puts us on a level with the birds, for even though they can flap and we can't, the ones that get high and stay up must soar just like us.

Soaring is not without its dangers and challenges, for it generally requires more wind and often presents more turbulence. However, you will learn to soar just like you learned most of your skills: with the guidance of an instructor and with gradual steps. We begin with ridge soaring, for it is the easiest to understand and exploit and it is found most anywhere there is a hill.

> *"Some folks see a bird's eye view, others haven't got a clue."*
> *— Jimmy Buffet*

Your Goal: *To understand the principles of ridge soaring and learn to exploit ridge lift efficiently.*

✓ The Story of Soaring

The definition of soaring is simple: *soaring* is staying aloft by using the lift generated in upward moving air. Paragliders are always descending with respect to the surrounding air at what is known as the sink rate, as figure 13-2 shows. If the air itself is moving upward as fast as your sink rate or better, then your descent through the air will no longer drop you toward the ground but hold you level or make you climb. That's soaring.

Figure 13-2: Effects of Vertical Moving Air

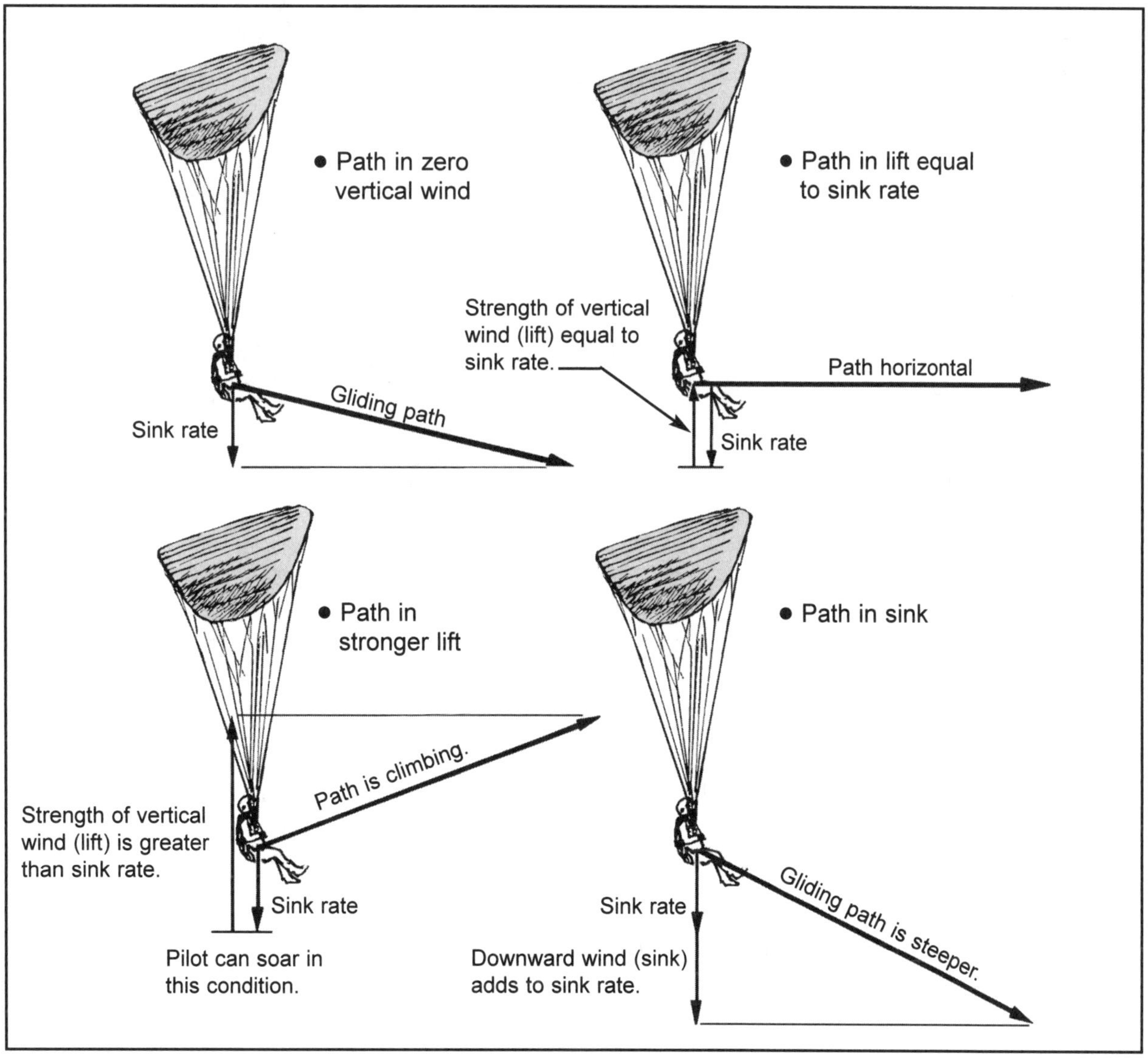

The first humans to soar were the early developers of aircraft near the turn of the century—Montgomery, the Wright brothers and Chanute's associates. They naturally started out with *ridge* or *slope soaring*. However, soaring was largely ignored once engines were added to the aircraft, so it wasn't until the 1920s that soaring really began to be explored by German sailplane enthusiasts.

Ridge soaring is using lift created by wind deflected upward by a ridge, hill, dune or mountain. (Buildings, tree lines and other obstructions also produce ridge lift that is usually too weak for paragliders to use.) Figure 13-3 shows the cross-section of various ridge shapes and the lift that is produced as air is deflected over them. The shaded areas indicate the position of lift strong enough to support a paraglider, known as the *soarable envelope*. Also shown are areas of possible turbulence.

Figure 13-3: Lift Produced on Various Hill Shapes

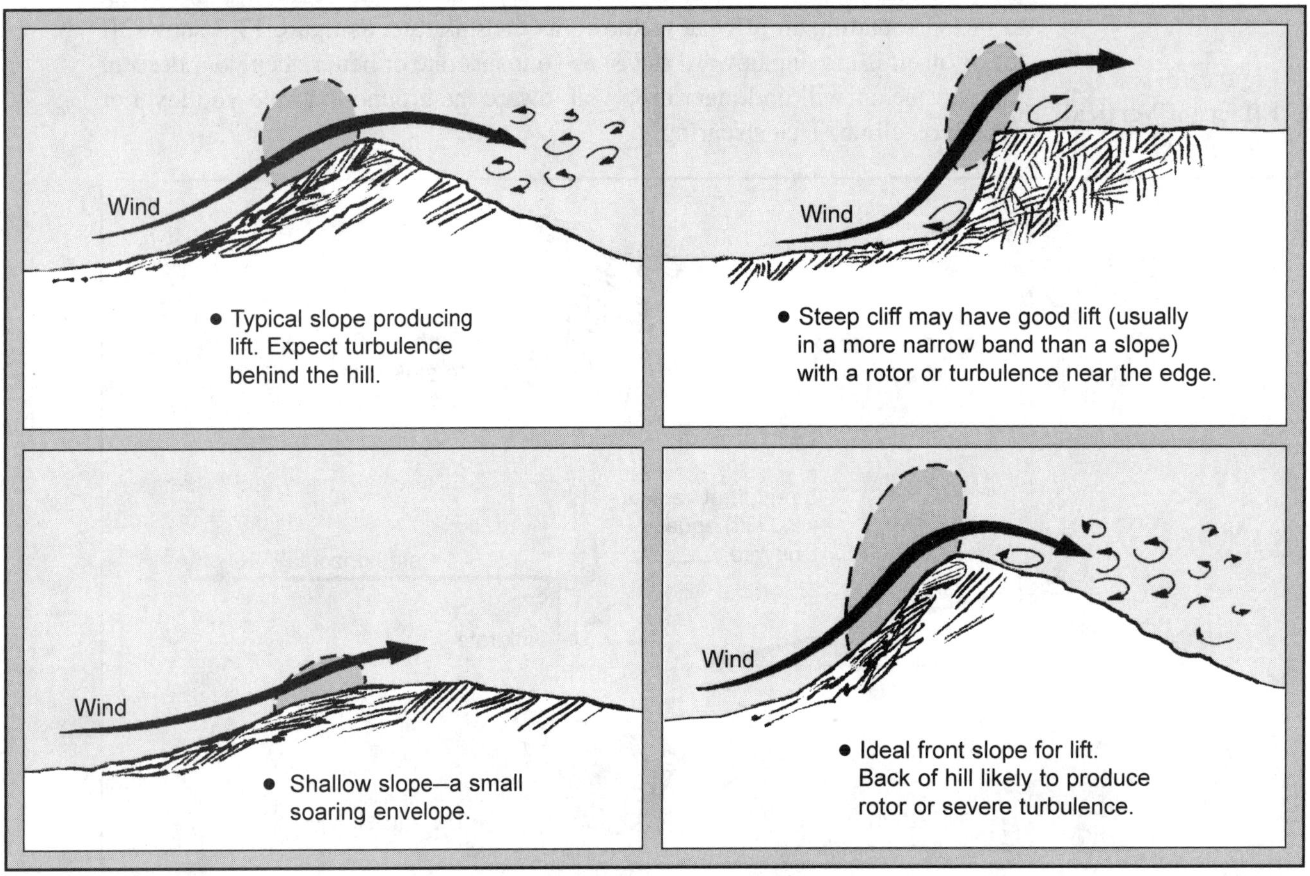

✓ *Soaring Qualifications*

How do you know when you are ready to begin soaring? Your instructor will most likely make you aware of the necessary skills appropriate for sites in your area, but here are some general pre-soaring requirements:

Soaring is good for your body and soul.

- Know how to clear traffic at launch
- Know the right-of-way rules.
- Be comfortable flying at higher altitudes.
- Be comfortable and competent to launch and fly in higher winds.

- Know how to handle turbulence.
- Know how to fly at minimum sink without stalling.
- Be able to perform efficient soaring turns of up to 180° heading change in higher winds.
- Be able to follow a precise ground track in wind by crabbing.

All of these qualifications should be acquired at the novice level and can be practiced at an appropriate high site before attempting to soar. By higher wind, we mean the wind strength needed to soar, which may be 12 mph (20 km/h) or less, depending on the site. Since some degree of texture or turbulence is often associated with moving air, you should also be comfortable flying in the "bumps".

Flying at minimum sink is efficient since your sink rate is the lowest possible, which allows you to utilize all the uprising air's energy. However, you know that minimum sink airspeed is near a stall, so you must fly a little faster—about trim or best glide speed— when you are near the terrain, which means right after take-off and until you climb at least 50 ft (15 m) above all obstructions.

✓ Soaring Turns

When you learn to ridge soar you will be making a series of passes along a ridge or hill. The turns you produce when you initiate your soaring pass and reverse directions are known as *soaring turns*. These turns are a combination of well-controlled and maximum efficiency turns. On one hand you want good control because you are near the terrain. On the other hand you must be efficient to exploit the lift.

Here's a general description of soaring turns. Be sure you have ample airspeed to avoid a stall just before the turn is initiated. Once the roll starts apply the control necessary to achieve a maximum efficiency turn. To end the turn, roll out with the appropriate increase in speed control added to the roll control. The roll-out leaves you flying parallel to the ridge with good control speed if you are low and close to the terrain. If you have plenty of clearance you should roll out to fly a bit slower—just above minimum sink speed.

A pilot makes a soaring pass close to the ground. Careful control is required in this situation.

You should always try to make your turns at the end of a pass in lift. Because of fluctuations in the wind and terrain differences as well as the presence of thermals, you will encounter areas of varying lift even in relatively smooth ridge lift conditions. Since we increase our sink rate in a turn, it pays to turn in the best lift. Also, the turn makes you linger in the lift. This technique is one of the secrets of the best ridge jockeys.

Note that your initial turn after launch should be automatically initiated with good airspeed if you have performed a good energetic launch. If your launch is slow (too much control pull) a turn to parallel the ridge may be unsafe for a stalled inside wing can occur. We'll show how to apply these turns when we discuss soaring.

✓ Ridge Soaring Steps

▲ 1. *The first thing to do is practice flying at minimum sink airspeed* from a point safely clear from launch to the landing field. Make all your turns maximum efficiency turns and practice losing as little altitude as possible. (During the landing pattern increase your speed to the normal approach speed.) You should become totally relaxed and comfortable with this slow speed flying before you attempt to soar. This may take as many as ten flights if your site is 1,000 feet (300 m) or less.

Ridge soaring is made easy by the presence of the slope that cries out, "The lift is here!"

▲ 2. *Next, on a smooth, light wind day try turning as soon after launch as you feel comfortable*, and fly parallel to the ridge. Repeat this procedure on subsequent flights until your instructor tells you that your turns are efficient and performed as soon after launch as necessary to stay in the *lift band*. The lift band is the soarable envelope of a slope as it extends along a ridge.

As figure 13-4 shows, on a long ridge the lift will be on the upwind side along the whole face. This is why you turn to parallel the ridge soon after takeoff. If you turn too far out you will miss the lift band as shown. If you turn too close you may get too close to the hill.

Caution: *Near a slope, the gradient or a turbulent gust can roll you towards the hill. Maintain ample clearance and airspeed to avoid this potentially dangerous occurrence.*

Figure 13-4: The Lift Band on a Long Ridge

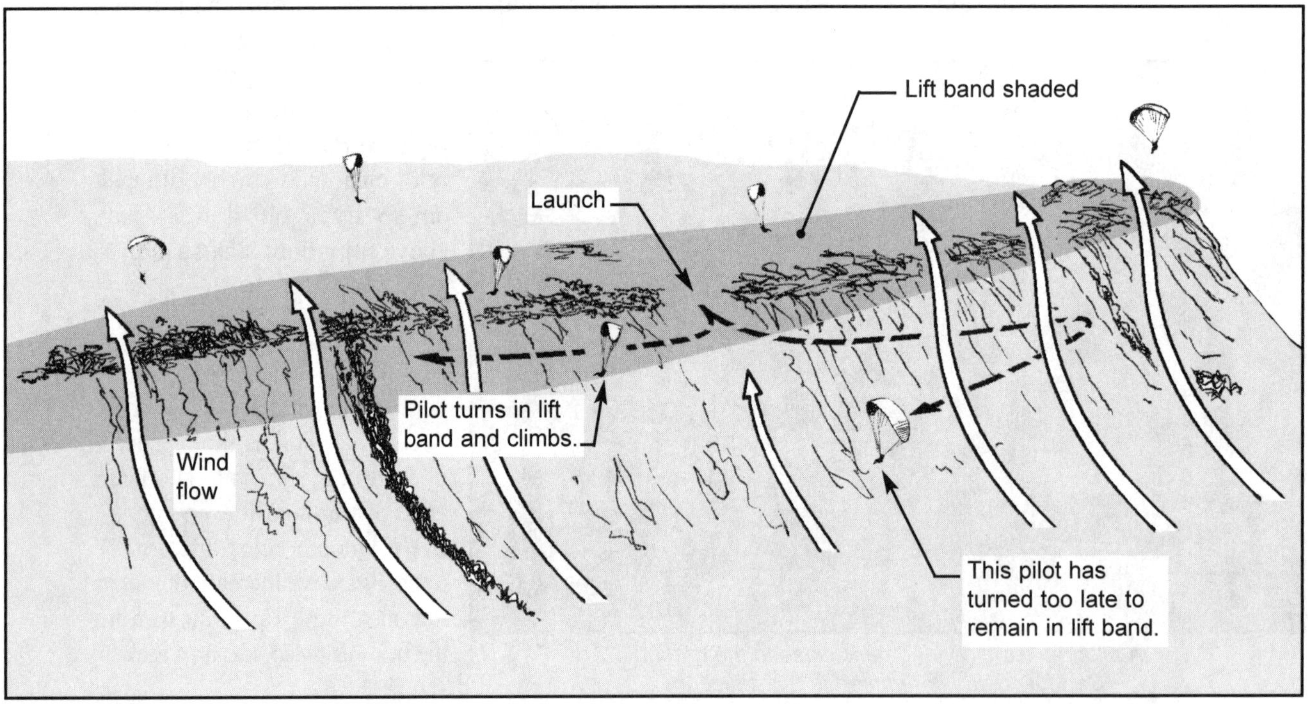

If your landing field is to one side of your takeoff, be sure to make your turn toward the landing field. Otherwise you may get caught away from it. Once you are flying comfortably along the ridge (still in light non-soarable wind), you can try a 180° turn to come back along the ridge if you have enough altitude. At all times be aware of your position with respect to the ridge and your landing field. You will be sinking, so you will have to move out as you descend. We call the passes along the ridge the *soaring pattern.*

Skill Check: *You are ready to try your soaring pattern flight in actual soarable conditions when your launch is well-controlled (not too slow), your turns are smooth and efficient, your airspeed is ample for control near the ridge and at minimum sink when flying away from it, and finally, you always arrive at your landing field with safe altitude.*

▲ 3. *Now you are ready to try the same flight plan in stronger winds.* At many sites the evening winds may become smooth and buoyant. This is the best condition for you to learn to soar. Coastal sites also offer ideal soaring training conditions, because the smooth, steady winds blowing off the sea let you fly in ample lift and reduce turbulence dangers. Be sure to use the advice of your instructor and other pilots concerning how smooth and strong the wind is. In any case, build your skills gradually by flying your soaring pattern in only slightly more wind than the previous experience.

As soon as you are stable in flight after launch, begin your first turn to parallel the ridge as before. Watch your ground track and maintain ample clearance from the ridge. If you don't begin to rise, keep to your pattern and fly to the landing field when necessary. You may find that you can make an extra pass or so in the lifting air, but be cautious of not being able to reach your landing field in the increased head wind.

If you begin to rise, maintain your position in the lift band and continue to parallel the ridge. Once you are well above the top, you may be able to move closer to the ridge to exploit the best lift. However, be extremely careful to avoid moving back into the venturi area above the ridge where the airflow is increased. Once you drift back into a venturi, you may not be able to come back out. If you detect more wind as you get higher, move forward away from the ridge to allow yourself an escape. It is better to err on the side of safety than performance at the risk of a very scary ride "down the drain" behind a ridge.

A flat-bottomed cumulus cloud in this shot promises the pilot some great thermal soaring.

Clear all your turns and make them smooth and shallow to minimize altitude loss. If you turn out of the lift band,

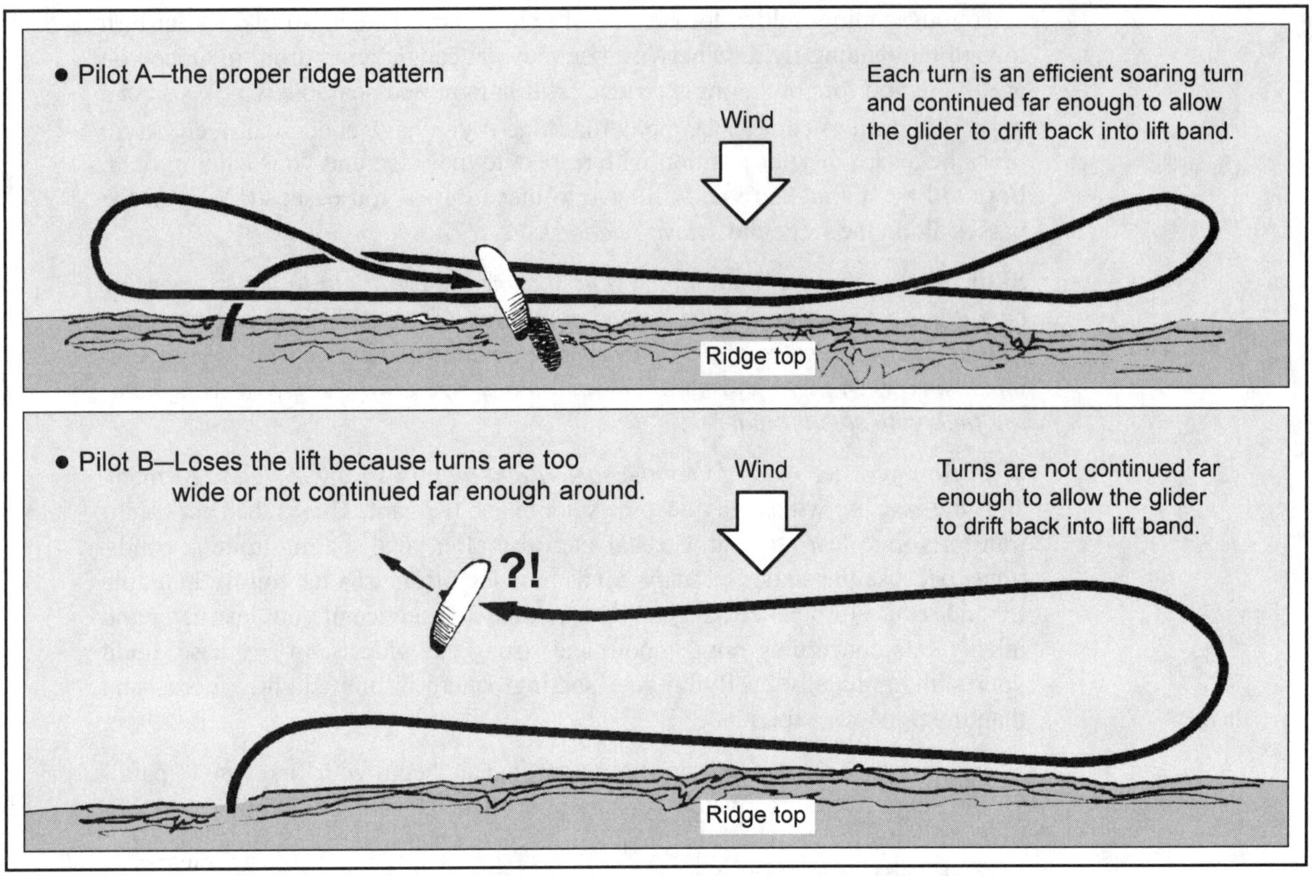

Figure 13-5:
Ridge Soaring Paths

continue turning a little more until your glider is aiming parallel to the ridge and you drift back into the hill. Once you are repositioned, readjust your heading to stop the drift and follow a path directly along the ridge. This pattern is shown in figure 13-5 as pilot A's path.

Maintain your position in relation to a ridge to stay in the lift band. If you stray in front, turn more parallel to the ridge. If you stray back, turn more perpendicular.

✓ *Crabbing on a Ridge*

This brings up a very important point concerning ridge soaring. Since the wind is blowing into the ridge, to fly parallel to the ridge you must be pointed a bit into the wind as shown in figure 13-5. This is crabbing, and the stronger the wind the greater your crab angle. (You could maintain the same crab angle in a stronger wind by speeding up, but then your sink rate would increase. At some point, as the wind increases, your crab angle is so great that you are aimed 90º to the ridge and are "parked".)

Because of this crabbing flight, the turns to reverse direction are not full 180º of *heading* change. However, if you simply change your *path* by 180º you may end up moving away from the ridge, as shown with pilot B's pattern. Pilot A holds the turn a little longer so that his path (not heading) changes a bit *more* than 180º and remains the same distance away from the hill. Note that he is still crabbing and his glider's heading actually changes *less* than 180º.

The secret of ridge soaring is staying in the best lift by controlling your path. The stronger the wind, the less you'll have to change your glider's heading to follow the ideal path of pilot A since your glider will point more into the wind. By paying attention to where you wish to be in relation to the ridge, you will crab almost automatically to follow the desired ground track, just as you do when flying in wind over flat ground.

Pro Tip: *The best lift on a ridge is usually from the edge of the top of the slope to about two wingspans upwind. As conditions get weaker the best lift is near the top of the slope (see figure 13-6).*

Figure 13-6: Best Lift Envelope in Light Conditions

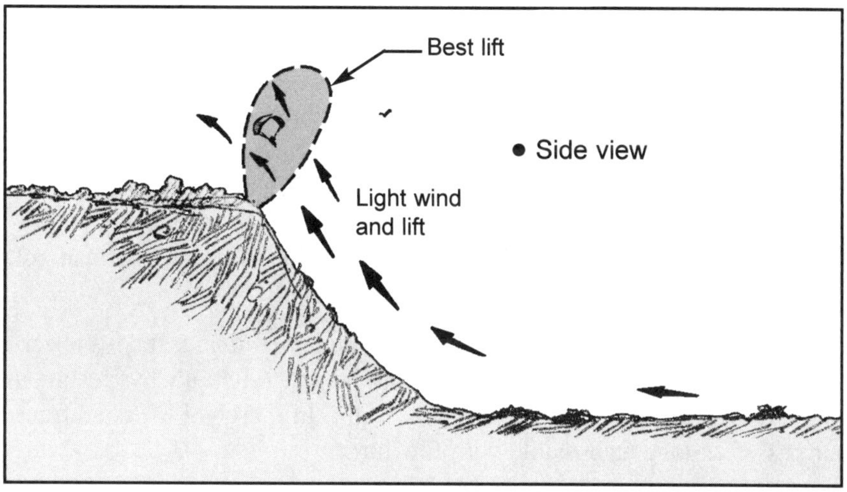

✓ Soaring Launches

One of the most important launch factors in soaring conditions is checking to clear traffic. Pilots tend to linger around launch, especially if conditions are light. Your first guiding principle before launching is to look both ways to make sure another glider isn't approaching the area you will occupy once you launch. If so, wait until the glider is moving away. In addition, announce your intentions clearly to the pilots in the air so they can give you clearance. You must learn to predict the progress and position of gliders in the air so you can judge when you have ample launch clearance. Be sure to allow a margin of safety. In a similar vein, once you are soaring, do not block the launch of other pilots laid out in position on the ground. Common courtesy allows a pilot five minutes or so to scratch up at launch, but if pilots are waiting to launch, don't out-stay your welcome.

On your launches you have been concentrating on getting away from the slope quickly and into the clear air. This is especially important for inexperienced pilots in windy (soaring) conditions. However, as your control abilities grow you can modify your technique somewhat to improve your chances of soaring in light lift conditions.

If the soaring prospects are just barely there, an efficient launch is of utmost importance. Such light conditions generally mean less turbulence to deal with near the hill (however, beware of thermals moving through). Perform the inflation and the first part of your launch as usual: *run hard with a perfect attitude* to avoid unnecessary diving. In fact the faster you are running at the end of your run, the more speed you have to pay off into altitude. Your goal should be to put energy into the system.

Next, as soon as you are balanced and clear of obstructions in the air, slow to the proper speed while turning to parallel the ridge. The slowing process helps the turn, and you will have good control since you started fast. The proper speed referred to is minimum sink in very smooth, light air. It is faster as necessary for more turbulence. If you begin to rise you can move in to the better lift. If you sink, move out and reach a safe landing area.

Caution: *In all cases concentrate on airspeed, and be ready to roll away from the hill if a surge lifts your outside wing.*

A pilot performs a soaring launch with a forward lean, knees bent and arms back.

✓ Ridge Soaring Problems

To help your learning progress smoothly, we list some potential problems that may crop up in your first soaring practice. Think about these as you make your first soaring attempts and you will learn quickly.

▲ 1. *Turning too late on takeoff.* Most beginners wait too long to turn and thus fly out of the lift. One note of caution: always turn in the direction of the wind on takeoff if it is crossing. The reason for this is there is a wind gradient close to the slope which can roll you into the hill.

▲ 2. *Turning too often.* If the lift is very light, it will take time for you to rise to maximum height. A turn always results in more altitude loss than straight minimum sink flight. In marginal lift conditions, make your passes as long as possible with few turns.

▲ 3. *Flying too far away from the ridge.* When you first learn to soar, you may be fearful of getting too close and thus not fly in the best lift. (The best lift is over the steepest part of the slope.) However, flying too close to the ridge is dangerous, so caution is warranted.

▲ 4. *Flying too slow or too fast.* Your best sink rate is at a speed just above a stall, but near the terrain this speed is not safe. If you do stall, you will lose plenty of altitude recovering. At worst, you will be blown back into the hill since you have very little control in a stalled condition. This, of course, can be quite dangerous. Flying too fast increases your sink rate and decreases your chance to soar. The best speed to use is trim speed down low and minimum sink speed once you are high.

▲ 5. *Gaps and breaks in the ridg*e. These are outlets for the wind so there is a lot of air rushing through them. Be ready for this and angle away from the ridge to cross them (see *Performance Paragliding* for details). Often you will notice no difference when crossing a gap, but be prepared for stronger winds.

▲ 6. *Getting too far back behind a ridge.* When the air is forced up over a ridge there is an area of faster moving wind just above the ridge due to the *venturi* effect when the wind is horizontal. Avoid this area by remaining in front of the crest of the ridge. (In upslope thermal winds this problem doesn't occur as much, but beginning soaring pilots do not have the experience to judge the difference in such conditions.)

▲ 7. *Getting too far from your landing area.* A new soaring pilot can get caught up in the ecstasy of staying aloft for the first time and drift too far away or wait too long in dying lift to reach the landing field. Remember that on some days lift can quickly shut off—especially in the evening—so keep in safe proximity to a good landing field.

An instrument is indispensable for efficient soaring. Make sure you're always in reach of a safe landing field

▲ 8. *Flying too far along a ridge when it tapers off or when you have a crossing tail wind.* Always try to fly upwind of your landing field in a crossing wind or make frequent upwind passes to make sure you can get back. Note your ground speed when flying in both directions on a ridge to detect a crosswind by a difference when going one way or the other. If you do experience a wind crossing the ridge, take great care in your positioning and look for an area turned more into the wind for better lift. Any time you note your groundspeed increasing along a ridge, it's time to turn around and head back the other way.

▲ 9. *Turning in sink.* Often the lift on a ridge varies due to the influence of thermals or the ridge contour itself. Try to linger in lift and fly faster through sink or areas of poor lift. Make all your turns in lift if possible and avoid turning in sink areas. Remember a turn increases your glider's sink rate.

Once you learn to ridge soar, you can perfect your skills by trying to imagine where the best lift is located and placing yourself accordingly. You can venture a distance away from your normal takeoff-to-landing flight path (but always remain in safe reach of a landing field) and begin to explore the sky. Our book *Performance Paragliding* provides a guide to more advanced ridge soaring techniques.

✓ Ridge Soaring Rules

Here we offer a rule summary to help you set safety limits in your early soaring practice.

Ridge Soaring Rules

- *Always* perform your mental launch list: CHAT – Canopy check, Harness check, Air check, Traffic check.
- *Always* perform a fast, controlled launch.
- *Always* turn to head into a crossing wind after launch.
- *Always* maintain ample control speed when flying close to a ridge to prevent unexpected turns and stalls.
- *Always* have a safe landing field well within reach.
- *Never* fly behind (downwind from) the top of the ridge when low.

✓ Ridge Soaring Lore

As you learn the basic skills of ridge soaring, you'll also learn that sites and conditions vary greatly. Soon you'll learn to judge when it's soarable at different sites and where to go at a specific site for the best lift. Let us investigate some of these matters here to speed up your learning.

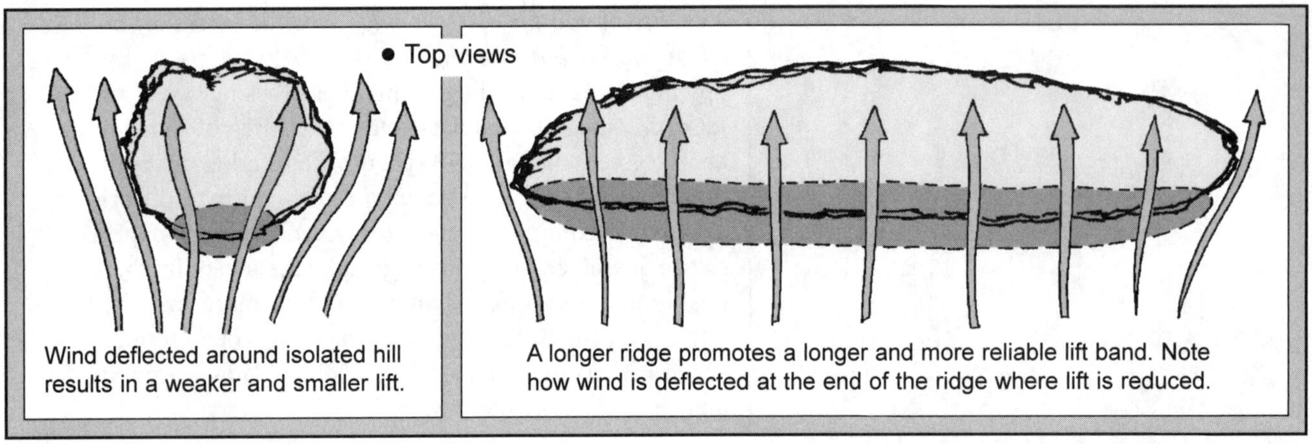

Figure 13-7: Ridge Lift is Dependent on Ridge Length

We'll simplify things by looking first at ridge, hill or mountain shapes, then add the wind factors.

▲ *Contours* — The first thing to note is how the cross-section of a slope is shaped. Previously, in figure 13-3, we showed several shapes of soarable hills. What you'll find in general is the steeper and higher the slope is, the more ridge lift it produces. Certain shapes that greatly encourage the air to move upward are also better lift producers, as shown in the figure. A vertical cliff, while producing good lift, will often have a more narrow lift band (and be less likely to give birth to thermals). A slope that is not too steep on top will create weaker ridge lift.

The length of a ridge is very important to lift production, as figure 13-7 shows. A rounded hill or mountain will allow the wind to pass *around* it rather than forc-

Figure 13-8: Crosswinds on a Slope

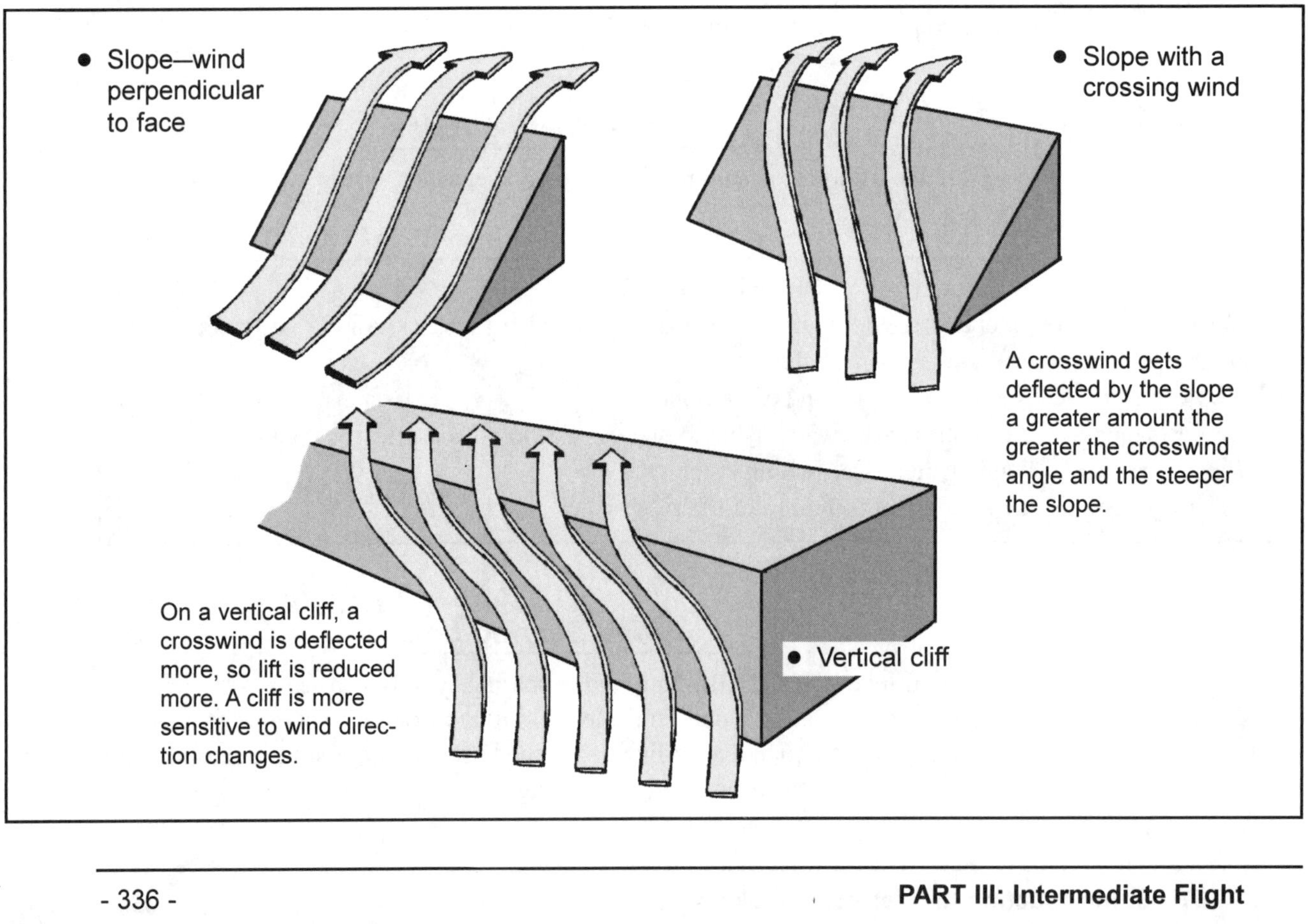

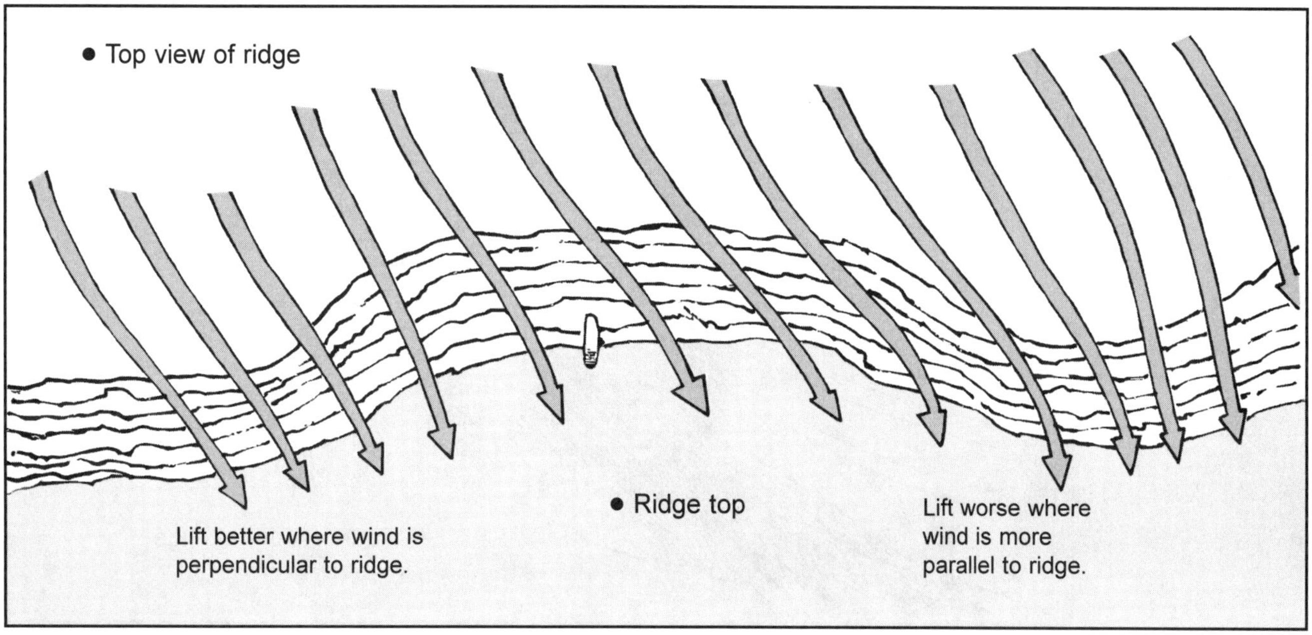

Figure 13-9: Best Lift Areas on a Ridge With Crossing Wind

ing the wind up and over. The longer ridge doesn't exhibit this problem, although on the ends of the ridge you should expect lift to dwindle as shown.

▲ *Wind Effects* — As the wind strikes a ridge, it is deflected upward but also to the side if it isn't exactly perpendicular to the face. This sideways deflection can reduce the amount of lift produced, as shown in figure 13-8. Note that the steeper slope is more sensitive to crossing winds. The rule quoted in the box is important to remember.

> The steeper the slope, the quicker lift dies off as the crosswind angle increases.

You can use this factor to your advantage by positioning yourself in areas of the ridge that face more directly into the wind, as shown in figure 13-9. Even when the wind is straight in, ridge contours can affect the lift production. As we show in figure 13-10, a belly or bowl in the ridge will deflect the air somewhat so

Figure 13-10: Lift factor in Bowls and Gaps

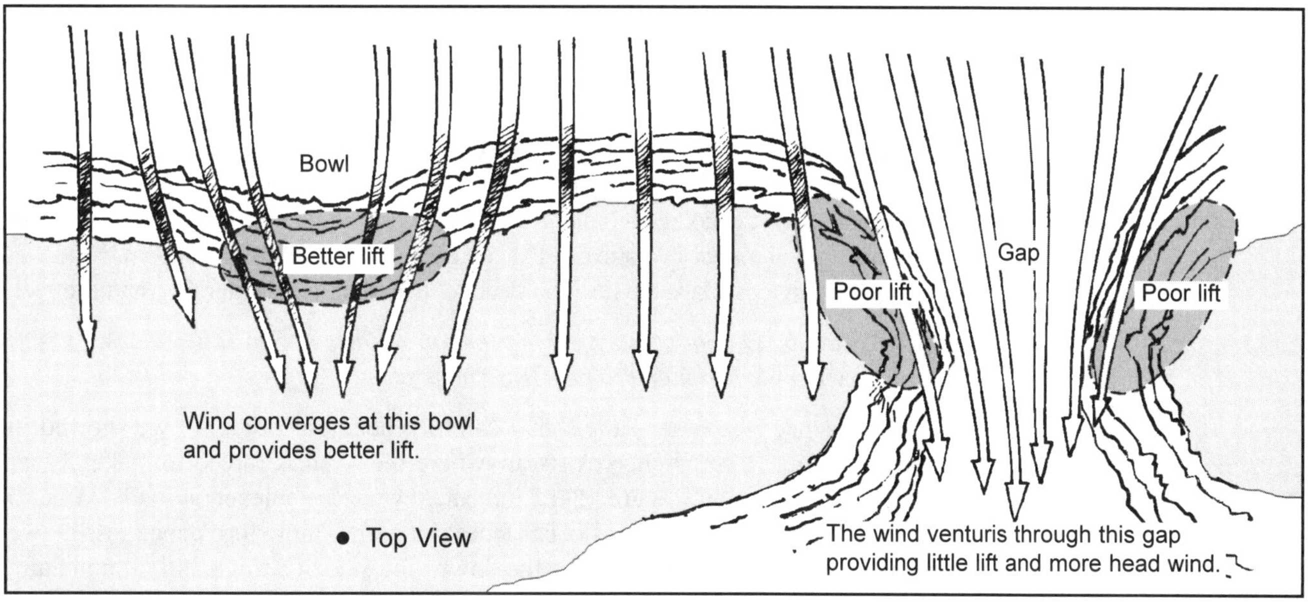

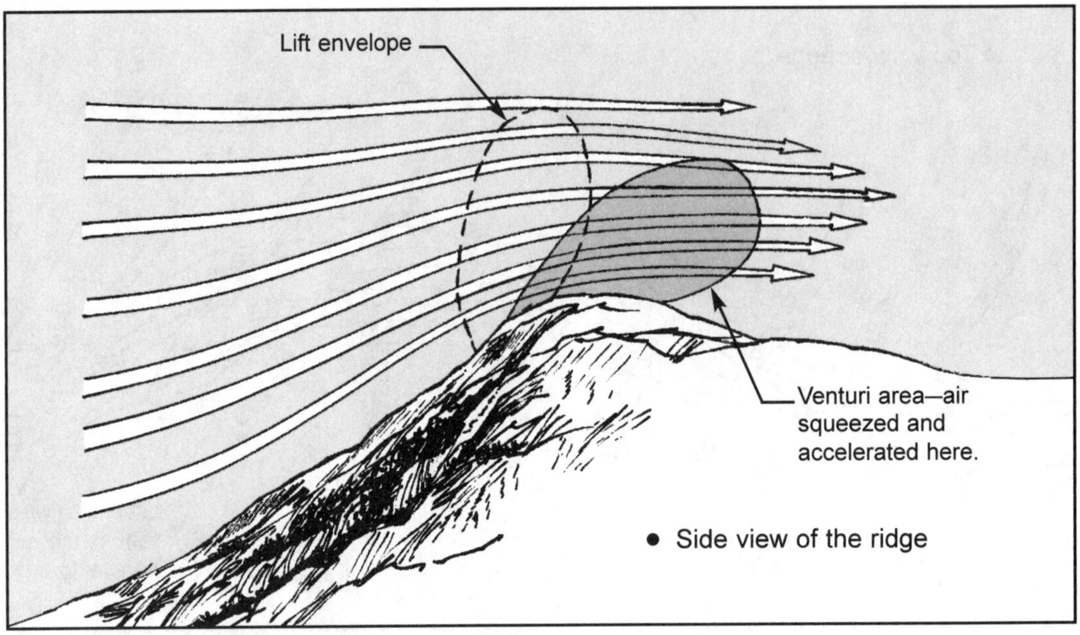

Figure 13-11: The Venturi Effect on a Ridge

it comes together and creates better lift at the back of the bowl. In lighter winds this is where you may find the best lift. In stronger winds, you will also find the strongest winds here, so you may wish to avoid the back of bowls if penetration is a problem. *Always* have a safe landing field well within reach.

Stronger winds do not necessarily mean better lift. At some point, as the wind increases in strength, you must fly faster than minimum sink and therefore will not get as high. In addition, as wind strength increases, heating effects such as anabatic (upslope) flow and thermals are suppressed so you can't float as high. In general, the lower the site height the higher winds you need to ridge soar. In any case, winds over 15 mph (25 km/h) are not ideal for soaring paragliders.

When a ridge opens into a gap, the air rushing through may result in little lift at the edges of the gap, as shown in figure 13-10. In fact, this gap area may exhibit strong wind due to the *venturi* effect as described in Chapter 9. Here the extra air coming together in the gap speeds up to escape, like water spewing through a breach in a dam. We discuss how to cross gaps in *Performance Flying*.

Another place where the air speeds up due the venturi effect is above the very ridge producing the ridge lift. As figure 13-11 shows, the air above the peak is forced to move faster as it encounters more air above it. This effect is similar to that of water moving around a rock in a stream. The areas of venturi and lift are shown. Note that the venturi area overlaps the lift. The stronger the wind, the greater the venturi effect.

⚠ **Caution:** *Do not venture behind the top of the slope in ridge soaring winds or you may be unable to penetrate back out!*

Turbulence is another danger that can lurk behind a ridge. As we showed in previous figures, turbulence exists anywhere the wind is forced to make abrupt changes of direction. Figure 13-12 indicates where turbulence is likely to be in soaring conditions. Note that steeper faces can harbor more turbulence. Again, we remind you to beware of flying on the downwind side of a ridge, hill or mountain.

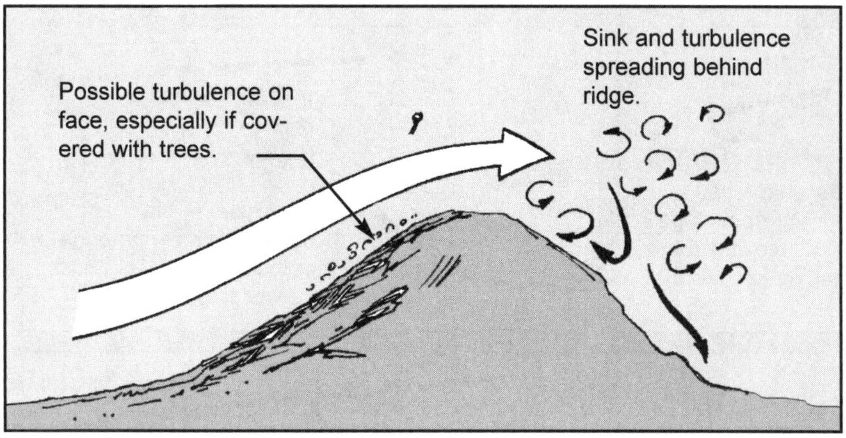

Figure 13-12: Turbulent Areas in Ridge Lift

Wind complicates all flying. However, without it we would be less likely to soar. There are times, however, when no detectable wind occurs at take-off, yet it is possible to stay up. What's happening? Once you realize that a typical glider minimum sink rate of 200 feet per minute is only 2.3 mph (3.6 km/h) you can see how light a straight up wind can be yet still allow soaring.

Pro Tip: *Watch the birds for soaring signs. Usually if the hawks and vultures are out, it is soarable for paragliders. Also, groups of swallows indicate soaring conditions, for they feed on the insects carried aloft in lifting currents.*

We provide a summary of where the best ridge soaring occurs below.

The Best Ridge Lift

- Above steeper slopes
- Above higher ridges or slopes
- Above ideal contours
- Above a smooth slope (without trees or obstructions)
- Above sections facing directly into the wind
- Where more wind occurs, such as in bowls (up to a point where penetration becomes a problem)
- When the air is more buoyant due to heating.

If you find stronger lift, try to linger in that area to climb higher.

Once you have mastered ridge soaring, you can move on to more experiences, but your ridge soaring skills will always be the mainstay of those "tricks of the trade" that make your flying exceptional. Besides, there is nothing like floating along on a swell of lift pushing along a ridge. This is your chance to cut loose and surf the sky!

Landing on Top

At some sites you can perform the neat trick whereby you land back up where you started. This makes retrieval easier and allows you to take multiple flights without lugging your glider back up the hill.

Landing on top is not without some added complications and thus some necessary precautions. The main problem is the turbulence and higher winds which can exist behind the lip of a ridge or hill, as explained in the previous section. Let us look at these matters more closely.

Your Goal: *To learn to judge the safety of landing on top for your skill level, and to perform controlled top landings.*

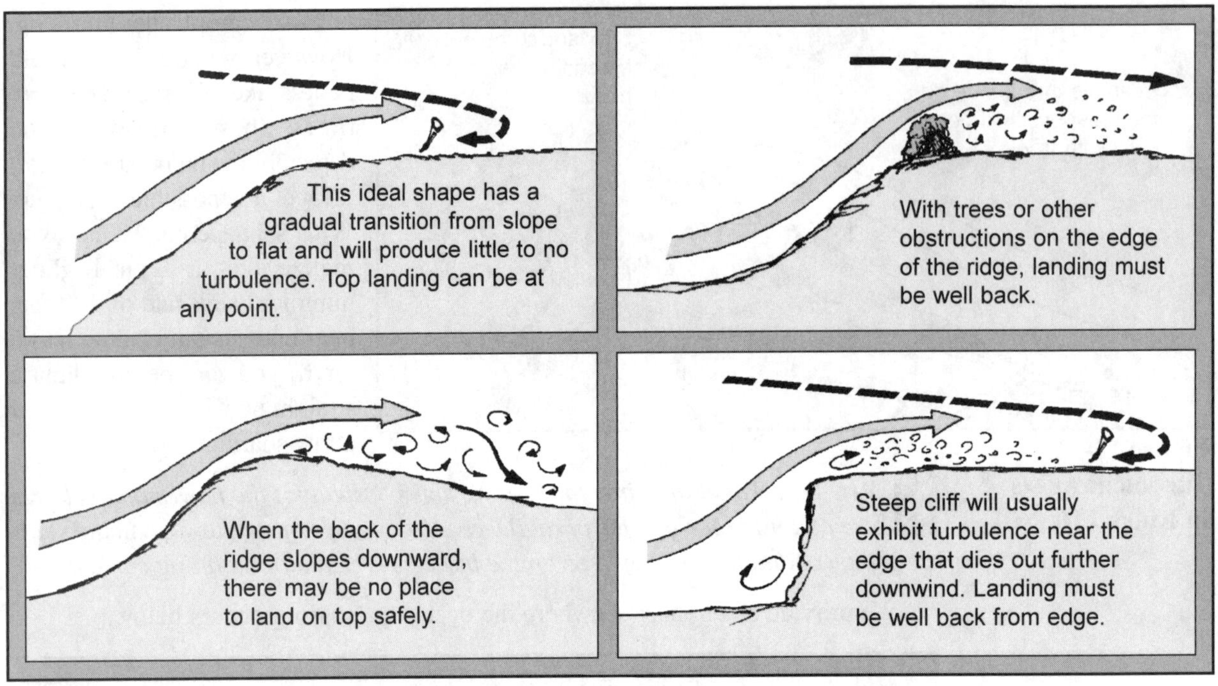

Figure 13-13: Ridge Shape Considerations When Landing on Top

✓ Safe Top Landings

In order for a top landing area to be safe, the top of the hill must be very rounded and smooth, the air itself must be very smooth (as in coastal conditions), or the landing must be far enough back from the edge of the ridge that turbulence is avoided. Figure 13-13 illustrates these concepts. For each different situation you need a different amount of height above the ridge to safely set up your landing.

In any event, at your skill level you should not top land without first observing other pilots performing such a feat. Learn the proper ground tracks and touchdown points from them. Ask for their guidance in determining a safe altitude for landing on top and where the turbulence danger areas are.

Figure 13-14: Pattern for Landing on Top

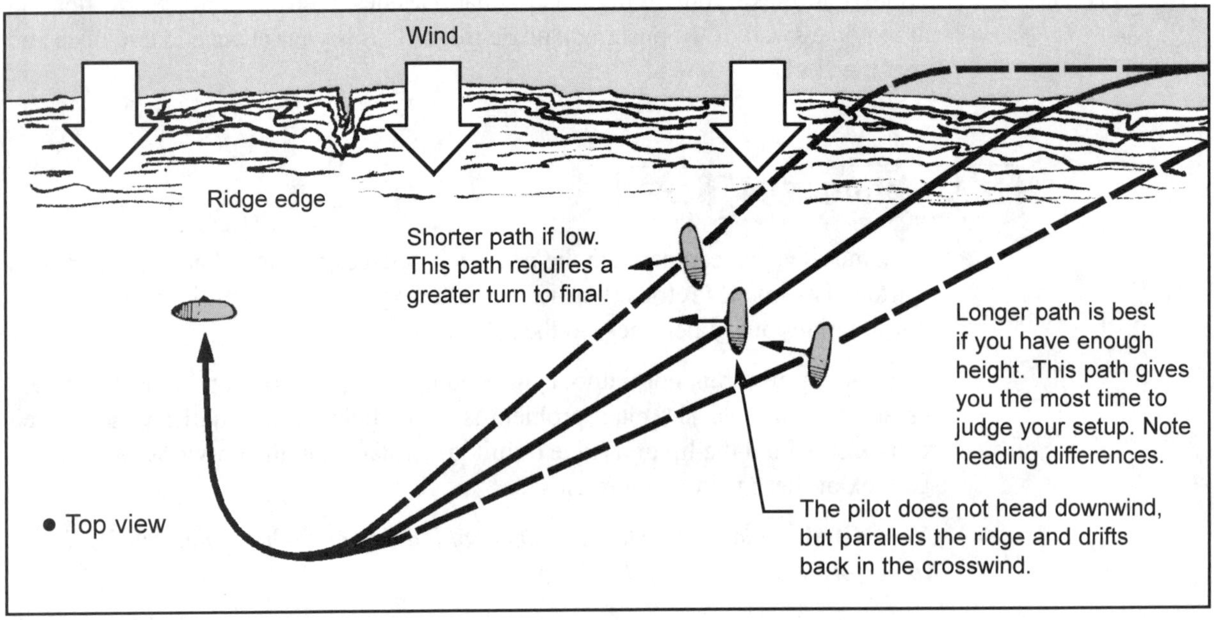

✓ The Crosswind Top Landing Approach

Naturally you must be soaring before you can land on top. To begin, you must position yourself to one side of the intended landing spot, as shown in figure 13-14. Then begin a crosswind approach that drifts you back behind the ridge face. Once you are back far enough, you turn to head into the wind and settle down for a landing.

Top landings are more complicated than bottom landings. Your setup and final skills must be perfected before you attempt them.

Notice that the longer you make your crosswind approach, the less you have to point your nose downwind, because the wind will have more time to drift you back to your turn position. This results in a smaller heading change being required at the turn, and also lets you see to judge the front of the ridge easier. For these reasons, we recommend that lesser experienced pilots use this crosswind approach.

Be careful of going back behind the ridge with too much altitude or you will have to fly back and forth in the potential rotor area too long as you lose height. Experienced pilots can give you a standard height above the ridge top for landing on top at a particular site. By all means maintain proper inflation pressure and control.

✓ The Downwind Top Landing Approach

More experienced pilots often use a direct approach for top landing, with the first leg being more or less downwind as shown in figure 13-15. The downwind leg can be initiated from a pass in either direction as shown. The advantage of such an approach is that it can pinpoint the exact landing spot more readily than the preceding one and doesn't require as much altitude.

Figure 13-15: Advanced Top Landing Pattern

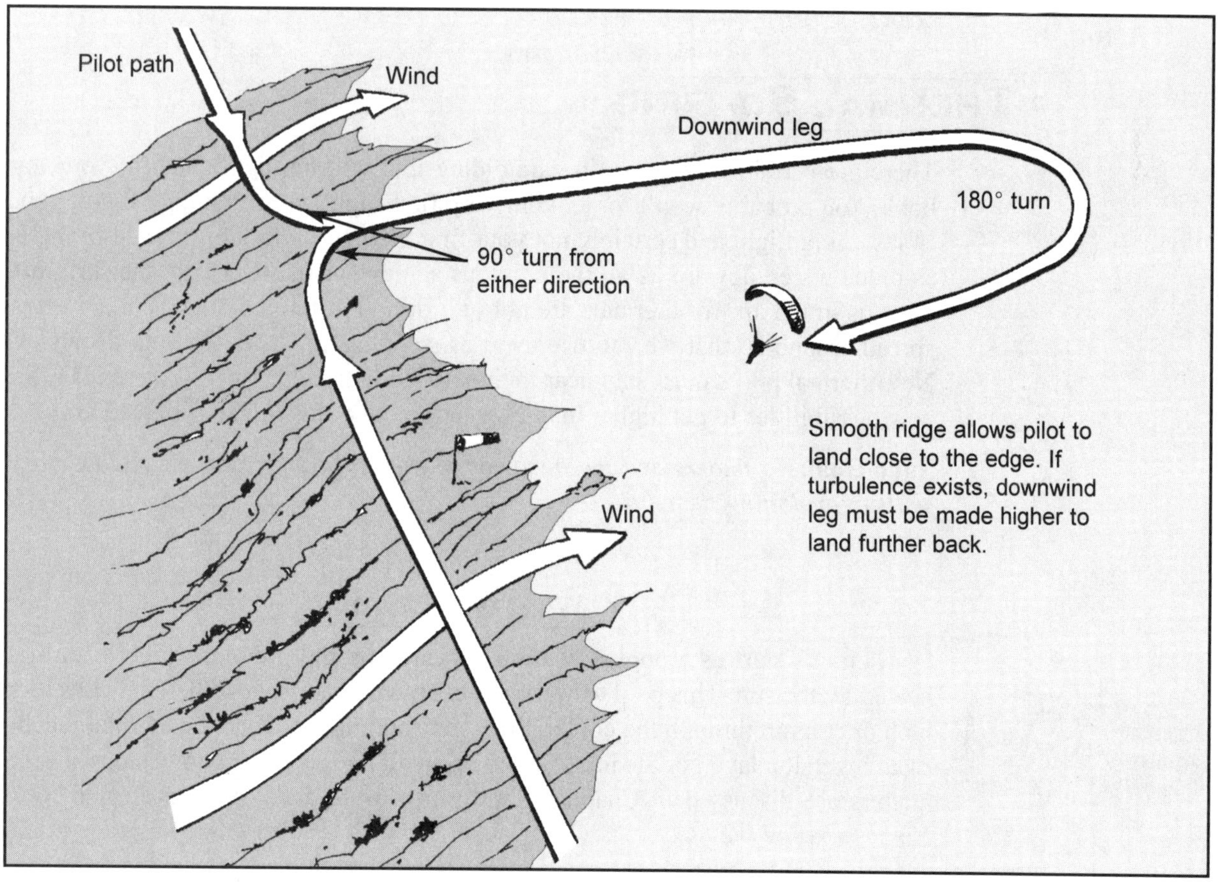

When making this approach, you must maintain good airspeed and be aware that you will be traveling over the ground at a high rate. Also, your turn will drift downwind a bit before it is completed and you will be performing a 180° turn close to the ground. There is a danger of misjudgement of the turn or a confusion of airspeed and groundspeed in the high winds close to the ground. This is a more advanced maneuver.

For both types of top landings, here are some essential guidelines:

Landing on Top

- Do not attempt this feat in strong winds. Smooth coastal sites tolerate more wind than inland sites. Ask local pilots what the limits of safety are.
- Carefully observe your ground track to position yourself properly. Be very careful of a possible venturi effect carrying you downwind farther than expected.
- Maintain good airspeed and pressurization to combat turbulence and wind gradient. Diving turns are useful to come down in the lift that can exist a bit back from the hill edge.
- If you overshoot you may pass to the front of the ridge, climb back up and try again, but this is possible *only at a very ideal site with a gently rounded lip.* At other sites it is better to land well back. Do not attempt a top landing until you can put your glider down very close to where you intend.

Some pilots use big ears during top landings to descend through the lift and potential turbulence. The problem with this technique is it warrants a landing with the ears on in soaring winds. See our warnings about big ears in Chapter 9.

THERMAL SOARING

There are certain milestones in paragliding that will be highlighted in your log book. You probably won't forget your very first flight, your first high flight, your first soaring flight and certainly not your first thermal flight. For thermal flying is as much a step upward as all the previous achievements. It is truly the skill that frees us in the air, for thermals are not only found in limited areas, but are often spread around so that we can use them as stepping stones to fly long distances. New thermal pilots must stay near their safe landing field, but still thermals offer new possibilities to get higher than ever before, and this in itself is freedom.

Your Goal: *To understand the behavior of thermals and learn the rudimentary skills of exploiting their lift.*

✓ *What is a Thermal*

Thermal are the stairway to heaven, literally.

Thermals start as a pool of warm air heated by the ground, which in turn is heated by the sun. This pool of warm air eventually comes together and rises as a blob or column through the cooler, drier surrounding air. Once the thermal reaches an inversion layer, or the moist air cools enough to form a cloud, it stops. These matters are discussed in Chapter 6 and much more detail appears in our book *Understanding the Sky*.

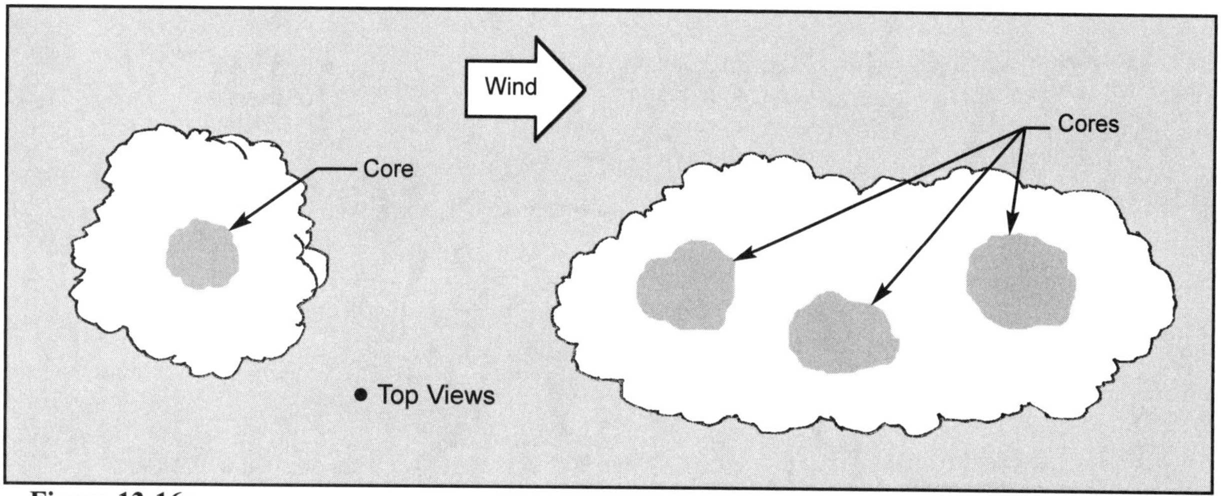

Figure 13-16: Thermal Cross-Sections

A thermal is often more or less circular in cross-section when viewed from above, as shown in figure 13-16. The center contains the fastest rising air and is known as the core. Larger thermals may have several cores or areas of stronger lift, especially in windy conditions.

> **Predicting Thermals**
> - Cumulus clouds forming. These clouds are created by thermals.
> - Variable winds on the ground. As thermals lift off they can change the wind direction for a few minutes to twenty minutes or more.
> - A very cool temperature at launch compared to the landing field. This is known as a good lapse rate (see Chapter 6).
> - The presence of circling, climbing birds and gliders. They are circling to remain in the thermal lift.

A thermal has mass—quite a bit of it to be sure—and will drift along and tilt with the wind, as figure 13-17 illustrates. If it is a bubble, it will drift with the wind as shown. One of the tricks of thermal soaring is to locate thermals in wind by imagining how they will drift from their source, or by noting where they are marked by another pilot circling.

One of the important skills you will learn when gaining thermal experience is how to recognize their presence on a given day. Some well-known clues are listed in the side box.

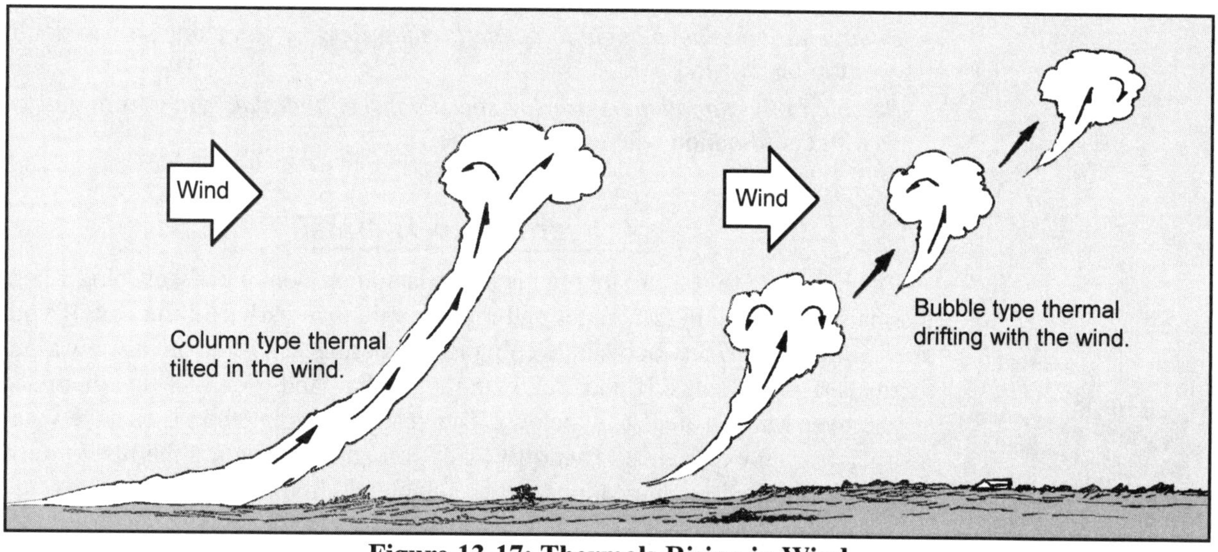

Figure 13-17: Thermals Rising in Wind

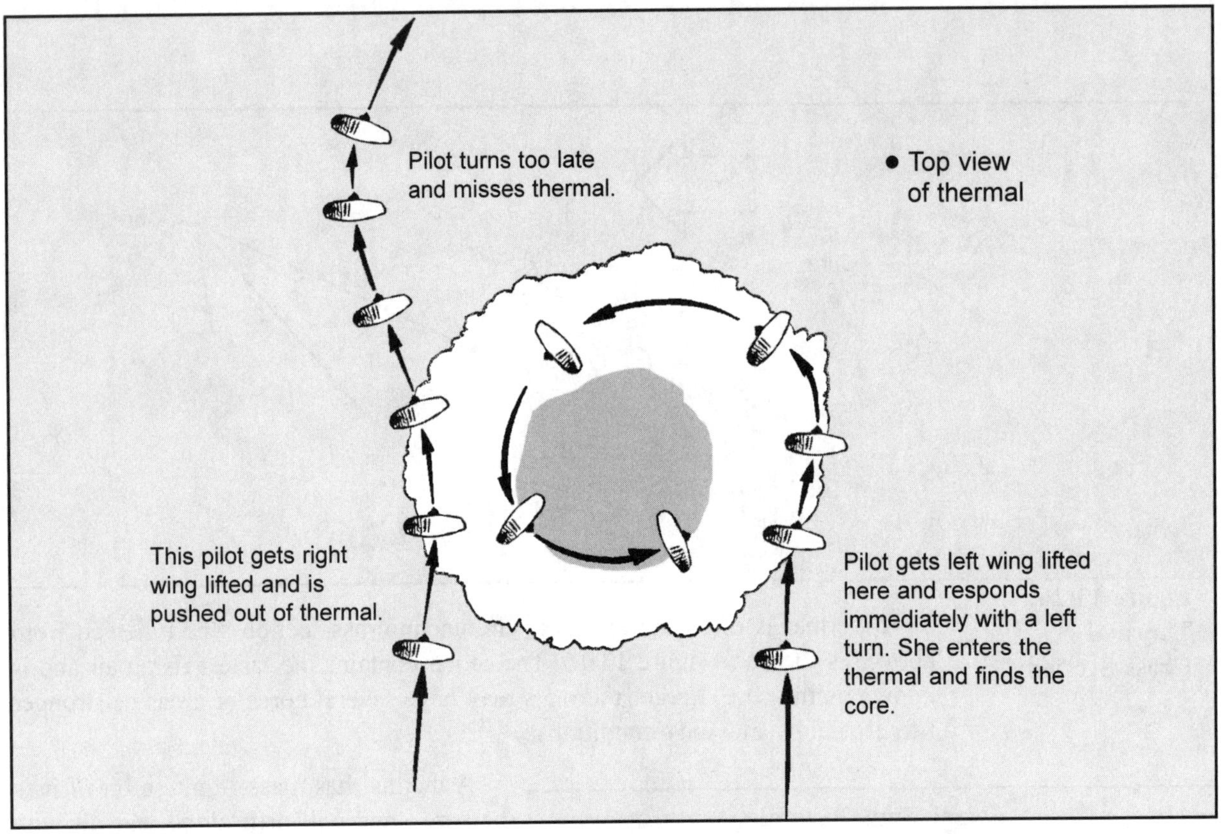

Figure 13-18: Responding to Thermal Lift

✓ Thermal Soaring Qualifications

Thermal soaring, or thermaling, is not too difficult on certain days, but since there are fewer visual clues to help you position yourself in the lift than with ridge soaring, thermaling sometimes takes longer to master. Also the skills required for thermaling are a bit more advanced. Here are some general pre-thermaling requirements:

- *Know the thermal right-of-way rules* (see Chapter 10).
- *Be comfortable turning 360s in both directions with varying degrees of bank angles*. We sometimes use very flat turns in thermals, but mostly we bank 20° to 30° with up to 45° used in tight, small cores.
- *Be able to perform maximum efficiency turns in turbulence*, making constant roll and pitch controls to keep from stalling or changing bank angle.
- *Be able to judge your angle to a safe landing field* as you circle so you don't drift too far in wind.
- *Be able to fly a good maneuvering speed between thermals* and to turn quickly with coordination when you meet one.

✓ Learning to Thermal

Thermals provide a wonderful gift of lift. They are a good reason to practice your 360 skills.

The best time to learn to thermal is in the late afternoon or early evening when thermals tend to be bigger, fatter and more gentle than earlier in the day. If you are learning on a ridge, buoyant evening ridge soaring will often have a few kind thermals rising in spots. If you are towing in the flat lands you will find thermals spread over warmer areas (see below). The best learning method is to have your instructor or an experienced friend fly with you and show you a thermal by circling in it. Radio communications are a valuable aid in this practice.

You have most likely experienced thermals before. They were those bumps that

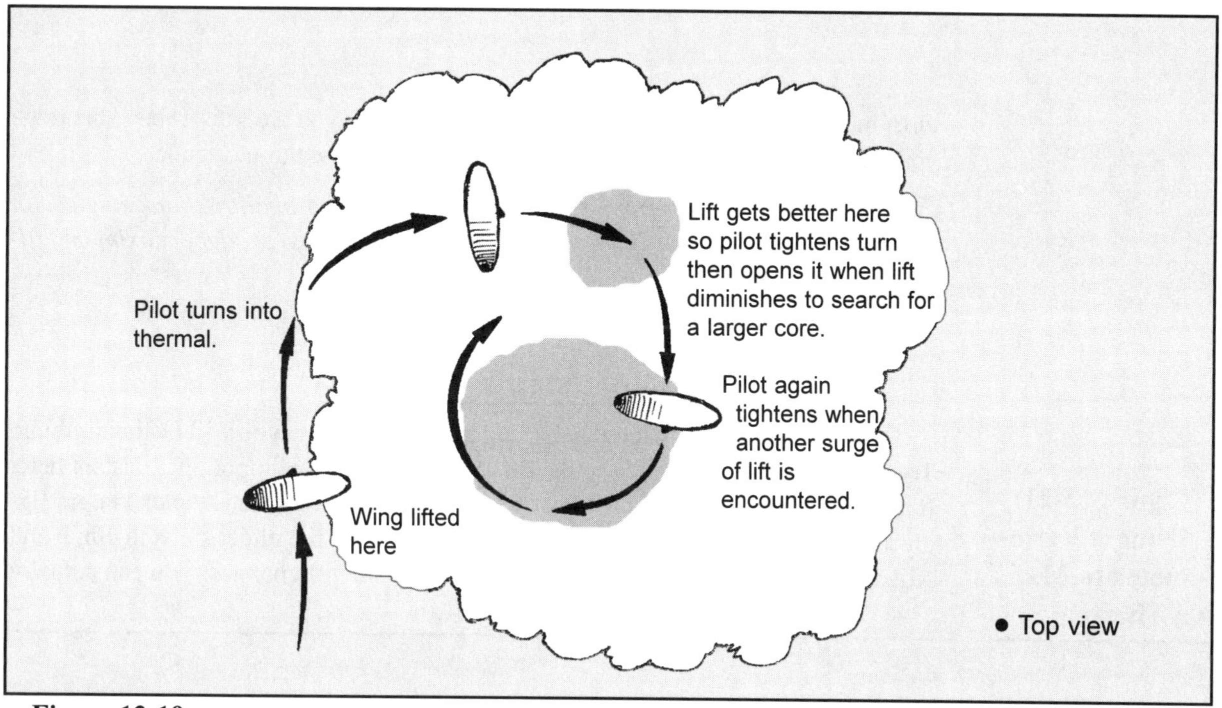

Figure 13-19: Searching for the Best Core

lifted your wing and felt like they wanted to change your flight direction. Most inexperienced pilots get tilted away from the thermals when their wings get lifted, as shown in figure 13-18. This then is your first clue to detecting thermals: a lifted wing.

The way to utilize thermal lift is to turn toward the lifted wing. This should be your number one thermaling rule.

Sometimes you will blunder directly into a thermal, in which case your entire glider lifts. Pull on a little control to maintain your pressurization, count off two seconds—hippopotamus one, hippopotamus two—and turn in the direction you feel the most lift, as shown in figure 13-19. This time delay is to allow you to make your circle in the core. When you become more experienced you will learn to use smaller thermals and turn in them sooner, but for now a two to three second delay is perfect.

Figure 13-20: Using Figure Eights to Thermal when Low

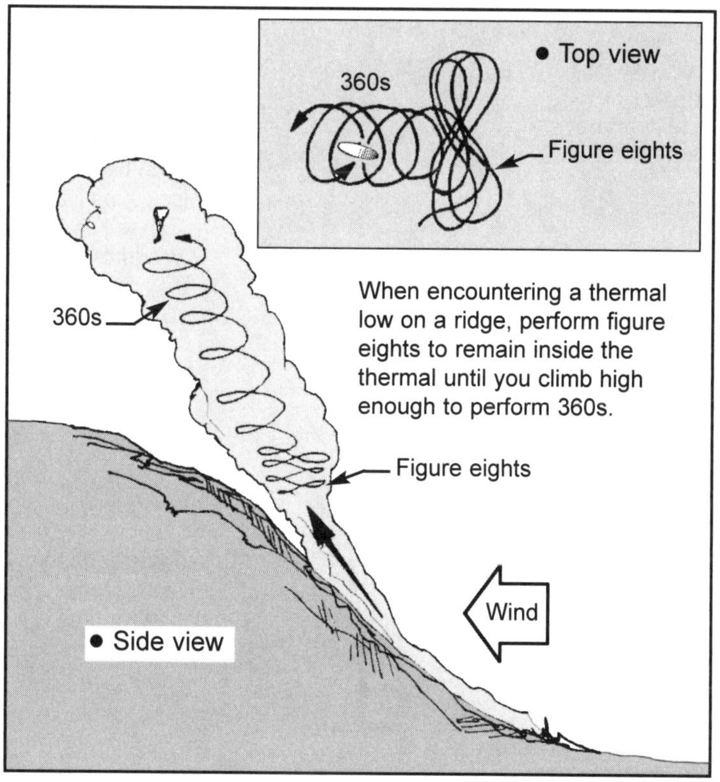

The way to thermal effectively is to remain in the best lift, so you must center your circles on the core, as shown in figure 13-19. To do this, tighten your turn when you feel surges of lift, then open it again to produce a smooth, regular turn. The idea is to shift your circle to the best core. *Performance Paragliding* provides many details on maximizing performance in a variety of thermal conditions.

If you are encountering thermals fairly low on a ridge you may not be able to turn 360s because of clearance (see figure 13-20). The trick here is to turn figure 8s in the thermal to remain in the lift, as shown. These turns must be fairly quick if the thermal is small. Once you are safely above the ridge you can begin circling as shown. Even advanced pilots use this technique.

Pro Tip: *Most beginning thermal pilots use too shallow bank angles. In order to use steeper bank angles, however, you must produce maximum efficiency turns and center in the core.*

✓ Using a Variometer

Once you have encountered a thermal and start turning, you will begin climbing. Try to concentrate on staying in the lift. A variometer is an important assistant here. It tells you when you are in the best lift by beeping faster (see Chapter 11). As figure 13-21 shows, your vario will indicate *up* soon after the glider starts to climb and will show a varying indication according to how the lift changes. You can actually

Figure 13-21: Using a Variometer to Map a Thermal

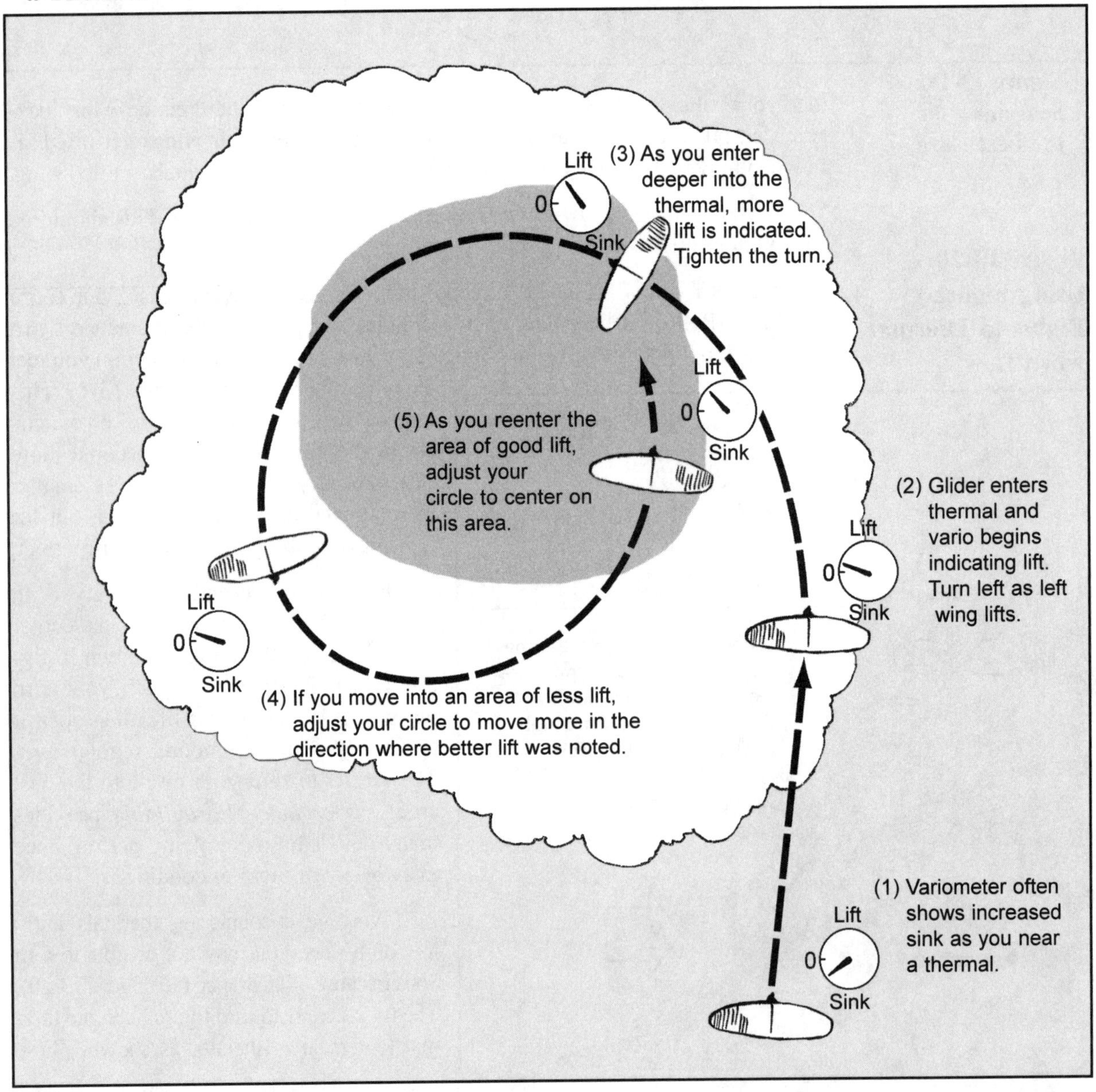

anticipate the vario by feeling the initial lift, but your vario will confirm how strong the lift is and help you center in the core.

Fishing for thermals requires knowledge, observation, lore and luck.

The most important things you can do to perform well in thermals are to turn efficiently and form an image of the thermal in your mind. A vario can help you "map" a thermal by indicating where you climb best. As you continue to thermal, note what part of the circle provides the best climb then shift your circle to encompass this area. You will have to flatten out a bit then tighten up again to perform this shift. A thermal is rarely a cohesive patch for long, so being able to shift with the thermal is the mark of a good thermal pilot. Mapping the thermal in your mind and using good turn controls are the necessary basics to thermaling skills.

Pro Tip: *To enhance your thermaling ability, learn to thermal with equal skill in both directions. On a hill or mountain, initiate your turns away from the ridge. In free air, turn toward the lifted wing.*

✓ Where to Find Thermals

Trolling for thermals is just like fishing. You can never be sure where the big one lurks, but your educated guesses help increase your percentages of soaring greatly. Here is where you should look to find thermals.

▲ 1. Along a ridge or high ground. Ridges collect thermals, as shown in figure 13-22, because the thermals tend to drift to the ridge and the ridge acts as a thermal trigger.

Figure 13-22: Thermal Collectors and Triggers

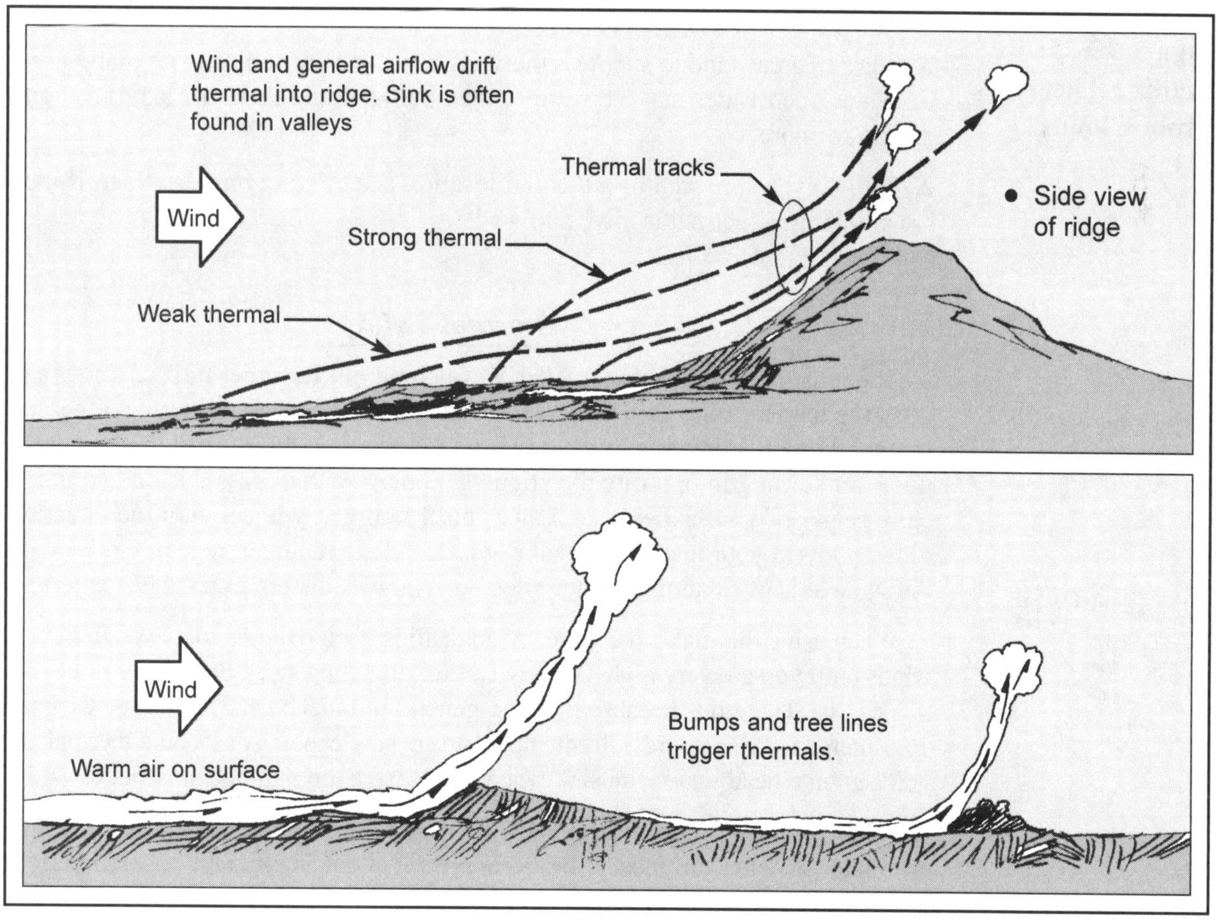

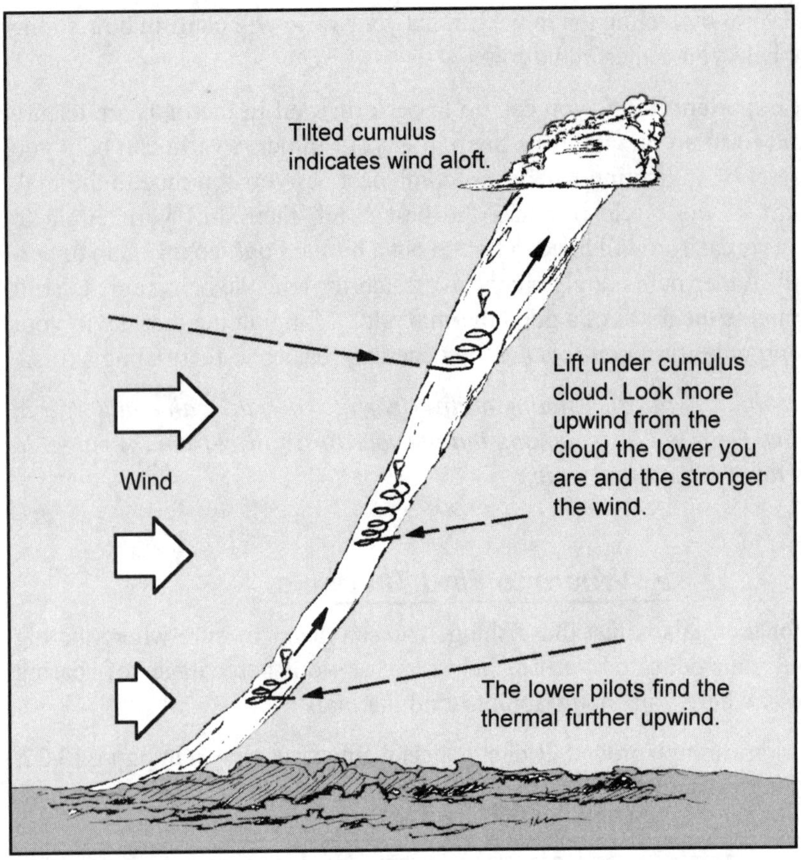

Figure 13-23: Finding Thermals From Clouds

▲ 2. In wind, look above hills, tree lines or any large object that can trigger a thermal since the moving air is deflected upward, as shown.

▲ 3. Look above dark plowed fields or dry areas. Rocky areas, quarries and towns often heat up well and produce great thermals. Crops can hold the air still to heat up well. Any area that is sheltered from the wind can produce a good thermal if the surface is dry.

▲ 4. Look under a cumulus cloud. If wind is blowing you will have to judge the thermal position as shown in figure 13-23. Generally the thermal will be more upwind of the cloud the lower you are.

▲ 5. Most sites have "house thermals", a more or less regular thermal production spot. The local pilots at a site can let you know where these spots are.

▲ 6. Damp areas tend to suppress thermals because they cool by evaporation, but in winter open water may be warmer than the land and thermals may form and drift in to shore.

▲ 7. Finally, a sure sign of a thermal location is a circling bird or glider. If they are climbing, a thermal is alive and well.

✓ Thermal Safety

The most important safety matter is watching out for other traffic. Know and obey the thermal rules (see Chapter 10), which mainly means entering a thermal turning the same direction as the others already in it, at the edge of their circle. If pilots are turning in opposite directions, you may wish to pass this one up in the early stages of your experience. Later you'll find that you can turn the direction of the pilots at your level and be all right. Be sure to maintain regular circles with the same axis of rotation as other pilots, or you will disrupt everyone's pattern.

Thermaling is like gambling in a gold rush town—fortunes are lost as soon as they're won.

When using thermals on a ridge, avoid drifting back over the top, even in lighter winds until you are very high. Be very careful of getting stuck behind the ridge, for this can be dangerous because of rotor, general turbulence and sink. (see Chapter 10, Emergency Procedures). Remember, strong sink can exist around a thermal, so you can hit a head wind with sink when you leave a thermal behind a ridge.

When you are flying in stronger thermals or in thermals with wind, you will encounter stronger turbulence. In these cases it is not as efficient or safe to circle flatly, for you will not be able to describe a precise circle since you will be constantly dropping a wing or falling out of the thermal. Use a steeper bank and more speed in these situations for maximum control.

If you do get kicked out of a thermal—known as locking out (not the same as a towing lockout)—it is best to control the canopy inflation and turn back into the thermal the same way you were originally circling. Finally, when landing in thermal conditions, be wary of turbulence near the ground due to thermals lifting off.

✓ *Thunderstorms*

When thermal conditions exist and there is ample moisture, *thunderstorms* are possible. In fact, often cloud suck develops under building thunderstorms. We call thermals clouds that keep on growing *overdevelopment*. As a thermal pilot you must always be aware of the dangers of thunderstorms and keep vigilant for their arrival on days where the clouds grow beyond the small cumulus size.

In *Performance Paragliding* we'll cover thunderstorm lore, avoidance and escape in much detail. Here we'll mention that their dangers are multiple—lightning, severe turbulence, extreme winds and if you get caught up in one, disorientation, hypoxia and freezing. Pilots have suffered from all these matters and unfortunately fatalities have occurred in all forms of aviation—from jet airliners to paragliding—directly due to flying in or around thunderstorms. Figure 13-24 shows a typical thunderstorm and its dangers.

Figure 13-24: Thunderstorm Dangers

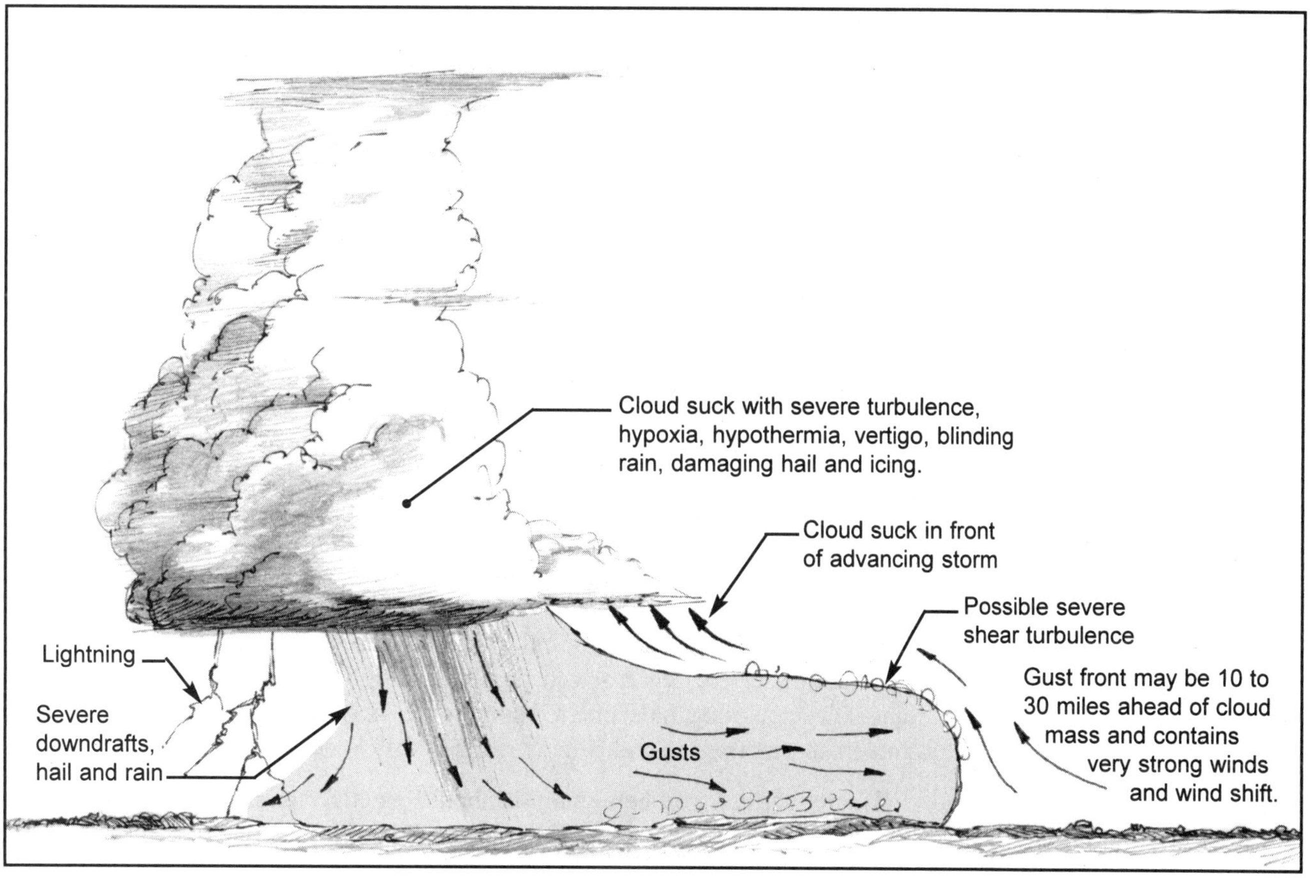

Small cumulus are created by individual thermals.

Inexperienced pilots should not be flying when thunderstorms are visible. As shown in figure 13-24, the gust front from a thunderstorm can extend many tens of miles or kilometers from the storm. The severe turbulence and violent winds of a gust front are very unfriendly to paragliders. One of the big problems with thunderstorms is that they may develop very fast and start pulling you up into them with lift that can exceed 2000 FPM (10 m/s)—see Cloud Suck below. If you see a thunderstorm while flying, consider landing or flying away from it as soon as possible. Follow the advice of safe expert pilots and don't try to second guess thunderstorms—they are often unpredictable.

Spread out cumulus can often generate cloud suck.

✓ Cloud Suck

The last danger we'll address is cloud suck. Sometimes, as you get nearer to the clouds, the lift gets stronger and more widespread. If you are not wary you can get sucked up into the clouds, which can be very frightening and dangerous. You can't see in the cloud, it is often quite turbulent, and it is bound to send you out of control with no visual reference.

Whenever the lift is over 500 FPM (2.5 m/s) and smooth you should expect cloud suck. If you aren't near a cloud you have no worries (but sometimes they form very fast above you). If a cloud is above you, widen your circle out to see if it is widespread lift. If not, you are not threatened for you can leave the lift when you wish, but be wary of a change as you gain altitude. If the lift is widespread, you should move to the edge of the cloud as you get higher. The stronger the lift and the closer you are to the cloud, the more urgent this move is.

✓ Escaping Cloud Suck

There are three primary ways of escaping cloud suck in a paraglider: using *big ears*, *B-line stalls* or *spirals*. We have previously discussed using big ears. In conjunction with a speed system, this maneuver is the safest for a lesser experienced pilot to use. Sink rate can be from 500 FPM (2.5 m/s) with 25% big ears pulled and no speed stirrup applied to about 900 FPM (4.5 m/s) with full big ears and full speed stirrup.

B-line stalls are easy to do but are not stable in turbulence and present more problems upon exit than big ears. Their description and practice are left for more advanced flying which we cover in *Performance Paragliding* .

We have introduced spirals in Chapter 9. They allow some of the fastest sink rates—1000 to 3000 FPM or 5 to 15 m/s—but they too have their problems. To get the extreme sink rates the spiral has to be steep which means high G forces, fast rotations, careful control and possible vertigo (disorientation). Many well-trained airplane pilots have entered spiral dives and lost control of their senses—so much so that the nickname "graveyard spiral" is well-known.

We are warning you here that you should practice spirals carefully by gradually adding to their steepness and number of turns. Remember also that different gliders behave differently in a spiral. By all means learn to do spirals as a safety escape technique well before you have to use them in an emergency, but do so under supervision. You will note that big ears will not provide enough sink rate to escape severe cloud suck—only spirals can do that.

✓ *Thermal Rewards*

Thermal soaring provides a much greater challenge than any other type of flying. It can be considered the peak of your soaring experience in normal conditions. Thermals are not found every day, but are so common that all experienced pilots use them on almost any soaring flight.

If you aren't warm on the ground you probably aren't dressed warmly enough for getting high in thermals.

Thermals can take us to incredible heights. Climbs over 20,000 feet (6,000 m) have been recorded. On a typical good day, gains of several thousand feet are the norm. Once you latch on to a good thermal, ride it like a rodeo bronc and it will carry you skyward for as long as it lasts. Your only job will then be to find the next thermal.

Pro Tip: *The key to good thermaling is to maximize the time you spend in lift and minimize the time in sink. Sink usually exists between thermals because what goes up must come down somewhere. So you fly faster between thermals to minimize your time in the sink. Also, the more capable you are of finding the next thermal the less altitude you will lose. The best course is to go to other gliders climbing. Beginning thermal pilots often seem to miss this obvious rule of thumb.*

SUMMARY

You have completed most of the learning that takes place to the intermediate level. There are many things you can still explore and delve into. Most notably, cross-country flying and competition attract many pilots. These subjects, along with flying in waves, convergence and inclement weather are covered in *Performance Paragliding*. This book also addresses maneuvers, parachute packing, speeds-to-fly, tandem flying, record setting, traveling with your glider and much more.

You learned to fly in slow steps at first. Then as you began to put all the controls and ideas together, learning progressed at a faster and faster pace. But once you learn to soar your experience takes off, for airtime is accumulated in large chunks rather than tiny bits. Soaring is the prize which you have been striving for. First ridge soaring, then thermal soaring, gives you the ability and privilege to command your time in the sky. Soaring unties your earthly fetters at long last and frees you for the next stage: traveling cross-country.

But before you can progress beyond what you have learned to date, you must pause and consider what you have achieved and how you did it. A clear assessment of our collective paragliding experience shows that the gradual approach and an attitude of safety is the method that produces long-time masterful pilots, for only a sound grounding in the basics will make future learning come readily. An intermediate pilot is a long way from an expert. Realize that, avoid intermediate syndrome, and you will end up being a better pilot in the long run.

Paragliding is a joy every step of the way as long as you respect the limits—both your own and those imposed by natural law. Keep your eyes open, ask plenty of questions, listen carefully, study all the material you can and you will continue developing. There is nothing to stop you. The dream has been realized; now reach for the sky.

Pro Tip: *In all your paragliding experiences perform this reality check frequently: ask yourself. "Am I having fun?" If the answer isn't a resounding yes, correct the situation.*

TEST YOURSELF (*Answers in Appendix II*)

1. Once you have learned to fly it is permanent like riding a bicycle and no further practice is required. *True or false*

2. Fear has no place in paragliding and should be ignored. *True or false*

3. One of the most dangerous phases a pilot can experience is the:
   a. Advanced stage
   b. Graduation from instruction
   c. Intermediate syndrome
   d. Soaring flight

4. When flying a new site you should:
   a. Consult the local pilots
   b. Make sure conditions are gentle
   c. Create a thorough flight plan
   d. Allow a margin of error
   e. Obey flying restrictions
   f. All of the above
   g. None of the above

5. If you are converging on another flying craft, the angle between your paths will:
   a. Remain unchanged
   b. Change uniformly

6. When learning to ridge soar some of the important skills to acquire are: (list all correct answers):
   a. Being able to control ground track in wind
   b. Judging strength of wind on takeoff
   c. Producing soaring turns
   d. Flawless forward launches
   e. Capability of flying in turbulence safely
   f. Competence at flying big ears
   g. Being able to fly safely at minimum sink speed

7. Getting too far behind a ridge when soaring is only dangerous for beginners. *True or false*

8. The increase of wind speed above a ridge or hill top is called the __________ effect.
   a. Drag
   b. Crosswind
   c. Colibri
   d. Venturi
   e. Coriolis
   f. Downwind

9. Thermal soaring requires the following skills: (list all that apply)
   a. High wind experience
   b. 360° turn abilities
   c. Efficient turning
   d. Crossed controls
   e. Handling of turbulence
   f. Knowledge of right-of-way rules
   g. a, c and d
   h. All of the above

10. The pilot's term for strong widespread lift below a cloud is ____________.

11. List at least three thunderstorm dangers: ______________, ___________, _______________.

12. List the three methods of escaping thunderstorm lift in order of descent rate (fastest to slowest): ______________, _____________, ______________

USHGA Intermediate Rating Checklist

General Description - The pilot has the knowledge and skills to fly most sites in mild to moderate soaring conditions, and to judge when the site and conditions are within the pilot's skill, knowledge, and experience level. The pilot understands the USHGA paragliding rating system as recommended operating limitations, and the FARs and other flying rules applicable to his/her flying (ridge rules, thermal right of way, FAR 103, aircraft sectional use and regulated airspace avoidance, etc.).

The pilot shall use good judgement and have a level of maturity commensurate with the rating.

Intermediate Rating - Foot/Tow Launch - Required Witnessed Tasks

A. Logged Requirements
   1) Must have logged a minimum of 30 flying days.
   2) Must have logged a total of at least 90 flights.
   3) Must have logged a minimum of 20 hours of solo airtime.

B. Demonstrated Skills and Knowledge
   1) Has received training in and/or understands the importance and significance of:
      a) Right-of-way rules.
      b) FAA Regulations and aircraft sectional charts
      c) Airspeed control, stalls, spins, and turbulence-induced collapses and recoveries.
      d) Canopy owners manual.
      e) USHGA Accident Report results currently in print.
   2) Can give verbal analysis of conditions on the hill, demonstrating knowledge of wind shadows, gradients, lift, sink, laminar air, turbulence and rotors, and the effect these items have on an intended flight path and turns.
   3) Must give a verbal flight plan for each observed flight.
   4) Must show thorough preflight of the harness, canopy, and backup reserve parachute.
   5) With each flight, demonstrates a method of establishing that the pilot is properly connected to the glider, with cleared lines and risers just prior to launch.
   6) All inflations/launches should be aggressive, confident, and with a smooth transition from running to flying. Flights with slow, unstable inflations/launches will not be considered adequate for witnessed tasks.
   7) For witnessed tasks, all landings must be safe, smooth, on the feet, and in control.
   8) Demonstrates the ability to differentiate airspeed from ground speed.
   9) Demonstrates linked 180° turns along a predetermined ground track showing smooth controlled reversals and proper coordination at various speeds and bank angles.
   10) Demonstrates 360° turns in both directions, and at various speeds and bank angles.
   11) Demonstrates symmetric and asymmetric tip folds (25% per side, 50% total) or some other method of canopy reduction for increased descent rate.
   12) Demonstrates one method to increase forward speed.
   13) Demonstrates proper surge control of canopy using properly timed control application.
   14) Gives a thorough verbal description of how to maintain directional control during, and correct for, a 50% asymmetric wing collapse.
   15) Explains characteristics of impending stall and impending spin.
   16) In 8-to-15 mph winds, demonstrates the ability to maintain airspeed at or near minimum sink during crosswind and upwind legs, without any evidence of stalls.
   17) Demonstrates 5 landings within 10 ft of a spot after flights requiring turns on approach.

18) Demonstrates proper airspeed control on landing approach when descending through a gradient.
19) Demonstrates proper airspeed for maximum distance flown into a significant head wind.
20) Demonstrates complete understanding of all Paragliding Tow Discussion Topics (for Tow rated pilots only).
21) Must pass the USHGA Intermediate Paragliding written exam.
22) Must agree to all the provisions of the USHGA standard waiver and assumption of risk agreement for the Intermediate rating and deliver an original signed copy to the USHGA office.

Recommended Operating Limitations for Intermediate Paraglider Pilots

A. Maximum base wind of 15 mph.

B. Maximum peak gusts to 18 mph.

C. Maximum gust rate of 5 mph in 5 seconds.

D. Avoid steep turns close to the ground.

E. Avoid application of either control beyond 3/4 of the way from full off to stall position.

F. Limit turns to bank angles recommended by the manufacturer, limit speed in turns to 2 times the straight line, control off, cruise speed, and smoothly exit any spiral turn that shows a tendency to steepen or accelerate.

G. Should initiate downwind turns only with 300 ft of clearance outward from the hill or ridge in winds above 15 mph, and 200 ft of clearance in winds above 10 mph.

H. Should not fly in thermals where peak climb rates exceed 500 fpm or where significant vertical cloud development exists.

I. Upon mastering the above skills, an Intermediate Paragliding Pilot should pursue new maneuvers, sites, and conditions with the guidance of a USHGA Certified Advanced Paragliding Instructor or Observer.

Flying with a friend lets you share the magic.

APPENDIX I - USEFUL ADDRESSES

NATIONAL ORGANIZATIONS

United States Hang Gliding Association
PO Box 1330
Colorado Springs, CO 80901 - USA
Tel: 719-632-8300
Fax: 719-632-6417
Web site: www.ushga.org
E-mail: ushga@ushga.org

Hang Gliding and Paragliding Association of Canada / Association Canadienne de Vol Libre
21593, 94a Ave.
Langley, B.C. V1M 2A5 - CANADA
Tel & Fax: 604-882-5090

British Hang Gliding and Paragliding Association
The Old Schoolroom
Loughborough Rd
Leicester LE4 5PJ - ENGLAND
Tel: 0533-612362
Fax: 0533-611323
Web site: www.bhpa.co.uk
E-mail: office@bhpa.co.uk

Hang Gliding Federation of Australia
PO Box 558
Tumut, NSW 2720 - AUSTRALIA
Tel: 02 6947 2888
Fax: 02 6947 4328
E-mail: hgfa@tpgi.com.au

South African Hang Gliding and Paragliding Association (SAHPA)
PO Box 1993
1685 Halfway House - SOUTH AFRICA
Tel: (011) 805-5429
Fax: (011) 805-5429
Web site: www.sahpa.co.za
E-mail: sahpa@paragliding.co.za

New Zealand Hang Gliding and Paragliding Association (NZHPA)
PO Box 3521
Richmond, Nelson - NEW ZEALAND
Web site: nzhgpa.org.nz

RESOURCES

Sport Aviation Publications
(Training manuals, videos and CD Roms for paragliding, hang gliding, ultralights and weather)
PO Box 43
Spring Mills, PA 16875 - USA
Tel/Fax: 814-422-0589
E-mail: pagenbks@lazerlink.com
Web site: www.lazerlink.com/~pagenbks

Books in Print:
Understanding the Sky
Towing Aloft
Hang Gliding Training Manual
Performance Flying
Powered Ultralight Flying
Powered Ultralight Training Course

Upcoming Books:
The Secrets of Champions
Performance Paragliding

Dixon's Airplay School and Flight Parks
(Paragliding lessons and instruction, Weather information)
PO Box 2626
Flagstaff, AZ 86003-2626 - USA
Tel: 520-526-4579
E-mail: Dixon@paraglide.com
Web site: www.paraglide.com

Adventure Productions
(Paragliding, hang gliding, weather and ultralight videos and CD roms)
6553 Stone Valley Dr.
Reno, NV 89523 - USA
Tel/Fax: 775-747-0175
Web site:www.adventurep.com - www.weathertofly.com

High Perspective
RR # 5, 865 Conc 7
Claremont, Ontario, L1Y 1AZ - Canada
Tel: 905-294-2536 Fax: 905-294-8495
E-mail: flyhigh@inforamp.net

APPENDIX II - ANSWERS TO CHAPTER TESTS

CHAPTER 2 - *1. Risers – 2. Leading edge – 3. D lines – 4. UV, heat, abrasion, jerking, tearing, moisture, stepping on...*

CHAPTER 3 - *1. Relative wind – 2. Less – 3. Same – 4. Add to – 5. a and d – 6. Varying control line pressure – 7. Angle of attack (or control pull), airspeed – 8. Loss of control, loss of altitude, loss of canopy shape.*

CHAPTER 4 - *1. Leading – 2. Knots, lineover, sticks, folded tips, line twists – 3. False – 4. False – 5. Tease leading edge in center up in wind, not too high in strong wind – 6. A – 7. Pull rear risers – 8. d*

CHAPTER 5 - *1. b, c and d – 2. Accelerating the canopy, frontal fold – 3. Rear – 4. Gentle, small, smooth, 2 inches of control, with floating hands, between 1 and 6 lbs (.5 to 3 kg) – 5. False – 6. Loading the canopy too soon and not having feet under you – 7. Other gliders landing (traffic) – 8. Pull gently down on right control and let up on the left control, carefully monitoring airspeed.*

CHAPTER 6 - *1. Differential, sharpness, frequency – 2. Noon to late afternoon – 3. False – 4. Downslope – 5. Cumulus – 6. Mechanical, shear, wake, thermal – 7. Downwind of solid objects – 8. Low*

CHAPTER 7 - *1. At horizon or briefly at canopy – 2. c – 3. True – 4. b – 5. Increased – 6. Getting dragged – 7. A*

CHAPTER 8 - *1. 4 – 2. Figure eight and aircraft (DBF) approach – 3. Figure eight – 4. Figure eight – 5. DBF – 6. Angles – 7. Turbulence, gusts, wind gradient, wing shadow, getting dragged – 8. e – 9. False – 10. False – 11. Uphill – 12. False!*

CHAPTER 9 - *1. Flying slower with control line pressure – 2. Controls about at the trim position – 3. Stall, spin – 4. With a timely addition of control when canopy 30° or more in front – 5. Low airspeed, lightening controls, wallowing – 6. Increased descent rate, no airspeed – 7. Clear your turn – 8. True – 9. Spin – 10. False – 11. Position and airspeed – 12. Correcting the turn – 13. False.*

CHAPTER 10 - *1. Pre-launch check – 2. One – 3. c, a, f, d, e, b – 4. False – 5. False – 6. The ridge – 7. A, B, C, D, restricted and prohibited airspace*

CHAPTER 11 - *1. Porosity, line stretch, line damage, canopy stretch, harness damage – 2. Tape tears less than 4 inches (10 cm) – 3. Every 100 hours, when stretched, when broken, when soaked in salt water – 4. False – 5. h – 6. Too rapid – 7. 6 months minimum*

CHAPTER 12 - *1. Stable – 2. Perpendicular – 3. Parallel – 4. e – 5. Parasite (form, profile), induced – 6. Span^2/area or thinness of a wing – 7. Yaw (vertical axis) – 8. Airfoil, shape and pendulum effect – 9. Increases – 10. Inside, spin*

CHAPTER 13 - *1. False – 2. False – 3. c – 4. f – 5. a – 6. a, b, c, e, g – 7. False – 8. d – 9. b, c, e, f – 10. Cloud suck – 11. Hypoxia, hail damage, freezing, lightning, disorientation, high winds, severe turbulence – 12. (1) Spirals, (2) B-lines, (3) big ears*

GLOSSARY - TERMS USED IN PARAGLIDING

AERODYNAMICS — The study of the movement of a body such as a paraglider wing through the air (from *aero* meaning air and *dynamics* meaning motion).
ACPULS — French glider certification standard institute.
ACTIVE FLYING — Continuously feeling the controls, keeping constant pressures and maintaining a steady canopy.
AGL — Abbreviation for Above Ground Level.
AIRCRAFT APPROACH — A landing approach consisting of downwind, base and final (upwind) legs (DBF).
AIRFOIL — A curved surface designed to generate lift when moving through the air.
AIR MASS — A large uniform volume of air such as behind a cold or warm front.
AIRSPEED — The velocity of the glider through the air.
AIRSPEED INDICATOR (or ASI) — An instrument for measuring airspeed.
ALTIMETER — An instrument for measuring altitude above a predetermined point.
ALTO — A prefix meaning high, referring to clouds from 6,500 to 20,000 ft.
ANGLE OF ATTACK — The angle the relative wind makes with the chord of an airfoil.
A RISERS — The most forward risers connecting to the A lines routing to the canopy's leading edge.
APPROACH — See LANDING APPROACH
ASPECT RATIO — A comparison of the span of a wing to its chord (fore and aft dimension).
ATTITUDE — The angle an airfoil's chord makes with the horizon.
BANK ANGLE — The angle the wing makes with the horizon in a roll.
BASE LEG — The crosswind leg of an aircraft approach between the downwind and final leg.
BEST GLIDE SPEED — The airspeed that provides the best glide ratio through the air.
BOXING THE FIELD — Flying around a field as a landing setup.
CAMBER — The amount of curvature in surfaces of an airfoil.
CANOPY — The material wing of a paraglider.
CAP CLOUD — A cloud above a high mountain caused by upslope flow.
CARABINER — An oval ring used to attach the harness to the risers.
CARDINAL SPEEDS — The primary speeds of a glider such as stall speed, minimum sink speed, best glide speed and best maneuvering speed.
CELLS — The separate volumes in a canopy defined by the ribs and the upper and lower surface.
CEN — European glider certification standard institute.
CENTER OF GRAVITY — The point along the chord of a wing where the weight of a pilot is located.
CENTER OF PRESSURE — The point along an airfoil or wing where all the aerodynamic forces (lift and drag) are centered or balanced.
CHORD — Measurement of an airfoil from the leading edge to the trailing edge.
CIRRUS CLOUDS — High clouds above 20,000 ft that are wispy in form.
CLOUD BASE — The bottom of cumulus or layered clouds which is the condensation or dew point.
CLOUD STREET — A line of cumulus clouds that indicates a "highway" of lift.
COLD FRONT — The leading edge of an advancing cool air mass.
CONE OF PENETRATION — The limit of reach you have in any direction at your present altitude.
CONTROL — Changing a glider's orientation in either the pitch, roll or yaw axes.
CONROLLED AIRSPACE — That part of airspace that is regulated. This is nearly all airspace in the U.S.
CONTROL LINES — Lines running from the toggles to the trailing edge of the canopy used to control the glider.
CONVECTION — Rising heat bubbles or columns in the air—thermals.
CONVERGENCE — A coming together of two air flows usually resulting in an upwelling (lift).
COORDINATED TURNS — A turn with the right amount of control applied so a slip or stall doesn't occur.
CORIOLIS EFFECT — A turning of the wind (to the right in the northern hemisphere) due to the earth's rotation.
CRABBING — Flying in wind with a different ground track from your heading due to the sideways push of the wind.
CROSSWINDS — Winds flowing in a direction across the flight path.
CUMULUS CLOUDS (CU) — A tumbled, cotton ball-like cloud created by rising thermal currents.
CUMULONIMBUS (CU-NIM) — An overgrown cumulus cloud that grows dark and turns into a thunderstorm.
DEEP STALL — Vertical, steady descent of the glider in a stalled state. Sometimes called parachutage.

DHV — German glider certification standard institute.
DIVE SYNDROME — The tendency for an inexperienced pilot to fly too fast on a first high flight due to the appearance that the ground is not moving compared to when flying lower.
DOWNSLOPE WINDS — Winds sliding down a slope that is cooled in the evening (also called catabatic winds).
DOWNWIND — Direction to which the wind is flowing.
DOWNWIND LEG — The initial segment of an aircraft landing approach flown in the downwind direction.
DRAG — The energy losses on a glider due to friction and deflection of the air.
DYNAMIC PRESSURE — The pressure felt on a wing due to the air's movement.
FIGURE 8 SETUP — A landing approach using figure 8 turns at the downwind side of the landing field.
FINAL LEG — The last (into the wind) portion of an aircraft type landing approach.
FLARING — Pulling the controls fully to raise the angle of attack abruptly on landing to stop forward motion.
FLIGHT PLAN — A set of expected flight procedures and paths planned before a flight.
FORM DRAG — Drag caused by the lines, harness and pilot exposed to the air.
FORWARD INFLATION — Inflating the canopy while facing into the wind, away from the canopy.
FRONT RISER — See A RISER
G FORCES (G's or GEES) — An apparent increase in weight due to curving flight.
GLIDE ANGLE — The angle between the glider path and the horizontal.
GLIDE PATH — The flight path of a glider.
GLIDER — An aircraft that remains flying through the energy of gravity only.
GLIDE RATIO — The ratio of the distance traveled forward to the distance dropped. This is erroneously used interchangeably with L/D (glide ratio and L/D are only identical in still air).
GROUND HANDLING — Floating and controlling a glider on the ground.
GROUND SPEED — The velocity of a glider over the ground. This is different form airspeed if any wind is present.
GROSS WEIGHT — Total weight of the glider and the heaviest allowed payload (pilot).
HANDLING — The quickness of reaction of a glider to a control input.
HARNESS — A suspension system that supports a pilot and attaches him to a canopy.
HEADING — The direction a glider points (this will be different from actual flight direction in a crosswind.
HEAD WIND — A wind from the front or opposite the heading.
HOUSE THERMAL — A thermal that is more or less regular at a site.
IMAGING — Forming a mental picture of a skill to be learned to review it in a relaxed manner.
INDUCED DRAG — Losses on an aircraft due to air escaping around the tips.
INFLATION — Lofting the canopy by pulling forward to fill it with air.
INITIATION POINT — The place in a landing approach when the downwind, base and final pattern is begun.
ISOBARS — Lines connecting points of equal pressure on a weather map. Wind generally follows isobars.
INSTABILITY — The tendency for any system (as a glider in flight) to go out of control or require a constant input to maintain equilibrium.
INTERMEDIATE SYNDROME — The tendency for pilots to become overconfident when reaching the intermediate level.
INVERSION — A layer of air where the temperature does not cool enough with height to continue to promote rising thermals.
KITING — Inflating the canopy and keeping it aloft in a wind with deft control.
LANDING APPROACH — The pattern used to set up a landing all the way to final.
LANDING SETUP — Positioning in the air to enter the landing approach or pattern.
LANDING ZONE (LZ) — An open area for landing.
LAPSE RATE — The temperature profile of the air from the ground up.
L/D — Lift-to-drag ratio which is the same as the glide ratio in still air.
LEADING EDGE — The most forward point of a wing.
LENTICULAR CLOUD — A flat "lens shaped" cloud formed by a wave.
LIFT — Uprising air used by pilots to soar, or the upward aerodynamic forces on an airfoil.
LIFT-TO-DRAG RATIO — See L/D
LIGHT AND VARIABLE — Light winds that change direction frequently due to thermals lifting off.
LOCAL WINDS — Winds caused by heating effects and developed over a limited area.
LOCKOUT — When towing, a turn to one side that is difficult to correct. When thermaling, falling out of a thermal at the edge.

LOG BOOK — A book used to list flights and achievements.
LOOMING — An apparent sudden expansion in size of an object as you approach it.
MANEUVERING SPEED — The airspeed to fly to allow the best glider response.
MAXIMUM GLIDE RATIO The best L/D performance possible on a glider.
MEAN CAMBER — The curve of a line drawn midway between the upper and lower surface on an airfoil.
MINIMUM SINK RATE — The slowest vertical descent speed possible.
MINIMUM SINK AIRSPEED — The airspeed to fly to allow the slowest glider descent.
MSL — Abbreviation for Mean Sea Level for indicating height above the average sea level.
MYLAR — Stiff polyester sheeting.
OVER-DEVELOPMENT — Cumulus cloud buildup and spreading that can lead to thunderstorms.
PARACHUTAGE — See DEEP STALLS
PARACHUTE BRIDLE — The webbing connection between a parachute and the harness.
PARASITIC DRAG — All forms of drag except induced drag.
PITCH — Movement of a glider where the front of the wing tilts up or down.
PLF — Parachute Landing Fall—a technique of falling in order to dissipate impact forces.
PORPOISING — Oscillating up and down in pitch due to overcontrolling.
PREFLIGHT — A careful integrity check of all glider parts before flying.
PRELAUNCH CHECK — A check of important points just before launch.
PROFILE — The curve of an airfoil.
PROFILE DRAG — Drag caused only by the wing.
QUICKLINKS — Metal connectors that fasten the risers to the suspension lines.
REAR RISERS — Risers connected to the rearmost suspension lines.
RELATIVE WIND — The apparent wind as the glider is flying. Since the glider is always moving in respect to the air around it the relative wind is different from the actual wind.
RESERVE PARACHUTE — A parachute carried by the pilot for emergencies.
RESTITUTION — Light widespread lift caused by cool evening air moving under air warmed by the sun throughout the day.
RESULTANT — The vector sum of lift and drag on an airfoil.
REVERSE INFLATION — Lofting the canopy from a position facing the canopy.
RIBS — Vertical cloth webs connecting to upper and lower surface of a canopy.
RIDGE SOARING — Same as slope soaring. Staying up on lift caused by air deflected over a ridge, hill or mountain.
RIGHT-OF-WAY — Rules determining who should give way during encounters with other air traffic.
RISERS — Webbing that connects from the harness to the suspension lines.
RISK MANAGEMENT — Making safety decisions by taking into consideration all factors that affect a flight.
ROLL — A control during which one wing tip raises or lowers.
ROOT — The center of a wing.
ROTOR — A standing eddy or swirl that forms close to and downwind from a solid object.
ROUNDOUT — Paying off speed to fly parallel to the ground during landing.
SCRATCHING — Working very light lift to remain aloft, usually without spare altitude.
SEA BREEZE — A local wind caused by differential heating between water and land.
SECTIONAL — An aviation map showing airports, airways, controlled airspace as well as ground features.
SHEATH — The outer protective coating of a suspension line.
SIMULATOR — A device to hook into with a harness and practice controls.
SINK — Falling air which makes the glider travel downward faster than normal.
SLIP — A falling to the inside of a turn due to insufficient control input.
SLOPE SOARING — See RIDGE SOARING
SOARABLE ENVELOPE — The area around a ridge producing enough lift to soar.
SOARING — Flight extended beyond the normal glide path of the glider.
SOARING TURNS — Efficient turns performed during ridge soaring while maintaining ample control.
SPAN — The total width of a glider from tip to tip.
SPEED STIRRUP — A system for lowering the angle of attack of a paraglider consisting of a foot bar attached to a line that routes to the A risers.
SPEEDS-TO-FLY — A method of flying to improve performance by speeding up in sink and slowing down in lift.

SPINS — The rapid uncontrolled turning of a glider due to a stall on one side.
SPIRAL DIVE — A coordinated, diving turn that a glider enters when held in a steep turn. High G's result because of the steep bank.
STABILITY — Tendency for a glider to return to level flight from any attitude or bank.
STAGING AREA — The area near a landing field you use to lose altitude.
STALL — A sudden loss of lift and increase in drag due to an excessive angle of attack.
STATIC PRESSURE — The pressure of air caused by the weight of air above the measuring point.
STRAPS — Webbing on a harness that secures the pilot into the seat (e.g. leg straps and chest straps).
STRATUS CLOUDS — Layered clouds that spread and usually suppress thermals.
STREAMER — A thin ribbon used as a wind indicator.
SUSPENSION LINES — The lines connecting risers to the canopy.
TAIL WIND — A wind from the rear or blowing in the heading direction.
TANDEM — Flying with two people on a glider.
TEMPERATURE INVERSION — See INVERSION
TEMPERATURE PROFILE — See LAPSE RATE
THERMAL — A warm bubble or column of air rising through the sky.
THERMAL SOARING — Climbing or remaining aloft by flying in thermal lift.
THUNDERSTORM — Excessive buildup of cumulus clouds until rain, lightning and other severe conditions occur.
TOGGLES — The handles at the end of control lines which the pilot holds during flight.
TOWING — Pulling a glider and pilot aloft with a tow rope and power source such as a winch.
TRAILING EDGE — The rearward most part of a wing or an airfoil.
TRIM SPEED — The speed a stable glider returns to when the controls are released.
TURBULENCE — Gusts or swirls of air encountered in flight.
UPPER SURFACE — The top surface of a canopy.
UPSLOPE WINDS — Winds flowing up a slope due to the sun's heating of the slope (also called anabatic winds).
UPWIND — A flight direction heading into the wind.
VARIOMETER — A device that indicates when the glider is rising or sinking and the rate at which this takes place.
VELOCITY — A measurement of the speed and direction of motion.
VENTURI — An increase in wind speed when air gets squeezed in a gap or over a ridge.
VERTIGO — Disorientation due to excessive turning or a disruption of visual reference.
VORTEX (TIP VORTICES) — The swirling of air at the tips of a glider, or any swirl of air.
WAKE TURBULENCE — The swirls left behind a glider or other aircraft.
WARM FRONT — The leading edge of an advancing warm air mass.
WIND DUMMY — A pilot who launches first to check conditions. A kinder term: wind technician.
WIND GRADIENT — Slowing of the wind as the ground is approached.
WIND INDICATOR — An instrument to measure the wind speed, or an object moved by the wind such as a wind sock.
WIND SHADOW — A still area close behind a large obstruction such as a building or tree line.
WIND SOCK — A cloth tube mounted on a pole to indicate wind direction.
WING LOADING — The weight-to-area ratio of an aircraft found by dividing the flying weight of the pilot plus the glider by the total sail area.
WINGOVER — A maneuver in which the glider follows a path like that of a bike riding up and back down a curved wall.
X-C — An abbreviation for cross-country.
YAW — Rotation of a wing about its vertical axis so that one wing moves forward and the other back.

INDEX

NOTES